THE SOLAR RADIATION AND CLIMATE EXPERIMENT (SORCE)

Mission Description and Early Results

Edited by
G. ROTTMAN, T. WOODS and V. GEORGE
University of Colorado, Boulder, CO, USA

Reprinted from *Solar Physics*, Volume 230, Nos. 1–2, 2005

Library of Congress Cataloging-in-Publication Data is available

ISBN-10 0-387-30242-5
ISBN-13 978-0-387-30242-3

Published by Springer,
P.O. Box 17, 3300 AA Dordrecht, The Netherlands.

Printed on acid-free paper

Printed in the Netherlands

TABLE OF CONTENTS

SOLAR RADIATION AND CLIMATE EXPERIMENT (SORCE) INSTRUMENTS

Total Irradiance Monitor (TIM)

Spectral Irradiance Monitor (SIM)

Solar Stellar Irradiance Comparison Experiment (SOLSTICE)

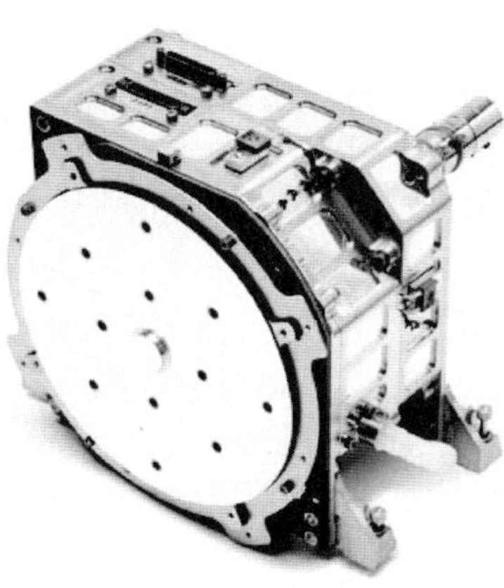

XUV Photometer System (XPS)

Solar Physics (2005) 230: 1–2

PREFACE

This volume on the Solar Radiation and Climate Experiment (SORCE) mission continues a *Solar Physics* tradition of special topical issues dedicated to major solar space missions. As one element of NASA's Earth Observing System, SORCE is a satellite carrying four instruments to measure the solar radiation incident at the top of the Earth's atmosphere. These observations are improving our understanding and generating new inquiry regarding how and why solar variability occurs and how it affects our atmosphere and climate.

The SORCE mission is a joint effort between NASA and the Laboratory for Atmospheric and Space Physics (LASP) at the University of Colorado. The mission is a Principal Investigator-led mission under the direction of Dr. Gary Rottman from LASP. LASP developed, calibrated, and tested the four science instruments before integrating them onto a spacecraft procured from Orbital Sciences Corporation.

The SORCE instruments include the Total Irradiance Monitor (TIM), the Spectral Irradiance Monitor (SIM), two Solar Stellar Irradiance Comparison Experiments (SOLSTICE), and the XUV Photometer System (XPS). The TIM instrument continues the precise measurements of total solar irradiance (TSI) that first began in 1978. SORCE also provides measurements of the solar spectral irradiance (SSI) from 1 to 2000 nm with its other instruments. The Sun has both direct and indirect influences on the terrestrial system, and SORCE's comprehensive total and spectral solar measurements are providing the requisite understanding of this important climate system variable.

SORCE was successfully launched from a Pegasus XL rocket on 25 January 2003. By early March 2003, all instrument doors were open and science operations had begun. The first validated science data were delivered approximately two months after the launch.

The SORCE satellite is orbiting the Earth every 95 minutes or 15 times daily. Ground stations are providing the communication links to the satellite two times each day. Science and mission operations are conducted from LASP's Mission Operations Center, which provides the computer hardware and software necessary to conduct spacecraft operational activities, including command and control of the satellite, mission planning, and assessment and maintenance of spacecraft and instrument health. The science operations include experiment planning, data processing and analysis, validation, and distribution of the finished data product. Within 48 hours of data capture, all instrument science data and spacecraft engineering data are processed to derive higher level science data products of the solar irradiance.

Included in this special *Solar Physics* issue are an overview of the mission, science objectives, detailed descriptions of the instruments and their calibrations, the spacecraft, operational procedures, data processing, and early science results. This volume has been an extremely worthwhile undertaking, and all authors would like to gratefully acknowledge the dedication of the *Solar Physics* referees and editors for their valuable suggestions and guidance. With their much appreciated assistance, the SORCE mission is thoroughly documented for future reference.

G. ROTTMAN, T. WOODS, and V. GEORGE (*Guest Editors*)
J. HARVEY, Z. ŠVESTKA, and O. ENGVOLD (*Editors*)

Solar Physics (2005) 230: 3–6

THE SOLAR RADIATION AND CLIMATE EXPERIMENT (SORCE) MISSION FOR THE NASA EARTH OBSERVING SYSTEM (EOS)

DONALD E. ANDERSON
NASA Headquarters, Washington, DC, U.S.A.
(e-mail: Donald.Anderson-1@nasa.gov)

and

ROBERT F. CAHALAN
NASA Goddard Space Flight Center, Greenbelt, MD, U.S.A.
(e-mail: Robert.F.Cahalan@nasa.gov)

(Received 5 July 2005; accepted 6 July 2005)

Abstract. The NASA Earth Observing System (EOS) is an advanced study of Earth's long-term global changes of solid Earth, its atmosphere, and oceans and includes a coordinated collection of satellites, data systems, and modeling. The EOS program was conceived in the 1980s as part of NASA's Earth System Enterprise (ESE). The Solar Radiation and Climate Experiment (SORCE) is one of about 20 missions planned for the EOS program, and the SORCE measurement objectives include the total solar irradiance (TSI) and solar spectral irradiance (SSI) that are two of the 24 key measurement parameters defined for the EOS program. The SORCE satellite was launched in January 2003, and its observations are improving the understanding and generating new inquiry regarding how and why solar variability occurs and how it affects Earth's energy balance, atmosphere, and long-term climate changes.

1. Introduction to the EOS Program

The concepts for the NASA Earth Observing System (EOS) program began in the 1980s as part of the NASA Earth Science Enterprise (ESE) through recommendations from the U.S. Global Change Research Program (USGCRP), the International Geosphere–Biosphere Program (IGBP), and the World Climate Research Program (WCRP). The driving motivation for the EOS program is the Earth science community concerns for potentially serious environmental changes, such as global warming, rising sea level, deforestation, desertification, atmosphere ozone depletion, acid rain, and reduction in biodiversity (King and Greenstone, 1999). Part of the EOS program research is to determine the actual changes in the environment both globally and on local scales and to access the contributions of human activity on the environmental changes as compared to natural variations. The EOS program has provided advanced and integrated scientific observing and data systems to address the hydrologic, biogeochemical, atmospheric, ecological, and geophysical processes that are important for improved understanding of the carbon cycle, water cycle, energy cycle, climate variability, atmospheric chemistry, and solid Earth science (King and Greenstone, 1999).

The series of satellites for the EOS program is based on a set of 24 key measurement objectives that will enable advances in understanding the long-term global changes of the solid Earth, its atmosphere, and oceans. These EOS measurements, as detailed in the *EOS Science Plan* (King, 2000), are grouped into five categories of atmosphere, solar radiation, land, ocean, and cryosphere. The early concept for these measurements was a series of satellites to provide observations over a period of 15 years. While the EOS program originally had several large satellites planned as flagship missions, the program restructured in the 1990s to have fewer large satellites and several small satellites with many of these smaller satellites being funded through the NASA Earth System Science Pathfinder (ESSP) program. Other important components of the EOS program are the archive and distribution of the EOS satellite data through the EOS Data and Information System (EOSDIS), interdisciplinary science research, calibration and validation, education and public outreach, and international cooperation (King and Greenstone, 1999).

2. SORCE Mission Contribution to the EOS Program

The total solar irradiance (TSI) and solar spectral irradiance (SSI) are two of the 24 EOS key measurements, and the Solar Radiation and Climate Experiment (SORCE) mission is now providing these solar irradiance measurements for the EOS program. The TSI is known to vary by a few tenths of a percent, and these small changes are considered a key climate-forcing component as related to Earth's energy budget (King, 2000; Pilewskie, Rottman, and Richard, 2005). The variations of the solar spectral irradiance are highly wavelength dependent, and the deposition of the solar irradiance into the Earth system is also strongly dependent on wavelength. The visible and infrared irradiance and its variation are important for radiation studies involving clouds and aerosols and their influence on climate changes (King, 2000; Pilewskie, Rottman, and Richard, 2005). The ultraviolet irradiance and its variation are important for atmospheric studies involving stratospheric chemistry, heating, and dynamics and the possible coupling to the lower atmosphere (King, 2000; Lean *et al.*, 2005). The accurate measurements of the TSI and SSI over periods of decades are important for establishing the solar influence on Earth's climate.

The SORCE mission is the merger of two EOS mission concepts. The original selection of instruments in 1989 for the EOS solar irradiance measurements included the Solar Stellar Irradiance Comparison Experiment (SOLSTICE) with the Principal Investigator (PI) being Dr. Gary Rottman at the Laboratory for Atmospheric and Space Physics (LASP) at the University of Colorado (CU) and the Active Cavity Radiometer Irradiance Monitor (ACRIM) with the PI being Dr. Richard Willson, who was at the Jet Propulsion Laboratory (JPL) at that time. The SOLSTICE includes several channels to measure the SSI, and the ACRIM is designed to measure the TSI. Both instruments were selected as "flight of opportunity" instruments without a satellite platform defined. After several iterations in studying various mission concepts, the ACRIM instrument was designed for its own small satellite called the ACRIMSAT, which launched in December 1999, and the SOLSTICE was being designed for a small satellite called the Solar and Atmospheric Variability Explorer (SAVE). NASA recognized in the mid 1990s that the second generation of solar irradiance instruments would need to be selected to follow the ACRIMSAT and SAVE missions and defined the new mission opportunity as the Total Solar Irradiance Mission (TSIM) that included requirements for measuring the TSI and limited bands of the SSI. Following a Phase A study of the TSIM concept, LASP with Dr. Gary Rottman as the PI was selected for the TSIM program. Partly because this TSIM concept included the Solar Irradiance Monitor (SIM) that was also planned for the SAVE mission, the SAVE and TSIM programs at LASP were integrated into a single mission and renamed the SORCE mission in 1999.

SORCE has four different instruments for measuring the solar irradiance in order to meet the SORCE mission objectives (Rottman, 2005). The Total Irradiance Monitor (TIM) measures the TSI (Kopp and Lawrence, 2005). The Spectral Irradiance Monitor (SIM) measures the near ultraviolet, visible, and near infrared SSI in the 200 to 2000 nm range (Harder *et al.*, 2005). The Solar Stellar Irradiance Comparison Experiment measures the far ultraviolet and middle ultraviolet SSI in the 115–320 nm range (McClintock, Rottman, and Woods, 2005). The XUV Photometer System (XPS) measures the soft X-ray (XUV) SSI in the 0.1–34 nm range (Woods, Rottman, and Vest, 2005).

The SORCE mission is a PI-led satellite program with a firm cost cap and is operated much like an ESSP or Small Explorer (SMEX) program. The NASA Goddard Space Flight Center (GSFC) provided the higher-level project oversight, and LASP provided the program management for the instrument development and for the subcontract to Orbital Science Corporation (OSC) for the spacecraft bus and the launch services on a Pegasus XL rocket. The SORCE satellite was successfully launched on 25 January 2003, and its mission will extend for at least 6 years. So far, the SORCE mission has completed two and a half very successful years of operations with the mission operations and data processing activities centered at LASP (Pankratz *et al.*, 2005).

3. Future of the Solar Irradiance Measurements

NASA has plans to transition many of the EOS key measurements into longer term observations using NOAA operational satellites. For example, the SORCE TIM and SIM measurements are currently planned on a series of satellites for the NOAA National Polar-orbiting Operational Environmental Satellite System (NPOESS). These NPOESS solar irradiance measurements are being referred to as the Total Solar Irradiance Sensor (TSIS), which is currently being procured by a contract from Northrop Grumman to LASP for the TIM and SIM instruments and a solar pointing platform. The first flight of the NPOESS TSIS is planned for a 2012 launch. Because of the large time separation between the SORCE launch in 2003 and the first NPOESS TSIS observations, NASA has considered a Solar Irradiance Gap Filler (SIGF) mission or "flight of opportunity" for additional solar irradiance measurements to bridge the potential gap between the SORCE and NPOESS measurements. One of these opportunities is the flight of the TIM instrument on the NASA Glory spacecraft with its launch now planned in 2008. Without the SIM instrument aboard the Glory satellite, additional opportunities are being explored for obtaining the solar spectral irradiance measurements in the 2008–2012 timeframe.

Acknowledgements

The SORCE program is supported under NASA contract NAS5-97045 to the University of Colorado. We are grateful to LASP and OSC for the highly successful SORCE mission.

References

Harder, J., Lawrence, G., Fontenla, J., Rottman, G., and Woods, T.: 2005, *Solar Phys.*, this volume.

King, M. D. (ed.): 2000, *EOS Science Plan*, NASA GSFC, Greenbelt, MD.

King, M. D. and Greenstone, R. (eds.): 1999, *EOS Reference Handbook*, NASA GSFC, Greenbelt, MD.

Kopp, G. and Lawrence, G.: 2005, *Solar Phys.*, this volume.

Lean, J., Rottman, G., Harder, J., and Kopp, G.: 2005, *Solar Phys.*, this volume.

McClintock, W. E., Rottman, G. J., and Woods, T. N.: 2005, *Solar Phys.*, this volume.

Pankratz, C. K., Knapp, B., Reukauf, R., Fontenla, J., Dorey, M., Connelly, L., and Windnagel, A.: 2005, *Solar Phys.*, this volume.

Pilewskie, P., Rottman, G., and Richard, E.: 2005, *Solar Phys.*, this volume.

Rottman, G.: 2005, *Solar Phys.*, this volume.

Woods, T. N., Rottman, G., and Vest, R.: 2005, *Solar Phys.*, this volume.

Solar Physics (2005) 230: 7–25

THE SORCE MISSION

GARY ROTTMAN
Laboratory for Atmospheric and Space Physics, University of Colorado, Boulder, CO 80303
(e-mail: gary.rottman@lasp.colorado.edu)

(Received 19 May 2005; accepted 20 May 2005)

Abstract. The Solar Radiation and Climate Experiment (SORCE) satellite carries four scientific instruments that measure the solar radiation at the top of the Earth's atmosphere. The mission is an important flight component of NASA's Earth Observing System (EOS), which in turn is the major observational and scientific element of the U.S. Global Change Research Program. The scientific objectives of SORCE are to make daily measurements of the total solar irradiance and of spectral solar irradiance from 120 to 2000 nm with additional measurements of the energetic X-rays. Solar radiation provides the dominant energy source for the Earth system and detailed understanding of its variation is essential for atmospheric and climate studies. SORCE was launched on January 25, 2003 and has an expected lifetime through the next solar minimum in about 2007. The spacecraft and all instruments have operated flawlessly during the first 2 years, and this paper provides an overview of the mission and discusses the contributions that SORCE is making to improve understanding of the Sun's influence on the Earth environment.

1. Introduction

The Solar Radiation and Climate Experiment (SORCE) was launched in January 2003. On board are four instruments that measure solar irradiance, both the total irradiance and the spectral irradiance at short X-ray wavelengths, and in the ultraviolet, visible, and infrared regions. The SORCE observations of solar irradiance represent the present state-of-the-art in observing the Sun and recording one of the primary climate system variables.

Solar radiation is the dominant energy input to the Earth system. Roughly 30% is scattered and reflected back to space, with the remaining 70% absorbed by the atmosphere, land, and ocean (Kiehl and Trenberth, 1997). This energy determines the temperature and structure of the atmosphere, warms the Earth surface, and sustains life. A delicate balance is maintained between incoming solar radiation, the Earth's albedo (fraction of radiation reflected back to space), and outgoing long-wave infrared radiation arising from a global mean temperature as altered by greenhouse gasses, clouds, and aerosols. Changes in solar irradiance will have both direct and indirect effects on the Earth climate system, and implications of a solar role are evident in many climate records (Lean *et al.*, 2005).

Observations of solar variability became possible only after attaining access to space, and therefore the observational record extends back only about 30 years.

The total solar irradiance or TSI, which is predominantly visible and infrared radiation, has shown a range of variation not exceeding a few tenths of one percent (Willson and Hudson, 1991). Based solely on radiative balance consideration the direct effect on global temperature from such a small solar variation is a change of only a fraction of a Kelvin. In parallel with these TSI observations, rockets and satellites were also making observations of very energetic radiation from the Sun (Friedman, 1961). This ultraviolet and X-ray radiation exhibited a much larger range of variation – factors of two to ten, and even more. Such highly variable short-wave radiation is easily reconciled with the small variation observed in TSI because the X-rays and ultraviolet (UV) radiation make up less than 1% of the Sun's total radiative output. These more energetic photons do not have access to the Earth surface and lower atmosphere; therefore, they do not have a direct effect on global surface temperature. However, because of their very important influence on the composition, temperature, and dynamics of the atmosphere, that in turn couples to the lower atmosphere, they may have important indirect influence on local and global climate (Haigh, 2001; Rind, 2002). Such indirect effects are complicated and their understanding requires models that accurately incorporate the complex and interrelated processes occurring throughout the Earth's atmosphere. Of course, the foundation for such models is reliable and accurate input data on the relevant climate system variables, including solar irradiance.

All of the terrestrial processes that absorb and scatter solar radiation do so in a very wavelength dependent manner (Meier *et al.*, 1997). In particular the photochemical processes occurring in the atmosphere can be strikingly wavelength dependent. For example, molecular oxygen is dissociated by a fairly broad continuum of radiation centered near 140 nm in the ultraviolet with additional strong and narrow absorption bands near 180 nm. The result is that while solar radiation near 140 nm is absorbed uniformly well above 120 km in the atmosphere (Nicolet and Peetermans, 1980), the penetration of radiation near 180 nm is far more irregular with one wavelength absorbed in the mesosphere and adjacent radiation, only a few nanometers away, penetrating an additional 20 km (Nicolet and Kennes, 1989). Further complications arise because the solar irradiance at these UV wavelengths varies considerably, and likewise in a very wavelength dependent fashion – from 5 to 20% variation over a few days to 10 to 70% variation over several years. These estimates of the Sun's variability are from observations obtained during the last few solar cycles (25 years), and over longer periods of time even larger variations cannot be ruled out. Solar irradiance at different UV wavelengths is absorbed and scattered by other major and minor constituents of the atmosphere, including ozone and water vapor, each in its own distinctive wavelength dependent way. Thus, in order to achieve a comprehensive understanding of the Sun's influence on the atmosphere, it is crucial that data sets of solar irradiance include detailed and specific wavelength information.

1.1. Early Measurement of Total Solar Irradiance

Historically mankind has long recognized the importance of understanding the Sun as the Earth's energy source. As new instruments and observing techniques were devised they were quickly applied to the Sun. The development of the telescope in the early seventeenth century brought the Sun under close scrutiny and discoveries rapidly followed – sunspots, filaments, prominences, faculae, etc. Developments and improvements in spectroscopy were applied to the Sun, and likewise led to discoveries of new atomic species and unexpected high states of ionization. Coronagraphs and other specialized instruments also have had great success in unraveling the mysteries of the Sun. Naturally as the science of radiometry progressed, attempts were made to record the radiative output from the Sun and to determine the amount of solar variability.

In the late nineteenth century Langley, Abbot, and others (Menzel, 1949; Abbot, 1948) pursued a very active research program, refining measurement techniques and deploying instruments to remote and high altitude observatories. These early ground-based observations concentrated on TSI, and the rigorous and careful programs continued well through the middle of the twentieth century. The observation is difficult at best, for a measurement taken at the ground must be corrected for atmospheric absorption and scattering. Rayleigh scattering (λ^{-4} dependence), aerosol scattering (unknown λ dependence), and molecular and atomic absorption (absorption lines, bands, and edges) all attenuate the solar radiation. In order to recover a top of the atmosphere (TOA) irradiance, the ground-based observations were carefully extrapolated, but the necessary adjustments were large. In the final analysis the estimated TSI values were deemed valid to only the order of a couple of percent. Details in the observational record of TSI although enticing were often misleading, for example dips as large as 5% seemed to accompany the passage of sunspots across the solar disk – a puzzling observation that could not be reconciled by the dark contrast and size of the sunspot on the solar disk (Eddy, 1983). In the final analysis there was general consensus that long-term variations of the Sun were probably less than the measurement error of a couple percent, and at this level the solar irradiance could be considered "constant." For a time the insolation was referred to as the "solar constant," a misnomer that was only retracted after space-based observations beginning in 1978 proved conclusively that TSI did indeed vary.

TSI devices have operated on a number of space missions since about 1978 (Willson, 1984, 1994; Lee, Barkstrom, and Cess, 1987; Hoyt and Schatten, 1993; Fröhlich *et al.*, 1997) and since that time they have observed almost three complete 11-year solar cycles. For these three cycles TSI values show a clear solar cycle variability of about 0.1%, with the higher levels coinciding with the maximum levels of sunspots. In fact, these observations seem to show quite conclusively that the dominant solar variability over the 11-year cycle is due to magnetic activity in the photosphere with a positive contribution originating in the bright faculae and a negative contribution arising from the dark sunspots (Foukal and Lean, 1988). The best fit to the TSI data is achieved with a faculae contribution roughly twice the sunspot darkening, giving a net variation of about 0.1% (Fröhlich and Lean, 1998, 2004).

Shorter-term variations of TSI are also apparent in the observational record, and their major cause is the passage of dark sunspots across the disk of the Sun. These appear as dips in TSI data of about 0.1% and last for several days as the sunspots and sunspot groups traverse the center of the solar disk. Since the associated faculae are more uniformly spread across the solar disk, they do not typically produce intermediate- and short-term increases to TSI as striking as the sunspot dips.

1.2. Early measurements of spectral irradiance

The solar spectrum has also been studied from the ground for many decades. This spectrum is rich in structure with features originating both at the Sun and additional telluric absorption features superposed. In general the telluric features, especially the weaker absorptions, can be identified and removed. However, some portions of the infrared are so overwhelmed by water vapor absorption bands that the true solar spectrum remained hidden, and only revealed after spectrometers were taken to high altitude aboard balloons and rockets. Atmospheric ozone is such an efficient absorber of radiation at wavelengths shorter than about 300 nm that no solar ultraviolet radiation penetrates to the ground. At still shorter wavelengths where ozone is not an effective absorber, the dominant atmospheric gases, nitrogen and oxygen, completely absorb the ultraviolet, extreme ultraviolet (EUV), and X-rays from the Sun. Indeed, the Earth's atmosphere provides a 100% effective shield against the harsh ultraviolet radiation. Figure 1 illustrates the solar spectrum as

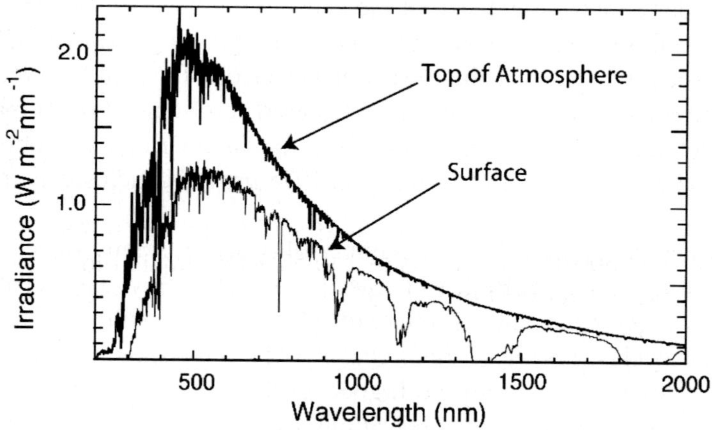

Figure 1. The solar spectrum (200 nm $< \lambda <$ 2000 nm) at the top of the Earth's atmosphere and for an overhead Sun after passing through the clear atmosphere.

seen above the Earth's atmosphere and then as observed at the surface. The difficulty of correcting the surface irradiance values to the top of the atmosphere is apparent.

An effective temperature of the Sun near 6000 K was determined by examining the visible portion of the solar spectrum (Menzel, 1949). Extrapolation of this black-body shape to short wavelengths suggested absent or extremely weak ultraviolet emission from the Sun. Meanwhile conflicting information from remote sensing studies of the Earth's upper atmosphere implied very high temperatures and wide spread ionization (Rense, 1961; Banks and Kockarts, 1973), facts that could only be reconciled by robust solar irradiance at short ultraviolet wavelengths. Moreover, some solar observations, especially related to the solar corona, implied very high temperatures within the solar atmosphere that would be accompanied with ultraviolet emission. In 1947 when sounding rockets first provided access to space, the issue was resolved when spectrometers recorded the ultraviolet emissions from the Sun – observing this robust radiation to be both intense and highly variable (Friedman, 1961).

The measurement of spectral irradiance improved steadily, beginning with multiple sounding rocket observations and progressing to long-duration observations from early satellites. The spectral coverage was expanded and observing techniques were refined, until slowly an understanding of the very energetic radiation emerged. In general, emission in the extreme ultraviolet and ultraviolet originates throughout the Sun's transition region and chromosphere, regions just above the photosphere where the effective temperature rises to about 10^4 K. The very shortest wavelength and most energetic X-rays originate in the highest layer of the solar atmosphere, the corona, where temperatures exceed 10^6 K.

In 1975 a compilation of review articles (White, 1975) summarized the prevailing understanding of the solar spectrum and its variations as derived from the first 20 years of space observations. Some of the conclusions at that time remain true today while others contained misinformation probably due to overly optimistic interpretation of the uncertainty in the observations, for example, a solar cycle variation of 20% at 300 nm.

Between 1975 and the launch of SORCE solar observations were further refined in their precision and accuracy. Today there is consensus on the following aspects: (1) the most energetic X-rays can vary by factors of 10 and larger (Woods *et al.*, 2004b), the EUV spectrum ($\lambda < 120$ nm) by factors of 2, and the UV spectrum ($\lambda < 300$ nm) by factors of 1 to 50% (Rottman, Floyd, and Viereck, 2004); (2) the integrated effect of UV radiation, $\lambda < 300$ nm, variability accounts for roughly 30% of the solar cycle variation of TSI (London and Rottman, 1989; Lean, 1989) with the remaining 70% attributed to the visible and infrared; and (3) with few exceptions (e.g., cores of strong Fraunhoffer absorption lines) the visible and infrared irradiance varies by only small fractions of 1% (Foukal and Lean, 1988).

2. The SORCE Mission

The SORCE program was developed between 1999 and launch in 2003 but originated in the early 1980's when NASA began developing an Earth Observing System (EOS) with a goal of determining the extent, causes, and regional consequences of climate change. EOS defined science and policy priorities based on recommendations of national and international programs including the Intergovernmental Panel on Climate Change (IPCC) and the Committee on Earth and Environmental Sciences (CEES)/Committee on the Environment and Natural Resources (CERN). In response to an Announcement of Opportunity extended in 1988, the Laboratory for Atmospheric and Space Physics submitted a proposal to build an instrument, the Solar Stellar Irradiance Comparison Experiment (SOLSTICE) that would measure solar ultraviolet irradiance for a period of 15 years. In response to the same opportunity, the Jet Propulsion Laboratory proposed an ACRIM III to measure the Total Solar Irradiance. NASA selected these two instruments, together with an additional 28 atmospheric, oceanic, and surface instruments, in early 1989. Beginning in 1991 the EOS program underwent major revisions, prompted by the need to make substantial budget reductions. ACRIM III slowly evolved to a small free-flying satellite, ACRIMSAT, which launched in 1999, while the SOLSTICE also evolved to a small free-flying satellite called the Solar and Atmospheric Variability Experiment, SAVE, with a planned launch in late 2002. Meanwhile other revisions to the EOS program reduced the 15-year mission life to a more realistic 5-year mission, with a possibility of three separate flights to achieve the 15-year objective. In 1997, NASA issued another Announcement of Opportunity in order to recompete the follow-on ACRIM III and that mission was called the Total Solar Irradiance Mission (TSIM). LASP proposed and was selected in 1999 to provide the TSIM. Recognizing that TSIM and SAVE were closely aligned in terms of mission scope, science objectives and timing, NASA combined the two into a single mission called the Solar Radiation and Climate Experiment and recommended a launch date in 2002.

The SORCE science objectives are to:

(1) Measure total solar irradiance (TSI) with sufficient precision and accuracy to produce a reliable record of short- and intermediate-term solar variations, and measure solar cycle variations as appropriate to the mission duration. Provide overlap with existent and future TSI observing programs. Provide a refined level of absolute accuracy (combined standard uncertainty) to establish a new and reliable benchmark for all TSI observations.

(2) Measure solar spectral irradiance for wavelengths between 1 and 2000 nm (the spectral range 27–115 nm was specifically eliminated in the original EOS SOLSTICE acceptance notification in 1989) with sufficient precision and accuracy to produce a reliable record of short- and intermediate-term solar variations. Provide a level of absolute accuracy (wavelength dependent) to determine longer-term solar variations.

(3) Calibrate and validate the SORCE data. Distribute SORCE data to the scientific community, with special emphasis on informing the atmospheric and climate communities and the solar physics community.

2.1. The SORCE instruments

The SORCE science objectives led to the measurement requirements as listed in Table I. SORCE achieves these measurement requirements with measurements made by four separate scientific instruments – the total irradiance monitor (TIM); the spectral irradiance monitor (SIM); the solar stellar irradiance comparison experiment (SOLSTICE); and the XUV photometer system (XPS). The instruments are described briefly here, and in detail in accompanying papers (Kopp and Lawrence, 2005; Harder *et al.*, 2005a; McClintock, Rottman, and Woods, 2005; Woods, Rottman, and Vest, 2005). The approximate mapping of science objective to corresponding instruments is: TSI measurement from TIM, UV spectral measurement from SOLSTICE, visible/near infrared measurement from SIM, and the EUV/X-ray measurement from XPS.

Figure 2 illustrates the spectral coverage of the SORCE instruments using as a backdrop the spectrum of the Sun as recorded on 21 April 2004. The bars at the bottom of the figure identify the wavelength range covered by the three SORCE spectral instruments, XPS, SOLSTICE, and SIM. To fill in the missing portions of the spectrum, data from the TIMED SEE instrument are inserted from 27 to 115 nm, and data from Thuillier *et al.* (2004) are used at wavelengths longer than about 2.4 μm. The solar spectrum of Figure 2 corresponds to the TOA spectrum of Figure 1, in a log–log format.

The SORCE spectral observations comprise more than 96% of the total irradiance. It is fortunate that the TIMED SEE instrument (Woods *et al.,* 2005) launched in 2001 fills the EUV wavelength gap between about 27 and 115 nm. The SORCE observations, supplemented by those of the TIMED SEE, are therefore providing the first time series of irradiance at most wavelengths, a near-simultaneous data set allowing detailed studies of how the Sun varies wavelength by wavelength as solar activity evolves throughout its 11-year cycle.

TABLE I

SORCE measurement requirements.

Measurement	Spectral range (nm)	Resolution (nm)	Precision (1σ)	Accuracy (1σ)
TSI	All wavelengths	N/A	10 ppm	100 ppm
UV spectral	$120 < \lambda < 300$	1	0.5%	5%
Visible/infrared	$300 < \lambda < 2000$	5–50	0.01%	0.05%
EUV/X-ray	$1 < \lambda < 34$	10	2%	20%

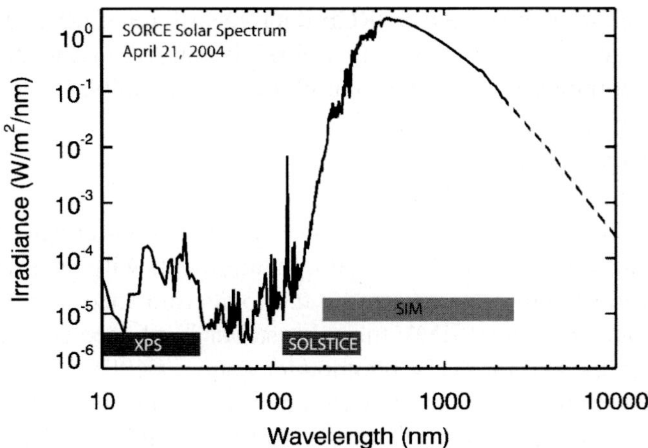

Figure 2. The solar irradiance spectrum in log–log format, with the wavelength range covered by the three SORCE spectral instruments, SIM, SOLSTICE, and XPS. The SORCE TIM measures the total solar irradiance or the integral over the entire spectrum.

2.1.1. *Total Irradiance Monitor*

The SORCE total irradiance monitor, called TIM, is a new and refined instrument that measures TSI. The measurement of TSI or radiant flux density requires an aperture to define the collection area in conjunction with a sensor to measure the incident power. The instrument must be sensitive uniformly to all wavelengths from the very energetic and short wavelength X-rays to the very longest infrared wavelengths. A bolometer, or a device that absorbs radiation and converts it to internal heat and sensed as a temperature increase in the collector, is such a simple device and is the sensor of choice used to measure TSI for over 100 years (Fröhlich, 2004). Consider a black metal disk with a temperature sensor placed behind an aperture. When the aperture is open and pointed at the Sun, the disk warms to an elevated temperature. When the aperture closes the disk cools and returns to its lower ambient temperature. The challenge is to place such a device in space, point it at the Sun and record TSI with an accuracy (combined standard uncertainty) of better than 100 ppm (0.001%) and a precision of a few parts per million (ppm). Of course the responsivity of the device must be very stable and, moreover, the observer must have knowledge of any changes in instrument responsivity in order to continually correct the solar observations and to retain a value of better than 100 ppm uncertainty in the measurement.

TIM is a four-channel electrical substitution radiometer, ESR (Lawrence *et al.*, 2000, 2003; Kopp and Lawrence, 2005). Each channel has a thin-wall conical bolometer with an integral heater and thermistor. The interior of each cone is extremely black using nickel phosphorous (NiP) black coating as an efficient absorber of radiation. The cones are used in pairs, one arbitrarily called the active cone and the other the reference cone. Electronics provide Joule heat (known voltage across the resistance of the cone's heater) to balance the two cones at a temperature slightly elevated to their surroundings. The pair of cones is pointed at the Sun, and although both have shutters over a precise aperture, at any time only one (the active cone) is open, with the result that solar radiation entering that cone is completely absorbed in its interior. The "balancing circuit" immediately reduces the Joule heat to that "active" cone in order to maintain its temperature balance to the "reference" cone, and the amount of heater power removed is precisely equivalent to the radiant power (watts) entering the shutter/aperture. Knowing the size of the aperture and the amount of power removed from the active cone provides a precise measurement of the solar radiant flux density (W m^{-2}) or irradiance. The TIM aperture is $\sim$0.5 cm^2 and the recorded power is therefore about 65 mW.

SORCE irradiance data are corrected to a distance of one astronomical unit (1 AU), although it is straightforward to adjust the data to true Earth distance. The SORCE data may also be extended to other locations throughout the solar system, but because the Sun's radiation is not isotropic, a model must be used to estimate the differences due to ecliptic latitude and longitude.

To achieve precision and accuracy measured in parts per million, all terms in the measurement equation that converts the instrument signal (current to the heater) to solar irradiance must be characterized with incredible detail. Some calibrations and characterizations are straightforward, although challenging, for example to measure the aperture to the order of 50 ppm. Other terms are far more difficult and elusive, for example the equivalence term that is the ratio of the bolometer thermal response to solar radiation absorbed to its response to Joule heat applied. Kopp and Lawrence (2005), Kopp, Heuerman, and Lawrence (2005), and Kopp, Lawrence, and Rottman (2005) provide extensive detail on the TIM design, its operations and calibrations, and first scientific results, respectively.

2.1.2. *Spectral Irradiance Monitor*

The measurement of solar spectral irradiance (SSI) is similar to the measurement of TSI with the important additional complexity that incoming radiant power density is first separated by wavelength before the detector records the solar signal. Usually the science requirement is to achieve a spectral resolution ($\Delta\lambda/\lambda$) on the order of 0.01 (resolving power greater than 100) or better, requiring a spectrometer type instrument. For an irradiance measurement this device usually includes an entrance aperture, followed by a dispersive element such as a grating or prism, and finally a means to refocus the dispersed radiation to an exit slit that establishes the bandpass of the instrument.

SIM is a newly developed prism spectrometer designed to measure solar irradiance throughout the visible and near infrared (Harder *et al.*, 2005a,b). The science objective of SIM is to make these measurements with a combined standard uncertainty of less than 0.05% and precision and long-term relative accuracy of 0.01%. Although SIM's spectral coverage extends to ultraviolet wavelengths as short as 200 nm, this region is the primary objective of SOLSTICE and only secondary to SIM. As mentioned earlier, the small TSI variations indicate that variations in the visible and near infrared do not exceed a fraction of one percent, and the SIM measurements now confirm this level of variability (Rottman *et al.*, 2005).

It is a challenge for a space-based spectrometer to provide a stable responsivity over many years on-orbit and to thus establish solar spectral irradiance variability at the level of 0.05%. SIM achieves this using only a single optical element – a Suprasil fused-silica prism with a concave front face and a convex rear surface aluminized for high reflectivity. The solar radiation enters an entrance slit and then is dispersed and refocused by the prism back to a set of exit slits. As the prism rotates the entire solar spectrum is recorded by a miniaturized version of the ESR used in TIM and described earlier. Because the SIM aperture is much smaller ($\sim 0.02\,\text{cm}^2$) than the aperture of TIM, and because furthermore the wavelength partitioning reduces the detectable light considerably, the power at the SIM ESR is only of the order of $10\,\mu$W. Such small signals are a challenge for a bolometer/ESR type detector, but nevertheless this realization is indeed the accomplishment of SIM (Harder *et al.*, 2005a,b).

In addition to the SIM ESR, four photodiodes – a combination of Si and InGAs – at four separate exit slits cover the spectral range 200 nm to 1 μm and provide rapid scans of portions of the spectrum. The ESR is the stable, absolute detector that is used to continually recalibrate the diodes. Harder *et al.* (2005a,b) provide a complete description of SIM's design and its operation and calibrations.

There are two completely independent optical channels in SIM. One is used on a daily basis and the second is used infrequently (approximately 1% duty cycle) to evaluate responsivity changes in the primary unit. In addition, there is a small, periscope device that can direct monochromatic radiation from either one of the two instruments into the other. The channel that is being calibrated has a diode pair to measure this incoming radiation and then move out of the beam to allow the light to pass to its prism. The test prism then refracts, transmits, and returns the light to the same diode pair, whereby its transmission is determined. The two SIM channels are symmetric and throughout the mission the transmissions of both prisms are repeatedly measured, tracking changes in instrument responsivity.

2.1.3. *Solar Stellar Irradiance Comparison Experiment*

SOLSTICE is a grating spectrometer that measures solar spectral irradiance ultraviolet wavelengths, $115 < \lambda < 320$ nm. The SOLSTICE measurements have a combined standard uncertainty of less than 5% (wavelength dependent), and a precision and long-term relative accuracy of better than 0.5%. McClintock, Rottman, and Woods (2005) and McClintock, Snow, and Woods (2005) provide a complete description of the SORCE SOLSTICE instrument and its calibrations.

This instrument is a second generation of the SOLSTICE (Rottman, Woods, and Sparn, 1993) currently flying on the Upper Atmosphere Research Satellite (UARS) which was launched in 1991. (The UARS mission is likely to end during the summer of 2005.) The SOLSTICE technique observes the Sun during daylight portions of the SORCE orbit, and then during nighttime portions it reconfigures itself, but uses the very same optics and detectors to observe bright blue stars. The large dynamic range between the stellar and solar flux is accommodated by changing only apertures (factor of 2×10^5) and integration times (factor of 10^3), both parameters in the measurement equation that are well calibrated and do not change during the mission. The repeated observation of the stars accomplishes two things. First, the stellar fluxes from main-sequence B and A stars are not expected to vary (Mihalas and Binney, 1981) and any changes in the SOLSTICE signal while observing the stars are unambiguously interpreted as changes in the instrument reponsivity, which is corrected accordingly. Second, both UARS and SORCE SOLSTICE establish the ratio of the solar to stellar flux that is independent of instrument responsivity. Future observations (perhaps up to thousands of years) can repeat these ratio measurements. Assuming that the stars do not vary, the ratios from the different "SOLSTICE" observers can be directly compared and thereby establish variations in the Sun's ultraviolet irradiance over any arbitrary time base.

SORCE carries two identical and redundant SOLSTICE units. Each instrument has two selectable detectors, one covering the spectral range 115–180 nm and the second 170–320 nm. In routine operations, one instrument observes the shorter wavelength range while the other observes the longer wavelengths, although periodically and infrequently this order is reversed. The duty cycling provides an estimate of changes in the detector efficiency, and the wavelength overlap of the two channels provides additional data validation. The full redundancy of the two SOLSTICE units insures that the SORCE science objective will be achieved in the event of a failure in either unit.

2.1.4. *XUV Photometer System*

XPS is a combination of filter photometers that measure solar irradiance from 0.1 to 34 nm with an additional channel at the important Lyman-α line at 121.6 nm. A very similar instrument flies on NASA's TIMED mission, launched in 2001 (Woods *et al.*, 1998). There are twelve silicon photodiodes, eight with metal films directly deposited on them, one with a 121 nm interference filter, and the remaining three are bare (Woods, Rottman, and Vest, 2005). The filter materials (Powell *et al.*, 1990), either metal coating or interference, establish the wavelength sensitivity (bandpass) and also block the long wavelength solar radiation that would overwhelm the relatively weak X-ray signal.

The 12 photodiodes/filters are packaged in a single unit with a filter wheel mechanism in front. As the wheel turns it places an open aperture, a blocked position, or a window (fused silica) in front of each diode. The open aperture allows the solar irradiance measurement, while in turn the blocked position provides a reading of the dark signal, and the window position provides a measure of the long wavelength leakage through the filter.

2.2. SORCE SPACECRAFT

The SORCE science objectives specify the measurement requirements of the four instruments as listed in Table I, and in turn, the proper functioning of the instruments places constraints on the SORCE spacecraft and operation system. The spacecraft (Sparn *et al.*, 2005) provides a stable three-axis platform capable of pointing the instruments at the Sun and at selected stellar targets. The spacecraft pointing is generally under star tracker control and is accurate on the order of one arc minute with knowledge better than 0.1 arcmin for subsequent ground processing of the data. The spacecraft uses solar cells and lithium-hydrogen batteries to collect and store power as required to operate all instruments and spacecraft subsystems throughout the orbit. The command and data-handling system collects science and engineering data from the instruments and telemeters it to the ground, and accepts command sequences from the ground to configure and operate the instruments. SORCE has a design lifetime of 5 years – with a goal of six – and this requirement has specified an orbit altitude in excess of 600 km. There was no hard requirement for the inclination of the orbit, and an inclination of 40° was chosen in order to avoid the somewhat harsher radiation of high-inclination orbits and to follow a ground track with suitable ground stations for data transfer (SORCE primarily uses Wallops Island, Virginia, USA and Santiago, Chile). The orbit period is about 97 min and SORCE completes 15 orbits per day. Once per day commands for the spacecraft and instruments are relayed to the satellite from the Mission Operations Center (MOC) at the Laboratory for Atmospheric and Space Physics (LASP), and on one or two ground station passes per day all data are transferred back to the MOC and from there into the LASP Science Operations Center for data processing.

At least 98% of all solar observing opportunities have provided data, and consequently time series extracted from the data base have negligible gaps. The science data processing algorithms (Pankratz *et al.*, 2005) have been refined and updated with the latest calibration files and all data processing is now routine. Reprocessing of the data is carried out on an *ad hoc* basis to implement improvements developed by the SORCE science team or the data processing team.

Although real-time data are examined as they arrive at the MOC, the vast majority of the data is in the playback mode and is processed roughly 24 h after the solar observation. Subsequently the data are examined, validated, and made available in a preliminary form to the scientific community from the LASP SORCE website: *http://lasp.colorado.edu/sorce/*. Additional examination, validation, and correction for instrument degradation are undertaken on an instrument-by-instrument basis, and this process takes from only a few days (as in the case of TIM and XPS) to several weeks (as in the case of SIM and SOLSTICE). Information about the data version, quality, and level of validation accompanies each data file as "header" information. Users of SORCE data are cautioned to pay careful attention to the quality, appropriateness, and reliability of each data set as expressed in these "metadata" files. Pankratz *et al.* (2005) provide details on the SORCE data processing system.

3. SORCE Accomplishments and Contributions

Irradiance measurements are unique in character for they qualify as both *in situ* and *remote sensing* observations. When considered as a climate system variable and used as input to terrestrial and planetary studies irradiance observations are indeed *in situ*. They are the appropriate *in situ* value not only at the satellite location, but also for all other locations across the top of the Earth's sunlit atmosphere. After adjustment to true solar distance, they also qualify as the *in situ* measurement of irradiance for all solar system objects seeing the same hemisphere of the Sun – objects near inferior conjunction and opposition. Measurements appropriate to other ecliptic latitudes and longitudes can be inferred from the measurements made at Earth with some ambiguity depending on the amount of solar variability and activity. Irradiance measurements also provide important information about the Sun, and when used in this context they may be considered a *remote sensing* observation. In effect, they equate to observing the Sun as a star. The extensive information on stellar variability (Radick *et al.*, 1998) corresponds directly to observations and inferences of solar irradiance variability.

Reducing the complex and structured Sun to a single point measurement without the advantage and insight of spatial information on the solar disk is a severe limitation for understanding the sources of irradiance variability, but it is essential for measuring the magnitude of the variability itself. With few exceptions, radiometry derived from solar images and spectroheliograms has never been of sufficient accuracy and duration for reliable detection and determination of solar variability. At X-ray and EUV wavelengths where the solar variations exceed factors of two, images have provided some insight (see for example analysis of the *Yohkoh* data by Acton, Weston, and Brunner, 1999), but not with the cadence, quality, and reliability of a true irradiance XUV instrument. At longer wavelengths, where the solar variations are on the order of, and less than one percent, space-based spectroheliograms do not achieve the precision and accuracy to adequately determine solar irradiance. It is unlikely in the foreseeable future that white-light imagers will have the instrument characterization and stability to achieve the required precision and accuracy of a few hundred parts per million of present day TSI devices. Nevertheless, to take full advantage of irradiance as a *remote sensing* observation of the Sun, and to improve our understanding of the physical processes of solar variability, additional, high spatial resolution images of the Sun are essential. Ideally these (spectral) images are comprehensive and include correlative information on temperature, density, and magnetic field intensity. Physical models that relate the irradiance observations to the physical condition of the Sun exist and are being refined (Fontenla *et al.*, 1999; Solanki and Unruh, 1998).

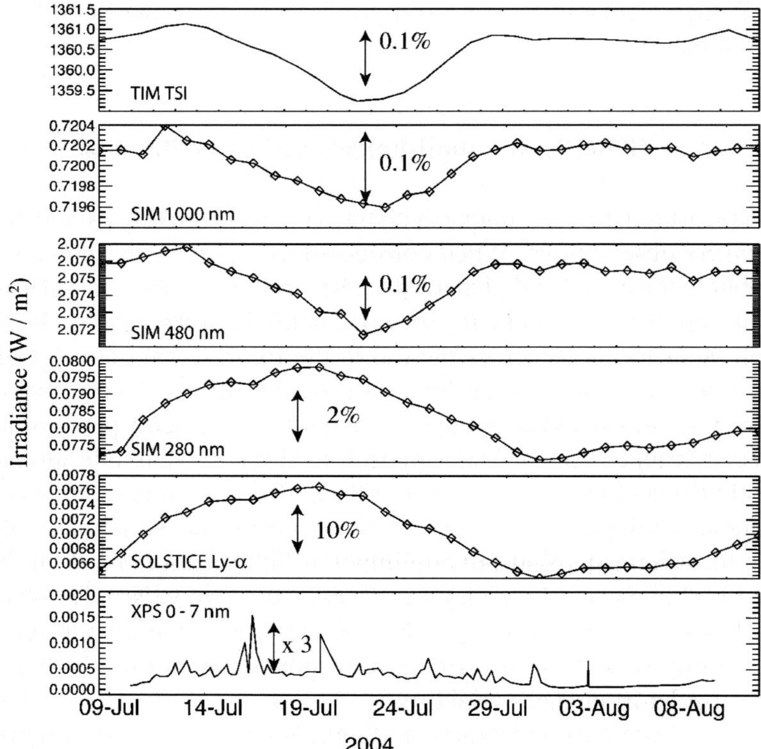

Figure 3. SORCE TSI and multi-spectral data for the time period 1 July 2004 to 1 August 2004. The *top panel* is TSI from TIM, *panels 2–4* are SIM data at 1000, 480, and 280 nm, respectively, *panel 5* is the SOLSTICE Lyman-α, and the *bottom panel* is the XPS 0–7 nm data. These are daily values of the SORCE data and higher time resolution data are available. In particular the XPS data could be presented in 5-min intervals and the data would be far more variable, perhaps by factors of 100 and larger during flares (Woods and Rottman, 2005).

The SORCE data are providing important constraints to the development of these models.

Figure 3 provides a typical example of the types of studies that are now possible with the SORCE data. Shown is a comparison of TSI and five spectral data sets, three from SIM, one from SOLSTICE, and one from XPS. The choice of data is an arbitrary 1-month interval chosen to illustrate the similarities (both striking and subtle) as well as the differences (both striking and subtle) that are present in the irradiance time series. These time series illustrate how the TSI and visible radiation decrease during this 1-month period (27-day solar rotation period) following the passage of a sunspot group across the solar disk. Meanwhile, the UV wavelengths do not carry the sunspot signature but rather are influenced by the active plage that accompanies the sunspot areas. Likewise the amplitude of the variability ranges from a fraction of 1% in the visible and near infrared, increasing to several percent in the ultraviolet and much larger factors in the X-rays. The time signatures are also quite different from a smooth and slowly varying shape in TSI and the visible to the far more rapid fluctuations in the energetic X-rays. The time series of Figure 3 demonstrate that it is complicated and difficult to postulate the behavior of the Sun at one wavelength from information at another.

The intent of this paper is to provide an overview of the SORCE mission, and it is beyond the scope to discuss and interpret the SORCE observations. Fontenla *et al.* (2004) compare SIM data and model calculations with interesting implications on how faculae contribute to the irradiance, especially in the infrared. Woods *et al.* (2004a) provide a study and comparison of irradiance data during the extremely active solar storms in October and November 2003, including the first TSI record of a solar flare. Rottman *et al.* (2005) and Lean *et al.* (2005) also provide initial comparisons of certain SORCE time series.

Understanding the Sun's influence on the terrestrial environment has been a long-term challenge, surrounded by uncertainty and controversy. Presently there is active debate regarding the extrapolation of today's solar observations back in time, for example back to the Maunder Minimum in the mid seventeenth century, and estimates of the "minimum" in the solar irradiance for that time differ by a factor of three (Lean, Skumanich, and White, 1992; Lean, Wang, and Sheeley, 2002; Foukal, North, and Wigley, 2004). A second controversy centers around the ability to determine the solar minimum TSI values during the two recorded solar minima in 1986 and 1996 (Willson, 1997; Willson and Mordinov, 2003; Fröhlich and Lean, 2004) with implications for possible very long-term solar variations. Certainly the improved observations of SORCE – including improved precision and accuracy together with comprehensive and complete wavelength coverage – will provide critical information to help resolve these standing controversies. Just as likely the SORCE data sets are expected to reveal further intricacies about subtle, and not so subtle variations of the Sun.

As the scientific community utilizes SORCE data, extensive interpretation of the observations and new understanding of solar and terrestrial variability will follow. Studies of the Earth atmosphere and climate will incorporate the data in model calculations and comparisons. Meanwhile, solar research will use the data to constrain models that calculate solar radiation and attempt to understand the generation of solar activity. The SORCE data will also identify and improve connections between emission at various wavelengths and between irradiance and solar phenomena, for example sunspots. Such studies improve the reliability of proxy relationships that have provided insight to past solar activity and emission, see for example Foukal and Lean (1988) and Woods *et al.* (2000). In addition to these anticipated uses of the SORCE data, serendipity will certainly provide new discoveries and unexpected results.

The SORCE data will be extensively used in present day solar and Earth science studies. The true test of the value of these data will be their continued use for future climate studies. By being accurately and reliably connected to the *Système*

International unit for irradiance, W m^{-2}, the SORCE data will form a basis for research conducted at all future times.

4. Summary and Conclusions

SORCE launched on January 25, 2003 and continues to function exceptionally well. After more than two years, all of the instruments on SORCE are fully operational and returning high-quality data. The spacecraft systems, which also continue to work extremely well, are fully redundant, and many of the instruments have redundancy also. To date there have been no issues or concerns that even suggest a switch to one of the redundant sides. Operations continue to go smoothly and all experiments are planned and executed efficiently. There have been no operational constraints placed on the spacecraft or any of the instruments.

SORCE had a requirement for an 18-month mission lifetime (as specified in NASA's Minimum Success Criteria for SORCE), and this has now been exceeded. The mission design lifetime is five years – with a goal of six – and there is every expectation that these milestones will be surpassed as well. The next solar minimum will occur in about 2007, and SORCE should continue operations through this minimum and into the rising phase of solar cycle 24.

The SORCE irradiance data will become part of the long-term climate record. The TIM data extend the TSI data that began with the NIMBUS-7 ERBS observations in 1978 and establish a new level of precision and accuracy. The SORCE TSI data are complementary to the ACRIMSAT and SOHO VIRGO TSI data, providing validation and redundancy. The SORCE SOLSTICE observations extend the UV irradiance record beginning with SBUV in 1978 and SME in 1981. SORCE has now had two years overlap with the UARS SUSIM and SOLSTICE observations, and in particular, the stellar comparison technique of SOLSTICE provides a direct and reliable method of tying together the UARS and SORCE data. The SORCE XPS data provide essential overlap and validation with similar data provided by the TIMED XPS, but in addition, SORCE has more extensive wavelength coverage and more frequent observations ($\sim$5 min vs. 90 min). The SORCE SIM observations, however, are quite different than these other three. SIM is providing new and unique data at visible and infrared wavelengths that have not been recorded before. The full wavelength coverage and the long duration of the SIM measurements will provide insight as to how the TSI variations are distributed in wavelength. These findings will have important implications regarding the atmospheric and climate response to solar variability. They will also improve and constrain models that describe the radiation processes and energy balance at the Sun.

There are plans to include the TIM and SIM instruments on the NOAA/DoD/NASA National Polar Orbiting Operational Satellite System, NPOESS. This series of satellites may have a first launch opportunity after 2010. The two irradiance instruments comprise the Total Solar Irradiance System, TSIS, presently scheduled on the second platform with a launch in 2013 (all of these dates are of course subject to change). The NPOESS SIM will have extended capability in the UV down to about 200 nm, but will not recoup the entire spectral range of SOLSTICE down below Lyman-α. The immediate problem is that it is unlikely that SORCE will remain operational for 10 or more years, and there is therefore a threat that a gap will occur in the irradiance record. The break in the data sets is troubling for two reasons, first, missing data can only be filled in with proxy or modeled data leading to ambiguity in the long-term record; and second, the two (or more) observational records that do not overlap must rely on their individual, inherent accuracies to bridge the break in the data – more than likely compromising the entire long-term climate record.

There is a plan to partially fill in a potential gap between SORCE and NPOESS by flying a TIM only (without SIM) on a NASA mission called *Glory*. This mission may be launched in 2008 (again subject to change) and so may overlap with SORCE, but it will take a fortuitous extension to its 3-year lifetime to continue operating into the NPOESS epoch. The *Glory* does not include SIM or any other spectral irradiance measurement capability, and will therefore leave the long-term spectral irradiance record in great jeopardy. Moreover, there is presently no plan that includes the far UV spectral irradiance after SORCE – a fact most troubling to maintaining the long-term UV data record, but also a great loss to process studies in atmospheric chemistry.

For the next 3–5 years SORCE will continue to make irradiance observations, extending the long-term climate forcing record, providing fundamental energy input measurements for atmospheric studies, and likely enabling important new discoveries related to atmospheric sciences and solar physics.

Acknowledgements

The realization of the SORCE Mission is a great accomplishment and a credit to all of the people who contributed including individuals at the University of Colorado, at Orbital Sciences Corp., throughout the NASA organization, and from the many subcontracting entities. Key individuals include T. Sparn, the LASP Program Manager; W. Ochs, the NASA GSFC Project Manager; R. Fulton, the Orbital Project Manager; D. Anderson, the NASA HQ Program Manager; and R. Cahalan, the NASA Project Scientist. Special recognition is extended to the professional staff and students at the Mission Operations Center and Science Operations Center at LASP who attend to the 24-7 care of the spacecraft and instruments with diligence and dedication.

References

Abbot, C. G.: 1948, *Smithsonian Miscellaneous Collections* **110**, Publication 3940.
Acton, L. W., Weston, D. C., and Brunner, M. E.: 1999, *J. Geophys. Res.* **104**, 14827.

Banks, P. M. and Kockarts, G.: 1973, *Aeronomy*, Academic Press, New York.

Eddy, J. A.: 1983, in B. M. McCormac (ed.), *Weather and Climate Responses to Solar Variations,* Colorado Association, University Press, Boulder, CO.

Fontenla, J. M., Harder, J., Rottman, G., Woods, T., Lawrence, G. M., and Davis, S.: 2004, *Astrophys. J.* **605**, L85.

Fontenla, J., White, O. R., Fox, P. A., Avrett, E. H., and Kurucz, R. L.: 1999, *Astrophys. J.* **518**, 1, 480.

Foukal, P. and Lean, J.: 1988, *Astrophys. J. Part I*, **328**, 347.

Foukal, P., North, G., and Wigley, T.: 2004, *Science* **306**, 68.

Friedman, H.: 1961, in William Liller (ed.), *Space Astrophysics*, McGraw-Hill, NY, 107.

Fröhlich, C.: 2004, in J. Pap, C. Fröhlich, H. Hudson, J. Kuhn, J. McCormack, G. North, W. Sprig, and S. T. Wu (eds.), *Solar Variability and Its Effects on Climate,* Geophysical Monograph Series, American Geophysical Union, Washington, DC, **141**, 97.

Fröhlich, C. and Lean, J.: 1998, *Geophys. Res. Ltrs.* **25**, 23, 4377.

Fröhlich, C. and Lean, J.: 2004, *Astron. Astrophys. Rev.* **12**, 4.

Fröhlich, C., Crommelynck, D., Wehrli, C., Anklin, M., Dewitte, S., Fichot, A., Frosterle, W., Jiménez, A., Chevalier, A., and Roth, H. J.: 1997, *Solar Phys.* **175**, 267.

Haigh, J. D.: 2001, *Science* **294**, 2109.

Harder, J., Lawrence, G., Fontenla, J., Rottman, G., and Woods, T.: 2005a, *Solar Phys.*, this volume.

Harder, J., Fontenla, J., Lawrence, G., Woods, T., and Rottman, G.: 2005b, *Solar Phys.*, this volume.

Hoyt, D. V. and Schatten, K. H.: 1993, *J. Geophys. Res.* **98**, 18895.

Kiehl, J. T. and Trenberth, K. E.: 1997, *Bull. Am. Meteorol. Soc.* **78**, 197.

Kopp, G. and Lawrence, G.: 2005, *Solar Phys.*, this volume.

Kopp, G., Heuerman, K., and Lawrence, G.: 2005, *Solar Phys.*, this volume.

Kopp, G., Lawrence, G., and Rottman, G.: 2005, *Solar Phys.*, this volume.

Lawrence, G. M., Rottman, G. J., Harder, J., and Woods, T.: 2000, *Metrologia* **37**, 407.

Lawrence, G. M., Kopp, G., Rottman, G., Harder, J., Woods, T., and Loui, H.: 2003, *Metrologia* **40** S78.

Lean, J.: 1989, *Science* **244**, 197.

Lean, J., Skumanich, A., and White, O.: 1992, *Geophys. Res. Lett.* **19**, 1595.

Lean, J., Wang, Y.-M., and Sheeley, Jr., N. R.: 2002, *Geophys. Res. Lett.* **29**(24), 77.

Lean, J., Rottman, G., Harder, J., and Kopp, G.: 2005, *Solar Phys.*, this volume.

Lee, R. B., Barkstrom, B. R., and Cess, R. D.: 1987, *Appl. Opt.* **26**, 3090.

London, J. and Rottman, G. J.: 1989, in J. Lenoble and J. Geleyn (eds.), *IRS'88: Current Problems in Atmospheric Radiation*, Deepak Publishing Co., pp. 472–473.

McClintock, W. E., Snow, M., and Woods, T. N.: 2005, *Solar Phys.*, this volume.

McClintock, W. E., Rottman, G. J., and Woods, T. N.: 2005, *Solar Phys.*, this volume.

Meier, R. R., Anderson, G. P., Cantrell, C. A., Hall, L. A., Lean, J., Minschwaner, K., *et al.*: 1997, *J. Atmos. Terr. Phys.* **59**, 2111.

Menzel, D. H.: 1949, *Our Sun*, The Blakiston Co., Garden City, NY.

Mihalas, D. and Binney, J.: 1981, *Galactic Astronomy: Structure and Kinematics,* W. H. Freeman and Co., San Francisco, CA.

Nicolet, M. and Kennes, R.: 1989, *Planet. Space Phys.* **37**, 459.

Nicolet, M. and Peetermans, W.: 1980, *Planet. Space Phys.* **28**, 85.

Pankratz, C., Knapp, B., Reukauf, R., Fontenla, J., Dorey, M., Connelly, L., *et al.*: 2005, *Solar Phys.*, this volume.

Powell, F. R., Vedder, P. W., Lindblom, J. F., and Powell, S. F.: 1990, *Opt. Eng.* **26**, 614.

Radick, R. R., Lockwood, G. W., Skiff, B. A., and Baliunas, S. L.: 1998, *Astrophys. J. Suppl. Ser.* **118**, 239.

Rense, W. A.: 1961, *Ann. NY Acad. Sci.* **95**, 33.

Rind, D.: 2002, *Science* **296**, 673.

Rottman, G. J., Floyd, L., and Viereck, R.: 2004, in J. Pap, C. Fröhlich, H. Hudson, J. Kuhn, J. McCormack, G. North, W. Sprig, and S. T. Wu (eds.), *Solar Variability and Its Effects on Climate,* Geophysical Monograph Series, American Geophysical Union, Washington, DC, **141**, 111.

Rottman, G. J., Woods, T. N., and Sparn, T. P.: 1993, *J. Geophys. Res.* **98** 10667.

Rottman, G., Harder, J., Fontenla, J., Lawrence, G., and Woods, T.: 2005, *Solar Phys.*, this volume.

Solanki, S. K. and Unruh, Y. C.: 1998, *Astron. Astrophys.* **329**, 747.

Sparn, T., Rottman, G., Kohnert, R., Anfinson, M., Holden, T., Boyle, B., *et al.*: 2005, *Solar Phys.*, this volume.

Thuillier, G., Floyd, L., Woods, T. N., Cebula, R., Holsenrath, E., Herse, M., *et al.*: 2004, in J. Pap, C. Fröhlich, H. Hudson, J. Kuhn, J. McCormack, G. North, W. Sprig, and S. T. Wu (eds.), *Solar Variability and Its Effects on Climate,* Geophysical Monograph Series, American Geophysical Union, Washington, DC, **141**, 171.

White, O. R.: 1975, *The Solar Output and Its Variation,* Colorado Association, University Press, Boulder, CO.

Willson, R. C.: 1984, *Space Sci. Rev.* **38**, 203.

Willson, R. C.: 1994, *The Sun as Variable Star, Solar and Stellar Irradiance Variations,* Cambridge University Press, Cambridge, UK.

Willson, R. C.: 1997, *Science* **277**, 1963.

Willson, R. C. and Hudson, H. S.: 1991, *Nature* **351**, 42.

Willson, R. C. and Mordinov, A. V.: 2003, *Geophys. Res. Lett.* **30**(5), 3.

Woods, T. N. and Rottman, G.: 2005, *Solar Phys.*, this volume.

Woods, T. N., Rottman, G., and Vest, R.: 2005, *Solar Phys.*, this volume.

Woods, T. N., Eparvier, F. G., Bailey, S. M., Solomon, S. C., Rottman, G. J., Lawrence, G. M., *et al.*: 1998, *SPIE Proc.*, **3442**, 180.

Woods, T. N., Tobiska, W. K., Rottman, G. J., Worden, J. R.: 2000, *J. Geophys. Res.* **105**, 27195.

Woods, T. N., Eparvier, F. G., Fontenla, J., Harder, J., Kopp, G., McClintock, W. E., *et al.*: 2004a, *Geophys. Res. Lett.* **31**, L10802.1.

Woods, T., Acton, L.W., Bailey, S., Eparvier, F., Garcia, G., Judge, D., *et al.*: 2004b, in J. Pap, C. Fröhlich, H. Hudson, J. Kuhn, J. McCormack, G. North, *et al.* (eds.), *Solar Variability and Its Effects on Climate,* Geophysical Monograph Series, American Geophysical Union, Washington, DC, **141**, 127.

Woods, T. N., Eparvier, F. G., Bailey, S. M., Chamberlin, P. C., Lean, J., Rottman, G. J., *et al.*: 2005, *J. Geophys. Res.* **110**, A01312.

Solar Physics (2005) 230: 27–53

SORCE CONTRIBUTIONS TO NEW UNDERSTANDING OF GLOBAL CHANGE AND SOLAR VARIABILITY

JUDITH LEAN
E. O. Hulburt Center for Space Research, Naval Research Laboratory, Washington, DC, U.S.A.

and

GARY ROTTMAN, JERALD HARDER and GREG KOPP
Laboratory for Atmospheric and Space Physics, University of Colorado, Boulder, CO, U.S.A.

(Received 5 May 2005; accepted 23 June 2005)

Abstract. An array of empirical evidence in the space era, and in the past, suggests that climate responds to solar activity. The response mechanisms are thought to be some combination of direct surface heating, indirect processes involving UV radiation and the stratosphere, and modulation of internal climate system oscillations. A quantitative physical description is, as yet, lacking to explain the empirical evidence in terms of the known magnitude of solar radiative output changes and of climate sensitivity to these changes. Reproducing solar-induced decadal climate change requires faster and larger responses than general circulation models allow. Nor is the indirect climatic impact of solar-induced stratospheric change adequately understood, in part because of uncertainties in the vertical coupling of the stratosphere and troposphere. Accounting for solar effects on pre-industrial surface temperatures requires larger irradiance variations than present in the contemporary database, but evidence for significant secular irradiance change is ambiguous. Essential for future progress are reliable, extended observations of the solar radiative output changes that produce climate forcing. Twenty-five years after the beginning of continuous monitoring of the Sun's total radiative output, the Solar Radiation and Climate Experiment (SORCE) commences a new generation of solar irradiance measurements with much expanded capabilities. Relative to historical solar observations SORCE monitors both total and spectral irradiance with significantly reduced uncertainty and increased repeatability, especially on long time scales. Spectral coverage expands beyond UV wavelengths to encompass the visible and near-IR regions that dominate the Sun's radiative output. The space-based irradiance record, augmented now with the spectrum of the changes, facilitates improved characterization of magnetic sources of irradiance variability, and the detection of additional mechanisms. This understanding provides a scientific basis for estimating past and future irradiance variations, needed for detecting and predicting climate change.

1. Introduction

A balance between incoming solar radiation (with peak flux near 500 nm) and outgoing radiation from the much cooler terrestrial surface (with peak flux near 10 μm) establishes Earth's global mean temperature (e.g., Pilewskie and Rottman, 2005). When this radiative balance is perturbed, for example by a change in solar radiation, atmospheric composition or surface reflectivity, Earth's surface temperature responds by seeking a new equilibrium. This response, which is in the range of 0.3 –1 °C per Wm^{-2} of forcing (Intergovernmental Panel on Climate Change, 2001) alters climate. Significant climate change can accompany even modest changes in global temperature. For example, during the last ice age 20 000 years ago, globally averaged temperatures were 5 °C cooler than at present; a response to forcing of 6.5 Wm^{-2} (Hansen, 2004).

As a part of NASA's Earth Observing System, the Solar Radiation and Climate Experiment (SORCE) seeks new understanding of the Sun's role in global change by measuring total and spectral solar irradiance, quantifying the solar sources of observed variations, and investigating the responses to these variations of Earth's climate and atmosphere (Rottman, 2005). Changes in solar electromagnetic radiation reaching Earth's surface perturb the radiative balance directly. Changes in solar UV radiation alter ozone and may have an indirect influence via coupling of the middle atmosphere with the surface by both radiative and dynamical processes (Haigh, 2001; Rind, 2002).

SORCE aims to specify daily total solar irradiance with an uncertainty of less than 100 ppm (0.01%) and repeatability of 0.001% per year, and daily solar ultraviolet irradiance from 120 to 300 nm with a spectral resolution of 1 nm, an uncertainty of better than ±5%, and repeatability of ±0.5%. These observations of total and UV spectral irradiance continue extant databases that now exceed, respectively, 27 and 14 years. SORCE is also making the first precise daily measurements of solar spectral irradiance between 0.3 and 2 μm with a goal of ±0.1% uncertainty and ±0.01% per year repeatability. These observations commence new databases of visible and near-IR solar irradiances. So that the SORCE observations may interface with the historical irradiance data and with future operational monitoring by the National Polar-orbiting Operational Environmental Satellite System (NPOESS), relationships must be identified with all concurrent, overlapping observations. Biases in absolute calibrations must be established and understood, as must differences in temporal trends among independent radiometers.

New understanding of the solar sources of the irradiance variations observed by SORCE permit improved models of contemporary irradiance variability. The parameterizations of the sunspot and facular sources in these models are the basis of reconstructions of past and future total and spectral irradiance changes. Reliable spectral irradiance time series enable more robust empirical and theoretical studies of Earth's surface, and ocean and atmospheric variability. Most studies conducted thus far have used proxies for solar variability, such as the 10.7 cm radio flux (e.g., Gleisner and Thejll, 2003; Labitzke, 2004), rather than the actual irradiances. Climate change simulations typically use total rather than spectral irradiance to specify solar forcing (e.g., Tett *et al.*, 2002; Meehl *et al.*, 2003). SORCE's spectral irradiance observations will enable more realistic climate model simulations for comparison with empirical evidence and projections of future change, in comparison with other climate influences.

2. Solar Influences on Global Change

New understanding of solar influences on global change is emerging from a variety of investigations. Empirical comparisons utilize databases that characterize the surface, ocean and atmosphere observationally; process studies seek to quantify mechanisms by which the Earth system responds to various forcings, and modeling simulations attempt to integrate the empirical evidence and the physical understanding. These efforts focus on three primary time scales – decadal variability in the current epoch of high quality observations, especially from space; centennial variability during the past millennium, for which actual temperature changes are known directly from measured records and indirectly from reconstructions (Jones and Mann, 2004); and centennial and millennial variability in the Holocene (e.g., Bond *et al.*, 2001).

2.1. Empirical Evidence

Comprehensive records now exist for a multitude of climate forcings (solar irradiance, greenhouse gas and volcanic aerosol concentrations), feedbacks (cloud cover and cloud properties, water vapor), internal oscillations (El Niño-Southern Oscillation/ENSO, North Atlantic Oscillation/NAO, Quasi-Biennial Oscillation/QBO) and climate itself (surface temperatures, rainfall, circulation patterns), including the overlying atmosphere (temperature, ozone, geopotential heights, winds). The records of the past 25 years sample a range of natural and anthropogenic radiative forcing strengths and internal modes. The period includes the El Chichón and Pinatubo volcanoes, almost three solar activity cycles (including the most recent cycle free of volcanic interference), a few major ENSO events, and significant increases in greenhouse gases, chlorofluorocarbons (CFCs), and tropospheric (industrial) aerosols.

A linear combination of solar, anthropogenic (combined greenhouse gases and industrial aerosols), volcanic and ENSO influences can account for approximately 50% of the observed variance in global surface temperature (as reported by GISS, the Goddard Institute for Space Studies) between 1979 and 2004. Figure 1 compares the relative strength of each influence, derived from multiple regression analysis of monthly data. Hansen *et al.* (2002) describe the datasets. The total solar irradiance is that modeled by Lean (2000), the surface temperatures are combined land and ocean records, volcanic aerosols are from Sato *et al.* (1993), and ENSO is depicted by the Multivariate ENSO Index of Wolter and Timlin (1998).

According to the correlation coefficients listed in Table I, ENSO, solar and volcanic influences account for, respectively, 1.5, 4, and 13% of the monthly mean global surface temperature variance over this 25-year period. The anthropogenic influence accounts for a significantly larger 43%. The surface warms 0.1 °C at solar cycle maxima (forcing of 0.2 Wm^{-2}) and 0.39 °C overall from anthropogenic

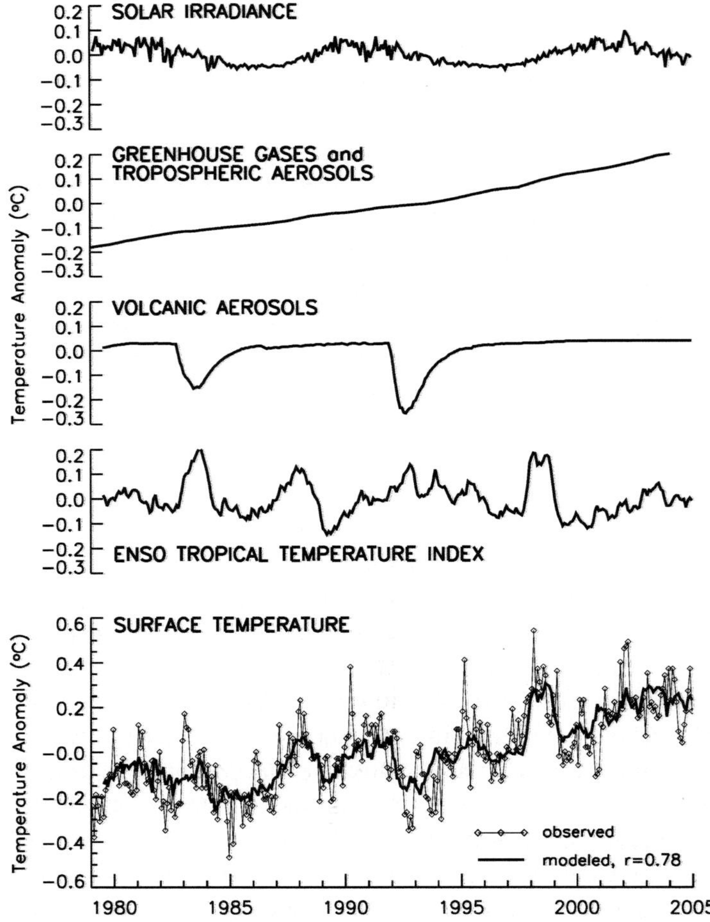

Figure 1. Comparison of different sources of variance in Earth's recent surface temperature, deduced from multiple regression analysis. The volcanic aerosols and ENSO indices are lagged by 6 months, and solar irradiance by 1 month, relative to the observed surface temperatures.

influences (forcing of $1\,\mathrm{Wm}^{-2}$). ENSO and volcanic activity produce episodic fluctuations that can exceed 0.2 °C. Douglass and Clader (2002) reported similar results for lower tropospheric temperatures measured by the microwave sounding unit.

The solar signal in surface temperature shown in Figure 1 is consistent with other detections of decadal solar effects in the ocean (White, Dettinger, and Cayan, 2003) and atmosphere (van Loon and Shea, 2000; Coughlin and Tan, 2004; Labitzke, 2004), and with several independent analyses that further explore the meridional and height dependences of various forcings. Overall, the troposphere is warmer, moister, and thicker during solar maximum, with a distinct zonal signature. The strongest response occurs near the equator and at mid latitudes (40–50 °) with

TABLE I

Comparison of solar and other contributions to variance in monthly mean global surface temperatures in recent decades and in the past century.

Process	1979–2004 (306 months)	1882–2004 (1469 months)
Solar irradiance	0.211	0.506
Anthropogenic gases	0.654	0.801
Volcanic aerosols	−0.361	−0.137
ENSO	0.124	0.278
Model (all of above)	0.778	0.846

Listed are the correlations of the reconstructed components of individual processes with surface temperatures using the time series shown in Figures 1 and 2. Squared values give explained variance.

subtropical minima (Gleisner and Thejll, 2003). The primary surface temperature expression of these changes is warming in two mid-latitude bands (increases of 0.5 K at 20 – 60° N and S) that extend vertically to the lower stratosphere where they expand equatorward (Haigh, 2003). The patterns suggest that solar forcing invokes dynamical responses in the troposphere, involving the Hadley, Walker, and Ferrel circulation cells (Kodera, 2004; van Loon, Meehl, and Arblaster, 2004).

The relative influences of solar and other climate forcings are less certain prior to the era of space-based observations. Figure 2 and Table I compare relative strengths of ENSO, solar, volcanic and anthropogenic influences on monthly global surface temperatures between 1882 and 2004 using the parameterizations deduced after 1979. Together these influences account for 72% of the observed variance. In this figure, the MEI ENSO index is extended prior to 1950 by the Japan Meteorological Index. The solar component is obtained by reducing the background component in the total solar irradiance reconstruction of Lean (2000) to be consistent with the recent model of Wang, Lean, and Sheeley (2005) in which the secular increase is 27% that of earlier irradiance reconstructions (e.g., Lean, 2000; Fligge and Solanki, 2000). With this new irradiance model, the secular solar-induced surface temperature increase of 0.06 °C since 1880 is more than a factor of 10 smaller than the 0.7 °C warming attributed to anthropogenic influences. Solar-related global warming since the seventeenth century Maunder Minimum is of order 0.1 °C, or less, which is smaller than suggested by previous studies in which reconstructed solar irradiance changes were larger (Lean, Beer, and Bradley, 1995; Crowley, 2000; Rind *et al.*, 2004).

Temperature responses to the solar cycle increase with altitude, from 0.1 K near the surface (Figure 1) to 0.3 K at 10 km, and 1 K around 50 km (van Loon and Shea, 2000). Accompanying the temperature changes is a solar cycle in global total ozone of ∼3% peak-to-peak amplitude. As with tropospheric climate, solar-induced

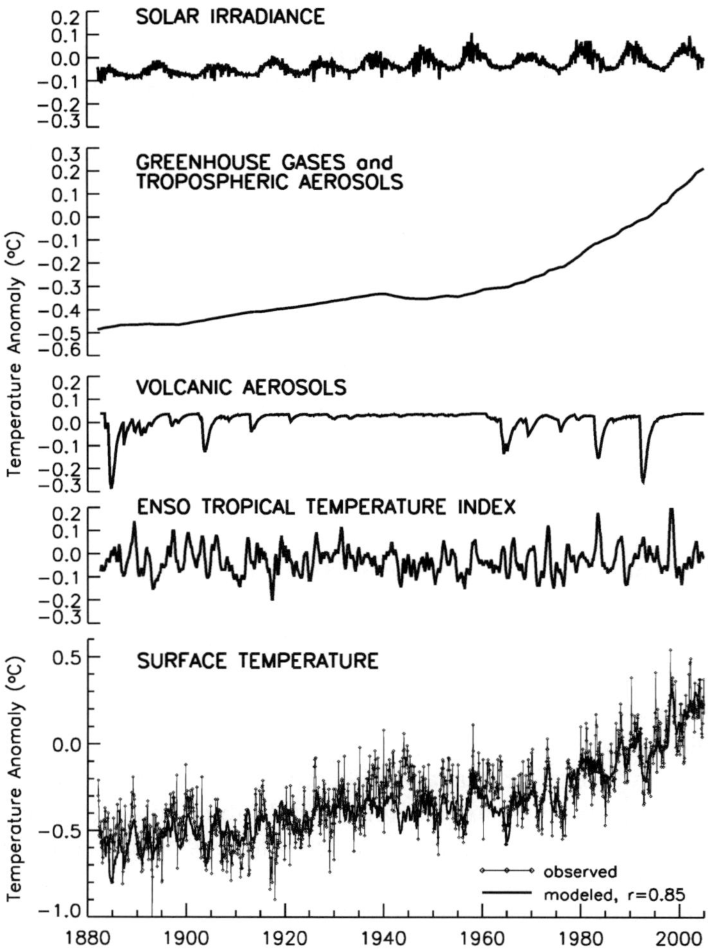

Figure 2. Shown are empirical estimates of the different sources of variance contributing to changes in the Earth's observed surface temperature during the past century, based on an extension of the multiple regression parameterizations in Figure 1.

changes in atmospheric temperature and composition occur simultaneously with anthropogenic effects and internal variability (e.g., QBO) (Jackman *et al.*, 1996; Geller and Smyshlyaev, 2002). Following the approach of McCormack *et al.* (1997) and Fioletov *et al.* (2002), Figure 3 illustrates the solar and other components of ozone variance extracted by statistical regression analyses of deseasonalized monthly total ozone from 1979 to 2004. Almost 80% of the total variance is explained by the combined effects of the QBO (9%), anthropogenic chlorofluorocarbons (39%), solar UV irradiance (42%) and volcanic ($\sim$1%) activity. In this figure the ozone record is the Version 8 TOMS merged ozone dataset constructed by Goddard Space Flight Center, the QBO is the 30 mb zonal wind from the National Weather Service

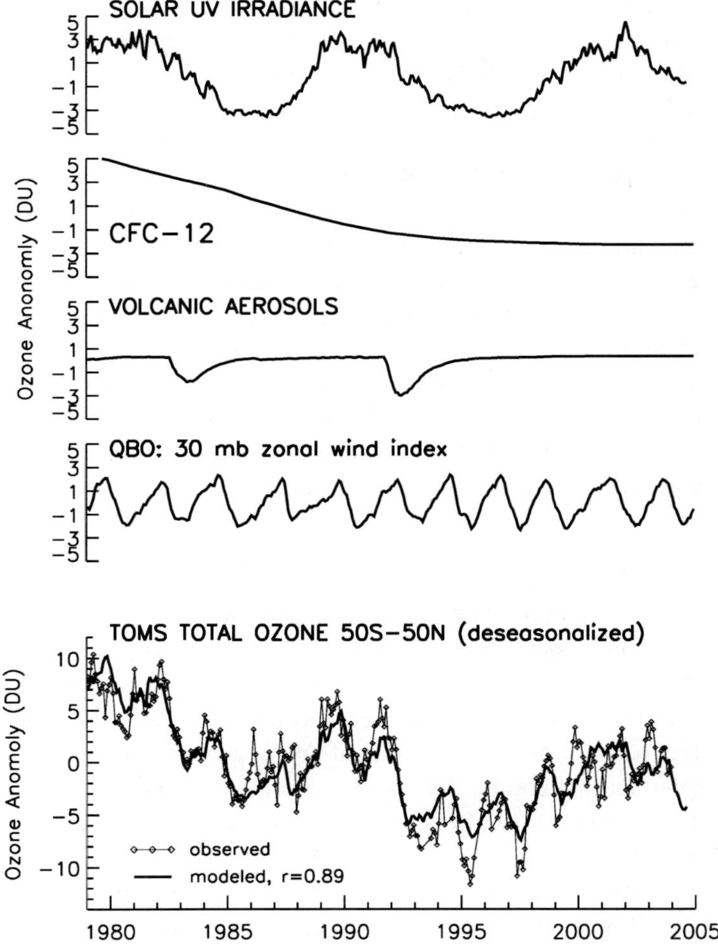

Figure 3. Comparison of difference sources of variance contributing to changes in observed, deseasonalized total ozone concentrations, deduced from multiple regression analysis.

Climate Prediction Center, and the UV irradiance is a band from 200 to 295 nm modeled by Lean (2000). The solar-induced ozone changes vary with geographical location and altitude in ways that are not clearly related linearly to the distribution of the forcing. For example, solar forcing appears to induce a significant and unexpected (from a modeling perspective) lower stratospheric response (Hood, 2003).

2.2. Mechanisms

At least three distinct mechanisms are surmised for climate's response to solar radiative forcing. Irradiance changes in the near-UV, visible, and near-IR spectrum can directly affect the Earth's radiative balance and surface temperature. Ultraviolet irradiance changes can alter the stratosphere whose coupling to the troposphere provides an indirect climate effect. Varying irradiance may interact with internal modes of climate variability (ENSO, NAO, and the QBO) and climate noise, triggering, amplifying or shifting the modes. Each mechanism is expected to have an individual geographical, altitudinal, and temporal response pattern.

An unequivocal determination of specific mechanisms has yet to be accomplished. As a result, alternate explanations are often proffered for common empirical evidence. For example, an apparent relationship between solar variability and cloud cover has been interpreted as a result of (1) sea surface temperatures altered directly by changing total solar irradiance (Kristjánsson *et al.*, 2002), (2) solar-induced changes in ozone (Udelhofen and Cess, 2001), (3) internal variability by ENSO (Kernthaler, Toumi, and Haigh, 1999), and (4) changing cosmic ray fluxes modulated by solar activity in the heliosphere (Usoskin *et al.*, 2004). In reality, different physical processes may operate simultaneously.

2.2.1. *Direct Surface Heating*

The near-UV, visible, and near-IR radiations that compose almost 99% of the Sun's total radiative output penetrate Earth's atmosphere to the troposphere and surface. Some 31% of the incident solar radiation is reflected back to space, the lower atmosphere absorbs 20%, and the surface and oceans absorb the remaining 49% (Kiehl and Trenberth, 1997). Geographical and seasonal inhomogeneities of this short-wave solar heating couple with land – ocean and cloud cover distributions to produce thermal contrasts that alter coupled land – atmosphere – ocean interactions (Rind and Overpeck, 1993; Meehl *et al.*, 2003). As a result, the regional response to solar forcing may be significant even when the net global change is modest. The heating is thought to stimulate vertical motions that involve the Hadley cell and affect monsoons. The response may depend on the background state of the climate system, and thus on other forcings such as greenhouse gases (Meehl *et al.*, 2003) and volcanic aerosols (Donarummo, Ram, and Stolz, 2002).

The empirical evidence suggesting a significant (0.1 K) surface temperature response to solar forcing (e.g., Figure 1), approximately in-phase with the solar cycle, is inconsistent with current understanding that oceanic thermal inertia strongly dampens (by a factor of 5) forcing at a period of the decadal solar cycle (Wigley and Raper, 1990). This suggests that the effect primarily involves the atmosphere and surface, but does not engage the deep ocean.

2.2.2. *Indirect Effects through the Stratosphere*

The Earth's atmosphere absorbs about 15 Wm^{-2} ($\sim$1%) of the Sun's radiant energy, in the ultraviolet portion of the spectrum. Solar UV radiation is more variable than total solar irradiance by at least an order of magnitude. It contributes significantly to changes in total solar irradiance (15% of the total irradiance cycle, Lean *et al.*, 1997) but is unavailable for direct forcing of climate because it does not reach the Earth's surface. Solar UV radiation creates the ozone layer (initially by photodissociating molecular oxygen in the atmosphere) and its effect on climate depends on the coupling of the stratosphere (where ozone primarily resides) with the troposphere (Haigh *et al.*, 2004). Both radiative and dynamical couplings are surmised.

Because ozone absorbs electromagnetic radiation in the UV, visible, and IR spectral regions, changes in ozone concentration alter Earth's radiative balance by modifying both incoming solar radiation and outgoing terrestrial radiation. Solar-driven radiative coupling effects of this type may influence not only surface temperature (Lacis, Wuebbles, and Logan, 1990) but dynamical motions such as the strength of the Hadley cell circulation, with attendant effects on, for example, Atlantic storm tracks (Haigh, 2001). Because ozone controls solar energy deposition in the stratosphere, its variations alter both the altitudinal temperature gradient from the troposphere to the stratosphere, and the latitudinal gradient in the stratosphere, from the equator to the poles. These changes are postulated to propagate surface-wards through a cascade of feedbacks involving thermal and dynamical processes that alter winds and the large scale planetary waves (Rind, 2002). Equatorial winds in the stratosphere appear to play an important role in this process because of their impact on wind climatology (Matthes *et al.*, 2004).

2.2.3. *Indirect Effects by Alteration of Internal Climate Variability*

Even with little or no global average response, solar radiative forcing may nevertheless influence climate by altering one or both of two main variability modes; ENSO (Neelin and Latif, 1998) and the NAO (Wallace and Thompson, 2002). Since the climate system exhibits significant "noise" the forcing may be amplified by stochastic resonance (Ruzmaikin, 1999). Also possible is the non-linear interaction of the forcing with existing cyclic modes. Such frequency modulation has been demonstrated on Milankovitch time scales (Rial, 1999).

Positive radiative forcing, including by solar variability, may suppress the frequency and occurrence of ENSO (Mann *et al.*, 2005) because of sea surface temperature gradients arising from the deeper thermocline in the west Pacific Ocean relative to the east. Solar UV irradiance changes may alter the high latitude stratospheric and the polar vortex, thereby affecting the NAO (Shindell *et al.*, 2003), which is observed to expand longitudinally to the Artic annual oscillation during solar maxima (Kodera, 2002). The phase of the quasi-biennial oscillation in stratospheric equatorial winds possibly modulates this interaction (Ruzmaikin and Feynman, 2002). That the phase of the QBO changes with the solar cycle (McCormack, 2003; Salby and Callaghan, 2004), as part of a pattern of non-linear stratospheric response to the 11-year cycle involving both the QBO and the SAO, underscores the complicated, multifaceted nature of solar influences on global change.

3. Solar Irradiance Variability

Empirical studies such as those in Figures 1–3, and theoretical investigations of climate processes and change, require reliable knowledge of solar spectral irradiance on multiple time scales. Continuous space-based measurements with adequate precision to detect real variations in total irradiance exist since 1978, in UV irradiance since 1991 and in visible and near-IR irradiance since 2003. For prior periods and future projections, irradiance variations are estimated using models that account for the changes observed in the contemporary era, in combination with proxies of solar activity recorded in the past and predicted for the future.

3.1. Observations

3.1.1. *Total Irradiance*

Four space-based instruments measure total solar irradiance at the present time. SORCE's Total Irradiance Monitor (TIM) (Kopp, Lawrence, and Rottman, 2005), together with the radiometers of the Variability of Irradiance and Gravity Oscillations (VIRGO) experiment on the Solar Heliospheric Observatory (SOHO), the ACRIM III on the Active Cavity Radiometer Irradiance Monitor Satellite (ACRIMSAT), and the Earth Radiation Budget Satellite (ERBS), contribute to a database that is uninterrupted since November 1978. Figure 4 compares three composite irradiance records obtained from different combinations of measurements. While the gross temporal features are clearly very similar, the slopes differ, as do levels at solar activity minima (1986 and 1996).

Secular trends differ among the three composite irradiance records because of different cross-calibrations and drift adjustments applied to individual radiometric sensitivities. The PMOD composite (Fröhlich and Lean, 2004) combines the observations by the ACRIM I on the Solar Maximum Mission (SMM), the Hickey–Friedan radiometer on Nimbus 7, ACRIM II on the Upper Atmosphere Research Satellite (UARS), and VIRGO on SOHO by analyzing the sensitivity drifts in each radiometer prior to determining radiometric offsets. In contrast, the ACRIM composite (Willson and Mordvinov, 2003), which utilizes ACRIMSAT rather than VIRGO observations in recent times, cross-calibrates the reported data assuming that radiometric sensitivity drifts have already been fully accounted for. For the Space Absolute Radiometric Reference (SARR) composite, individual absolute irradiance measurements from the shuttle are used to cross-calibrate satellite records (Dewitte *et al.*, 2005).

Solar irradiance levels are likely comparable in the two most recent cycle minima when absolute uncertainties and sensitivity drifts in the measurements are assessed (Fröhlich and Lean, 2004). The upward secular trend of 0.05% proposed by Willson and Mordvinov (2003) may be of instrumental rather than solar origin. This irradiance "trend" is not a slow secular increase but a single episodic increase between

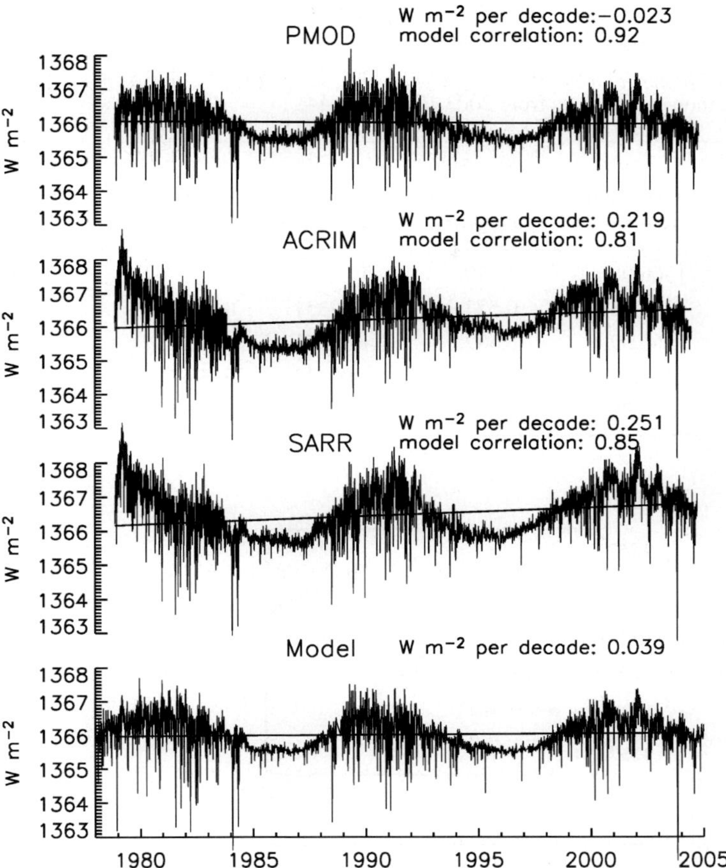

Figure 4. Shown in the *upper three panels* are different composite records of total solar irradiance during the era of space-based monitoring. For quantitative comparison, the slopes of the time series are computed from 7538 daily values between November 1978 and June 2004. Compared in the *bottom panel* is a model of total solar irradiance calculated from parameterized sunspot and facular influences.

1989 and 1992 that is present in the Nimbus 7 data. Independent, overlapping ERBS observations do not show a comparable increase at this time (Lee III *et al.*, 1995). The trend is absent in the PMOD composite, in which total irradiance at successive solar minima is constant to better than 0.01%. Although a long-term trend is present in the SARR composite, the increase of 0.15 $\mathrm{Wm^{-2}}$ between successive solar activity minima (in 1986 and 1996) is not significant because the uncertainty is $\pm 0.35\ \mathrm{Wm^{-2}}$.

SORCE's TIM observations, shown in Figure 5, aim at reducing such instrumental uncertainties in the long-term irradiance record. Table II compares TIM's mean irradiance and standard deviation with each of the irradiance time series in Figure 4 for the duration of the SORCE mission thus far. On average, TIM measures

TABLE II

Compared are absolute values, standard deviations, and trends of TIM observations with the three composite irradiance records and the empirical model in Figure 4, during the SORCE mission thus far (415 common daily values from 2003.15 to 2004.44).

TSI record	Mean value (Wm^{-2})	Standard deviation (Wm^{-2})	Ratio to TIM	Correlation with TIM	Slope (Wm^{-2} per year)
TIM	1360.98	0.579	1	1	−0.182 (0.013%)
PMOD	1365.76	0.573	1.00351	0.9964	−0.134 (0.010%)
ACRIM	1366.09	0.582	1.00375	0.9829	−0.261 (0.019%)
SARR	1366.73	0.577	1.00423	0.9967	−0.191 (0.014%)
Model	1365.95	0.479	1.00365	0.9634	−0.069 (0.005%)

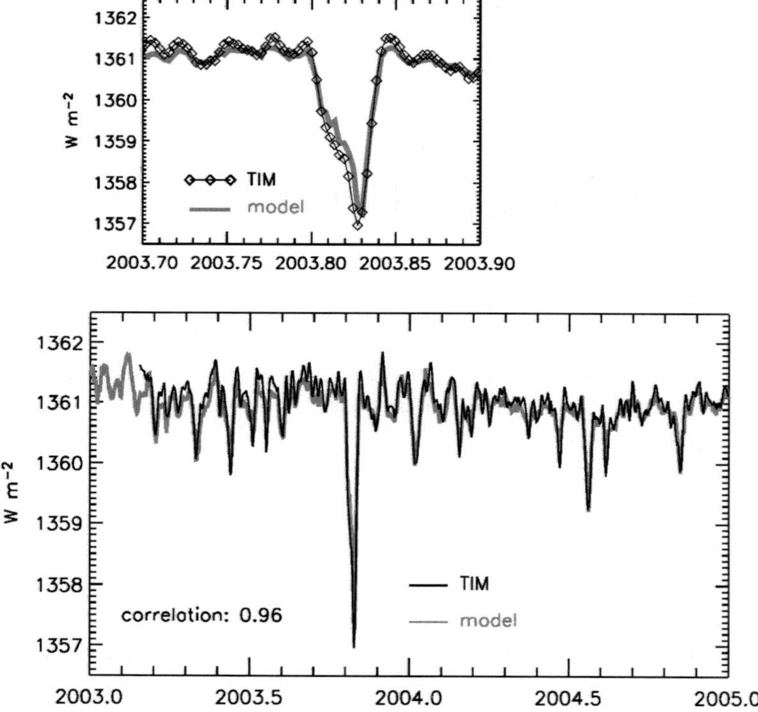

Figure 5. Shown are the TIM daily mean observations (symbols with *dark line*), compared with an empirical variability model developed from sunspot and facular influences (Lean, 2000). In the *upper panel* the observations and model are compared during the period of high solar activity in October 2003. In the *lower panel* the observations and model are shown for the duration of the SORCE mission thus far.

absolute solar irradiances 5.2 Wm^{-2} (0.4%) lower than the other radiometric time series. This difference is an order of magnitude larger than the combined uncertainties claimed for the respective measurements (e.g., ±0.01% for SORCE; ±0.025% for SARR) and the discrepancy is under investigation.

Standard deviations in Table II, which primarily reflect true solar irradiance variations, are on average comparable (~0.04%) in TIM and the PMOD, ACRIM and SARR composites records. The notably different irradiance trends from 2003.15 to 2004.44 likely arise from residual instrumental drifts in the reported measurements. During this time of overall decreasing solar activity with the approach of solar minimum, TIM's downward slope is 1.55 times that of the PMOD composite, but 0.57 times that of the ACRIM composite.

3.1.2. *Spectral Irradiance*

With its Spectral Irradiance Monitor (SIM, Harder *et al.*, 2005) and Solar Stellar Irradiance Comparison Experiment (SOLSTICE, McClintock, Rottman, and Woods, 2005), SORCE monitors the Sun's spectral irradiance almost simultaneously across ultraviolet, visible, and near-IR regions, for the first time from space on a daily basis with sufficient precision to detect real changes.

The overlap in time and wavelength of the SORCE SOLSTICE UV measurements with those made since 2000 by the EUV Grating Spectrometer on the Thermosphere Ionosphere Mesosphere Energetics and Dynamics (TIMED) spacecraft (Woods *et al.*, 2005) extends spectral irradiance information to extreme ultraviolet wavelengths. The X-ray photometer systems on SORCE (Woods and Rottman, 2005) and TIMED complete the spectral coverage. In the UV spectrum, SORCE continues the spectral irradiance observations made since October 1991 by an earlier SOLSTICE instrument and the Solar Ultraviolet Spectral Irradiance Monitor (SUSIM), both flown on the Upper Atmosphere Research Satellite (Woods *et al.*, 1996).

The comprehensive spectral coverage of the SORCE instruments provides unprecedented characterization of solar irradiance variability. As expected, variations occur at all wavelengths. The comparisons in Figure 6 and Table III illustrate the changes in the solar spectrum accompanying the increase in solar activity from 17 to 30 October 2003. During this time, the Sun's visible surface, shown in Figure 7, evolved from being almost sunspot free, to having significant sunspot coverage. At the same time, total solar irradiance is seen in Figure 5 to decrease by 4 Wm^{-2} (0.3%). The middle panel in Figure 6 shows the corresponding decreases in spectral irradiance energy, while the lower panel shows the fractional changes. Maximum energy changes occur at wavelengths from 400 to 500 nm, whereas fractional changes, listed numerically in Table III are greatest at UV wavelengths, where the energy change is, however, considerably smaller.

Radiation in the UV spectrum has a notably different temporal character during solar rotation than the spectrum above 300 nm, as the time series in Figure 8 illustrate. The standard deviation of the 200–300 nm time series in Figure 8 is 0.15%,

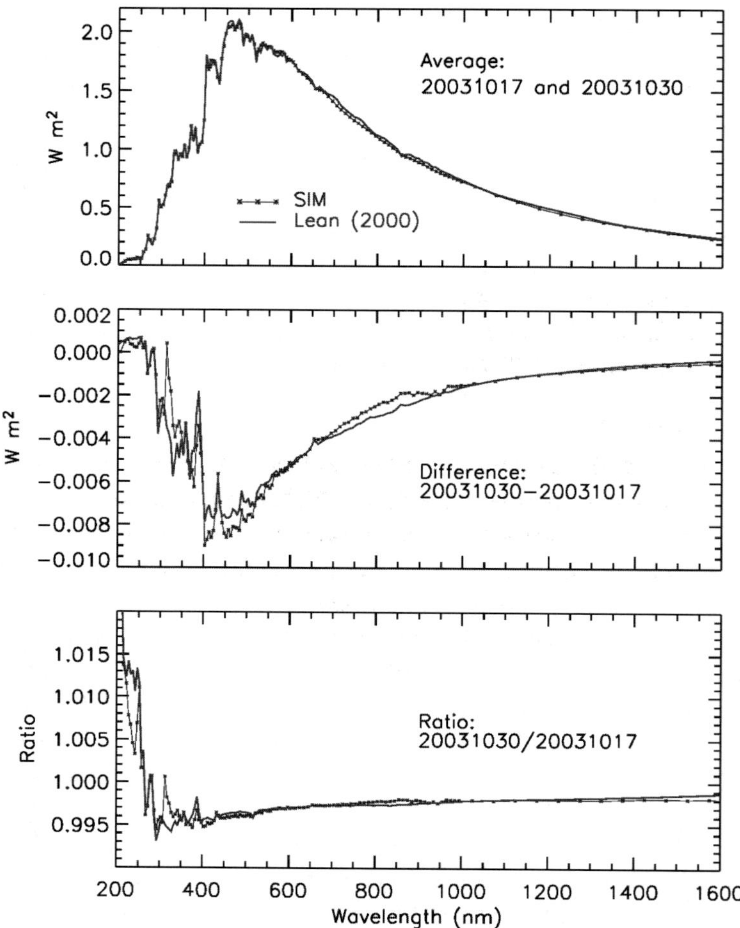

Figure 6. The solar spectral irradiance measured by SIM on SORCE, shown in the *upper panel*, is the average of two spectra, on 17 and 30 October 2003. Spectral irradiance changes caused by significant solar activity on 30 October, relative to quieter conditions on 17 October, are shown in the *middle panel* as energy differences, and the *lower panel* as fractional changes. A model of the irradiance variations caused by sunspots and faculae is compared with the SORCE observations.

decreasing to 0.04% for radiation in the wavelength band from 400 to 700 nm, and to 0.03% at 1000 to 1600 nm. These differences reflect the different solar origins of irradiance variability since the observations relate to emission from a range of temperatures and structures within solar atmosphere.

3.2. Models

3.2.1. *Present*

Two decades of solar observations and analysis have established the primary roles of sunspots and faculae in causing solar irradiance to vary (Fröhlich and Lean, 2004).

TABLE III

Variations in spectral irradiance bands during the strong solar rotation of October 2003, observed by SORCE and estimated from a model of facular and sunspot influences.

Spectral band (nm)	SORCE rotation 30 Oct 2003/ 17 Oct 2003	Model rotation 30 Oct 2003/ 17 Oct 2003	Model solar cycle 1989/1986	Model secular change 1713/1986 (Wang, Lean, and Sheeley, 2005)	Model secular change 1713/1986 (Lean, 2000)
200–300	0.9990	0.9993	1.013	0.9957	0.9864
315–400	0.9959	0.9956	1.002	0.9990	0.9968
400–700	0.9965	0.9967	1.0008	0.9995	0.9983
700–1000	0.9979	0.9975	1.0004	0.9997	0.9990
1000–1600	0.9980	0.9982	1.00025	0.9998	0.9994

The modeled changes are also given for the solar cycle (Lean *et al.*, 1997) and Maunder Minimum, for which the recent estimates of Wang, Lean, and Sheeley (2005) are compared with earlier estimates of Lean (2000).

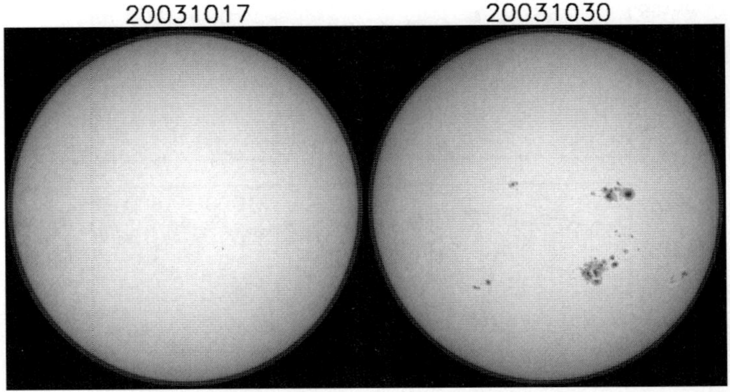

Figure 7. Continuum-light solar images made by the Michelson Doppler Imager instrument on SOHO are compared for 17 and 30 October 2003, the two days of the solar spectra compared in Figure 6.

Sunspots deplete the Sun's local emission so that their presence on the disk reduces irradiance, especially in the visible and infrared spectral regions. TIM and SIM record irradiance fluctuations that are the net effect of sunspot-induced reductions and facular-induced enhancement. The two influences compete continually as active regions emerge, evolve and decay on the solar surface, altering the relative strengths and phase of the sunspot and facular emissions (Rottman *et al.*, 2005).

SORCE instruments observed the dramatic effects of active regions on irradiance when sunspot darkening increased dramatically and faculae brightening more modestly from 17 to 30 October 2003 (Figures 7 and 9). Total solar irradiance decreased 4 $\mathrm{W m^{-2}}$ (Figure 5), as exceptionally large sunspots transited the Earth-facing solar disk (Figure 7). As Figure 6 shows, the sunspots depleted the entire solar spectrum at wavelengths between 350 and 1600 nm.

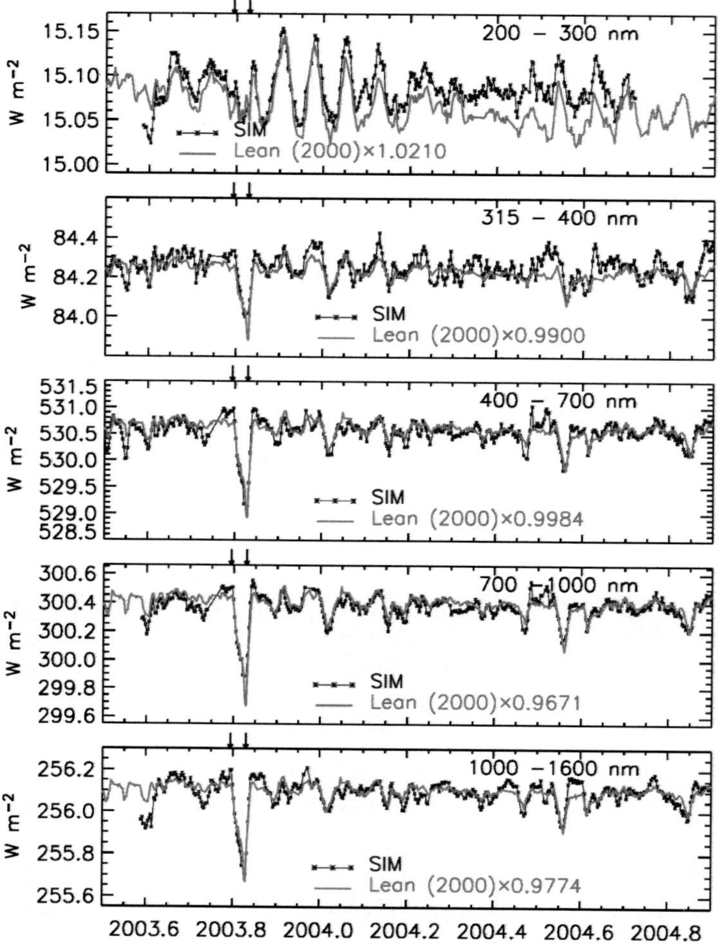

Figure 8. Compared are SORCE observations (*symbols*) and empirical variability model values (*solid line*) of irradiance in the: (a) middle ultraviolet, 200–300 nm, (b) near-UV, 315–400 nm, (c) visible, 400–700 nm, (d) visible, 700–1000 nm, and (e) near-IR, 1000–1600 nm wavelength bands. SORCE observations are made by SIM, except at wavelengths between 200 and 210 nm, which are made by SOLSTICE. The SORCE time series have been detrended by subtracting a 30-day running mean, to remove known instrumental drifts not yet incorporated in the data reduction algorithms. The model time series are scaled by the values shown in each panel to agree with the SIM absolute scale. Table III lists fractional changes of the time series from 17 to 30 October 2003 (indicated by the *arrows*).

Models that quantify the sunspot and facular influences on solar irradiance have been developed using a variety of approaches. Typically, the sunspot component is calculated directly using information about the location and size of all sunspots on the disk, obtained from visible images (Lean *et al.*, 1998), magnetograms (Krivova *et al.*, 2003), or images at other wavelengths (Preminger, Walton, and Chapman, 2002). More diverse is the approach for estimating brightness enhancements in

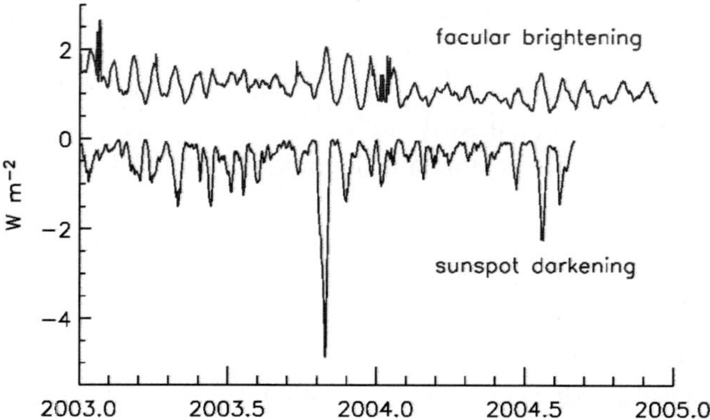

Figure 9. Shown are variations in the bolometric faculae brightening (*upper curve*) and sunspot darkening (*lower curve*) that together produce changes in total solar irradiance observed by TIM. The combination of these competing effects produces the modeled irradiance in Figure 5. Linear scalings that determine the relative contributions of the sunspots and facular are obtained from multiple regression of these time series with the PMOD composite in Figure 4.

faculae, which have lower contrasts and are more widely dispersed over the solar disk than are sunspots. Because photospheric faculae usually underlie bright chromospheric active regions, Ca II K images are often used (Lean *et al.*, 1998; Walton, Preminger, and Chapman, 2003). So too are fluxes of the Ca II and Mg II chromospheric emission lines (Lean, 2000; Fröhlich and Lean, 2004). More physical approaches for modeling irradiance are also being developed using magnetograms, precise solar photometry (Kuhn, Lin, and Coulter, 1999), and spectral synthesis techniques that seek to represent the intensity of a range of disk features as functions of wavelength and disk position (Fontenla *et al.*, 1999).

A model that linearly combines sunspot and facular effects is seen in Figure 5 to provide close (but imperfect) tracking of total irradiance variations recorded by TIM during October 2003, and in the SORCE mission thus far, accounting for 92% of observed variance. The model uses an estimate of sunspot darkening calculated explicitly from information about sunspot areas and locations recorded from white-light solar images (archived by NOAA in the National Geophysical Data Center), together with adopted center-to-limb functions and bolometric contrast (Lean *et al.*, 1998). Facular brightening is represented by the Mg II chromospheric index (Viereck *et al.*, 2004). Multiple regression of the sunspot darkening and facular proxy time series with the PMOD total solar irradiance composite (Figure 4) establishes the relative strengths of the sunspot and facular influences, as shown in Figure 9. According to the model, the 4 Wm^{-2} total irradiance reduction in October 2003 is the net effect of a 5 Wm^{-2} irradiance depletion by sunspot darkening, compensated by a 1 Wm^{-2} enhancement in bright facular emission.

The quantitative comparison summarized in Table II suggests that on average the total irradiance model underestimates the strength of both the facular and sunspot influences on solar rotation time scales. This is evident visually in Figure 5, where the model irradiance is seen to be slightly lower than the observations during facular increases, and slightly higher during times of large sunspot reductions. It is also evident when compared statistically with all three composite time series during the past 26 years; the model's standard deviation of 0.038% is uniformly smaller than that of the observations (0.042% for PMOD, 0.051% for ACRIM, 0.053% for SARR). Model revisions are in progress to increase the modeled variance by revising the bolometric sunspot and facular parameterizations, and to investigate the model's underestimate of the overall downward total irradiance trend from 2003.15 to 2004.44 (Table I), possibly related to the Mg II facular index.

Relative spectral irradiance changes modeled by incorporating the spectral dependence of the sunspot and facular contrasts (Lean, 2000) are shown in Figure 6. The overall agreement is surprisingly good, since the modeled wavelength dependence is based on limited measurements of sunspot contrasts (Allen, 1981) and theoretical calculations of sunspot and facular contrasts (Solanki and Unruh, 1998). Nevertheless distinct differences are evident. From 17 to 30 October 2003, the model predicts energy changes that are smaller than observed at wavelengths from 400 to 500 nm, and larger than observed at wavelengths from 700 to 1000 nm. Nor are the variations of the spectral features in the region 300 to 400 nm modeled exactly. A particular deficiency is in the spectral region near 1.6 μm where the models appear to underestimate facular brightness (Fontenla *et al.*, 2004). By clearly delineating deficiencies in spectral irradiance models, the SORCE measurements are facilitating model improvements, such as in the wavelength dependence of the sunspot and facular contrasts and their relative temporal influences.

3.2.2. *Past*

Observations and models of irradiance (such as those shown in Figures 4–6 and 8) provide a scientific basis for reconstructing past solar irradiance. Proxy indicators of solar activity such as the *aa* index (Lockwood and Stamper, 1999), cosmogenic isotopes in tree-rings and ice-cores (Baud *et al.*, 2000), and the range of variability in Sun-like stars (Baliunas and Jastrow, 1990) place current solar activity levels in a broader context. The irradiance reconstructions of Hoyt and Schatten (1993), Lean, Beer, and Bradley (1995), Lean (2000), Lockwood and Stamper (1999), and Fligge and Solanki (2000) assume that longer term irradiance variations are larger than during the 11-year cycle, since the proxies suggest that the Sun is capable of a greater range of activity than witnessed during recent times. With this approach, total irradiance during the seventeenth century Maunder Minimum is reduced in the range of 0.15–0.4% (2–5 $\mathrm{W m^{-2}}$) below contemporary cycle minima values. Table IV summarizes different estimates.

New studies (Lean, Wang, and Sheeley, 2002; Foster, 2004; Foukal, North, and Wigley, 2004; Wang, Lean, and Sheeley, 2005) raise questions about the proper interpretation of the proxies, and suggest that long-term irradiance changes are a

TABLE IV

Compared are estimates of the reduction in total solar irradiance during the Maunder Minimum relative to contemporary solar minimum.

Reference	Assumptions and technique	Maunder Minimum irradiance reduction (global climate forcing) from contemporary minimum (Wm^{-2})
Schatten and Orosz (1990)	11-Year cycle extrapolation	~0.0 (0)
Lean, Skumanich, and White (1992)	No spots, plage, network in Ca images	1.5 (0.26)
Lean, Skumanich, and White (1992)	No spots, plage, network, and reduced basal emission in cell centers in Ca images non-cycling stars	2.6 (0.45)
Hoyt and Schatten (1993)*	Convective restructuring implied by changes in sunspot umbra/penumbra ratios	3.7 (0.65)
Lean, Beer, and Bradley (1995)	Non-cycling stars	2.6 (0.45)
Fligge and Solanki (2000)*	Combinations of above	4.1 (0.72)
Lean (2000)	Non-cycling stars (revised solar stellar calibration)	2.2 (0.38)
Foster (2004) Model No. 1	Non-magnetic Sun estimates by removing bright features from MDI images	1.6 (0.28)
Foster (2004) Model No. 3	Extrapolated from fit of 11-year smoothed total solar irradiance composite	0.8 (0.14)
Solanki and Krivova (2005)	Accumulation of bright sources from simple parameterization of flux emergence and decay	2.2 (0.38)
Wang, Lean, and Sheeley (2005)*	Flux transport simulations of total magnetic flux evolution	0.5 (0.09)

The solar activity cycle of order 1 Wm^{-2} is superimposed on this decrease. The climate forcing is the irradiance change divided by 4 (geometry) and multiplied by 0.7 (albedo). Reconstruction identified by * extend only to 1713, the end of the Maunder Minimum.

factor of 3 – 4 less (see Table IV). A reassessment of the stellar data has been unable to recover the original bimodal separation of (lower) Ca emission in non-cycling stars (assumed to be in Maunder Minimum type states) compared with (higher) emission in cycling stars (Hall and Lockwood, 2004) which underpins the Lean, Beer, and Bradley (1995) and Lean (2000) irradiance reconstructions. Long-term instrumental drifts may affect the *aa* index (Svalgaard, Cliver, and Le Sager, 2004) on which the Lockwood and Stamper (1999) irradiance reconstruction is based.

Nor do long-term trends in the *aa* index and cosmogenic isotopes (generated by open flux) necessarily imply equivalent long-term trends in solar irradiance (which track closed flux) according to simulations of the transport of magnetic flux on the Sun and propagation of open flux into the heliosphere (Lean, Wang, and Sheeley, 2002; Wang, Lean, and Sheeley, 2005).

Past solar irradiance has recently been reconstructed on the basis of solar considerations alone, without invoking geomagnetic, cosmogenic, or stellar proxies. From the identification of bright faculae in solar visible images made by the Michelson Doppler Imager (MDI) on SOHO, Foster (2004) estimates that removing all bright faculae reduces solar irradiance by 1.6 Wm^{-2} (see Table IV). This estimate of the irradiance of the "non-magnetic" Sun is consistent with an earlier estimate of Lean, Skumanich, and White (1992), who inferred a reduction of 1.5 Wm^{-2} from a similar analysis of solar Ca K images and fluxes (removal of all network but no alteration of basal cell center brightness). Both the Foster (2004) and Lean, Skumanich, and White (1992) approaches suggest that were the Maunder Minimum irradiance equivalent to the "non-magnetic" Sun, then the irradiance reduction from the present would be about half that of earlier estimates (see Table IV).

Using a quite different approach, Wang, Lean, and Sheeley (2005) also suggest that the amplitude of the background component is significantly less than has been assumed, specifically 0.27 times that of Lean (2000). This alternate estimate emerges from simulations of the eruption, transport, and accumulation of magnetic flux since 1713 using a flux transport model with variable meridional flow (Wang, Lean, and Sheeley, 2005). Both open and total flux variations are estimated, arising from the deposition of bipolar magnetic regions (active regions) and smaller-scale ephemeral regions on the Sun's surface, in strengths and numbers proportional to the sunspot number. The open flux compares reasonably well with the geomagnetic and cosmogenic isotopes which gives confidence that the approach is plausible. A small accumulation of total flux (and possibly ephemeral regions) produces a net increase in facular brightness which, in combination with sunspot blocking, permits the reconstruction of total solar irradiance shown in Figure 10. The increase from the Maunder Minimum to the present-day quiet Sun is ~ 0.5 Wm^{-2} (Table IV), i.e., about one-third the reduction estimated for the 'non-magnetic" Sun.

Based on current physical understanding, the most likely long-term total irradiance increase from the Maunder Minimum to current cycle minima is therefore in the range 0.5–1.6 Wm^{-2}. The larger amplitude secular irradiance changes of the initial reconstructions are likely upper limits. Figure 11 shows modeled changes in the spectral irradiance bands of Figure 8 that correspond to the Wang, Lean, and Sheeley (2005) flux transport simulations. The model changes were obtained by using a background component 27% of that adopted in the spectral irradiance reconstructions of Lean (2000), which extend from 1610 to the present. In Table III are estimated Maunder Minimum reductions in the bands, compared with that of Lean (2000).

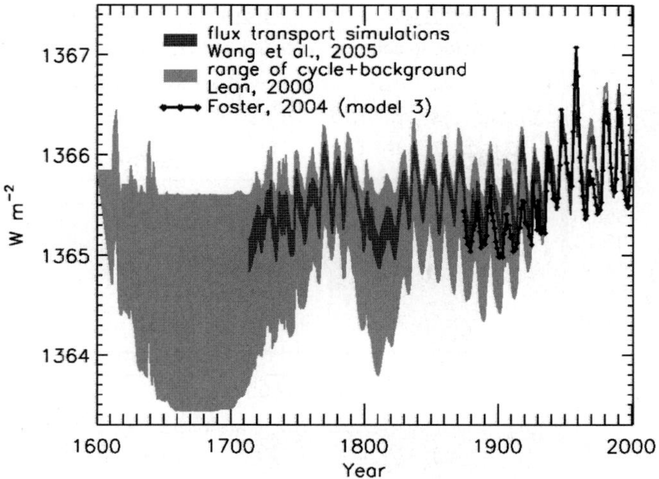

Figure 10. Shown as the upper envelope of the shaded region are total solar irradiance variations arising from the 11-year activity cycle. The lower envelope is the total irradiance reconstructed by Lean (2000), in which the long-term trend was inferred from brightness changes in Sun-like stars. In comparison are recent reconstructions based on solar considerations alone. That of Wang, Lean, and Sheeley (2005) uses a flux transport model to simulate the long-term evolution of the closed flux that generates bright faculae.

3.2.3. *Future*

Solar irradiance is expected to continue cycling in response to the 11-year activity cycle. Figure 12 suggests a possible scenario for the next few decades, based on a linear relationship of annual mean irradiance with the 10.7 cm flux (Lean, 2001). Predictions of the 10.7 cm flux, made by Schatten (2003) use a precursor approach that invokes solar dynamo theory to forecast cycle maxima from the strength of the Sun's polar fields at minima.

The prediction in Figure 12 has large uncertainty, in part because total solar irradiance is not linearly related to solar activity. Rather, its amplitude is the net effect of sunspot darkening and facular brightening, both of which vary with solar activity. Notably, total solar irradiance was as high in cycle 23 as in the prior two cycles, even though solar activity was not. Additionally, the irradiance database is too short for the detection or understanding of long-term solar irradiance trends that may also affect future radiative output.

That current levels of solar activity are at overall high levels, according to both the sunspot numbers and cosmogenic isotopes, may imply that future solar irradiance values will not exceed significantly those in the contemporary database. Spectral synthesis of the cosmogenic isotope record confirms that solar activity is presently peaking, and in 2100 will reach levels comparable to those in 1990 (Clilverd *et al.*, 2003). Projections of combined 11-, 88-, and 208-year solar cycles also suggest that solar activity will increase in the near future, until 2030, followed by decreasing

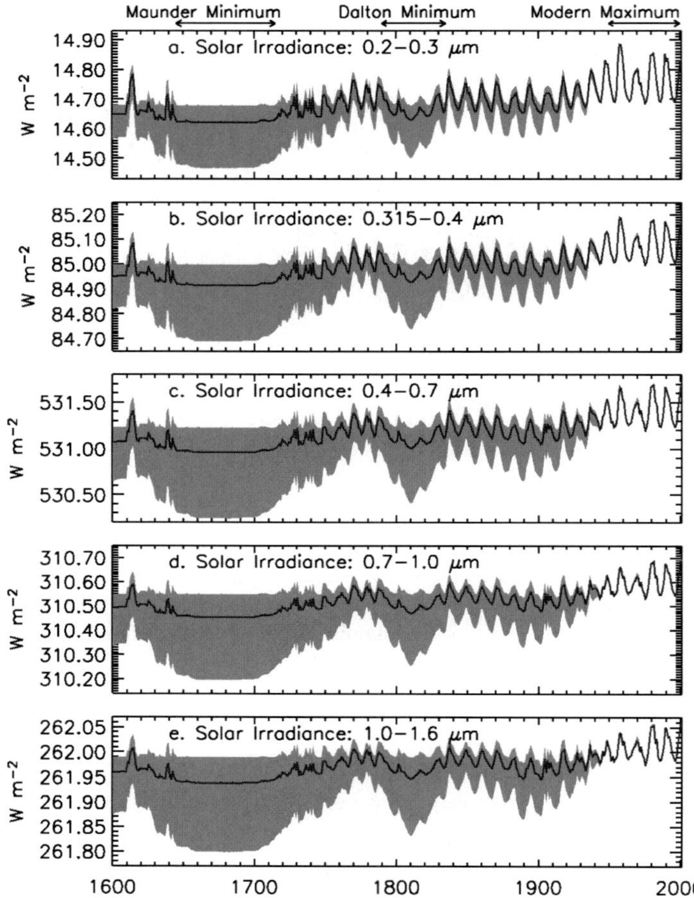

Figure 11. The shaded region shows the range of spectral irradiance variations in five wavelength bands, from the 11-year activity cycle alone to the estimate of Lean (2000), in which the long-term trend was inferred from brightness changes in Sun-like stars. In comparison are new irradiance reconstructions based on solar considerations alone, by Wang, Lean, and Sheeley (2005), using a flux transport model to simulate the long-term evolution of the closed flux that generates bright faculae.

activity until 2090 (Jirikowic and Damon, 1994). In contrast, a numerical model of solar irradiance variability which combines cycles related to the fundamental 11-year cycle by powers of 2 predicts a 0.05% irradiance decrease during the next two decades (Perry and Hsu, 2000).

SORCE's irradiance observations during the upcoming activity minimum (predicted for 2007) and cycle 24 activity maximum (predicted for 2010) will provide unique observations crucial for understanding the relationship of irradiance to solar activity, and for clarifying activity minima levels that may help resolve the controversy introduced by the different composite time series in Figure 4. Noting that radiometer sensitivities degrade most quickly during the beginning of the mission

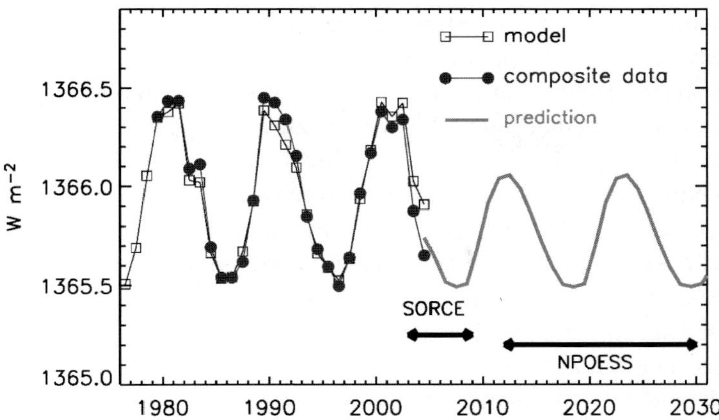

Figure 12. Predicted variations of total solar irradiance are shown during upcoming solar cycles, based on the Schatten (2003) predicted 10.7 cm flux. Also indicated is the expected epoch of operational monitoring by NPOESS, during which time solar activity may be notably less than during the present.

and that the NPOESS observations will commence in the declining phase of cycle 24, it will be essential to have independent monitoring by heritage radiometers during that time period.

4. Summary

Increasingly sophisticated statistical studies of high-fidelity climate, atmospheric, and solar variability time series in recent decades are contributing new knowledge of the Sun's influence on global change. Empirical evidence indicates surface and lower tropospheric temperature changes of order 0.1 K (peak-to-peak) associated with the solar activity cycle. The solar signal strength grows with altitude, to 1 K at 50 km. Changes in rainfall patterns in tropical regions also exhibit solar cycle periodicities, as do atmospheric ozone concentrations. Multiple regression analysis suggests that the solar influence on global change from solar minimum to maximum is comparable to anthropogenic effects over the same 5-year interval. Superimposed on both the cycling solar signal and the monotonically increasing anthropogenic influence are additional variations arising from internal variability and volcanic eruptions. When the contemporary empirical relationships are extended to the past 120 years, the solar influence on global surface temperature, consistent with current understanding of plausible secular irradiance change, is likely an order of magnitude smaller than the net warming from increasing concentrations of greenhouse gases and tropospheric aerosols.

The apparent surface temperature response to the solar activity cycle is inconsistent with current assumptions that the thermal inertia of the ocean attenuates the decadal solar forcing by a factor of 5. Efforts are underway to conduct more realistic response scenarios, by extending the upper boundary of general circulation models to above the stratosphere, improving ozone chemistry parameterizations, and using spectral, rather that total, irradiance inputs. Simulations with these expanded models suggest that solar effects on ozone and winds may propagate into the troposphere, and may influence the NAO. Recent approaches have used a regional model to explore ENSO responses to direct solar forcing and a stratospheric model to simulate changes in QBO phase caused by the UV irradiance cycle, but these mechanisms have yet to be included in general circulation models.

Reliable solar irradiance time series are crucial for both empirical and model investigations. Thus far, most studies have used the total solar irradiance record which extends since late 1978. SORCE observations by TIM continue this database. A critical, independent radiometric assessment of the space-based datasets is necessary to resolve radiometric differences of order 5 Wm^{-2} between SORCE and prior observations. Also needed is radiometric assessment of sensitivity drifts in the Nimbus 7 radiometer whose measurements between 1989 and 1992 are the basis for surmising that solar irradiance has increased between the two recent solar minima in 1986 and 1996. In comparison, overlapping ERBS observations and model simulations based on sunspot and facular influences do not show such an increase. The possibility of mistaking instrumental effects for real secular irradiance change emphasizes the need for continuing, overlapping measurements by independent radiometers to obtain a properly cross-calibrated long-term record of solar irradiance.

SORCE's SIM measurements realize the first continuous monitoring of variations in the near-UV, visible, and near-IR regions of the solar spectrum. This new database of spectral irradiance changes will enable more realistic investigations of the mechanisms of climate responses to solar forcing. The SORCE spectral irradiance observations are being examined to better quantify the sources of irradiance variability and the spectral partitioning of the variations. Empirical models of present and past solar irradiance variations, which utilize parameterizations of sunspot and facular influences, are being revised. Differences already detected include the apparent underestimation of the facular and sunspot influences, especially of the facular brightness in the near-IR regions. More physical approaches for modeling solar irradiance variations are also underway and may lead to better understanding of plausible longer term changes such as during the Maunder Minimum. SORCE's observations during the upcoming activity minimum (in 2007) and cycle 24 maximum (2010) will provide crucial benchmark data.

Acknowledgements

NASA and ONR funded this work. The efforts of the SORCE team in acquiring and reducing the data are appreciated, especially the help of Chris Pankratz. Ken Schatten provided solar cycle predictions. Bill Livingston and Rodney Viereck provided proxies to update the irradiance variability model. Mike Lockwood provided recent estimates of long-term irradiance changes. Gratefully acknowledged are discussions with Claus Fröhlich, David Rind, and Yi-Ming Wang. Additional data were obtained from the MDI, NGDC, GISS, GSFC, NOAA, and JMA websites.

References

Allen, C. W.: 1981, *Astrophysical Quantities*, 3rd edn, The Athlone Press, University of London.
Baliunas, S. and Jastrow, R.: 1990, *Nature* **348**, 520.
Baud, E., Raisbeck, G., Yiou, F., and Jouzel, J.: 2000, *Tellus* **52B**, 985.
Bond, G., Kromer, B., Beer, J., Muscheler, R., Evans, M. N., Showers, W., Hoffmann, S., Lotti-Bond, R., Hajdas, I., and Bonani, G.: 2001, *Science* **294**, 2130.
Clilverd, M. A., Clarke, E., Rishbeth, H., Clark, T. D. G., and Ulich, T.: 2003, *Astron. Geophys.* **44**, 2025.
Coughlin, K. and Tan, K. K.: 2004, *J. Geophys. Res.* **109**, D21105.
Crowley, T.: 2000, *Science* **289**, 270.
Dewitte, S., Crommelynck, D., Mekaoui, S., and Joukoff, A.: 2005, *Solar Phys.* **224**, 209.
Donarummo, J., Jr., Ram, M., and Stolz, M. R.: 2002, *Geophys. Res. Lett.* **29**, 75.
Douglass, D. H. and Clader, B. D.: 2002, *Geophys. Res. Lett.* **29**, 33.
Fioletov, V. E., Bodeker, G. E., Miller, A. J., McPeters, R. D., and Stolarski, R.: 2002, *J. Geophys. Res.* **107**, D22.
Fligge, M. and Solanki, S. K.: 2000, *Geophys. Res. Lett.* **27**, 2157.
Fontenla, J., White, O. R., Fox, P. A., Avrett, E. H., and Kurucz, R. L.: 1999, *Astrophys. J.* **518**, 480.
Fontenla, J. M., Harder, J., Rottman, G., Woods, T., Lawrence, G. M., and Davis, S.: 2004, *Astrophys. J.* **605**, L85.
Foster, S. S.: 2004, Ph.D. thesis, School of Physics and Astronomy, Faculty of Science, University of Southampton.
Foukal, P., North, G., and Wigley, T.: 2004, *Science* **306**, 68.
Fröhlich, C. and Lean, J.: 2004, *Astron. Astrophys. Rev.* **12**, 273.
Geller, M. and Smyshlyaev, S.: 2002, *Geophys. Res. Lett.* **29**, 5.
Gleisner, H. and Thejll, P.: 2003, *Geophys. Res. Lett.* **30**, 942.
Haigh, J. D.: 2001, *Science* **294**, 2109.
Haigh, J. D.: 2003, *Philos. Trans. R. Soc.* **361**, 95.
Haigh, J. D., Austin, J., Butchart, N., Chanin, M.–L., Crooks, S., Gray, L. J., Halenka, T., Hampson, J., Hood, L. L., Isaksen, I. S. A., Keckhut, P., Labitzke, K., Langematz, U., Matthes, K., Palmer, M., Rognerud, B., Tourpali, K., and Zerefos, C.: 2004, *SPARC Newslett.* **23**, 19.
Hall, J. C. and Lockwood, G. W.: 2004, *Astrophys. J.* **614**, 942.
Hansen, J.: 2004, *Sci. Amer.* **290**, 68.
Hansen, J., Sato, M., Nazarenko, L., Ruedy, R., Lacis, A. et al.: 2002, *J. Geophys. Res.* **107**, 4347.
Harder, J., Lawrence, G., Fontenla, J., Rottman, G., and Woods, T.: 2005, *Solar Phys.*, this volume.
Hood, L.: 2003, *Geophys. Res. Lett.* **30**, 10.
Hoyt, D. V. and Schatten, K. H.: 1993, *J. Geophys. Res.* **98**, 18895.
Intergovernmental Panel on Climate Change, Third Assessment Report: 2001.
Jackman, C., Flemming, E., Chandra, S., Considine, D., and Rosenfield, J.: 1996, *J. Geophys. Res.* **101**, 753.
Jirikowic, J. L. and Damon, P. E.: 1994, *Clim. Change* **26**, 309.
Jones, P. and Mann, M.: 2004, *Rev. Geophys.* **42**, RG2002.
Kernthaler, S. C., Toumi, R., and Haigh, J. D.: 1999, *Geophys. Res. Lett.* **26**, 863.

Kiehl, J. T. and Trenberth, K. E.: 1997, *Bull. Amer. Meteorolog. Soc.* **78,** 197.
Kodera, K.: 2002, *Geophys. Res. Lett.* **29**, 1218.
Kodera, K.: 2004, *Geophys. Res. Lett.* **31**, L24209.
Kopp, G., Lawrence, G., and Rottman, G.: 2005, *Solar Phys.*, this volume.
Kristjánsson, J. E., Staple, A., Kristiansen, J., and Kaas, E.: 2002, *Geophys. Res. Lett.* **29**, 22.
Krivova, N. A., Solanki, S. K., Fligge, M., and Unruh, Y. C.: 2003, *Astron. Astrophys.* **399**, L1.
Kuhn, J. R., Lin, H., and Coulter, R.: 1999, *Adv. Space Res.* **24/2**, 185.
Labitzke, K.: 2004, *J. Atmos. Solar-Terrest. Phys.* **66**, 1151.
Lacis, A. A., Wuebbles, D. J., and Logan, J. A.: 1990, *J. Geophys. Res.* **95**, 9971.
Lean, J.: 2000, *Geophys. Res. Lett.* **27**, 2425.
Lean, J.: 2001, *Geophys. Res. Lett.* **28**, 4119.
Lean, J., Beer, J., and Bradley, R.: 1995, *Geophys. Res. Lett.* **22**, 3195.
Lean, J., Wang, Y.-M., and Sheeley, N. R., Jr.: 2002, *Geophys. Res. Lett.* **29**, 77.
Lean, J., Skumanich, A., and White, O.: 1992, *Geophys. Res. Lett.* **19**, 1595.
Lean, J. L., Rottman, G. J., Kyle, H., Woods, T. N., Hickey, J. R., and Puga, L. C.: 1997, *J. Geophys. Res.* **102**, 29939.
Lean, J. L., Cook, J., Marquette, W., and Johannesson, A.: 1998, *Astrophys. J.* **492**, 390.
Lee III, R. B., Gibson, M. A., Wilson, R. S., and Thomas, S.: 1995, *J. Geophys. Res.* **100**, 1667.
Lockwood, M. and Stamper, R.: 1999, *Geophys. Res. Lett.* **26**, 2461.
Mann, M. E., Cane, M. A., Zebiak, S. E., and Clement, A.: 2005, *J. Climate* **18**, 447.
Matthes, K., Langematz, U., Gray, L. L., Kodera, K., and Labitzke, K.: 2004, *J. Geophys. Res.* **109**, D66101.
McClintock, W. E., Rottman, G. J., and Woods, T. N.: 2005, *Solar Phys.*, this volume.
McCormack, J.: 2003, *Geophys. Res. Lett.* **30**, 6.
McCormack, J. P., Hood, L. L., Nagatani, R., Miller, A. J., Planet, W. G., and McPeters, R. D.: 1997, *Geophys. Res. Lett.* **24**, 2729.
Meehl, G. A., Washington, W. M., Wigley, T. M. L., Arblaster, J. M., and Dai, A.: 2003, *J. Clim.* **16**, 426.
Neelin, J. D. and Latif, M.: 1998, *Phys. Today*, December.
Perry, C. A. and Hsu, K. J.: 2000, *PNAS* **97**, 12433.
Pilewskie, P. and Rottman, G.: 2005, *Solar Phys.*, this volume.
Preminger, D. G., Walton, S. R., and Chapman, G. A.: 2002, *J. Geophys. Res.* **107**, 1354.
Rial, J. A.: 1999, *Science* **285**, 564.
Rind, D.: 2002, *Science* **296**, 673.
Rind, D. and Overpeck, J.: 1993, *Quat. Sci. Rev.* **12**, 357.
Rind, D., Lonergan, P., Lean, J., Shindell, D., Perlwitz, J., Lerner, J., and McLinden, C.: 2004, *J. Clim.* **17**, 906.
Rottman, G.: 2005, *Solar Phys.*, this volume.
Rottman, G., Harder, J., Fontenla, J., Woods, T., White, O., and Lawrence, G.: 2005, *Solar Phys.*, this volume.
Ruzmaikin, A.: 1999, *Geophys. Res. Lett.* **26**, 2255.
Ruzmaikin, A. and Feynman, J.: 2002, *J. Geophys. Res.* **107**, D14.
Salby, M. and Callaghan, P.: 2004, *J. Clim.* **17**, 34.
Sato, M., Hansen, J. E., McCormick, M. P., and Pollack, J. B.: 1993, *J. Geophys. Res.* **98**, 22987.
Schatten, K. H.: 2003, *Adv. Space Res.* **32**, 451.
Schatten, K. H. and Orosz, J. A.: 1990, *Solar Phys.* **125**, 179.
Shindell, D. T., Schmidt, G. A., Miller, R. L., and Mann, M. E.: 2003, *J. Clim.* **16**, 4094.
Solanki, S. K. and Krivova, N. A.: 2004, *Solar Phys.* **224**, 197.
Solanki, S. K. and Unruh, Y. C.: 1998, *Astron. Astrophys.* **329**, 747.
Svalgaard, L., Cliver, E. W., and Le Sager, P.: 2004, *Adv. Space Res.* **34**, 436.

Tett, S. F. B., Jones, G. S., Stott, P. A., Hill, D. C., Mitchell, J. F. B., Allen, M. R., Ingram, W. J., Johns, T. C., Johnson, C. E., Jones, A., Roberts, D. L., Sexton, D. M. H., and Woodage, M. J.: 2002, *J. Geophys. Res.* **107**, 4306.

Udelhofen, P. N. and Cess, R. D.: 2001, *Geophys. Res. Lett.* **28**, 13.

Usoskin, I. G., Marsh, N., Kovaltsov, G. A., Mursula, K., and Gladysheva, O. G.: 2004, *Geophys. Res. Lett.* **31**, 16.

van Loon, H. and Shea, D. J.: 2000, *Geophys. Res. Lett.* **27**, 2965.

van Loon, H., Meehl, G. A., and Arblaster, J. M.: 2004, *J. Atmos. Solar-Terrest. Phys.* **66**, 1767.

Viereck, R. A., Floyd, L. E., Crane, P. C., Woods, T. N., Knapp, B. G., Rottman, G., Weber, M., and Puga, L. C.: 2004, *Space Weather* **2**, S10005.

Wallace, J. M. and Thompson, D. W. J.: 2002, *Phys. Today*, February.

Walton, S. R., Preminger, D. G., and Chapman, G. R.: 2003, *Astrophys. J.* **590**, 1088.

Wang, Y.-M., Lean, J. L., and Sheeley, N. R., Jr.: 2005, *Astrophys. J.* **625,** 522.

White, W. B., Dettinger, M. D., and Cayan, D. R.: 2003, *J. Geophys. Res.* **108**, 3248.

Wigley, T. M. L. and Raper, S. C. B.: 1990, *Geophys. Res. Lett.* **17**, 2169.

Willson, R. C. and Mordvinov, A. V.: 2003, *Geophys. Res. Lett.* **30**, 3.

Wolter, K. and Timlin, M. S.: 1998, *Weather* **53**, 315.

Woods, T. N. and Rottman, G.: 2005, *Solar Phys.*, this volume.

Woods, T. N., Prinz, D. K., Rottman, G. J., London, J., Crane, P. C., Cebula, R. P., Hilsenrath, E., Brueckner, G. E., Andrews, M. D., White, O. R., VanHoosier, M. E., Floyd, L. E., Herring, L. C., Knapp, B. G., Pankratz, C. K., and Reiser, P. A.: 1996, *J. Geophys. Res.* **101**, 9541.

Woods, T. N., Eparvier, F. G., Bailey, S. M., Chamberlin, P. C., Lean, J., Rottman, G. J., Solomon, S. C., Tobiska, W. K., and Woodraska, D. L.: 2005, *J. Geophys. Res.* **110**, A01312.

Solar Physics (2005) 230: 55–69

AN OVERVIEW OF THE DISPOSITION OF SOLAR RADIATION IN THE LOWER ATMOSPHERE: CONNECTIONS TO THE SORCE MISSION AND CLIMATE CHANGE

PETER PILEWSKIE, GARY ROTTMAN, and ERIK RICHARD
Laboratory for Atmospheric and Space Physics, University of Colorado, Boulder, CO, U.S.A.
(e-mail: pilewskie@lasp.colorado.edu)

(Received 31 March 2005; accepted 16 August 2005)

Abstract. Solar radiation is the primary energy source for many processes in Earth's environment and is responsible for driving the atmospheric and oceanic circulation. The integrated strength and spectral distribution of solar radiation is modified from the space-based Solar Radiation and Climate (SORCE) measurements through scattering and absorption processes in the atmosphere and at the surface. Understanding how these processes perturb the distribution of radiative flux density is essential in determining the climate response to changes in concentration of various gases and aerosol particles from natural and anthropogenic sources, as is discerning their associated feedback mechanisms. The past decade has been witness to a tremendous effort to quantify the absorption of solar radiation by clouds and aerosol particles via airborne and space-based observations. Vastly improved measurement and modeling capabilities have enhanced our ability to quantify the radiative energy budget, yet gaps persist in our knowledge of some fundamental variables. This paper reviews some of the many advances in atmospheric solar radiative transfer as well as those areas where large uncertainties remain. The SORCE mission's primary contribution to the energy budget studies is the specification of the solar total and spectral irradiance at the top of the atmosphere.

1. Radiative Energy Budget Overview

Solar radiation is the Earth's primary source of energy, exceeding by four orders of magnitude the next largest source, radioactive decay from the Earth's interior (Sellers, 1965). At the top of the atmosphere (TOA) the radiative balance between the incoming solar radiation and outgoing scattered solar radiation (the difference being the absorbed solar radiation) and outgoing (emitted) infrared radiation defines the radiative effective temperature of the planet, which is approximately 255 K. This temperature is derived from the albedo of the Earth (defined as the ratio of the incident to reflected solar radiation) and the incident TOA total solar irradiance and it is much lower than the average surface temperature of the planet of about 288 K. Therefore, the TOA view alone reveals nothing about processes such as the atmospheric greenhouse effect, or the disposition of radiative energy within the atmosphere. Furthermore, radiative balance is achieved only over large time and spatial domains. Thermal gradients induced by local radiative imbalance drive the Earth's atmospheric and oceanic circulation.

A complete understanding of the Earth's radiation budget, and the nature by which radiative imbalance drives weather and climate change must begin by establishing the boundary condition at the top of the atmosphere. Uncertainties in either the absolute magnitude of solar irradiance or in its spectral distribution will increase as radiation propagates through the atmosphere and interacts with constituent gases and particulates. The measurement of total solar radiation from space commenced nearly thirty years ago with Nimbus 7 and continues today with the Solar Radiation and Climate Experiment (SORCE) which was launched in January, 2003 (Rottman, 2005). SORCE is a suite of four instruments providing an unprecedented level of absolute accuracy and spectral coverage: the Total Irradiance Monitor (TIM) measures total (spectrally integrated) solar irradiance (TSI) with a goal of 100 ppm uncertainty (Kopp, Lawrence, and Rottman, 2005); the Spectral Irradiance Monitor (SIM) covers the spectral range from 300 to 2000 nm with 1–30 nm (wavelength-dependent) spectral resolution, and with a goal of 0.1% uncertainty (Harder *et al.*, 2005). SIM provides the first continuous sampling of spectral irradiance from space over the visible and near-infrared wavelengths which interact most strongly with Earth's lower atmosphere and surface; the Solar Stellar Irradiance Comparison Experiment SOLSTICE is the second generation instrument first flown on UARS and covers the spectral range from 120 to 300 nm with 1 nm resolution and 5% uncertainty (McClintock, Snow, and Woods, 2005); and the XUV Photometer System (XPS) measures between 1–40 nm with 10 nm resolution and 20% uncertainty (Woods and Rottman, 2005). Since the focus of this paper is on radiative processes in the lower atmosphere, TSI and spectral irradiance from 300 to 2000 nm (that is, the SIM spectral range) is given highest consideration. Lean *et al.* (2005) address indirect climate effects through the radiative processes in the stratosphere in a separate SORCE paper in this volume. Those tropospheric radiative processes that contribute the greatest uncertainty to the radiative energy budget, and therefore toward understanding and predicting climate change, will be emphasized here.

Several fates await a photon of solar radiation as it enters the atmosphere: it may continue unimpeded to the Earth's surface where it will be either absorbed or scattered back into the atmosphere; it may be scattered by molecules, clouds, or aerosol particles; or it may be absorbed in the atmosphere by molecules, clouds, or aerosol particles. Scattering and absorption processes depend upon the wavelength of incident light as well as the composition, size, and shape of the constituent particulates. Figure 1 shows a MODTRAN4 (Berk *et al.*, 2000) simulation (over the SORCE SIM spectral range) of the spectral optical thickness (defined as the logarithm of atmospheric transmittance) for the entire atmospheric column for all relevant atmospheric constituents (excluding clouds) individually and for their collective spectral effect. For wavelengths shorter than 0.7 μm, extinction is dominated by continuum molecular (Rayleigh) scattering and aerosol scattering and absorption. Ozone absorption contributes in the ultraviolet (Huggins bands near 300 nm in Figure 1) and in the mid-visible (Chappiuis bands) spectral regions. In the near-infrared water

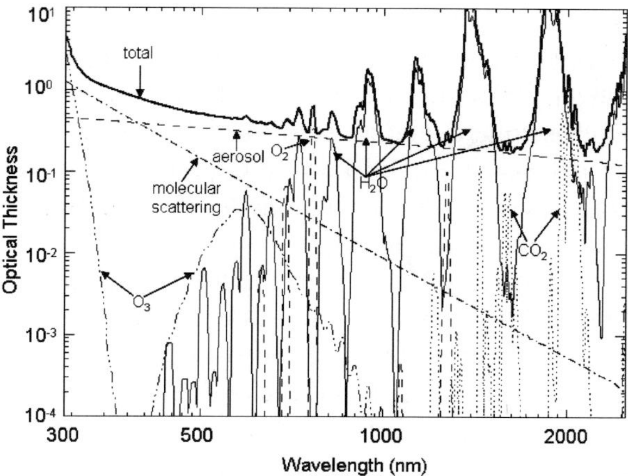

Figure 1. Spectral optical thickness for various constituents (labeled) and their combined effect, using MODTRAN4 mid-latitude summer atmospheric profile with surface visibility of 23 km. This spectral region is roughly equivalent to the SORCE SIM spectral range.

vapor is the dominant absorbing species, with smaller contributions from O_2 and CO_2.

The processes which modify incoming solar radiation as it propagates through the atmosphere, as illustrated on the left side of Figure 2 (from Kiehl and Trenberth, 1997), are of importance for the energy budget of the Earth, and therefore drive weather and climate. From the extraterrestrial solar irradiance of 342 W m^{-2}

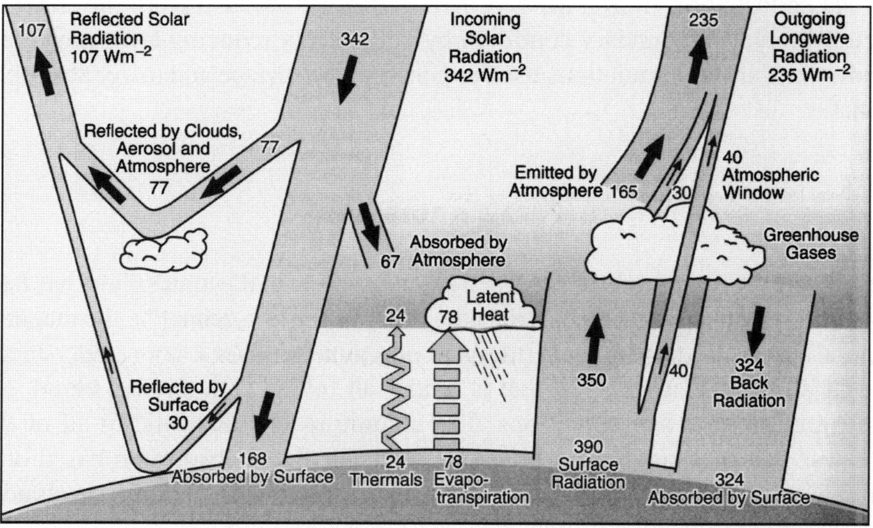

Figure 2. The Earth's annual global mean energy budget, from Kiehl and Trenberth (1997). Values are global and annual averages in units of W m^{-2}.

(annually and globally averaged) incident at the top of the atmosphere about $77\,\mathrm{W\,m^{-2}}$ are reflected to space by clouds, aerosol particles, and atmospheric gases. Approximately $30\,\mathrm{W\,m^{-2}}$ are reflected by the surface, $67\,\mathrm{W\,m^{-2}}$ are absorbed within the atmosphere, and an additional $168\,\mathrm{W\,m^{-2}}$ are absorbed by the surface. Radiative balance is established by the emission of terrestrial radiation (right side of Figure 2) and this outgoing longwave radiation determines the equilibrium temperature of the Earth-atmosphere system.

The TOA radiation budget depicted in Figure 2 is relatively well understood from satellite observations such as those made by Nimbus-7 (Ardanuy, Stowe, and Gruber, 1991), the Earth Radiation Budget Experiment, ERBE (Ramanathan *et al.*, 1989), the Clouds and the Earth's Radiant Energy System, CERES (Wielicki *et al.*, 1996), and most recently, SORCE. The knowledge of irradiance terms for the TOA energy budget far exceeds that for the surface and lower atmosphere (Li, Moreau, and Arking, 1997). For example, the accuracy and annual variability in total solar irradiance is known to within a few tenths of a percent and for top of the atmosphere albedo, about 1%. By contrast, the range of estimates for the surface radiative energy budget vary by more than 10% between satellite remote sensing and model simulations, and it is difficult to quantify the absolute uncertainties associated with the individual terms (Li, Moreau, and Arking, 1997). The partitioning of radiative energy throughout the atmospheric column and at the surface relies on radiative transfer modeling, general circulation models (GCM), remote sensing from space, and *in situ* observation in the atmosphere (primarily airborne) and at the surface. Ultimately, these methods generally suffer from sparse spatial and/or temporal coverage. It should be noted here that SORCE improvements to the radiation budget depicted in Figure 2 will come not only from the obvious contribution from TIM TSI measurements, but also from the SIM TOA spectral irradiance boundary condition by collectively reducing error propagation in radiative transfer simulations used to generate the surface and lower atmosphere estimates.

2. Solar Absorption

Quantifying the magnitude of absorbed solar radiation in the atmosphere has been a controversial topic during the last decade. What began as a debate on the magnitude of radiation absorbed by clouds migrated into concern over a poor understanding of solar absorption in general, that is, under all (cloud free, broken cloud, complete overcast, etc.) sky conditions. The ubiquitous characteristic in all of these cases was that measured absorption exceeded calculated absorption based on the best estimates of model input parameters (for example, in the case of cloud absorption, cloud thickness, droplet size, etc.). Because the Earth's albedo has been well constrained from space-based observations (Wielicki *et al.*, 2005), this "excess" absorption would have to replace absorption at the surface in energy budget models (for example, Figure 2). Precisely where energy is deposited in the vertical column determines the lapse rate of temperature, atmospheric stability, and heating profiles. A perturbation in solar absorption due to clouds, aerosols, and gases, would have a subsequent effect on cloud formation, maintenance, and dissipation. In general, the distribution of water in all phases and the global hydrological cycle is closely linked to the partitioning of solar absorption between the surface and the atmosphere (Ramanathan *et al.*, 2001). What follows is a brief background of the problem and the current status based on knowledge gained from the most recent experiments and analysis. It will be shown that improvements in quantifying cloud and aerosol absorption have been made through spectral rather than broadband observations. This underscores the significance of the SORCE SIM measurements for interpreting spectral data via radiative transfer modeling and for quantifying spectral absorption in the context of the radiative energy budget.

2.1. Absorption by clouds

The earliest airborne measurements of cloud absorption typically (but not always) exceeded the best model estimates of cloud absorption (see Stephens and Tsay, 1990, for a review of the cloud absorption anomaly prior to 1990). Cloud absorption was determined by measuring the net (downwelling minus upwelling) solar broadband irradiance at the top and base of the cloud and taking the difference to derive the flux divergence or absorption. Because of large uncertainties associated with the commonly used broadband, thermopile sensors, sampling error introduced by horizontal flux divergence and cloud advection, as well as general problems related to the differencing of two large terms (the cloud top and base net irradiance), reports of excess absorption were often dismissed – a position somewhat justified by the occasional report of *negative* absorption.

A more compelling argument was made by Stephens and Tsay (1990) for a cloud absorption anomaly based on the analysis of spectral reflectance measurements. In the near-infrared, absorption by a water droplet is directly proportional to the product of droplet radius and absorption coefficient. For high orders of multiple scattering the probability of absorption in a cloud layer increases to an amount approximately proportional to the square root of the probability of absorption by a single droplet (Twomey and Bohren, 1980). As reported by a number of investigators (for example, Twomey and Cocks, 1982, 1989; Stephens and Platt, 1987; and Foot, 1988), droplet size inferred from visible and near-infrared reflectance measurements exceeded direct measurements of cloud droplet size by as much as 40%. Thus, there was growing concern that something was lacking in theoretical models of cloud absorption. The most direct explanation, that the bulk absorption coefficients of liquid water and ice were grossly in error, seemed implausible. Nevertheless, that was the conclusion drawn by Rozenberg *et al.* (1974) upon examination of data made on-board the Soviet Cosmos satellites.

Before the 1990's the strongest evidence for the existence of a cloud absorption anomaly came from the poor agreement between the remote sensing-derived cloud droplet size and that measured *in situ*. The impact on the absorbed radiation in the global radiation budget (67 W m^{-2} in Figure 2), however, was less certain. This uncertainty was due primarily to the poor quality of measurements of broadband cloud absorption. New studies in the 1990's suggested that such an impact was substantial, perhaps as great as 25 W m^{-2} (Cess *et al.*, 1995). In order to explain the energy budget in the tropical "warm pool" region of the western Pacific, Ramanathan *et al.* (1995) deduced that clouds must absorb considerably more than previously assumed. In that same region Pilewskie and Valero (1995), using data from improved pyroelectric detector broadband radiometers (Valero, Gore, and Giver, 1982) deployed from collocated aircraft above and below cloud layers, concluded that absorption by clouds exceeded most model estimates. These findings stimulated the staging of a number of experimental (for example, the Department of Energy ARESE I and ARESE II airborne field campaigns in 1995 and 2000, respectively) and theoretical studies to examine cloud absorption. The present level of understanding of the cloud absorption "anomaly" based on results from these focused studies is summarized here.

One source of the discrepancy between measurement and theory was the relatively primitive level of radiative transfer models employed by the general circulation models used in some of the prior studies (Li *et al.*, 2003). With more sophisticated treatments of radiative transfer, specifically, line-by-line calculations of gas absorption, the more recent comparisons revealed overlap between measured and modeled cloud absorption within their respective levels of uncertainty (Valero *et al.*, 2003). Investigations into cloud absorption led to improvements in the spectroscopy of water vapor (Giver, Chackerian, and Varanasi, 2000) and the contributions from trace gases such as NO_2 (Solomon *et al.*, 1999), but the integrated absorbed energy fell short of explaining a possible excess of 25 W m^{-2} globally (Bennartz and Lohmann, 2001). Improved airborne sampling strategies eliminated instantaneous errors due to horizontal flux divergence in real, three-dimensional clouds (Marshak *et al.*, 1997). In general, while absorption in three-dimensional clouds has been shown to exceed plane parallel layers for certain geometries, other cloud geometries have exhibited less absorption when three-dimensional radiative transfer was utilized. To date there has been no conclusive evidence that observed three-dimensional cloud water distributions lead to greater (or less) absorption than equivalent cloud water contents distributed uniformly over horizontal layers.

Over the past decade the overall uncertainties in measured and modeled absorption have been reduced. A persistent and unresolved issue, however, is that models still appear to be biased toward less absorption compared to observations, even though in the most recent studies they agree within 10%. Although the uncertainty in the globally averaged absorbed solar irradiance is likely not as large as the proposed 37% (25 W m^{-2} out of 67 W m^{-2}), uncertainties in cloud absorption must be reduced even further. Our ability to predict future climate hinges upon our understanding the present state. If a climate forcing (see Section 3) of 4 W m^{-2}, due to a doubling of CO_2 from pre-industrial levels, is to be used as a benchmark for climate change then it is a reasonable goal to expect that the terms in the radiative energy budget be known to within 4 W m^{-2}. Wendisch *et al.* (2005) argued that the usual method of employing flux divergence is flawed because errors in solar irradiance propagate to unwieldy levels when deriving the absorption in an atmospheric layer. This occurs in addition to errors associated with spatially inhomogeneous surface reflectance, and to model errors from poorly quantified input parameters such as cloud water, water vapor, and cloud geometry. Spectrally resolved observations appear to be the most promising solution to deriving cloud absorption from observations. Improvements in absolute spectral irradiance monitors (Pilewskie *et al.*, 1998) have generated a high level of agreement between measured and modeled spectral irradiance and albedo (Figure 3), providing strong evidence that cloud absorption can be modeled accurately. In an experiment where similar instrumentation was employed Feingold *et al.* (2005) report an unprecedented agreement between airborne, surface, and satellite derived cloud droplet radii.

Utilization of SORCE SIM TOA spectral irradiance data in cloud radiation models is expected to lead to further improvement in comparisons to spectral measurements. For example, cloud spectral albedo (such as in the lower panel in Figure 3)

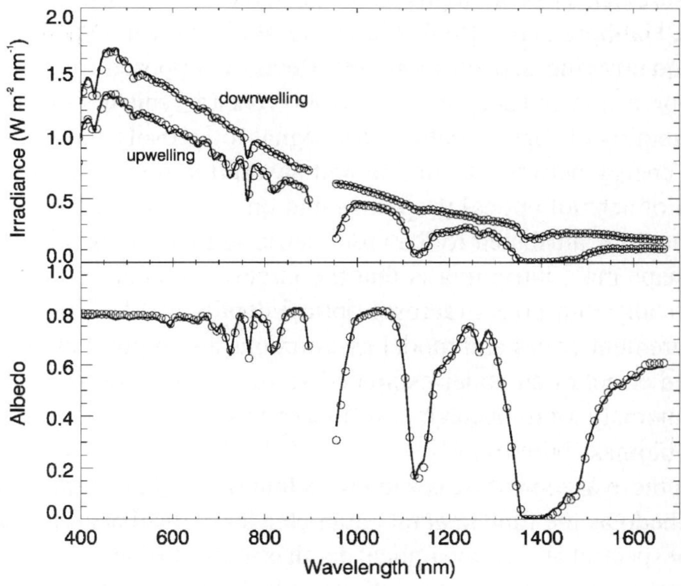

Figure 3. Top panel: Upwelling and downwelling solar spectral irradiance measured (*solid curve*) from above a cloud layer during the DOE ARESE II field experiment. Modeled irradiance values are indicated by *circles* and are within 3% of measurements across the spectrum. *Lower panel*: Same as above, for albedo, the ratio of upwelling to downwelling irradiance. The mean cloud droplet radius which produced this best fit was 8 μm, in close agreement to *in situ* measured values.

generally showed a higher level of agreement than either the upwelling or downwelling spectral irradiance because spectral anomalies in the TOA irradiance used for model input would cancel in the ratio (albedo). The need for highly accurate incident TOA solar spectra is even more crucial for spaceborne remote sensors lacking a direct solar reference spectrum. For example, the Earth Observing System Moderate Resolution Imaging Spectrometer (MODIS) includes a 3.7 μm band which is used for the retrieval of cloud droplet size. Uncertainties in the incident solar irradiance over the MODIS 3.7 μm band have been shown to propagate into a retrieved droplet radius error of between 1 and 2 μm (Platnick, 2003). This issue also argues for extended spectral coverage as the SORCE SIM is limited to wavelengths less than 3 μm.

2.2. Absorption by aerosols

Similar to the case for absorption by clouds, inferring the amount of absorption in aerosol layers has been afflicted with difficulties and has often led to larger absorption than standard accepted levels. In a number of investigations the single scattering albedo, defined as the ratio of scattering cross-section to extinction (absorption plus scattering) cross-section, needed to bring calculated solar irradiance in agreement with measurements was determined to be smaller than either *in situ* optical measurements or commonly accepted values used in climate modeling (see for example, Halthore *et al.*, 1998; Mlawer *et al.*, 2000; Russell *et al.*, 2002). One problem in quantifying aerosol radiative effects, as opposed to clouds, relates to radiometric precision and accuracy. Aerosol layers are typically much thinner than clouds. Attributing change in radiometric signal to aerosols and then partitioning the residual energy between scattering and absorption relies not only on *a priori* assumptions of aerosol optical properties, but on a level of instrumental accuracy greater than can be attributed to the radiometric sensors applied in some of these studies. Perhaps most intriguing is that the largest "anomalies" have occurred in cases of optically thin layers (aerosol optical depths < 0.1). This raises concern about measurement errors and model assumptions and input. For very thin layers the integrated effect of an under-estimated aerosol absorption is less of a concern than if the anomaly were occurring in thicker layers, say in regions of industrial pollution or biomass burning.

Based on these arguments it is suggested that rather than identifying the single scattering albedo as the fundamental parameter for aerosol absorption, absorption coefficient or spectral absorption optical depth is more appropriate in distinguishing the spectral behavior of aerosol absorption and in determining the absolute absorbed energy. What is further revealing when examining aerosol absorption optical depth (rather than single scattering albedo) is that the range of absorbing aerosols is more tightly constrained than what is typically considered in radiative transfer models. Take, for example, two of the primary absorbing species, carbonaceous aerosols and mineral dust. Spectral single scattering albedo for each species is very different: for most mineral dusts, single scattering albedo is a minimum in the ultraviolet and increases monotonically to near-unity in the mid-visible. Black carbon single scattering albedo is nearly opposite in behavior, reaching a maximum in the ultraviolet and falling with wavelength throughout the visible. However, measurements of the wavelength dependence of black carbon and dust absorption optical depths can be closely approximated by inverse power laws. An exponent of -1 is indicative of black carbon absorption (Bergstrom, Russell, and Hignett, 2002). Desert dust and dust/pollution mixtures result in larger exponents (Fialho, Hansen, and Honrath, 2005) from two to three (Figure 4). Thus the prospects of quantifying the impact of aerosol particles on the atmosphere's radiative energy budget as well as the identifying aerosol direct radiative effects on climate forcing (defined in the following section) may be more promising than, say, determining their influence on cloud radiation and various feedback processes associated with the hydrological cycle. Toward this end, the strong reliance on radiative transfer modeling for the interpretation of field measurements and for quantifying the deposition of radiative energy in the atmosphere makes it imperative that the spectral distribution of radiant energy entering the atmosphere be known to a high degree of accuracy and precision as in the SORCE SIM observation. It is therefore anticipated that the improvements in quantifying the TOA incident spectral irradiance will lead to better understanding of radiative energy deposition in the lower atmosphere and surface.

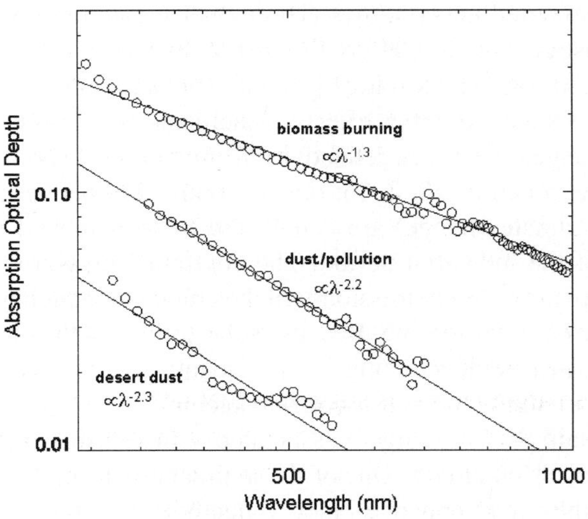

Figure 4. Aerosol absorption optical depth for biomass burning aerosol (from the SAFARI, 2000 experiment; Pilewskie *et al.*, 2003), complex dust/industrial pollution aerosol (from ACE-Asia; Bergstrom *et al.*, 2004), and desert dust (from PRIDE; Reid *et al.*, 2003). Lines are regression fits indicating power-law dependence of absorption optical depth with wavelength.

3. Climate Change and Climate Forcing

The fundamental parameter adopted by the Intergovernmental Panel on Climate Change (see IPCC, 2001) for determining long-term climate change is *forcing*, defined as a perturbation of the Earth's TOA energy budget (Hansen *et al.*, 1998) and quantified in units of $W\,m^{-2}$. Climate *sensitivity* is defined as the change in global mean temperature to an induced climate forcing. Forcing has supplanted sensitivity as the fundamental quantity for studying climate change because many forcing agents are so poorly quantified. Sensitivity, on the other hand, is considered to be constrained by the paleoclimate record of temperature response to induced forcings (Hansen *et al.*, 1998). It is recognized, however, that the equilibrium concept of forcing is limited (Board on Atmospheric Sciences and Climate, 2005). For example, it does not account for the highly non-linear response of the hydrological system (see Stephens, 2005, for a review on cloud-climate feedbacks). Nor does it adequately account for regional climate response and variability. Nevertheless, forcing is a useful conceptual tool for comparing different climate models and for comparing the influences of various forcing agents. For example, the response of the climate system in this context is independent of the forcing agent: the resultant temperature response due to a $0.1\,W\,m^{-2}$ solar forcing is equivalent to the same forcing from, for example, black carbon aerosol.

The key contributors to climate forcing as identified by the IPCC, 2001 are shown in Figure 5. The bars depict the global mean radiative forcing in 2000 relative to 1750, the beginning of the industrial age. The associated error bars indicate their estimated uncertainties. The radiative forcing induced by the well-mixed greenhouse gases CO_2, CH_4, N_2O, and halocarbons is considered to be known most accurately, followed by forcing by ozone. The remaining contributors, namely various aerosol species, indirect effects of aerosols on clouds, aviation-induced cirrus cloud, changes in surface reflectivity from land use, and direct and indirect solar forcing (see Lean *et al.*, 2005, on the connection between solar irradiance variability and climate change) are considered to be less well understood since the beginning of the industrial period. The continued measurement of TSI from SORCE and by planned future missions will be crucial to reducing the large relative uncertainty in direct solar forcing. Because solar forcing is the only variable which can be directly measured, reducing its uncertainty is vital to predicting climate response to other natural and anthropogenic agents.

The least certain of all forcings is associated with aerosols and in particular, the influence of aerosols on clouds. One of the indirect effects of aerosols on radiative forcing is that they may enhance cloud reflectivity (Twomey, 1974). Pollution, for example, contributes additional nuclei upon which water condenses, leading to a condensed water mass consisting of a larger concentration of smaller drops. The reflectance of a cloud increases when the water is divided into more droplets because the total surface area of scatterers increases. The effect of aerosols on

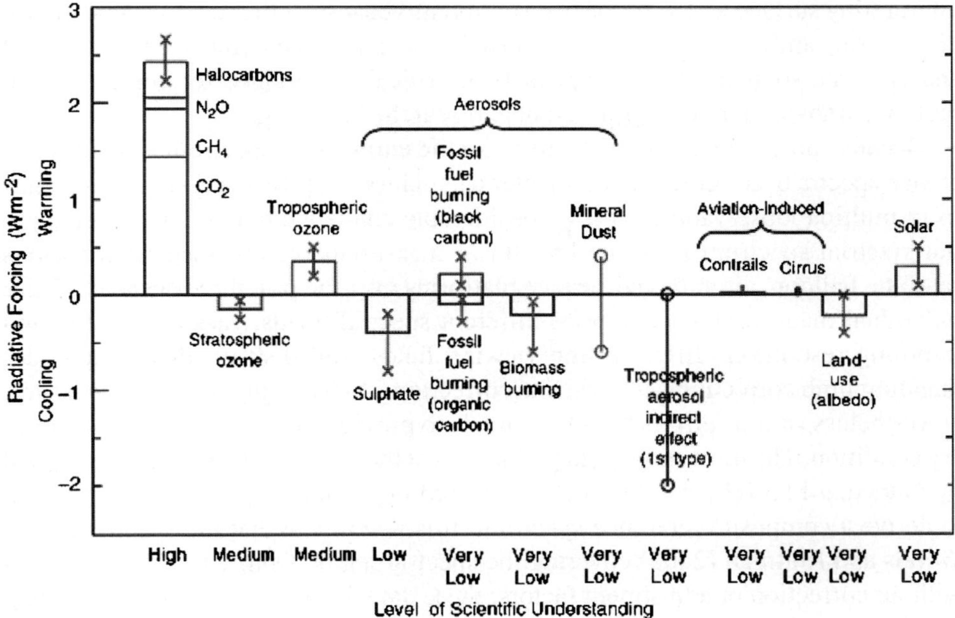

Figure 5. Radiative forcings for various natural and anthropogenic sources since 1750, along with their uncertainties. From IPCC, 2001.

clouds is potentially much larger than their direct radiative forcing because cloud droplets interact with radiation much more strongly than the nuclei upon which they form. Various other indirect aerosol effects have been proposed, including influences on cloud formation (Ackerman *et al.*, 2000) and maintenance (Albrecht, 1989), and a possible reduction of cloud reflectance through broadening of the cloud droplet distribution (Liu and Daum, 2002). The challenge will be to quantify these effects globally and reduce the uncertainties in order to better predict long-term climate change. Efforts underway using surface and space-based remote sensing and process modeling have been reviewed by Schwartz (2004). Further improvement in interpreting the significance of indirect aerosol radiative forcing in the context of climate forcing in general will be aided by reduction in uncertainty in direct solar forcing through continuation of the TSI record by SORCE TIM and future missions.

4. Solar Reference Spectrum for Radiative Transfer Modeling

The accuracy to which TOA incident solar irradiance is known would suggest that it does not contribute to the uncertainty in the surface radiative budget. However, the spectral distribution of solar irradiance is less well known than the integrated irradiance. Uncertainties in specific spectral bands may become important when interpreting surface and airborne measurements because of the spectral dependence of scattering and absorption processes in the atmosphere. Thus, the need for standard and accurate solar irradiance spectra is of critical importance for computational radiative transfer applications in climate research.

Because no single instrument can cover the entire solar spectrum, standard reference spectra used for radiative transfer calculations are, by necessity, composites from multiple observations. It is a considerable challenge to compile a single extraterrestrial spectrum from the breadth of measurements obtained from various satellite, balloon, aircraft, and surface platforms over the past three decades. These individual measurements comprise different spectral bands, varying spectral and sampling resolution, differing Sun-viewing fields (full disk vs. disk center) demanding limb corrections, telluric line corrections, to name just some of the issues. Nevertheless various efforts have been made to provide this essential model boundary condition. The most recent and perhaps most thorough study to date is described by Gueymard (2004). A total of 23 measured or modeled spectra were analyzed to derive a composite reference spectrum. It is noted here that the spectrum from Woods and Rottman (2002) covering the spectral region from 1 to 280 nm is used without correction or adjustment factors; these data derive from sensors measuring in the ultraviolet and extreme ultraviolet identical to those flying in the SORCE mission (Woods and Rottman, 2005; and McClintock, Snow, and Woods, 2005). Because SORCE launched just prior to completion of the reference spectrum, SIM data have yet to be analyzed and applied in this context. However, a reanalysis of the Gueymard (2004) spectrum will likely take into account new SIM, SOLSTICE, and XPS spectra obtained at different solar cycle levels during the SORCE mission. In addition, because the reference spectrum derived by Gueymard (2004) is tied to the integrated solar irradiance, future modifications will be influenced by the new measurement from TIM as well.

5. Summary and Conclusions

The combination of the various solar irradiance measurements by the SORCE instruments will be important for the continuing studies of the radiative energy budget. Improvements in quantifying the disposition of solar radiative energy in Earth's atmosphere relies upon a high degree of accuracy in the TOA incident TSI as well as its spectral distribution, provided by SORCE TIM and SIM, respectively. Clouds and aerosols contribute the largest uncertainties in the radiative energy budget, primarily due to absorption. Recent advances in quantifying both aerosol and cloud absorption have occurred because of improved observational methods using spectral rather than broadband observations. Interpretation of these spectral data relies on the accurate knowledge of TOA spectral irradiance that is being provided by SORCE SIM.

Unraveling the contributions to global climate change from various natural and anthropogenic forcing agents is one of the key problems in climate science. Only the radiative forcing from the greenhouse gases is known with sufficient accuracy to enable reliable climate prediction. Large uncertainties remain in forcing due to aerosols, most significantly in their effect on changing the scattering properties of clouds. In this context of climate change, although the forcing due to solar variability is small, its uncertainty is relatively large. Extending the 30-year TSI data record, currently with SORCE, and with future planned missions will contribute toward reducing that uncertainty and will lead to a more comprehensive understanding of climate change.

Acknowledgment

This research was supported by NASA contract NAS5-97045.

References

Ackerman, A. S., Toon, O. B., Stevens, D. E., Heymsfield, A. J., Ramanathan, V., and Welton, E. J.: 2000, *Science* **288**, 1042.

Albrecht, B. A.: 1989, *Science* **245**, 1227.

Ardanuy, P. E., Stowe, L. L., and Gruber, A.: 1991, *J. Geophys. Res.* **96**, 18 537.

Bennartz, R. and Lohmann, U.: 2001, *Geophys. Res. Lett.* **28**, 4591.

Bergstrom, R. W., Russell, P. B., and Hignett, P.: 2002, *J. Atmos. Sci.* **59**, 567.

Bergstrom, R. W., Pilewskie, P., Pommier, J., Rabbette, M., Russell, P. B., Schmid, B., Redemann, J., Higurashi, A., Nakajima, T., and Quinn, P. K.: 2004, *J. Geophys. Res.* **109**, 4467.

Berk, A., Anderson, G. P., Acharya, P. K., Chetwynd, J. H., Bernstein, L. S., Shettle, E. P., Matthew, M. W., and Adler-Golden, S. M.: 2000, *MODTRAN4 User's Manual*, Air Force Research Laboratory, Space Vehicles Directorate, Air Force Materiel Command, Hanscom AFB, MA.

Board on Atmospheric Sciences and Climate: 2005, *Radiative Forcing of Climate Change: Expanding the Concept and Addressing Uncertainties*, National Academies Press, Washington, DC, pp. 63–74.

Cess, R. D., Zhang, M. H., Minnis, P., Corsetti, L., Dutton, E. G., Forgan, B. W., Garber, D. P., Gates, W. L., Morcrette, J.-J., Potter, G. L., Ramanathan, V., Subasilar, B., Whitlock, C. H., Young, D. F., and Zhou, Y.: 1995, *Science* **267**, 496.

Feingold, G., Furrer, R., Pilewskie, P., Remer, L. A., Min, Q., and Jonsson, H.: 2005, *J. Geophys. Res.*, in press.

Fialho, P., Hansen, A. D. A., and Honrath, R. E.: 2005, *J. Aerosol Sci.* **36**, 267.

Foot, J. S.: 1988, *Q. J. R. Meteor. Soc.* **114**.

Giver, L. P., Chackerian, C., and Varanasi, P.: 2000, *J. Quant. Spectrosc. Radiat. Transfer* **66**, 101.

Gueymard, C. A.: 2004, *Solar Energy* **76**, 423.

Halthore, R. N., Nemesure, S., Schwartz, S. E., Imre, D. G., Berk, A., Dutton, E. G., and Bergin, M. H.: 1998, *Geophys. Res. Lett.* **25**, 3591.

Hansen, J., Sato, M., Lacis, A., Ruedy, R., Tegen, I., and Mathews, E.: 1998, *Proc. Natl. Acad. Sci.* **95**, 12753.

Harder, J., Rottman, G., Woods, T., White, O., and Lawrence, G.: 2005, *Solar Phys.*, this volume.

Intergovernmental Panel on Climate Change (IPCC): 2001, *Climate Change 2001: The Scientific Basis. Contribution of Working Group I to the Third Assessment Report of the Intergovernmental Panel on Climate Change*, J. T. Houghton *et al.* (ed.), Cambridge University Press, Cambridge.

Kiehl, J. T. and Trenberth, K. E.: 1997, *Bull. Am. Meteor. Soc.* **78**, 197.

Kopp, G., Lawrence, G., and Rottman, G.: 2005, *Solar Phys.*, this volume.

Lean, J., Rottman, G., Harder, J., and Kopp, G.: 2005, *Solar Phys.*, this volume.

Li, Z., Moreau, L., and Arking, A.: 1997, *Bull. Amer. Meteor. Soc.* **78**, 53.

Li, Z., Ackerman, T. P., Wiscombe, W., and Stephens, G. L.: 2003, *Science* **302**, 1151.

Liu, Y. and Daum, P. H.: 2002, *Nature* **419**, 580.

Marshak, A., Davis, A., Wiscombe, W., and Cahalan, R.: 1997, *J. Geophys. Res.* **102**, 16 619.

McClintock, W., Snow, M., and Woods, T.: 2005, *Solar Phys.*, this volume.

Mlawer, E. J., Brown, P. D., Clough, S. A., Harrison, L. C., Michalsky, J. J., Kiedron, P. W., and Shippert, T.: 2000, *Geophys. Res. Lett.* **27**, 2653.

Pilewskie, P. and Valero, F. P. J.: 1995, *Science* **267**, 1626.

Pilewskie, P., Goetz, A. F. H., Beal, D. A., Bergstrom, R. W., and Mariani, P.: 1998, *J. Geophys. Res.* **25**, 1141.

Pilewskie, P., Pommier, J., Bergstrom, R., Gore, W., Howard, S., Rabbette, M., Schmid, B., Hobbs, P. V., and Tsay, S. C.: 2003, *J. Geophys. Res.* **108**, 8486.

Platnick, S.: 2003, SORCE Science Meeting, Dec. 4–6, Sonoma, California.

Ramanathan, V., Cess, R. D., Harrison, E. F., Minnis, P., Barkstrom, B. R., Ahmad, E., and Hartmann, D.: 1989, *Science* **243**, 57.

Ramanathan, V., Crutzen, P. J., Kiehl, J. T., and Rosenfeld, D.: 2001, *Science* **294**, 2119.

Ramanathan, V., Subasilar, B., Zhang, G., Conant, W., Cess, R. D., Kiehl, J., Grassl, H., and Shi, L.: 1995, *Science* **267**, 499.

Reid, J. S., Kinney, J. E., Westphal, D. L., Holben, B. N., Welton, E. J., Tsay, S. C., Eleuterio, D. P., Campbell, J. R., Christopher, S. S., Colarco, P. R., Jonsson, H. H., Livingston, J. M., Maring, H. B., Meier, M. L., Pilewskie, P., Prospero, J. M., Reid, E. A., Remer, L. A., Russell, P. B., Savoie, D. L., Smimov, A., and Tanre, D.: 2003, *J. Geophys. Res.* **108**, 8586.

Rottman, G.: 2005, *Solar Phys.*, this volume.

Rozenberg, G. V., Malkevich, M. S., Malkova, V. S., and Syachinov, V. L.: 1974, *Izv. Acad. Sci. USSR, Atmos. Oceanic Phys.* **10**, 14.

Russell, P. B., Redemann, J., Schmid, B., Bergstrom, R. W., Livingston, J. M., McIntosh, D. M., Hobbs, P. V., Quinn, P. K., Carrico, C. M., Rood, M. J., Öström, E., Noone, K. J., von Hoyningen-Huene, W., and Remer, L.: 2002, *J. Atmos. Sci.* **59**, 609.

Schwartz, S. E.: 2004, *16th International Conference on Nucleation and Atmospheric Aerosols*, Kyoto, Japan, July 26–30.

Sellers, W.: 1965, *Physical Climatology*, The University of Chicago Press, Chicago, IL.

Solomon, S., Portmann, R. W., Sanders, R. W., Daniel, J. S., Madsen, W., Bartram, B., and Dutton, E. G.: 1999, *J. Geophys. Res.* **104**, 12047.

Stephens, G. L.: 2005, *J. Climate* **18**, 237.

Stephens, G. L. and Platt, C. M. R.: 1987, *J. Climate Appl. Meteor.* **26**, 1243.

Stephens, G. L. and Tsay, S. C.: 1990, *Q. J. R. Meteorol. Soc.* **116**, 671.

Twomey, S.: 1974, *Atmos. Environ.* **8**, 1251.

Twomey, S. and Bohren, C. F.: 1980, *J. Atmos. Sci* **37**, 2086.

Twomey, S. and Cocks, T.: 1982, *J. Meteorol. Sci. Japan* **60**, 583.

Twomey, S. and Cocks, T.: 1989, *Beitr. Phys Atmosph.* **62**, 172.

Valero, F. P. J., Gore, W. J. Y., and Giver, L.: 1982, *Appl. Opt.* **21**, 831.

Valero, F. P. J., Pope, S. K., Bush, B. C., Nguyen, Q., Marsden, D., Cess, R. D., Simpson-Leitner, A. S., Bucholtz, A., and Udelhofen, P. M.: 2003, *J. Geophys. Res.* **108**, 1029.

Wendisch, M., Barker, H., Pilewskie, P., and Heintzenberg, J.: 2005, *Atmos. Res.*, submitted.
Wielicki, B. A., Barkstrom, B. R., Harrison, E. F., Lee III, R. B., Smith, G. L., and Cooper, J. E.: 1996, *Bull. Amer. Meteorol. Soc.* **77**, 853.
Wielicki, B. A., Wong, T., Loeb, N., Minnis, P., Priestley, K., and Kandel, R.: 2005, *Science* **308**, 825.
Woods, T. N. and Rottman, G.: 2005, *Solar Phys.*, this volume.
Woods, T. N. and Rottman, G. J.: 2002, *Comparative Aeronomy in the Solar System*, M. Mendillo *et al.* (eds.), American Geophysical Union Monograph.

Solar Physics (2005) 230: 71–89

THE SORCE SPACECRAFT AND OPERATIONS

THOMAS P. SPARN, GARY ROTTMAN, THOMAS N. WOODS, BRIAN D. BOYLE, RICHARD KOHNERT, SEAN RYAN and RANDALL DAVIS
Laboratory for Atmospheric and Space Physics, University of Colorado, Boulder, CO, U.S.A.
(e-mail: sparn@lasp.colorado.edu; rottman@lasp.colorado.edu)

ROBERT FULTON
Orbital Sciences Corporation, Dulles, VA, U.S.A.

and

WILLIAM OCHS
NASA Goddard Space Flight Center, Greenbelt, MD, U.S.A.

(Received 7 June 2005; accepted 11 June 2005)

Abstract. The Solar Radiation and Climate Experiment, SORCE, is a satellite carrying four scientific instruments that measure the total solar irradiance and the spectral irradiance from the ultraviolet to the infrared. The instruments were all developed by the Laboratory for Atmospheric and Space Physics (LASP) at the University of Colorado, Boulder. The spacecraft carrying and accommodating the instruments was developed by Orbital Sciences Corporation in Dulles, Virginia. It is three-axis stabilized with a control system to point the instruments at the Sun, as well as the stars for calibration. SORCE was successfully launched from the Kennedy Space Center in Florida on 25 January 2003 aboard a Pegasus XL rocket. The anticipated lifetime is 5 years, with a goal of 6 years. SORCE is operated from the Mission Operations Center at LASP where all data are collected, processed, and distributed. This paper describes the SORCE spacecraft, integration and test, mission operations, and ground data system.

1. Introduction

One of the observational goals of NASA's Earth Science Enterprise is the precise measurement of solar irradiance. The Laboratory for Atmospheric and Space Physics (LASP) at the University of Colorado was selected by NASA to manage, develop, and operate a dedicated spacecraft carrying four scientific instruments to measure solar irradiance as part of the NASA Earth Observatory System (EOS). The mission, called the Solar Radiation and Climate Experiment (SORCE), is a PI-led mission with Dr. Gary Rottman as the Principal Investigator, and it represents the merging of two Earth Science Enterprise missions: the Solar and Atmospheric Variability Explorer (SAVE) and the Total Solar Irradiance Mission (TSIM). SORCE is a science-driven mission to continue the solar ultraviolet spectral irradiance and total irradiance data sets. SORCE was launched into a low Earth orbit on 25 January 2003 by a Pegasus XL launch vehicle and is designed to operate for a minimum of five years.

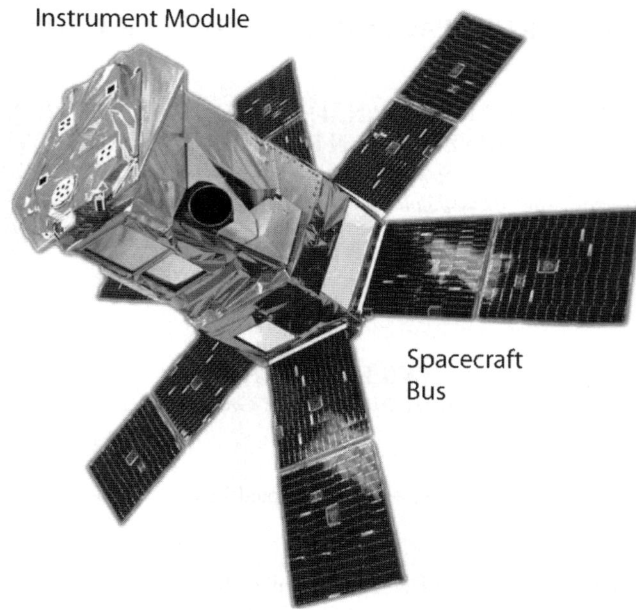

Figure 1. The SORCE spacecraft.

The SORCE mission is the culmination of numerous mission design approaches. After studying the possible flight opportunities, a small and highly capable spacecraft on a dedicated small launch vehicle was identified as the best platform to conduct the science mission. This optimal approach properly balanced performance, cost, risk, and schedule.

Figure 1 is an actual photograph of SORCE as it was being prepared for launch (background is removed). The hexagonal portion to the right, including the deployed solar arrays, is referred to as the spacecraft bus module. The more irregular, hexagonal portion extending toward the upper left contains the five scientific instruments together with the sun sensors and star trackers, and is referred to as the instrument module (IM). The entire spacecraft is encased in multilayer thermal blankets, and several exposed panels serve to passively cool internal electronic components. The left-most flat top surface contains the apertures for the instruments and it is this face that is pointed at the Sun and at selected calibration targets.

The spacecraft bus was developed and provided by Orbital Sciences Corporation (OSC) in Dulles, Virginia. It is based on a "common bus" approach that incorporates a modular design with substantial use of flight-proven hardware and software from previous successful programs. It is redundant for high reliability ($R > 0.85$ for a 6-year mission) and supports autonomous operations. The spacecraft uses GaAs solar arrays with nickel–hydrogen batteries in the power system, reaction wheels, sun sensors and star trackers in the attitude control system, and S-band communication for uplink and downlink. OSC was also responsible for the integration and test

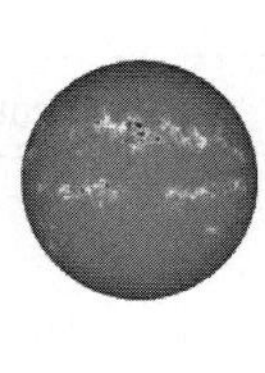

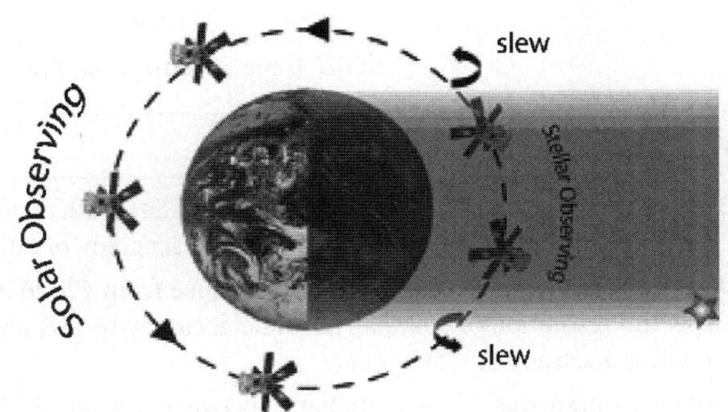

Figure 2. Typical SORCE orbit.

phase, and supported the launch activities. SORCE was boosted into low Earth orbit by a Pegasus XL space launch vehicle.

LASP operates and maintains the satellite from a control center on the University of Colorado, Boulder, campus. The LASP Mission Operations Center (MOC) plans observations, uplinks spacecraft and instrument commands and observation loads, and receives and analyzes all telemetry data. The Science Operations Center (SOC) at LASP analyzes, archives, and distributes the results to the scientific community.

The SORCE orbit altitude is approximately 640 km with an inclination of 40°. This particular orbit was selected because it met all of the science objectives for solar pointing and night-time stellar calibration. The altitude insures that the mission lifetime will far exceed the 6-year goal and the orbit inclination sends the spacecraft directly over suitable NASA ground stations. Moreover, this orbit matches the launch capabilities of the Pegasus to the mass of the SORCE satellite with adequate margin, and the radiation environment of the orbit is relatively benign simplifying the parts selection and shielding for both the spacecraft and instruments.

Figure 2 depicts the nominal orbital observation sequences used by SORCE. During the Sun-lit portion of the orbit, the SORCE instruments make solar observations. The spacecraft is then pointed toward stars during the eclipse portion of the orbit for stellar calibrations.

An overview of the science goals, mission design, spacecraft design and implementation, launch, ground operations, and flight performance from the mission's first 2 years are provided in this paper.

2. SORCE Objectives and Instruments

The Solar Radiation and Climate Experiment provides the solar/terrestrial research community with precise measurements of solar radiation. These measurements are critical to studies of the Sun, its effect on the Earth system, and its influence on humankind. SORCE's four instruments measure solar radiation incident at the top

TABLE I
SORCE measurement requirements.

Measurement objective	Instrument
Daily measurements of total solar irradiance (TSI), with an absolute accuracy of 0.01% and a long-term relative accuracy of 0.001% year^{-1}.	TIM
Daily measurements of the solar UV irradiance from 120 to 300 nm, with a spectral resolution of 1 nm, an absolute accuracy of 5%, and a long-term relative accuracy of 0.5% year^{-1}.	SOLSTICE
Daily measurements of solar irradiance between 300 and 2000 nm, with a spectral resolution ($\Delta\lambda/\lambda$) of at least 1/30, an absolute accuracy of 0.05%, and a long-term relative accuracy of 0.01% year^{-1}.	SIM
Daily measurements of the solar UV irradiance 1–34 nm, with a spectral resolution of 10 nm, an absolute accuracy of 20%, and a long-term relative accuracy of 2%.	XPS

of the Earth's atmosphere. Data obtained by SORCE are used to model the Sun's output and to explain and predict the effect of the Sun's radiation on the Earth's atmosphere and climate. In addition, the SORCE measurements address relevant questions from the U.S. Global Change Research Panel: (1) How does the Sun's output vary and what is the impact on terrestrial climate, and (2) What aspects of solar variability are influencing the stratospheric ozone layers?

SORCE primary objectives are to improve our understanding of how and why the variability occurs at the Sun and how the variable irradiance affects our atmosphere and climate and to use this knowledge to estimate past and future solar behavior and climate response (Rottman, 2005). The SORCE measurement requirements specify the instruments as listed in Table I, and these in turn determined the mission design.

LASP developed the four instruments to meet the scientific measurement requirements. The instruments are the Total Irradiance Monitor (TIM; Kopp, Heuerman, and Lawrence, 2005; Kopp and Lawrence, 2005), the Spectral Irradiance Monitor (SIM; Harder *et al.*, 2005a,b), the Solar Stellar Irradiance Comparison Experiment (SOLSTICE; McClintock, Rottman, and Woods, 2005; McClintock, Snow, and Woods, 2005), and the XUV Photometer System (XPS; Woods, Rottman, and Vest, 2005).

SOLSTICE and XPS are improved versions of similar devices that are flying on UARS (Rottman, Woods, and Sparn, 1993) and TIMED (Woods *et al.*, 2005). In order to meet the 5-year reliability requirements, the SOLSTICE and SIM are fully redundant. This is accomplished using two identical copies of the SOLSTICE instrument and a fully redundant SIM channel.

A single microprocessor unit (MU) controls the instruments and handles the command and telemetry interface between the spacecraft and ground, as well as all instrument commanding and sequencing. The resulting architecture provided a clean interface between the spacecraft bus and the instrument module (IM), permitting the instruments and the MU to be calibrated, integrated, and tested as a unit at LASP before delivery as a single unit to Orbital Sciences Corporation.

The spacecraft's fine Sun sensor is directly attached to the SIM housing and the star trackers are on the instrument deck. This assures an instrument reference for all pointing, places the alignment requirements on the IM, and insures the flight system tracks the selected target.

3. SORCE Spacecraft

The SORCE spacecraft is based on the Orbital Sciences Corporation LEOStar-2™ spacecraft bus design. This design was first used for the OrbView-4 commercial remote sensing program and subsequent adaptations were implemented for the SORCE and GALEX spacecraft (Martin *et al.*, 2005). (GALEX launched on 28 April 2003 at NASA's Kennedy Space Center on a Pegasus XL launch vehicle.)

SORCE is the first redundant version of the LEOStar-2 bus and features nearly complete redundancy throughout all subsystems (single battery). This produces a high reliability design in a small Pegasus-class spacecraft, that meets the 5-year requirement for the SORCE science mission. The SORCE technical capabilities are summarized in Table II.

The SORCE satellite uses a 3-axis stabilized, zero momentum design that provides precision pointing and attitude knowledge during science observations. The satellite bus includes: S-band communications compatible with NASA's ground network, science data storage with capacity for more than 24 h of spacecraft data, six solar array wings for power generation and reaction wheels, star trackers, Sun

TABLE II

SORCE spacecraft capabilities.

Mass	290 kg
Power	348 W peak output
	Fixed GaAs solar arrays
	23 Ahr nickel hydrogen batteries
Communications	Redundant S-band transceivers
	2 kbps command uplink
	1.5 Mbps data downlink
	4 kbps TDRSS downlink
Inertial pointing	3-axis stabilized design
	Slew rate $>1^\circ$ s^{-1}
	Knowledge <36 arcs
	Control <60 arcs
Data system	1024 Mbit on-board storage (24 h)

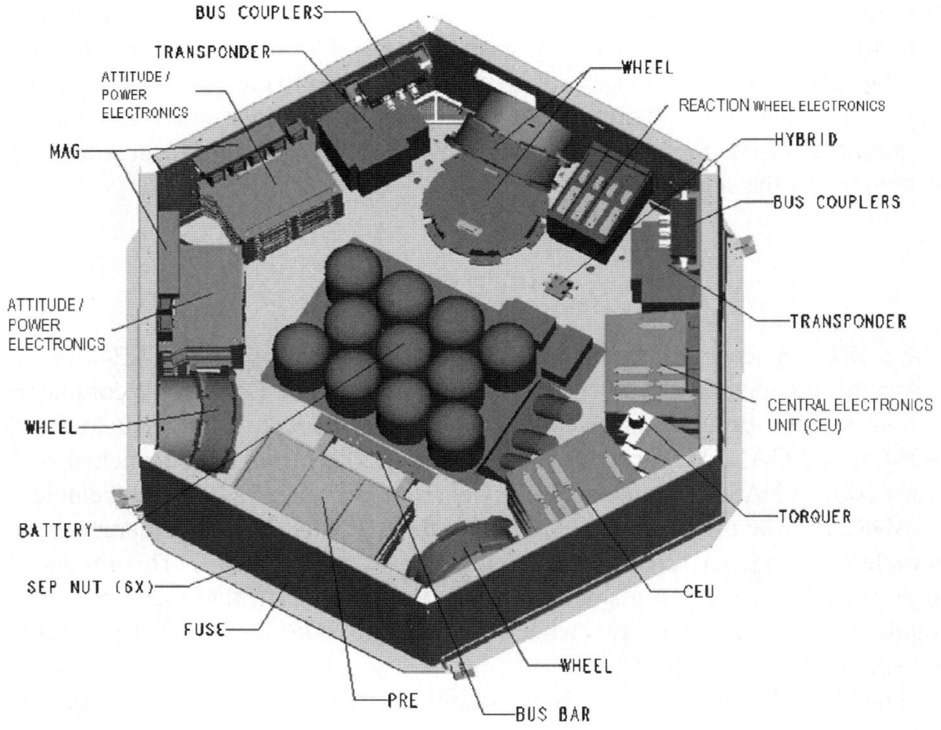

Figure 3. The SORCE spacecraft bus.

sensors, magnetometers, and torque rods for attitude control. Engineers and scientists from NASA, LASP, and OSC worked together throughout the design phase to optimize the hardware components and to maximize performance and reliability.

The SORCE spacecraft consists of two integrated modules: the satellite bus and the instrument module that contains the instruments and their associated electronics. The bus module structure is a robust, hexagonal modular unit of aluminum honeycomb. The bus components are configured within the structure as shown in Figure 3.

The instrument module (IM) features a graphite-epoxy optical bench assembly to provide a thermally stable mounting platform for the SORCE instruments and the two star trackers. The optical bench is kinematically supported by structural elements of the IM as shown in Figure 4. The design's stiff structure provides a stable platform for the instruments, which is essential for science missions requiring fine precision pointing performance, rapid slew rates, and short settling times.

A key strategy in optimizing the production schedule was the parallel integration effort of the IM and spacecraft bus. These two units were developed and integrated separately at the OSC and LASP facilities prior to their system level integration and environmental testing. Orbital Sciences Corporation developed and fabricated

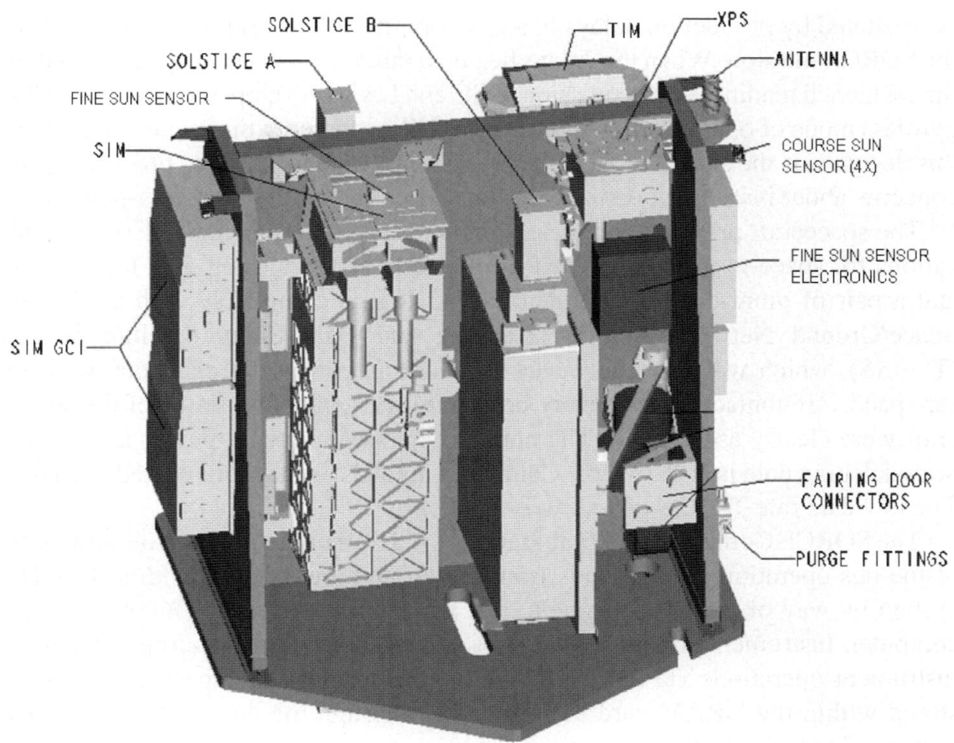

Figure 4. The SORCE instrument module.

the IM structure and then delivered it to LASP for the physical integration of the instrument suite.

The SORCE power system generates, stores, and distributes electrical power to the spacecraft bus and to the science instruments. The spacecraft generates electrical power through 66 strings of high efficiency GaAs solar cells bonded to the flat surface of the graphite-epoxy and aluminum honeycomb array panels. There are six deployable solar array wings, each consisting of two foldable panels. The wings are arranged radially near the aft of the spacecraft. After launch, the wings were deployed shortly after separation from the Pegasus XL launch vehicle. They will remain fixed throughout the on-orbit operations.

Attitude control is accomplished by a combination of reaction wheels, two star trackers, a fine sun sensor, a control processor, torque rods, and magnetometers. The physical location of each of these components is determined by not only functionality but also with consideration for magnetic field interference, field-of-view obstruction, alignment error, and mechanical and thermal considerations. Of particular interest is SORCE's unique "gyroless" design in the Attitude Control System (ACS), where the two star trackers are used for rate information in lieu of a more traditional gyroscope or Inertial Reference Unit-based (IRU) design. This design was necessitated by production delays in a developmental IRU originally baselined for the SORCE mission. When the delays began to threaten the master schedule and ultimate launch readiness of the mission, OSC and LASP developed and analyzed this gyroless mode of operations. A significant enabling factor in the implementation of this design was the ability of the solar instruments to view the Sun, precluding any concerns about instrument damage during spacecraft contingency Sun-pointing.

The spacecraft provides completely redundant radio frequency (RF) communications for space-to-ground and space-to-space interfaces using dual transceivers and a pair of omni-directional antennas. SORCE is compatible with the NASA Space/Ground Network and the Tracking and Data Relay Satellite System (TDRSS), which was especially beneficial during early-orbit checkout allowing for spacecraft contacts nearly every orbit until the health and status of the spacecraft were clearly ascertained. The nominal, high data rate output of science and housekeeping data is at 1.5 Mbps. Command uploads are accommodated at 2 kbps. The low data rate TDRSS downlink is at 4 kbps.

The SORCE Command and Data Handling (CDH) subsystem controls all aspects of the bus operations. Commands from the ground are received within the CDH system by way of the RF subsystem, and acted upon by the RAD6000 on-board computer. Instrument commands are passed directly to the IM microprocessor for instrument operations via the 1553 data bus. Instrument and spacecraft data are stored within the DRAM card and processed through the downlink card for RF transmission to the ground. The spacecraft has several modes of operation to support the solar measurements required by the SORCE mission, the stellar calibrations required for the instruments, and contingency operations in the event of a mission anomaly.

Thermal control of the spacecraft and the science instruments is optimized through judicious selection and use of passive and active thermal control elements including heaters and multi-layer insulation thermal blankets. The spacecraft is built to maintain temperature control of all elements sensitive to temperature variations. The instrument temperatures are maintained through a stable thermal environment providing proper pointing stability and calibration.

4. Preparation for Launch

4.1. Integration and Test

System engineering considered the challenges of integration and test (I & T) early and factored them into the system architecture. Key aspects of the SORCE development approach included a parallel and independent instrument and spacecraft bus build, having the spacecraft contractor lead the observatory I & T effort, and testing the observatory using the flight operations system (OASIS). SORCE implemented the modular design and independent system philosophy of the instrument module

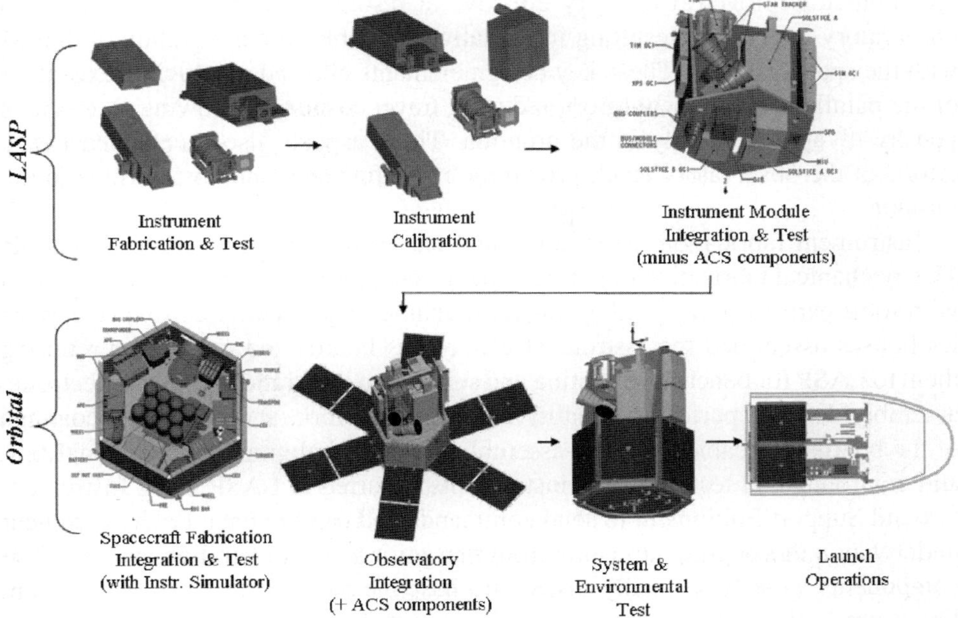

Figure 5. Instrument module and spacecraft bus integration flow.

(IM) prior to the selection of the bus contractor. The modular IM concept allowed the independent development, integration, and test of the instrument module and spacecraft bus prior to observatory integration. The two discrete structural elements (bus and IM) are separable modules. The successful execution of the SORCE integration and test program depended upon independent operational design, flight software autonomy between the IM and the spacecraft bus, and performing observatory I & T with OASIS. The early delivery of the IM structural module to the LASP facilities in Boulder, Colorado, allowed the I & T of the instrument suite at the same time that bus avionics and associated components were integrated and tested within the bus structure module at OSC's facilities in Dulles, Virginia. Figure 5 shows this independent integration and test flow.

Key aspects of the parallel development include independent microprocessor units for both modules, and the exchange of high fidelity interface simulators to test command, data, and telemetry interfaces. Both the instrument simulator (ISIM) and the spacecraft bus command and telemetry simulator (CTSIM) contained PowerPC processors similar to the RAD6000 processors used as the flight computers. They also contain MIL-STD 1553B interface cards operated with a pre-release of the 1553 flight software modules. The simulators and 1553 drivers for each simulator were delivered early in the development phase allowing the integration and test of the IM at LASP to be accomplished with the spacecraft bus "in the loop" and the bus integration at OSC with the instruments "in the loop". Exchange of the simulators proved to be very effective at resolving interface issues prior to observatory integration resulting in a relatively trouble-free integration of the IM with the spacecraft bus. These key design elements allowed an efficient execution of the parallel development effort, reducing travel costs, and allowing a "test it as you fly it" approach early in the program. These aspects also accelerated I & T efforts at the observatory level, providing both efficiency and cost-savings to the mission.

Instrument fabrication, calibration, and test were the responsibility of LASP. The mechanical fabrication of the instruments occurred in LASP's instrument shop with some parts manufactured by outside machine shops. NASA-certified electronics houses assembled the instrument electronics boards and cables and returned them to LASP for bench-level testing and subsystem integration. LASP's electronic assembly division performed quality inspection, rework, and polymeric coatings of the boards and cables. Optical assembly, focus and alignments, functional test, and environmental testing of the instruments occurred at LASP using Instrument Ground Support Equipment to send commands and capture data. Each instrument underwent a thorough qualification program prior to IM integration and test. The components were functionally tested and baseline performance was established. Environmental testing, including vibration and thermal vacuum testing, followed. Photometric calibrations with traceability to NIST standards were conducted at the Synchrotron Ultraviolet Radiation Facility (SURF-II) in Gaithersburg, Maryland, and LASP completed the performance verification at the instrument level. Detailed information of the design, calibration, and performance of the individual instruments can be found in the accompanying papers on TIM (Kopp, Heuerman, and Lawrence, 2005; Kopp and Lawrence, 2005; Kopp, Lawrence, and Rottman, 2005), SOLSTICE (McClintock, Rottman, and Woods, 2005; McClintock, Snow, and Woods, 2005), SIM (Harder *et al.*, 2005a,b: Rottman *et al.*, 2005), and XPS (Woods and Rottman, 2005; Woods, Rottman, and Vest, 2005).

Orbital Sciences Corporation fabricated, qualified, and vacuum baked the IM structure prior to delivery to LASP for the integration of the instruments. LASP and OSC worked together to install fasteners for the IM harness, survival heaters, and grounding straps for the subsystem electronics boxes. ACS sensors, such as the solar sensor and the star trackers, were not integrated onto the IM structure until it was returned to OSC for observatory integration.

The flight operations team participated early in the IM integration phase. The MU and CTSIM were crucial to the development and test of the OASIS flight operations and ground test procedures. Integration of the IM began with the installation of the instrument MU. OASIS commands and procedures were developed and the 1553 communications interface tested in this configuration. A systematic integration of each instrument followed the establishment of the command and telemetry interface and the verification of OASIS command procedures. The flight operations team executed test procedures throughout performance testing and environmental testing of the IM. The parallel development and test of the IM allowed the flight operations team to verify instrument test procedures at observatory level testing, prior to the delivery of the integrated IM.

Assembly of the spacecraft bus occurred in parallel at OSC. Spacecraft bus subsystems underwent a thorough unit-level qualification program prior to the bus I & T. The subsystems were functionally tested and their baseline performance established. Environmental testing included vibration, shock, and thermal vacuum. The solar array vendor performed acoustic testing on the solar array panels prior to their delivery to OSC for observatory integration. Comparisons of the post-environmental performance to the pre-environmental baseline verified the successful qualification of the subsystems in the spacecraft bus.

Integration and test of the spacecraft bus began with the mechanical installation of harnesses and electrical subsystems. Performance testing of the spacecraft bus followed the completion of safe-to-mate tests and electrical checks. Integration of the ACS sensors and the ISIM through test cables followed. The development of detailed test procedures for bus verification occurred in parallel using a SORCE bench top simulator referred to as the "flat-sat". The flat-sat was built around a RAD6000 computer using engineering model subsystems of the spacecraft bus. OSC used heritage operations software MAESTRO to integrate and test the spacecraft bus. The spacecraft bus underwent a rigorous integration and functional test program verifying the bus integrated performance and instrument interfaces, prior to the return delivery of the integrated IM. The transition to testing SORCE with the flight operations software, OASIS, came during observatory level testing.

Observatory I & T occurred at OSC's Spacecraft Manufacturing Facility in Virginia inside a class 10K clean tent residing in a class 100K clean room with controlled access to each. Orbital Sciences Corporation led the combined efforts of LASP and OSC through the observatory I & T activities. The LASP engineers and flight operations staff, along with OSC's integration and test personnel, formed a cohesive team. A LASP integration engineer relocated to Dulles, Virginia to provide continuous support throughout observatory I & T. The LASP flight operations team complemented I & T with near continuous on-site support for transitioning from MAESTRO to OASIS, executing observatory level testing, and gaining familiarity with operations of the spacecraft bus.

Observatory integration began with the installation of the ACS harnesses on the instrument module following the IM's post-ship functional verification. Integration of the IM onto the spacecraft bus followed the electrical safe-to-mate tests and initial verification of star tracker and solar sensor alignments. Installation of the coarse sun sensors and solar arrays completed observatory integration.

LASP and OSC jointly performed the initial alignments and integration of the ACS components onto the IM. The alignment of the six optical channels (TIM, two SIM, two SOLSTICE, and XPS), the fine Sun sensor, and the two star trackers all fell within a range of ± 10 arcmin, and this alignment held throughout the entire test phase and was verified on orbit. The field-of-view of each instrument is much larger, and instrument responsivity was calibrated in the laboratory over the entire field of view. It is of course important to have precise knowledge of the pointing for ground data processing to correct for alignment offsets, and this information is provided by the spacecraft attitude system as determined from the star trackers and fine sun sensor. On a typical target the RMS of the pointing errors is on the order of 2–3 arcs.

Details of these pointing corrections are included in the individual instrument papers (TIM: Kopp, Heuerman, and Lawrence, 2005; SIM: Harder *et al.*, 2005b; SOLSTICE: McClintock, Snow, and Woods, 2005; and XPS: Woods, Rottman, and Vest, 2005).

The observatory underwent a thorough functional and environmental test program following integration. The transition from MAESTRO to OASIS for testing occurred at this point. The baseline performance of the observatory was established in the flight configuration using ground support stimulus for the instruments and for the ACS sensors. The normal environmental test sequence followed. It included vibration tests, pyro-shock tests, thermal vacuum tests, and a 5-day mission simulation test. Key training and experience with the spacecraft functions and operations were provided early to the LASP engineering and flight operations staff through their on-site participation in all aspects of the spacecraft system testing.

4.2. Launch Operations

Following shipment from OSC's Virginia facility in October 2002, the SORCE spacecraft arrived at the Kennedy Space Center Multi-Payload Processing Facility for final launch preparations. The spacecraft was functionally tested and certified for flight. Flight simulations (interface tests) were performed to validate the spacecraft's compatibility with the Pegasus XL launch vehicle prior to and after the physical mate of the spacecraft to the launch vehicle.

OSC's Pegasus XL launch vehicle is an air-launched, internally guided, 3-stage solid rocket capable of launching up to 1000 pounds to low Earth orbit. The Pegasus is mated to its L-1011 carrier aircraft and dropped at approximately 38 000 feet. The vehicle free falls for about 5 s, its delta wing providing lift, before firing its first-stage rocket motor. From drop to insertion into orbit takes approximately 11 min and is depicted in Figure 6. The SORCE observatory was launched at 3:15 pm (EST) on 25 January 2003 into a 645 km altitude, 40° inclination orbit with no orbit maintenance required during the mission.

4.3. Program Management

As a Principal Investigator-led mission, the PI, Gary Rottman, is responsible for the mission success and scientific integrity over the entire SORCE project life cycle. LASP was given full responsibility for procuring the spacecraft and support services necessary to meet the mission requirements. This PI-led mission, featuring a true

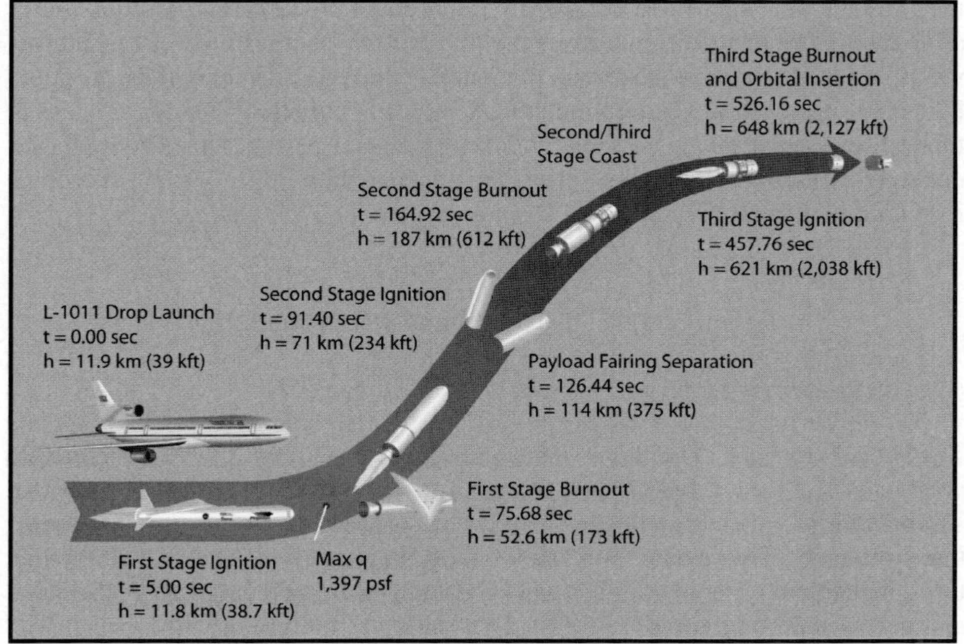

Figure 6. The SORCE flight, from drop to insertion into orbit.

partnership between NASA, LASP, and OSC, has successfully achieved the science goals of the SORCE mission.

The management of cost, schedule, and technical progress within each mission element is critical to successful development and test in preparation for launch. For this reason, SORCE created five focused Integrated Product Development Teams (IPDTs) overseen by a single Integrated Project Team (IPT). The IPT is comprised of LASP, OSC, and NASA members. Chaired by the SORCE Program Manager, the IPT coordinates all mission-level activities, interfaces, and trade-offs among mission elements. The IPT includes the PI, SORCE Program Manager, LASP Lead Systems Engineer, OSC Program Manager, OSC Lead Systems Engineer, and NASA Mission Manager. This organization is the basis for ensuring that the management processes achieve the SORCE science objectives. The IPDTs managed the individual mission elements to ensure smooth interfaces and execution. Each IPDT consisted of a core team of technical specialists, augmented when needed by specialty disciplines.

This management structure was successfully implemented by the SORCE team, and was used throughout the entire SORCE program. The close communication between LASP engineering staff and OSC spacecraft engineering staff, as well as NASA mission management, allowed this process to effectively negotiate and solve all of the management problems quickly and efficiently. Monthly IPDT meetings and weekly telecons, in addition to other IPDT meetings depending on the phase of the program, were conducted to insure excellent communication and coordination of all elements. Micro-management was avoided, and responsibility and accountability was distributed to each of the team leads for all of the technical areas of the program. Financial responsibility and accountability was held at the IPT and the transfer of funds from one IPDT to another optimized financial performance. SORCE was delivered on-time and under budget and has operated with 100% success, exceeding its required on-orbit life and Level 1 requirements.

5. Spacecraft Operations and Data Processing

5.1. Mission operations

The SORCE Mission Operations Center, MOC, is at LASP's multi-mission satellite operations facility located on the campus of the University of Colorado, Boulder. The operations system uses the same hardware, software, procedures, and personnel that simultaneously operate QuikSCAT and ICESat satellites. The LASP facility, the information systems within it, and the flight team personnel meet the same stringent standards that apply to NASA's own satellite operations centers. The operations team is composed of three elements: planning and scheduling, real-time operations, and science data processing.

The planning and scheduling team is composed of a mix of professionals and students. The professional team members are responsible for developing and maintaining the expert scheduling system software. The planning and scheduling software is implemented using the latest Java-based version of OASIS-PS. This software package is required to ingest station contact, orbit, and experiment requests from the science team and to produce daily command loads for the instruments. On a typical day the scheduling system plans for approximately 650 science experiments and 150 spacecraft activities. The students assigned to the planning and scheduling team assist in software development and generate the daily stored command loads for the observatory.

The real-time operations team for SORCE consists of LASP professionals and students. After two years on-orbit, there are 17 professional and 15 student members of the LASP flight team spread across several missions. Prior to being certified as a student command controller, students undergo an extensive summer-long training program culminating in a series of written and practical exams. Upon completion of the class, the students become certified to assist in the real-time operations of the observatory. The real-time operations team performs the final steps of the planning and scheduling process by collecting and reviewing the data products generated by the planning and scheduling team and making the final preparations for uplink to the spacecraft.

On a typical day, the operations team has two uplink/downlink sessions with the spacecraft. Each contact lasts anywhere from 8 to 10 min and is staffed by one professional flight controller and one student command controller. On the first contact of the day, a 24-h command sequence is loaded for each of the instruments. The spacecraft is also commanded to download all of the observatory state of health information and a portion of the science data. The second pass of the day is reserved for loading the spacecraft commands required to support the science operations for the next 24 h, as well as downloading the remainder of the science data and the state of health information since the first contact.

In addition to supporting the uplink of commands and downlink of data, the real-time operations team is also responsible for monitoring the state of health of the observatory. Any out-of-limit conditions, either flagged in real-time by OASIS CC or through analysis of stored data, are immediately investigated. In addition, after each contact a set of plots are created for several "critical" telemetry points. These plots are available on time ranges from 4 to 48 h and provide a big picture view of the health of the observatory.

A separate set of plots is generated daily, focusing on individual subsystems. These plots cover the last 36 h of operations and are reviewed by student subsystem teams with specialized training to identify subsystem-level anomalies. These teams also prepare weekly reports that summarize all activities of their respective subsystems. Every three months, the subsystem teams review all of the data from the past year. This is a comprehensive review of all telemetry items and is used to identify long-term trends in the data.

All data downlinked from the spacecraft are limit checked and stored in a database for access by scientists and engineers. The science operations team takes the raw data from this database and generates science data results for use by collaborators and the general public. Raw data and science data products are archived at the Distributed Active Archive Center (DAAC) located at the Goddard Space Flight Center (GSFC). See Pankratz *et al.* (2005) for additional information.

The planning and scheduling and real-time operations team report directly to the Mission Operations Manager. It is the responsibility of the Mission Operations Manager to verify that both teams and the ground network are meeting the mission requirements. Information on the status and performance of the operations system is provided to the scientists, and science data processing and management teams meet on a weekly basis, or as needed during critical events.

5.1.1. *On-Orbit Checkout*

The initial orbit check-out (IOC) of the SORCE spacecraft and instruments was performed per a pre-planned sequence of events within the first 30 days after launch. OSC engineering staff provided early flight operations support to the LASP Mission Operations Center at the University of Colorado, remaining on-site during bus and instrument commissioning activities that extended until 90 days after launch.

The OSC-led bus commissioning activities occurred in the first 4 weeks of spacecraft operations to certify that all spacecraft systems were operating nominally.

Some modifications to table values for ACS gains were performed to optimize the overall spacecraft pointing performance.

Instrument commissioning was led by the LASP engineering team and instrument scientists. These activities began 6 days after launch. Each instrument required 1 day to complete its checkout activities and the entire IM was commissioned by day 13. During the first 30 days on orbit, the spacecraft was commanded to solar pointing with a 2° offset in order to minimize SOLSTICE instrument degradation. On the 30th day of the mission, with the outgassing period complete, the spacecraft was commanded to solar pointing with zero offset signaling the start of normal instrument operations. Initial experiments included all instrument and pointing calibrations in order to establish the on-orbit baseline.

As the mission progressed toward routine science mission operations, the LASP flight operations team assumed total control of the spacecraft activities including operation and performance analysis. The previously defined milestone for the completion of the IOC occurred as scheduled, 90 days after launch. OSC staff is available to the SORCE mission for the remainder of the 5-year mission, ensuring complete support for future operational needs for software enhancements, anomaly resolution, and continuous performance data review.

5.1.2. *Ground Data System*

The SORCE Ground Data System (GDS) consists of the facilities and hardware required to launch, commission, and operate the SORCE observatory. The Mission Operations Center is located at LASP and is the center of the GDS.

A diagram of the SORCE Ground System is shown in Figure 7. SORCE uses a combination of commercial and government-provided tracking stations to support flight operations. Two ground stations are used on a routine basis, and one additional station was used for early-orbit support. The stations used routinely are Wallops Ground Station (WGS) (with the 11-meter antenna designated as prime and the 9-meter antenna designated as backup) located at Wallops Island, Virginia (NASA) and the 9-meter Santiago Ground Station (AGO) at the University of Chile located in Santiago, Chile. Additionally, the 10-meter Hartebeesthoek Ground Station (HBK) located in Johannesburg, South Africa, was used for additional early-orbit support during the first 30 days of operations. SORCE also utilized the Space Network (SN) for early-orbit support and continues to use the SN as a backup to the ground services emergencies. The MOC is connected to the remote tracking stations via NASA's closed IONet. This network provides a secure and reliable network for real-time operations.

NASA's Data Services Management Center (DSMC), located in White Sands, New Mexico, provides ground station scheduling. The SORCE program uses the same interfaces to the DSMC that were developed to support the SNOE, QuikSCAT, and ICESat missions (all of which are also operated from LASP's MOC). A conflict-free schedule is delivered to the MOC once per week and typically covers a time period of at least 10 days into the future. The MOC and Ground Network use this

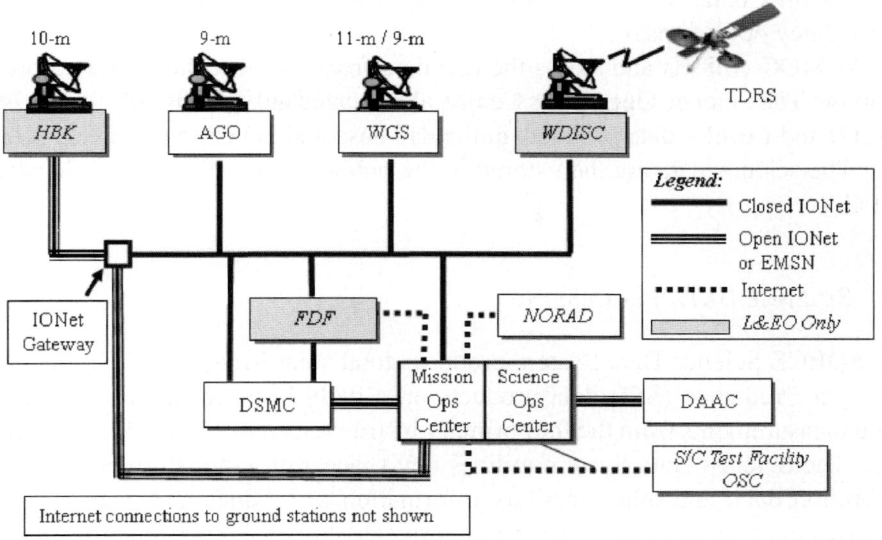

Figure 7. SORCE operations support network.

schedule to plan the appropriate activities and assets necessary to support nominal operations.

The Flight Dynamics Facility (FDF), located at NASA's GSFC, was used during launch and early-orbit operations to provide the initial knowledge of the SORCE orbit. During the first 2 weeks after the SORCE launch, the FDF processed tracking data from the ground stations, generated spacecraft ephemeris information, and provided the ephemeris to the MOC, DSMC, tracking stations, and SN. The task of orbit prediction and determination for SORCE has since shifted from the FDF to the MOC that uses the NORAD Two-Line Element (TLE) sets. Tracking data are still provided by the tracking stations to the FDF after every contact. In doing so, the FDF retains enough information that it could step in and provide ephemeris data should a problem arise with the orbit predictions provided by NORAD.

The Standard Autonomous File Server (SAFS) is a computer system for buffering telemetry data received at ground stations and distributing those data post-pass to the operations facilities. For SORCE, the SAFS collects files of telemetry data recorded at the WGS tracking station and then forward the files to the MOC. Post-pass data files from AGO and HBK are sent directly to the MOC at LASP. All data transmitted via the SN are forwarded in real time.

The SORCE observatory typically downlinks 140 Mbytes of data per day at 1.5 Mbs^{-1}. As a result, two contacts are scheduled per day during routine operations to support data downlink and command uplink. In addition to the high rate downlink, there is also a 4 kbps downlink rate that is used with the SN only. Due to the low downlink data rate available through the SN, its use is generally restricted to contingency operations.

The MOC collects and stores the raw data from the observatory into a project database. The Science Operations Center, also located at LASP, is able to access the Level 0 and Level 1 data products in the database and generate processed science data. The science data are then stored in the publicly accessible DAAC located at GSFC.

5.2. Science Data Processing

The SORCE Science Data System produces total solar irradiance (TSI) and spectral solar irradiance (SSI) data products on a daily basis, which are formulated using measurements from the four primary instruments onboard the SORCE spacecraft. The Science Data System utilizes raw spacecraft and instrument telemetry, calibration data, and other ancillary information to produce and distribute a variety of data products that have been corrected for all known instrumental and operational factors. SORCE benefits from a highly optimized object-oriented data processing system in which the software itself determines the versions of data products at runtime. This unique capability facilitates optimized data storage and CPU utilization during reprocessing activities by requiring only new data versions to be generated and stored. For more information refer to Pankratz *et al.* (2005) for a complete discussion of the SORCE data processing system, its design, implementation, operation, and details on how to access SORCE science data products.

6. Conclusions

SORCE has been operating successfully for over two years. The spacecraft has performed exceptionally well with no hardware anomalies and continues operations on the primary side of all redundant components. Similarly, the instruments are operating flawlessly and have returned over 35 Gb of data used to create the high quality science data products. Over 99.3% of all spacecraft and instrument data have been captured. The SORCE mission is testimony to the ability for a PI-led mission to be highly successful through partnerships of a university, an industry partner, and NASA centers. SORCE adheres to the successful tradition of other PI mode missions including solar missions TRACE (Handy *et al.*, 1999), ACRIMSAT (*http://www.acrim.com/*), and RHESSI (Lin *et al.*, 2002).

SORCE continues to fulfill its science objectives of recording the total solar irradiance and spectral irradiance on a daily basis. In general, these data exceed expectations in terms of accuracy and precision. In addition to the normal solar irradiance observations, SORCE observed the transit of Venus on 8 June 2004, has observed the Moon to obtain albedo and phase function information at ultraviolet wavelengths, monitored the intense solar activity during October/November 2003 including the first TSI observation of a class X17 flare (Woods *et al.*, 2004), and observed comet Temple I during the Deep Impact event in July 2005 (the UV emission was apparently too faint to be detected by SOLSTICE). The mission design lifetime is 5 years – with a goal of 6 – and there is every expectation that these milestones will be surpassed.

Acknowledgements

This research was supported by NASA contract NAS5-97045. We are grateful to the NASA Goddard Space Flight Center for their support in both the development of SORCE and its continued operations, to Orbital Sciences Corporation for their commitment to quality spacecraft engineering, and to the LASP engineering and mission operations staff for their dedication to the SORCE project.

References

Handy, B. N., Acton, L. W., Kankelborg, C. C., Wolfson, C. J., Akin, D. J., Bruner, M. E., Caravalho, R., Catura, R. C., Chevalier, R., et al.: 1999, *Solar Phys.* **187**, 229.

Harder, J., Lawrence, G., Fontenla, J., Rottman, G., and Woods, T.: 2005a, *Solar Phys.*, this volume.

Harder, J., Fontenla, J., Lawrence, G., Woods, T., and Rottman, G.: 2005b, *Solar Phys.*, this volume.

Kopp, G. and Lawrence, G.: 2005, *Solar Phys.*, this volume.

Kopp, G., Heuerman, K., and Lawrence, G.: 2005, *Solar Phys.*, this volume.

Kopp, G., Lawrence, G., and Rottman, G.: 2005, *Solar Phys.*, this volume.

Lin, R. P., Dennis, B. R., Hurford, G. J., Smith, D. M., Zehnder, A., Harvey, P. R., Curtis, D. W., Pankow, D., Turin, P., Bester, M., *et al.*: 2002, *Solar Phys.* **210**, 3.

Martin, D. C., Fanson, J., Schiminovich, D., Morrissey, P., Friedman, P. G., Barlow, T. A., Conrow, T., Grange, R., Jelinsky, P. N., et al.: 2005, *Astrophys. J.*, **619**, L1.

McClintock, W. E., Rottman, G. J., and Woods, T. N.: 2005, *Solar Phys.*, this volume.

McClintock, W. E., Snow, M., and Woods, T. N.: 2005, *Solar Phys.*, this volume.

Pankratz, C. K., Knapp, B., Reukauf, R., Fontenla, J., Dorey, M., Connelly, L., and Windnagel, A.: 2005, *Solar Phys.*, this volume.

Rottman, G.: 2005, *Solar Phys.*, this volume.

Rottman, G., Harder, J., Fontenla, J., Woods, T., White, O., and Lawrence, G.: 2005, *Solar Phys.*, this volume.

Rottman, G. J., Woods, T. N., and Sparn, T. P.: 1993, *J. Geophys. Res.* **98**, 10667.

Woods, T. N. and Rottman, G.: 2005, *Solar Phys.*, this volume.

Woods, T. N., Eparvier, F. G., Bailey, S. M., Chamberlin, P. C., Lean, J., Rottman, G. J., Solomon, S. C., Tobiska, W. K., and Woodraska, D. L.: 2005, *J. Geophys. Res.* **110**, A01312.

Woods, T. N., Eparvier, F. G., Fontenla, J., Harder, J. Kopp, G., McClintock, W. E., Rottman, G., Smiley, B., and Snow, M.: 2004, *Geophys. Res. Lett.* **31**, L10802.

Woods, T. N., Rottman, G., and Vest, R.: 2005, *Solar Phys.*, this volume.

Acknowledgements

References

highly reflective, reducing intrinsic heating. Each aperture is 0.76 cm thick, providing good thermal conductivity via an Indium gasket to the low-stress mounts to reduce thermal gradients. The 60° knife edge, with bevel facing the instrument interior, provides a precision area that is robust during machining and also prevents directly incident sunlight from reflecting off the specular interior aperture surfaces into the instrument. A resistive thermal device (RTD) mounted to the aperture mounting plate allows aperture area corrections due to temperature fluctuations after the apertures are characterized for their coefficient of thermal expansion. The 10.4 cm between the aperture and the cavity reduces both input stray radiation and corrections for shutter thermal emission, but comes at the expense of increased diffractive losses. Black baffles between the apertures and the cavities reduce stray light from off-axis objects such as the Earth.

2.5. Shutters

Each ESR has an independent shutter, located immediately sun-ward of the precision aperture. These bi-stable shutters open or close in 10 ms. The thin aluminum shutters are low mass and mounted with low friction ball bearings. Accelerated life tests in thermal vacuum conditions exceed the SORCE mission lifetime of 5 years for cycles of the primary cavity's shutter by greater than a factor of 2. Gold plating on each shutter's interior surface reduces thermal emission into the instrument interior, while a thermistor embedded in each shutter allows correction for its thermal emission. Shutter open (10.0–11.0 ms) and close (9.8–10.8 ms) times are characterized and accounted for in the phase sensitive analysis.

2.6. Digital control electronics maintain thermal balance

A 16 MHz Analog Devices TSC21020F–20MB/833 digital signal processor (DSP) performs all major instrument functions: thermally balancing the ESRs; regulating instrument temperature; maintaining shutter timing; and interfacing commands and telemetry with the spacecraft microprocessor. The DSP operates three 100 Hz AC servo bridges, which thermally balance the two paired ESRs and regulate instrument temperature. To maintain thermal balance, the DSP applies pulse-width modulated power to each cavity at 100 Hz via a field programmable gate array.

Since the solar irradiance is predictable to better than 1% during a given week, the estimated jump in replacement power is applied by the DSP at each shutter transition – a process called feedforward. This feedforward allows operation of the servo systems at high gain while preventing servo saturation during shutter transitions, and reduces sensitivity to uncertainty in the servo gain. By anticipating the decrease/increase in electrical power needed to maintain cavity temperature as a shutter opens/closes, the DSP's feedforward reduces the overshoot that would occur if the servo system only reacted to measured changes. The use of feedforward

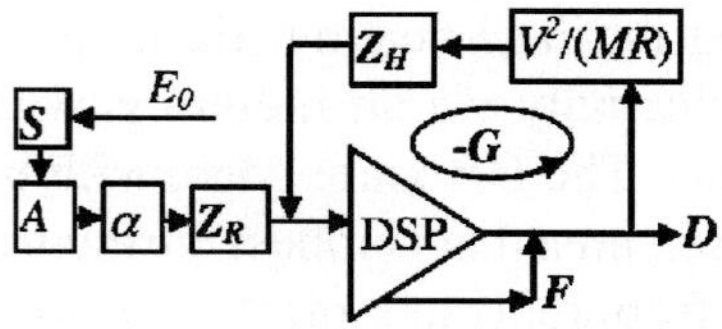

Figure 2. TIM signal transfer diagram. This signal transfer diagram illustrates the conversion from input solar irradiance E_0 to data numbers D. Variables in boxes are out/in ratios at some frequency. The servo loop gain is $-\boldsymbol{G}$. The digital signal processor (DSP) adds a known feedforward signal $\boldsymbol{F}$. The ratio of thermal impedances $\mathbf{Z_H}/\mathbf{Z_R}$ is the equivalence ratio. Bold symbols represent phasors.

essentially increases the effective servo loop gain and reduces the measurement uncertainty due to loop gain fluctuations. The instrument's signal transfer diagram including this DSP is shown in Figure 2.

To reduce the servo gain's sensitivity to operating temperature fluctuations, the thermistor bridge resistors are slightly asymmetrical so that the gain is at a maximum with respect to temperature at the nominal cavity operating temperature.

2.7. Phase-Sensitive Detection Reduces Noise

In ground processing, phase-sensitive detection of the electrical power applied to maintain ESR thermal balance gives the measured TSI. This method is similar to a Fourier analysis of a finite length of data. The TIM algorithm uses a filter that analyzes data in 400-s sections; this is long enough to benefit from some smoothing in the filter, much shorter than the time of an orbit, and comparable to time scales of short-term TSI variations, such as those due to solar oscillations.

The phase-sensitive detection method analyzes only changes in the applied ESR power at, and in phase with, the shutter fundamental frequency (0.01 Hz, or 100-s period). This method greatly reduces sensitivity to thermal drifts, $1/f$ noise, and parasitic thermal emission from the heat sink (which will be out-of-phase with the shutter). This detection method also means that the non-equivalence of heater and radiant power only needs to be known at the shutter frequency, making its calculation much simpler. Similar techniques are planned for the Scripps National Institute of Standards and Technology Absolute Radiometer (Scripps NISTAR) and have been used in an ambient temperature prototype (Rice, Lorentz, and Jung, 1999). The shutter frequency was selected near the minimum in the ESR-mounted thermistors' noise power spectrum, where noise levels reach a value of $\sim$1 ppm.

2.8. Equivalence Need Only Be Calculated at the Shutter Fundamental

The equivalence ratio, Z_H/Z_R, accounts for differences between electrical replacement heater power and absorbed radiant power. This ratio would be unity in an ideal

ESR; however, delays in thermal propagation for input electrical vs. radiant power can make this term differ significantly from unity, particularly at higher frequencies.

In the TIM ESR design, with the high diffusivity of the silver in the cavities and with both heater and radiant power inputs being nearly co-spatial at the cone end of the ESR, this ratio is within a few ppm of unity for signals in-phase with the shutter. The potentially large equivalence deviations at higher frequencies are not relevant, since one advantage of the phase sensitive detection method employed in the TIM analysis is that the instrument equivalence only needs to be known at the shutter frequency. This makes the calculation of the equivalence much easier and reduces uncertainties in this term.

Appendix A details the calculation of the TIM's equivalence ratio based on an algorithm developed in terms of known parameters of the ESRs.

2.9. TIM INSTRUMENT PERIPHERALS PROVIDE CORRECTIVE CAPABILITIES AND DIAGNOSTICS

The TIM ESRs, apertures, and shutters are the core of the instrument. Other items on the instrument help maintain thermal stability, maintain cleanliness, or give diagnostic information. A central heat sink maintains instrument temperature and stability both during a shutter's 100-s period and during the solar-illuminated and eclipsed portions of the 95-min SORCE orbit. A vacuum case with two vacuum doors maintains cleanliness of the instrument interior during integration and launch. Multi-layer aluminized mylar blanketing provides thermal isolation from the spacecraft's local environment. Multiple thermistors provide knowledge of various instrument temperatures, and are used in estimating thermal background corrections. Photodiodes monitor the cavity interiors for changes in reflectance, which might be indicative of a change in the absorptive NiP layer. A detector electronics board on the rear of the instrument contains the voltage references in close proximity to the ESRs to reduce the corrections for electrical leads supplying power to the cavity resistive heaters.

3. Relative Standard Uncertainties as Designed

Table I summarizes the combined standard uncertainty, σ, expected for the instrument as designed. Individual uncertainties are based on calibrations of prototypes, capabilities of state-of-the-art calibration facilities, analyses, and calculations using parameter uncertainties. (The actual uncertainties for the flight TIM are described by Kopp, Heuerman, and Lawrence (2005).) The dominant uncertainties are in the aperture area A and the cavity absorption α. The individual component uncertainties are assumed independent. Their quadrature sum gives a combined relative standard uncertainty of less than 100 ppm.

TABLE I

TIM uncertainty budget summary as designed (uncertainties are 1 standard deviation).

Factors/corrections	Size [ppm]	σ [ppm]
Distance (f_{AU})	33 537	0.1
Velocity (f_{Dopp})	57	0.7
Shutter waveform ($\boldsymbol{S}$)	100	1
Aperture (A)	1 000 000	55
Reflectance ($1 - \alpha$)	200	54
Servo gain ($\boldsymbol{G}$)	16 000	0
Standard (V^2)	1 000 000	7
Non-linearity	1 000 000	6
Standard R and leads	1 000 000	17
Equivalence ($\mathbf{Z_H}/\mathbf{Z_R}$)	7	22
Dark signal	1800	2
Scattered light	100	14
Repeatability (noise)		1
RSS total		84

4. TIM Measurement Equation

The TIM measurement equation follows from the signal transfer diagram shown in Figure 2. Scalars represent the input irradiance time series E_0, aperture area A, standard voltage V, standard resistance R, and a time series of output data numbers D. The bold-type terms are complex phasor components representing the amplitude and phase of sinusoidal variations as a function of frequency. These complex numbers describe shutter transmission $\boldsymbol{S}$, thermal impedances $\boldsymbol{Z}$, and servo gain $-\boldsymbol{G}$ within the instrument. TIM data and servo response are analyzed at frequencies between 10^{-4} and 50 Hz, but the irradiance is determined at the 0.01 Hz shutter fundamental. The data numbers D are output at rates up to 100 Hz, providing a maximum of 10^4 numbers per shutter cycle.

In ground data processing, the time series D is frequency analyzed and smoothed using a boxcar filter of the same period as the shutter, which eliminates sensitivity to changes at the shutter frequency. Four successive applications of this filter provide a nearly Gaussian weighting to the data within 200 s of a desired time (see Figure 3). Weighting by sinusoidal components at the shutter frequency produces the complex phasor $\boldsymbol{D}$ which provides knowledge of the changes occurring at, and in-phase with, the shutter fundamental. This data processing filter has the benefits of Fourier analysis at the shutter frequency while being able to analyze data within a 400-s interval.

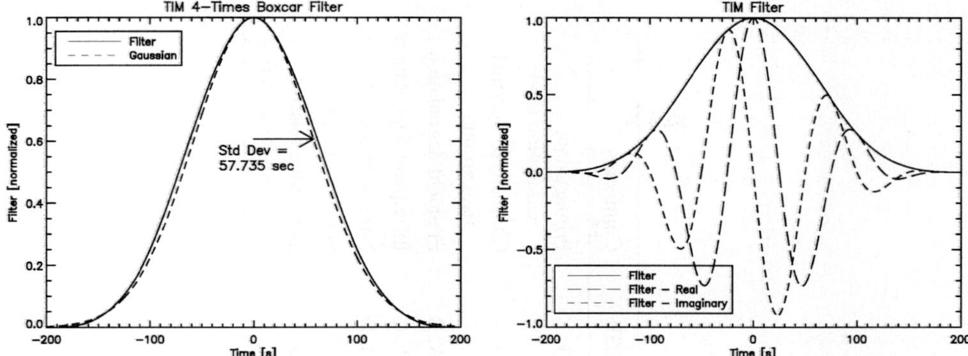

Figure 3. TIM data processing filter. This four-time repeated 100-s boxcar filter weights data with a near-Gaussian having total length of 400 s and a standard deviation of 57.7 s (*left plot*). Data in this range are Fourier analyzed to give changes at, and in-phase with, the TIM shutter fundamental (*right plot*).

The phasors $\boldsymbol{D}$ are converted to measured irradiances via

$$E_{\text{meas}} = \frac{V^2}{MR} \cdot \frac{1}{\alpha\, A f_{\text{corr}}} \cdot \text{real}\left[-\frac{\boldsymbol{Z}_{\mathbf{H}}}{\boldsymbol{Z}_{\mathbf{R}}} \cdot \frac{1}{\boldsymbol{S}} \cdot \left(\boldsymbol{D} + \frac{\boldsymbol{D} - \boldsymbol{F}}{\boldsymbol{G}} \right) \right], \tag{1}$$

which corrects for the complex servo system gain, shutter timing, equivalence $\boldsymbol{Z}_{\mathbf{H}}/\boldsymbol{Z}_{\mathbf{R}}$, applied feedforward values $\boldsymbol{F}$ at the shutter frequency, and scalar absorptance of the cavity α. Four correction factors are combined in f_{corr} to account for the spacecraft's distance to the Sun, Doppler shifts due to spacecraft velocity, cavity responsivity degradation, and pointing effects. The flight standard voltage V and the standard resistance R determine the standard watt, V^2/R. The fixed scalar $M = 64\,000$ converts data numbers D to duty cycle $q = D/M$; the standard voltage is applied to the active cavity's resistive heater at 100 Hz by a pulse width modulator.

Observations of empty space during the eclipse portion of each orbit provide a measurement of the instrument's thermal background. These thermal background measurements are modeled by four instrument temperature monitors. This empirical model, computed using instrument temperatures measured during actual solar measurements on the daytime side of the orbit, estimates the effective thermal signal appropriate for the solar observations, $E_{\text{dark\_est}}$. The reported TSI value is then the difference

$$E_0(t) = E_{\text{meas}}(t) - E_{\text{dark\_est}}(t). \tag{2}$$

5. Observation Modes and Data Products

The TIM's observation modes are summarized in Table II.

TABLE II

TIM observation mode summary.

Observation mode	Frequency	Duration	Target	ESR	Comments
Normal	Every orbit, daylight side	400 s per meas., continuous	Sun	B	Primary data acquisition mode
Dark	Every orbit, eclipse	400 s per meas., continuous	Dark space	B	Characterize thermal background
Degradation A	1%	1 orbit/week	Sun and dark space	A, B	Frequent degradation
Degradation C	0.5%	1 orbit/2 weeks	Sun and dark space	A, C	Infrequent degradation
Degradation D	0.2%	1 orbit/4 weeks	Sun and dark space	B, D	Infrequent degradation
Gain	Bi-weekly	4 orbits	NA	All	Servo calibration
FOV map	Every 6 months	15 orbits	Sun, 5 × 5 raster, 5′ steps	B	Determine relative sensitivity to FOV

In SORCE's low Earth orbit of 640 km and 40° inclination, the TIM generally observes the Sun for the 'daytime' portion of every orbit; this is the 'Normal' mode. These sunlight periods last from 50 to 75 min of the orbit's 95-min duration. In this mode, the TIM uses its primary ESR, shuttered at the instrument's 100-s period, to acquire irradiance measurements of the Sun. A feedforward value appropriate for the expected solar irradiance level is used.

On the remaining portion of each orbit, during which the Sun is eclipsed by the Earth, the TIM acquires measurements of dark space using the same operational configuration as for the Sun but with a lower feedforward value. These measurements provide knowledge of the thermal, or 'dark,' contributions from the instrument that are used in correcting the measurements of the Sun.

At a 1% duty cycle, amounting to one orbit every week, the primary and secondary ESR are used simultaneously to measure TSI. This lesser-used secondary ESR has a lower rate of solar exposure, providing a stable monitor by which long-term variations in the primary ESR can be corrected. Similarly, the third and fourth ESRs are used at 0.5% and 0.2% duty cycles to monitor changes in the more frequently used ESRs. These modes are known as 'Degradation' modes. This degradation correction approach was first applied to TSI measurements by Willson (1979) and Willson *et al.* (1981) using the three cavities in the Active Cavity Radiometer Irradiance Monitor.

TIM servo gain is calibrated during a 6-hour period every 2 weeks (Kopp, Heuerman, and Lawrence, 2005). During this mode no solar observations are acquired. The instrument shutters remain closed, and each ESR's response to a square-wave electrical heater transition is measured. From this response, the servo gain is determined and monitored throughout the mission.

Every 6 months the SORCE spends several orbits performing a field of view (FOV) map to determine instrument sensitivity to pointing. This calibration is a 5×5 grid with 5-arc min spacings centered on the Sun. Measurements with each SORCE instrument at every grid position determine that instrument's pointing sensitivity and changes with time.

The TIM's primary data products are TSI in units of W m^{-2} reported at a constant distance of 1 AU from the Sun as well as the value measured at the top of the Earth's atmosphere. These are used for long-term studies of the Sun's output and for Earth climate modeling, respectively. Both daily and 6-hourly averages are reported. Reported uncertainties are based on the instrument's combined standard uncertainty and on the standard deviation of the Sun's output during the time period.

TSI values are computed at a 50-s cadence, from which the reported daily and 6-hourly values are computed by averaging valid data. Each measurement requires 400 s of data, or four complete shutter cycles, so these high-cadence values are not independent. The high cadence values are a research product and are useful for studying short-term solar features responsible for irradiance variations.

TIM TSI data are processed within a few days of acquisition and are available to the public after regular, frequent updates. Data are versioned such that any change to the processing code or calibration parameters, such as when new degradation or gain data are applied, causes an increase in data version number. This configuration control links any data set with the associated parameters used in its generation.

The TIM data are available in ASCII text format online through the SORCE web site (*http://lasp.colorado.edu/sorce*) as well as through the NASA DAAC (*http://daac.gsfc.nasa.gov/upperatm/sorce/*).

6. Summary

The TIM design includes two significant improvements over previously flown TSI radiometers: (1) The use of phase sensitive detection lowers sensitivity to noise and thermal drifts and improves knowledge of the equivalence, and (2) NiP provides robust absorptive cavity interiors to withstand long-term exposure to solar radiation. This is the first space flight design intended to achieve 100 ppm combined standard uncertainty for an ambient temperature radiometer, and is made possible by exacting calibrations at the component level and extensive system design and analysis.

Appendix A: TIM Equivalence Calculation

In the TIM signal transfer diagram (Figure 2), measured signals from the combined heater power DV^2/MR and absorbed radiant power αE_0 are proportional to constants Z_H and Z_R, respectively. These (complex) thermal impedances Z are physical properties of the ESR and characterize its thermal response to input power. Both thermal impedances are frequency-dependent. The equivalence ratio, Z_H/Z_R, gives the conversion between replacement heater power and absorbed radiant power; this ratio would be unity in an ideal ESR. In the TIM ESR design, with the high diffusivity of the silver in the ESR and with both heater and radiant power inputs being nearly spatially co-located at the cone end of the ESR, this ratio is within a few ppm of unity for signals in-phase with the shutter.

The TIM's equivalence ratio, Z_H/Z_R, is calculated based on an algorithm developed in terms of known parameters of the ESRs. The computation proceeds by solving the thermal diffusion equation for power input at cavity position x, giving a Green's function solution, and then averaging the Green's function over the spatial distributions of input power. Thus, the three parts of the algorithm to calculate the equivalence ratio are the thermal Green's function $Z(x)$, and the normalized spatial distributions of electrical heat input $f_H(x)$ and radiant heat input $f_R(x)$. ESR parameters are varied over their uncertainty limits in a Monte Carlo program to calculate corresponding values of the equivalence ratio. The average of this ensemble is the final estimate of the equivalence ratio and the standard deviation of the ensemble is the uncertainty in the ratio.

Heat flow from the cone to the rest of the cavity can be treated as a one-dimensional (axial) problem because the four thermistors near the center of the ESR average out azimuthal temperature variations and because the initial power inputs are nearly azimuthally invariant. This approximation was verified numerically with two-dimensional finite element solutions and found to contribute less than 0.1% error to the calculated thermal impedances and hence has only a sub-ppm effect on the equivalence.

Another approximation is that heat from the replacement heater (embedded in the ESR's external surface) and heat from incident radiation (on the inside of the ESR) are taken to both have the same thermal impedance once the heat flows into the wall of the ESR. The relaxation time through this silver wall is on the order of 1 ms and the relaxation time through the inside NiP absorptive coating is only a few microseconds, which justifies this approximation for the TIM ESRs, having about 2-s net thermal response times.

The main difference in the two thermal impedances is from the heater wires, which are insulated from the silver wall in which they are embedded and have a thermal delay of about $\tau_{\text{wire}} \approx 20$ ms, as determined from finite element models. At the shutter fundamental of 0.01 Hz, this wire delay causes an amplitude attenuation of <1 ppm and a phase shift of $0.07°$, creating about a 10 ppm shift in the equivalence ratio. Neglecting the microsecond NiP delay and the relaxation through the wall, the Green's functions for the heater wire and the radiative thermal impedances are then

$$Z_{\text{wire}} = Z(x)/(1 + i\omega\tau_{\text{wire}}), \tag{A.1}$$

and

$$Z_{\text{NiP}} = Z(x), \tag{A.2}$$

where $Z(x)$ describes the flow along the axial coordinate x in the wall of the ESR. Again, $Z(x)$ is the temperature signal at the thermistors divided by the power input at position x along the cavity. Because unit power is applied, $Z(x)$ is the desired thermal impedance between position x and the thermistors.

The use of sinusoidal signals (appropriate for the TIM's phase sensitive method) converts the diffusion equation from a partial differential equation in time and space to an ordinary differential equation in space. This is where the phase sensitive method of the TIM enables a much simpler estimate of the equivalence than allowed by more traditional, DC-subtraction based radiometers; with the phase sensitive detection method, the equivalence is only needed at one frequency, and the large equivalence variations at high frequencies are not relevant.

The heat flow (diffusion) equation for sinusoidal variations in the temperature of the cylinder portion of the cavity is

$$\frac{d^2Z}{dx^2} = -\beta^2 Z, \tag{A.3}$$

where the complex wavenumber β (Equation (A.5)) characterizes both thermal conduction and thermal radiation from the outer surface of the ESR wall. The solutions of Equation (A.3) are the circular functions $\cos(\beta x)$ and $\sin(\beta x)$. Similarly, for the cone portion of the cavity the diffusion equation is

$$\frac{\mathrm{d}^2 Z}{\mathrm{d}x^2} + \frac{1}{x}\frac{\mathrm{d}Z}{\mathrm{d}x} = -\beta^2 Z. \tag{A.4}$$

The basic solution of Equation (A.4) that is finite at the tip of the cone ($x = 0$) is the Bessel function $J_0(\beta x)$. For both the cone and cylinder portions of the cavity, β is defined by

$$\beta^2 = -\left(\frac{4\varepsilon\sigma T_0^3}{kW} + i\frac{2\pi f}{D}\right), \tag{A.5}$$

with D the effective thermal diffusivity of the cavity (mostly silver), f the frequency of interest, $4\varepsilon\sigma T_0^3$ the incremental surface radiant conductivity at operating temperature T_0, k the thermal conductivity of the ESR's silver, and W the wall thickness of the cone and cylinder portions of the ESR cavity. Thus, distributed radiation from the exterior surface is included in the formalism, assuming the emissivity is constant along the cavity. At the TIM shutter fundamental, $f = 0.01$ Hz, $\beta^2 \approx -(0.0001129 + i0.03696)\,\mathrm{cm}^{-2}$ and $\beta \approx (0.1357 - i0.1361)\,\mathrm{cm}^{-1}$.

The actual thermal impedance Green's function $Z(x)$ is constructed as linear combinations of the basic solutions by matching the boundary conditions at the junction of the cone and cylinder, including the conductivity Γ of the cavity support structure at this junction. Then, by averaging $Z(x)$ over the (normalized) distributions $f(x)$ of input power, the equivalence ratio is

$$\frac{Z_\mathrm{H}}{Z_\mathrm{R}} = \frac{1}{1 + i\omega\tau_\mathrm{wire}} \frac{\int Z(x) f_\mathrm{H}(x)\mathrm{d}x}{\int Z(x) f_\mathrm{R}(x)\mathrm{d}x}. \tag{A.6}$$

Using the dimensions shown in Figure A1 and matching boundary conditions, relative solutions of the diffusion Equations (A.3) and (A.4) are, for inputs $x \leq B$,

$$Z(x, x_\mathrm{T}) = J_0(\beta x)\cos(\beta C - \beta x_\mathrm{T}). \tag{A.7}$$

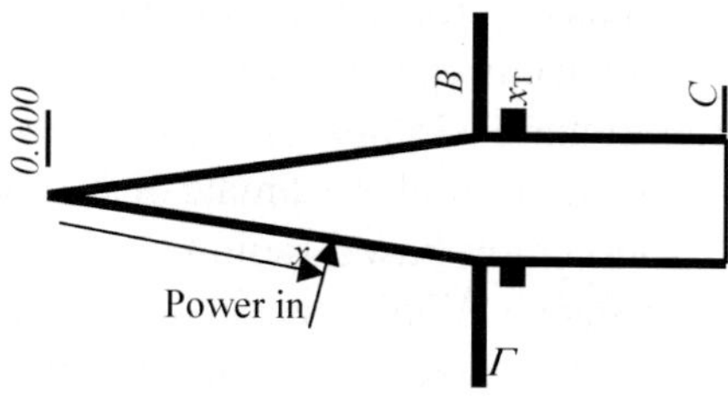

Figure A1. ESR cavity. With coordinate $x = 0$ at the apex of the cone portion of the ESR, the cone has slant length B, the ESR total length is C, and the thermistors are at x_T. The mechanical support, the heater leads, and the thermistor leads at B provide a thermal conduction of Γ to the surrounding instrument heat sink. Consider power input at some position x. The Green's function $Z(x, x_\mathrm{T})$ is then the temperature/power ratio between points x and x_T.

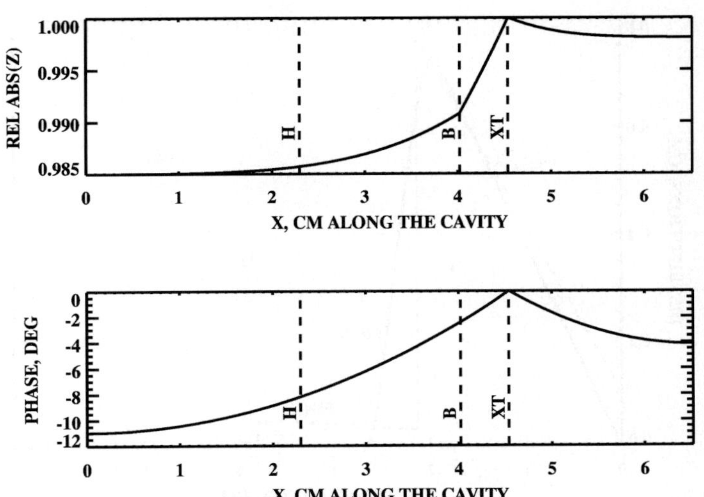

Figure A2. Thermal impedance. Relative thermal impedance to the thermistor as a function of input distance along the cavity for nominal parameters. B marks the cone/cylinder junction. XT is the location of the thermistors. H marks the nominal end of heat input.

For heat inputs on the cylinder ($x \geq B$), let x_L and x_G be the lesser and greater, respectively, of input position x and thermistor position x_T. For these positions x, the relative transfer impedance is given by

$$Z(x, x_T) = \cos(\beta C - \beta x_G) \times \\ \times\{\cos(\beta B - \beta x_L)J_0(\beta B) + \sin(\beta B - \beta x_L)(J_1(\beta B) - \xi J_0(\beta B))\}, \tag{A.8}$$

where $\xi = \Gamma/2\pi\beta BkW \sin(\theta)$, with Γ the thermal conductivity of the cavity support located at $x = B$ and θ the half angle of the cone. The thermal impedance, $Z(x, x_T)$, of Equations (A.7) and (A.8) are plotted in magnitude and phase in Figure A2. The functions are even in the argument βx.

The heater distribution in x increases linearly along the axial coordinate x from x_1 to x_2, being from the wire wound resistor on the external portion of the cone and heating very nearly the same region around the apex as the incoming radiant power. The normalized heater distribution is

$$f_H(x) = \frac{2x}{x_2^2 - x_1^2}, \quad x_1 \leq x \leq x_2. \tag{A.9}$$

In the actual TIM ESRs, there are about 1.5 turns of high pitch heater winding at the end nearest the cylinder to bring the wire (under the outer copper/gold plating) up to a low-resistance copper terminal strip and then to the heater lead connection at $x = B$. This extra heating is added numerically to the distribution and the entire distribution is normalized.

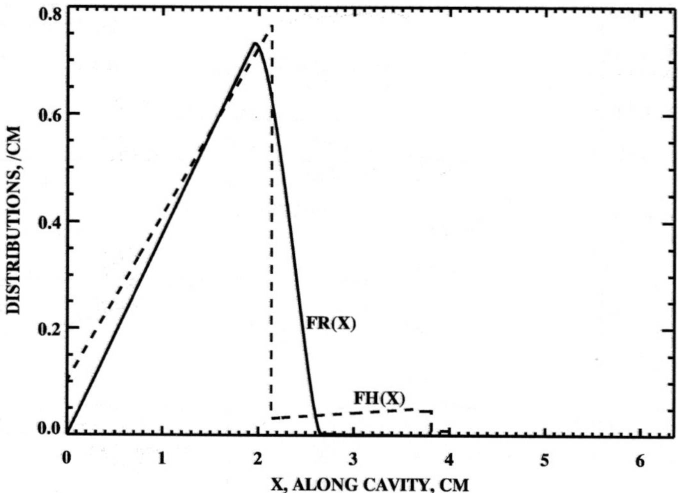

Figure A3. Heat distributions. Normalized distributions of heater power (FH) and absorbed radiant power (FR), given the nominal parameters of Table AI.

The incident radiant power distribution is nominally in a uniform cylindrical beam having a diameter defined by the precision aperture, but with fuzzy edges due to the finite size of the Sun. By design, this incoming solar beam has approximately the same uniform distribution as the electrical heater power, as shown in Figure A3. However, approximately 1.6% of the incident light scatters diffusely from the initial contact point, and this broadens the radiant heater distribution. For calculating equivalence, only the first bounce matters because the second bounce is reduced nearly another 2 orders of magnitude. For a total hemispherical Lambertian reflectance η at the first bounce, the distribution becomes

$$f_{\mathrm{RAD\_DIFFUSE}}(x) = (1-\eta)\frac{2x}{H^2} + \eta\, f_{\mathrm{DIF}}(x), \tag{A.10}$$

where H is the length of the illuminated area measured axially. $f_{\mathrm{DIF}}(x)$ is the first bounce scatter distribution derived from the Lambertian scattering and the conical geometry, and is given by

$$f_{\mathrm{DIF}}(x < H) = \\ = -\frac{H^2 - (H-2x)(x+Y) + H(H-3x-Y)\cos(2\theta) - 4xY\sin(\theta)}{2\tan(\theta)H^2Y} \tag{A.11}$$

and

$$f_{\mathrm{DIF}}(x > H) = \\ = \frac{H^2 - Hx + 2x(x-Y) + 2HY\cos^2(\theta) + H(H-3x)\cos(2\theta)}{2\tan(\theta)H^2Y}. \tag{A.12}$$

TABLE A1

Parameters and their uncertainties for equivalence calculations (all uncertainty distributions are assumed uniform and rectangular to the ±limits, and all are assumed independent.).

		Nominal	Distribute	
Equivalence Model Parameters				
K	Thermal conduction, silver	4.27	±10%	$\mathrm{W\,c^{-1}\,K^{-1}}$
B	Slant distance, cone vertex to cylinder	4.06	±0.2	cm
XT	Slant distance, vertex to thermistor	4.54	±0.2	cm
C	Slant distance, vertex to cone mouth	6.35	±0.2	cm
D	Diffusivity of silver + copper plate	1.6	±10%	$\mathrm{cm^2\,s^{-1}}$
EPS	IR emissivity of the cone exterior	0.04	±0.02	
W	Wall thickness	0.07	±0.02	cm
Tau	Zero-order pole time constant	220	±10%	s
C_H	Specific heat of the cone/cylinder	0.26	±10%	$\mathrm{J\,g^{-1}\,K^{-1}}$
	Mass of the cone assembly	15.78	±10%	g
H	Slant distance to edge of light beam	2.3	±0.1	cm
Tau_wire	Time constant of heater wire	20	±10	ms
Heater Winding Geometry				cm
	Uncertainty in the vertex position		±0.1	cm
S0	Start of cone winding	0.329	±0.01	cm
R0	Radius of winding at the start	0.058	±0.01	cm
S1	End of fine pitch winding, start fast pitch	2.465	±0.01	cm
R1	Radius of fine pitch winding at the end	0.435	±0.01	cm
S2	End fast pitch, start leads	4.012	±0.01	cm
R2	Radius at end of fast pitch	0.73	±0.01	cm
S3	Copper strap start	4.141	±0.01	cm
S4	Copper strap end	4.217	±0.01	cm
S5	Heater lead terminal point	4.349	±0.01	cm
η	NiP black total hemispherical reflectance	0.02	±0.01	

θ is the cone half angle and the auxiliary length Y is defined as

$$Y = \sqrt{H^2 + x^2 - 2Hx\cos(2\theta)}. \tag{A.13}$$

The total integral over dx of this formula is slightly less than one because of some escape out the front of the cone. This escaped fraction is accounted for separately by the calibration cone absorptance factor α. Because of this, numerical calculations must renormalize the distribution's area to one.

Based on this formalism, the total reflected loss out the front will be $1 - \alpha \approx 1.23\%$ of η, a result consistent with non-sequential ray trace studies. For example, a Lambertian reflectance $\eta = 1.6\%$, consistent with bidirectional reflectance

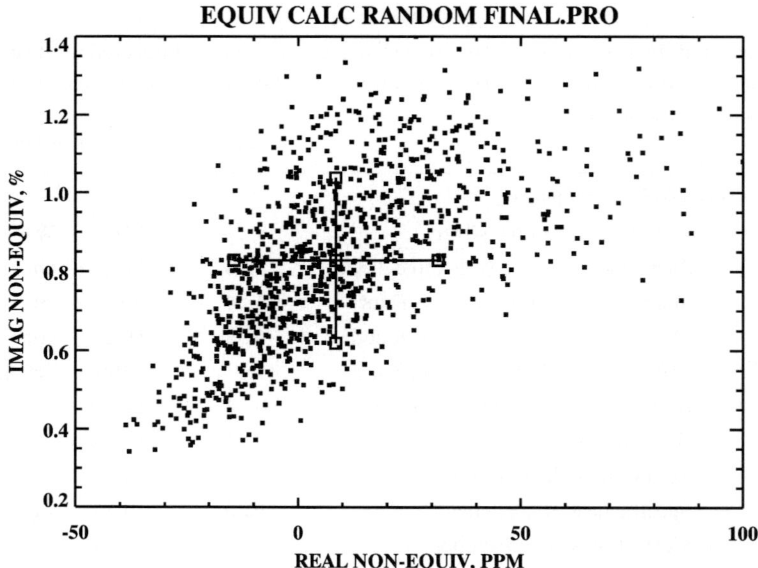

Figure A4. Equivalence estimate. Monte Carlo calculation of the "non-equivalence", $Z_H/Z_R - 1$. The cross in the center indicates the average and ± 1 standard deviation of the ensemble.

distribution function measurements of NiP, would give a total ESR reflection loss $1 - \alpha = 250$ ppm, close to the calibrated values for the TIM ESRs.

There should also be a thermal IR radiation term included in the definition of β to account for IR losses out the mouth of the ESR. In the analytic solution (A.5), β characterizes the constant emissivity of the gold-plated exterior of the cone, but neglects the effect of thermal radiative emission of the cone from the interior surface. The total radiative conduction in the front is

$$\Gamma_{\text{IR\_mouth}} < \pi R_{\text{mouth}}^2 \times 4\sigma T_0^3 \approx 0.0011\,\text{W}\,\text{K}^{-1}. \tag{A.14}$$

The measured total conductivity of the cone support is $\Gamma_{\text{total}} \approx 0.016\,\text{W}\,\text{K}^{-1}$, so the interior IR conductivity is approximately 7% of the total. This radiative conduction is distributed proportional to the view factor from the interior to the outside. Analysis shows it to come mostly from the centimeter of the cylinder nearest the mouth of the cavity. This is removed as far as possible from the heat distribution functions within the cone, so does not provide a significant effect on the equivalence ratio.

Inclusion of this distributed IR gives the diffusion equation a wavenumber β that varies slightly with x, requiring a numerical rather than an analytic solution. Numerical solutions obtained using finite element methods and sparse matrices ($18\,000 \times 18\,000$) verified the analytic model and showed that the addition of the IR distributed conductivity changes the equivalence ratio by less than 1 ppm. The distributed IR radiation is therefore neglected in the determination of Z_H/Z_R.

Additionally, in the Monte Carlo calculations, the support conduction Γ is assigned an uncertainty of $\pm 10\%$, which accommodates the distributed 7% IR loss near the mouth of the cavity.

The analytic result enables a Monte Carlo uncertainty analysis of the equivalence ratio. The parameterized thermal impedance given by Equations (A.7) and (A.8), and the distribution of heater power (Equation (A.9)) and of absorbed radiation (Equations (A.11) and (A.12)) yields an equivalence ratio given the set of parameters and their uncertainties listed in Table AI.

From the distributions of parameters listed in Table AI, an ensemble of 1000 equivalence ratios was computed at the shutter fundamental, giving the average and standard deviations shown in Figure A4. This gives an equivalence ratio for the TIM of

$$Z_{\mathrm{H}}/Z_{\mathrm{R}} = (1.000008 \pm 0.000023) + i(0.0083 \pm 0.0021). \quad \text{(A.15)}$$

Acknowledgements

This research was supported by NASA contract NAS5-97045. Informal reviewers from NIST and NASA provided much useful advice. LASP engineers are greatly acknowledged for their major contributions to the instrument design and function.

References

Foukal, P.: 2003, *EOS Trans. AGU* **84**, 22, 205.

Gundlach, J. H., Adelberger, E. G., Heckel, B. R., Swanson, and H. E.: 1996, *Phys. Rev. D*, **54**, R1256.

Kopp, G., Heuerman, K., and Lawrence, G.: 2005, *Solar Phys.*, this volume.

Kopp, G., Lawrence, G., and Rottman, G.: 2003, *SPIE Proc.* **5171**, 14.

Lawrence, G. M., Rottman, G., Harder, J., and Wood, T.: 2000, *Metrologia* **37**, 407.

Lawrence, G. M., Kopp, G., Rottman, G., Harder, J., Woods, T., and Loui, H.: 2003, *Metrologia* **40**, S78.

Lean, J., Beer, J., and Bradley, R.: 1995, *Geophys. Res. Lett.* **22**, 3195.

Pang, K. D. and Yau, K. K.: 2002, *EOS Trans. AGU* **83**, 43, 489.

Rax, B. G., Lee, C. I., and Johnston, A. H.: 1997, *IEEE Trans. Nuclear Sci.* **44**, 1939.

Rice, J. P., Lorentz, S. R., and Jung, T. M.: 1999, in: 10th Conference on Atmospheric Radiation, 28 June–2 July, Madison, Wisconsin (Preprint volume).

Spreadbury, P. J.: 1991, *IEEE Trans. Instrum. Meas.* **40**, 343.

Willson, R. C.: 1979, *J. Appl. Opt.* **18**, 179.

Willson, R. C., Gulkis, S., Janssen, M., Hudson, H. S., and Chapman, G. A.: 1981, *Science* **211**, 700.

Woods, T., Rottman, G., Harder, G., Lawrence, G., McClintock, B., Kopp, G. *et al.*: 2000, *SPIE Proc.* **4135**, 192.

Additionally, in the Monte Carlo calculation the [illegible] conductor [illegible] is assigned an uncertainty of [illegible], which accommodates the [illegible] near the mouth of the cavity.

The measured results [illegible] Monte Carlo uncertainty analysis of the equivalence ratio. The parameters [illegible] given by Equations [illegible] and the distribution [illegible] Equation (A.[illegible]) and [illegible] absorbed radiation (Equations A.13 and A.14) [illegible] equivalence ratio [illegible] parameters and their uncertainties listed in Table A.[illegible]

From the distributions of parameters listed in Table [illegible], an ensemble of 1000 equivalence ratios was computed as described above, [illegible] the average and standard deviation shown in Figure A.4. This gives an equivalence ratio for the ERM of

[illegible] (A.16)

Acknowledgements

The research was supported by NASA contract [illegible]. Informal reviews from NIST and NASA provided much useful advice. [illegible] are greatly acknowledged for their major contributions to the instrument design and function.

References

[illegible]

Solar Physics (2005) 230: 111–127

THE TOTAL IRRADIANCE MONITOR (TIM): INSTRUMENT CALIBRATION

GREG KOPP, KARL HEUERMAN and GEORGE LAWRENCE
Laboratory for Atmospheric and Space Physics, University of Colorado, Boulder, CO 80309, U.S.A.
(e-mail: kopp@lasp.colorado.edu)

(Received 7 February 2005; accepted 13 May 2005)

Abstract. The calibrations of the SORCE Total Irradiance Monitor (TIM) are detailed and compared against the designed uncertainty budget. Several primary calibrations were accomplished in the laboratory before launch, including the aperture area, applied radiometer power, and radiometer absorption efficiency. Other parameters are calibrated or tracked on orbit, including the electronic servo system gain, the radiometer sensitivity to background thermal emission, and the degradation of radiometer efficiency. The as-designed uncertainty budget is refined with knowledge from the on-orbit performance.

1. Introduction

The Total Irradiance Monitor (TIM) is an ambient temperature electrical substitution solar radiometer designed to achieve 100 parts per million (ppm) combined standard uncertainty in total solar irradiance (TSI). The TIM contains four electrical substitution radiometers (ESRs), which are electrically heated to maintain constant temperature while a shutter modulates sunlight through a precision aperture and into an ESR's absorptive cavity. The modulation in electrical heater power needed to maintain an ESR's temperature as its shutter modulates incident sunlight determines the radiative power absorbed by that ESR's cavity. Phase sensitive detection of this heater power, combined with knowledge of the aperture area over which the sunlight is collected, yields TSI in ground processing.

Meeting the design uncertainties presented by Kopp and Lawrence (2005) requires several precision calibrations of components and subsystems in the TIM. The most fundamental calibrations, such as the aperture area, ESR power applied, and cavity absorptivity are ground calibrations. Other items are directly calibrated on orbit. These include the servo system gain and the instrumental thermal infrared contribution to the measured signal. All calibrations are tracked for changes on orbit, including cavity absorptivity. After ground processing of these on-orbit calibrations or calibration changes, the resulting instrument data are updated to include the latest calibration values, many of which can be applied retroactively.

2. Ground Calibrations

Ground calibrations of spacecraft instruments are generally more accurate than possible in flight, as ground-based precision calibration facilities are not constrained by mass, power, or vibrations typical of launch environments. Ground calibrations are most appropriate for stable components, enabling transfer of the calibration to the in-flight instrument unchanged by age or launch vibrations.

2.1. Cavity Absorptance α is Near 0.9998

Cavity reflectance 1-α is measured using laser scans mapping the cavity interiors at six wavelengths spanning the spectral peak of the emitted solar flux. These spatially resolved measurements are supplemented by spatially integrating calibrations at mid-infrared wavelengths to extend the spectral coverage. The effective cavity reflectance is the average of the reflectance measurements at individual wavelengths weighted by the solar energy distribution given by Lean (2000).

A spatial map of cavity reflectance from a two-dimensional laser scan is averaged over the region of the cone illuminated by sunlight to obtain an effective reflectance at that laser wavelength. The solar limb-darkened profile is accounted for when computing these averages, and is a $\sim$0.1 ppm effect. Spatial maps, such as that shown in Figure 1, are acquired at each of six laser wavelengths (457, 532, 633, 830, 1064, and 1523 nm) spanning the primary solar spectrum.

The spatially resolved laser measurements are supplemented by a broad-beam laser calibration at 10.6 microns to extend the reflectance calibrations to the mid-infrared. NIST measurements of select cavities from 2 to 20 microns using an FTIR, described by Hanssen *et al.* (2003), fill in the large spectral gaps between the discrete laser wavelengths, and show that the cavity reflectance varies smoothly with wavelength through the mid-infrared.

A spline fit of the reflectance interpolates between the discrete laser wavelength calibrations. This fit is constrained at long wavelengths by estimating a reflectance at 100 μm that maintains a smoothly decreasing slope so the fitted long wavelength reflectance never exceeds unity, and at short wavelengths by the shortest-wavelength reflectance measurement. Since the Sun emits relatively little power at these far-infrared and ultraviolet wavelengths, the uncertainty in the cavity reflectance from this estimate is low. The reflectance calibrations for the primary TIM cavity are shown in Figure 2.

The effective solar-weighted reflectances for the four TIM cavities are very low, with values of 169, 139, 307, and 360 ppm. Reflectance uncertainties are 14%, including nearly equal estimated measurement uncertainty and statistical portions. These ground-based measurements of absorptance are tracked for relative changes on orbit, as described in Section 3.3.

2.2. Aperture Area is Corrected for Diffraction and Thermal Variations

The areas A of the diamond-turned aluminum apertures were measured at NIST using the non-contact geometric aperture calibration facility described by Fowler,

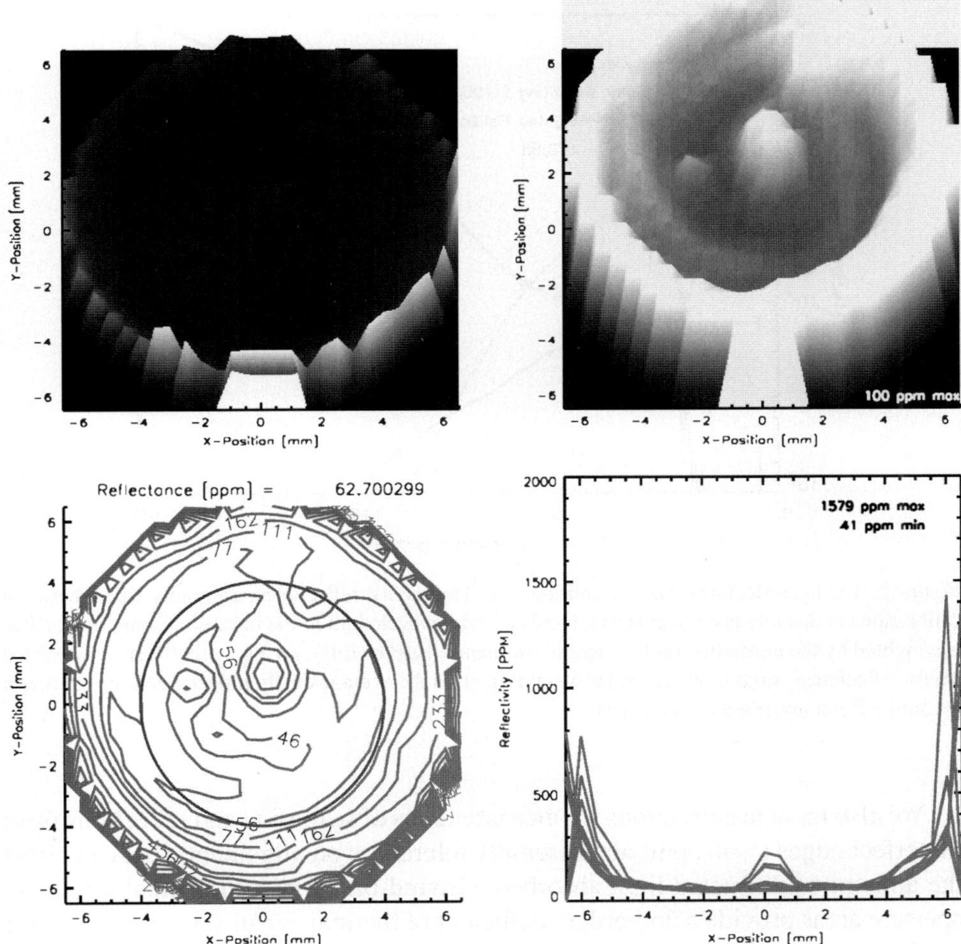

Figure 1. Cavity reflectance spatial map. Maps of the cavity interiors at six wavelengths spanning the solar spectrum peak are used to calibrate the cavity absorptance.

Saunders, and Parr (2000). The four TIM apertures have geometric areas 0.49928, 0.49938, 0.49936, and 0.49926 cm^2 with relative standard uncertainties of 25 ppm. This aperture area is corrected for temperature changes, diffraction, and bulk modulus expansion to the space environment.

Because of the precision with which NIST can measure geometric aperture areas, the dominant TIM aperture area uncertainty is the diffraction correction. This correction is proportional to the energy-weighted average wavelength $\langle\lambda\rangle$ of the solar spectrum and corresponds to a relative TIM diffraction correction of 430 ppm using $\langle\lambda\rangle = 947$ nm, based on measured solar spectra reported by Lean (2000). Shirley (1998, 2000) of NIST advises including 10% (43 ppm) of the diffraction correction as uncertainty.

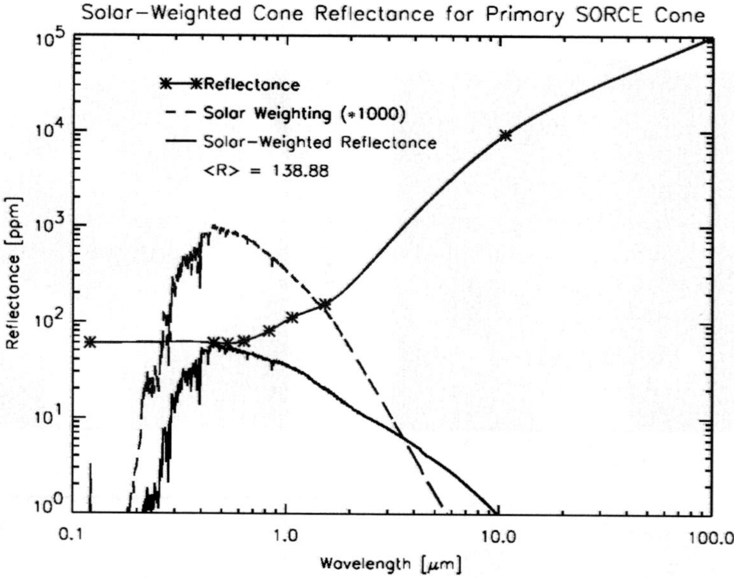

Figure 2. Cavity reflectance spatial calibrations. The cavity reflectance, smoothly fitted between calibrations at discrete laser (*asterisks*), increases with wavelength in the infrared. Cavity reflectance is weighted by the normalized solar spectral irradiance (*dashed line*), giving a relative solar-weighted cavity reflectance (*solid line*); this solar weighting gives the average cavity reflectance used to correct for sunlight not absorbed by the cavity.

We also include corrections for measurements of scattering and reflections from imperfect edges (~40 ppm) and assembly tolerances on alignments that can affect the amount of diffracted light absorbed. Ground-based measurements of witness aperture areas provide a 2nd order coefficient of thermal expansion appropriate for the flight apertures' aluminum. A platinum resistive thermal device (RTD) provides the flight apertures' temperature used in thermal corrections to the area, which are known to about 12 ppm. Pressure changes between ground calibrations and flight operations increase the flight aperture area by <1 ppm and are corrected. These corrections yield a net uncertainty in aperture area of 55 ppm, as summarized in Table II.

2.3. Standard Watt Comes from Stable Voltage and Resistance References

Changes in the electrical heater power applied to the ESRs directly compensate variations of the absorbed radiant power. This electrical power is purely resistive, and is produced by pulse width modulating a voltage standard reference through a standard reference resistor embedded in each ESR. There is no on-orbit monitor of these voltage or resistance references, as no low-mass, space-certified meter with

10^{-6} absolute accuracy is available. Instead, the TIM relies on stable voltage and resistance references that are calibrated and characterized prior to launch. Changes on orbit may be tracked by comparing simultaneous TSI measurements with two ESRs, which have different heater resistors and use different voltage references.

The standard watt relies on a 7.1 VDC reference voltage from temperature-stabilized Linear Technology LTZ1000 Zener diodes applied across resistive windings of encapsulated wire, and on pre-flight ground calibrations using an 8.5-digit HP3458A meter under temperature-controlled conditions. Both the standard voltage and the standard resistors were monitored and found stable throughout instrument assembly, environmental testing, and spacecraft integration and test, with the final calibrations being 3 months prior to launch.

Pre-flight calibrations of the two voltage references give

$$V_1 = 7.166434\,(1 - 0.201404 \times 10^{-6}\,T) \quad (1)$$

$$V_2 = 7.120490\,(1 - 0.112085 \times 10^{-6}\,T) \quad (2)$$

for temperature T in degrees Celsius. These temperature dependences are fairly linear across a 50 °C temperature range, as shown in Figure 3. Pre-flight measurements of the flight standard voltages have been stable to <1 ppm in spite of six qualification temperature cycles from −35 to +50 °C. Over 3 years of continual operation, five laboratory copies of the standard voltage circuits have changed in relative voltage by -0.9 ± 0.7 ppm/year (see Figure 4), consistent with the previous

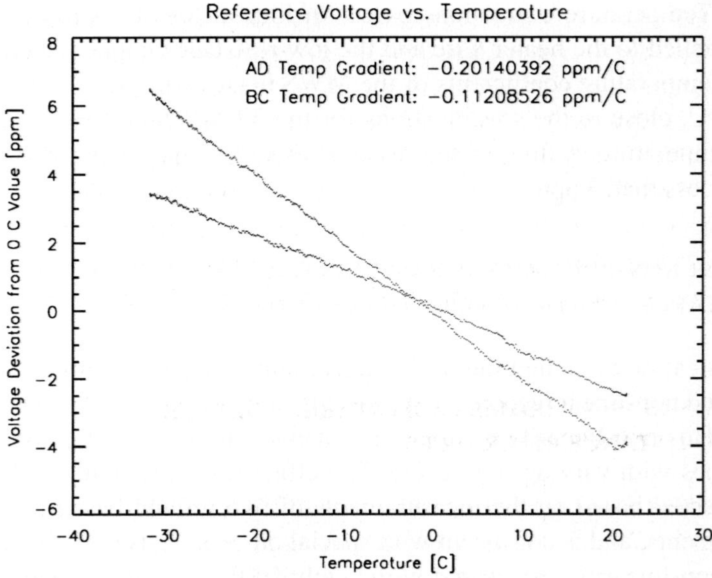

Figure 3. Temperature dependence of standard voltage. The temperature dependence of the LTZ1000 voltage references used in the TIM is linear from −30 to +20 °C.

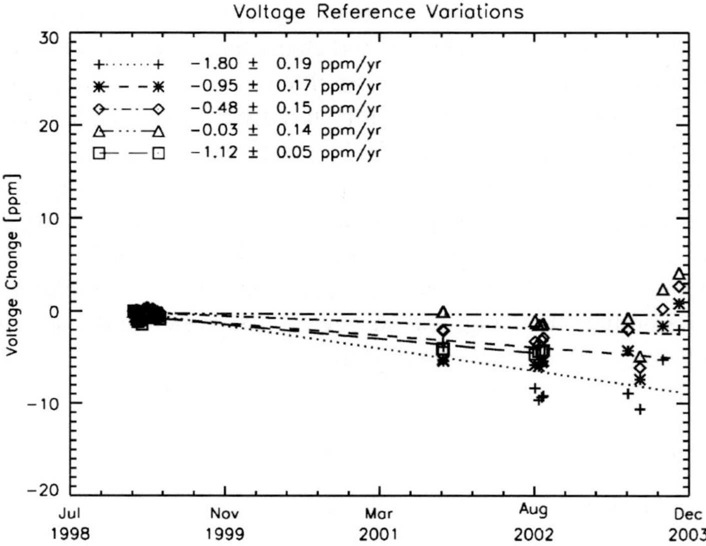

Figure 4. Witness voltage unit stability. Several ground-based LTZ1000 voltage references change by only 1 ppm/year.

studies (Spreadbury, 1991). The flight calibrations of the TIM voltage references are updated for systematic behavior of these five plus eight additional units.

The effective heater resistances for the four ESRs are 543.9689, 538.4464, 546.3407, and 537.9084 Ω at the cavity's operating temperature of 30.8 °C (see Figure 5). Temperature corrections based on four different instrument temperatures are applied to the heater wire and the low-resistance copper electrical leads. Measured temperature coefficients of the MWS heater wire itself are in the range 8–11 ppm/°C, close to the specifications for the 39-MWS-800-HML wire. Qualification temperature cycling of the resistor references has changed their relative resistances less than 3 ppm.

2.4. Power Non-Linearity is Based on Ground Calibrations of Nearly Identical Flight-Like Units

While most calibrations met the design uncertainty budget, the flight TIM had an unanticipated non-linear response with varying pulse width in the power applied to the ESRs. This non-linearity is due to the changes in the applied pulse width rise and fall times with varying duty cycle. The effect was only noticed after launch, and is corrected based on the measurements of two ground-based but flight-like TIM instruments, and is consistent with special on-orbit tests with the flight unit.

Power non-linearities measured in the eight ESRs from the two ground-based TIM units are similar but not identical, as shown in Figure 6. Variations between the four ESRs within an individual instrument are smaller than those in different units.

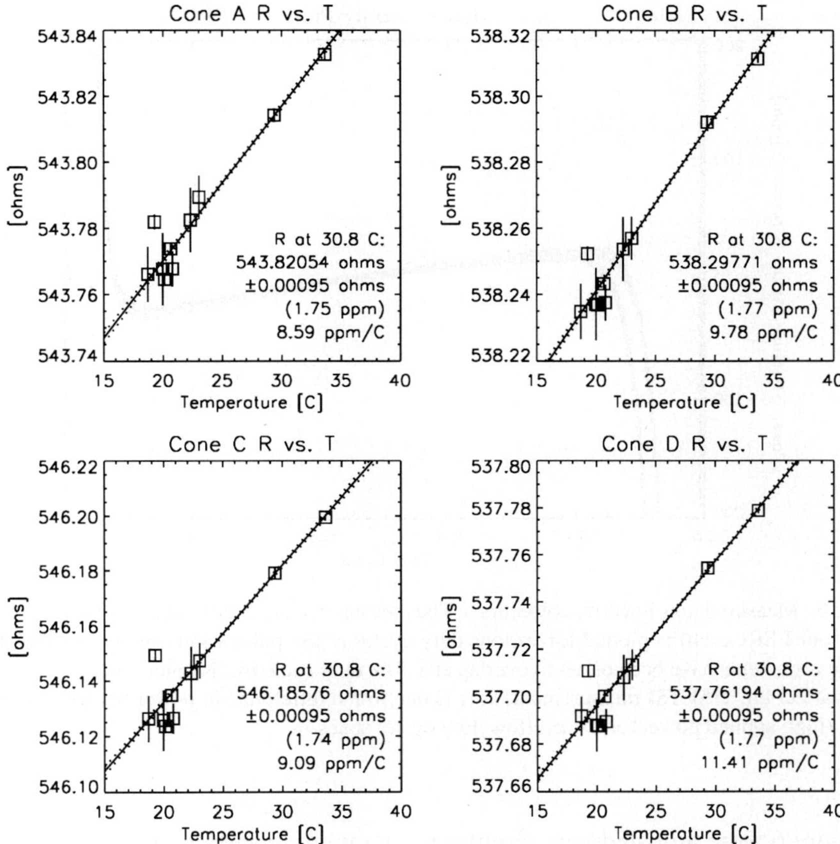

Figure 5. ESR heater resistance calibrations. The effective heater resistance is derived from ground-based calibrations at different temperatures. The contributions from low-resistance leads is removed to determine the effective ESR heater resistance.

The variations between these instruments or between their individual ESRs help determine the uncertainty by which we can expect a ground-based non-linearity determination to correctly apply to the flight unit. The TIM is operated at upper (shutter closed) power levels that are generally below the non-linear region at high duty cycles in the curve in Figure 6; however, the lower (shutter open) power levels used are affected by the non-linear region at low duty cycles, so the knowledge of this non-linearity is important.

Two flight tests help refine these ground-based non-linearity corrections for the flight unit. In the first test, simultaneous solar irradiance observations were made over the course of a day using both the primary and secondary ESRs. While the secondary ESR remained at a nominal power level to track small changes in the Sun's output power, the primary ESR was scanned from the lower limit of its power range to the upper limit. The characteristic non-linearity corrections shown

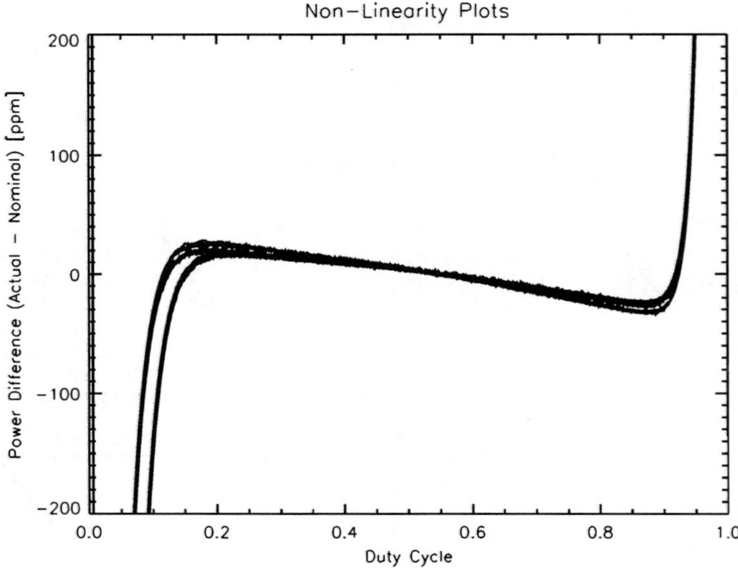

Figure 6. Measured non-linearity corrections. The measured non-linearity of two ground-based TIM units (four ESRs each) is plotted for various duty cycles of the pulse width modulated ESR power. The measurements have been offset to overlap at a 50% duty cycle for this plot. Such vertical offsets do not affect the TIM TSI measurement, as it is due to the difference in power between the shutter closed (high applied power) and open (low duty cycle) states.

in Figure 6 were adjusted very slightly to maintain a nearly constant difference in measured TSI between the two ESRs, meaning that the primary ESR's non-linearity is corrected. The second test comes from intermittent simultaneous observations of the Sun over the course of a year as its measured irradiance varies by ±3.3% due to changes in the Earth–Sun distance. If there are no non-linearities in the applied ESR power, the difference between the two ESRs due to solar distance variations would be zero. The non-linearity corrections shown in Figure 7 include small adjustments to the calibration measurements in Figure 6 that make the flight ESR results consistent with the expected responses for both of these on-orbit tests.

With these linearity adjustments from two ground-based TIM instruments and the on-orbit tests with the flight instrument, we estimate the uncertainty attributable to the flight unit's linearity to have a systematic uncertainty of ∼150 ppm and a measurement uncertainty as shown in Figure 7 that can be as high as ∼110 ppm. This net uncertainty of ∼186 ppm dominates the instrument's uncertainty budget.

2.5. Shutter Waveform Has <1 ppm Effect on In-Phase Signal

The shutter moves between a transmission of unity when open, and a transmission <3 ppm when closed. Shutter operation times are 10 ms. For the TIM's 100-s shutter

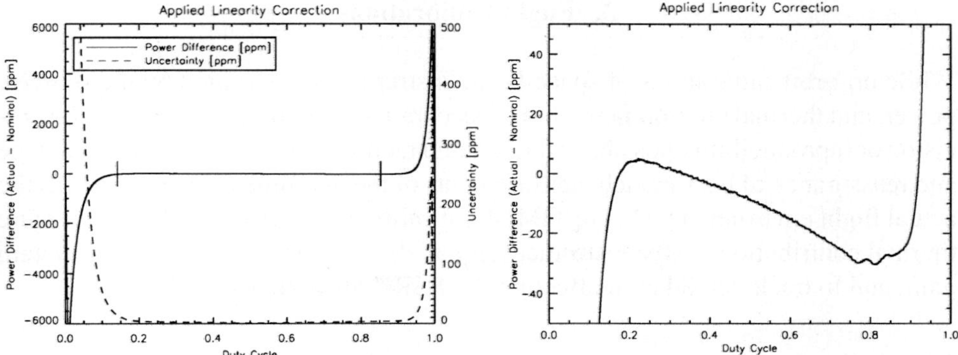

Figure 7. Applied non-linearity correction. The non-linearity of the flight unit is corrected by the plotted amount for various duty cycles of the pulse width modulated ESR power. The *plot on the left* includes vertical markings indicating the nominal power levels used by the primary TIM ESR. This plot also shows the uncertainty in the linearity correction (*right axis*), as estimated from the variations in the linearity measurements plotted in Figure 6. The *plot on the right* is an expanded view of the linearity correction.

period, these shutter operation times produce a relative correction <1 ppm in the phase sensitive analysis of the data. This effect, though small, is accounted for in data processing.

2.6. Pointing Sensitivity Estimated to be Low

A pre-launch field of view (FOV) analysis, based on cavity reflectance maps and variations in the cavity region illuminated as pointing changes, shows <100 ppm relative response up to and past 10 arcmin off-axis. This pointing sensitivity is at least a 2nd order effect, so for the worst case 2 arcmin SORCE pointing errors, no significant TIM pointing effect is expected.

2.7. Equivalence

The equivalence ratio $\mathbf{Z}_H/\mathbf{Z}_R$ comes largely from model calculations, with generous uncertainties in the model parameters, and differs from unity by 7 ppm with an uncertainty of 22 ppm (based on calculations by Lawrence *et al.*, 2000 and described in detail by Kopp and Lawrence, 2005) at the shutter fundamental frequency. The larger non-equivalence at higher harmonics may be used to further constrain the equivalence models. The uncertainty in the equivalence is low because the TIM's phase-sensitive detection only relies on knowledge of the equivalence at one (the shutter fundamental) frequency, and this is a low frequency where the equivalence is very nearly unity.

3. Flight Calibrations

While on-orbit calibrations of space-borne instruments are limited by the sources, power, and thermal environment of the spacecraft and are frequently not of the precision or reproducibility possible using ground facilities, they do offer the advantage and reassurance of post-launch measurements of the instrument performance in the actual flight environment. On the TIM, flight calibrations are conducted to measure thermal contributions to the instrument signal, determine the in-flight servo system gain, and to track degradation affecting the ESRs' absorptance.

3.1. Measurements of Dark Space Provide the Instrument's Thermal Background (Dark Correction)

The instrument's thermal-background ("dark") signal is measured by observing dark space during the eclipsed portion of each orbit. The dark signal is fitted to four instrument temperatures to model the contributions from different portions of the instrument. The instrument temperatures during solar measurements are used to estimate the background contribution to the solar signal at actual observing times.

The dark signal data numbers are converted to irradiances using the TIM measurement equation without applying Doppler, solar distance, or pointing corrections, as these have no relevance for instrument thermal background. These dark irradiances are fitted to four instrument temperatures, which form the basis vectors for modeling the observed dark irradiances with the linear combination

$$D_{\mathrm{Dark}} \approx \sum_{J=1}^{4} C_{\mathrm{J}} T_{\mathrm{J}}^{4}, \tag{3}$$

where the C_{J} are determined for best fit and T_{J} are the four Kelvin temperatures of the cavity, aperture plate, pre-baffle, and shutter. Since the basis vectors in Equation (3) are highly correlated, the Singular Value Decomposition (SVD) method of Press *et al.* (1993) is used to isolate individual temperature dependences of the dark signal. The coefficients C_{J} from the SVD fit have little physical significance and vary sufficiently that the SVD fit is computed daily using a 7-day running fit to background measurements, rather than presuming the C_{J} remain constant with time. The thermal background during the actual solar observations is then estimated by applying this temperature dependence to the daytime portion of the orbit and using actual instrument temperatures measured during the solar observations.

Dark signals of roughly $-3.15\,\mathrm{W/m^2}$ are measured. (The negative value is due to the *loss* of energy from the cavities into space when the shutter is opened, so the dark correction increases the measured TSI.) This value varies due to thermal fluctuations in the instrument by $0.1–0.2\,\mathrm{W/m^2}$ between orbit sunset and sunrise,

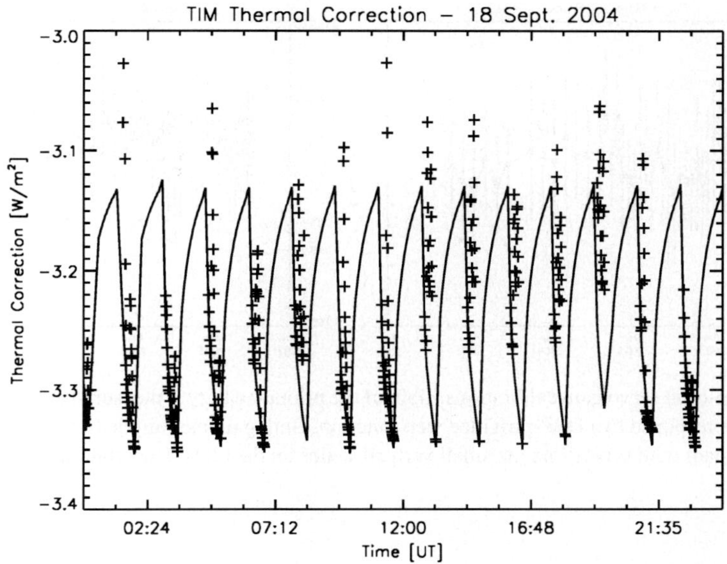

Figure 8. Dark signal observations and model. A fit of measured instrument temperatures to measured signals of dark space ('+' *signs*) gives a model of the thermal contributions of the instrument signal. This model estimates the thermal contribution at the temperatures relevant during solar observations (*solid line*).

as shown in Figure 8, giving a relative effect during the orbit of ~100 ppm. The uncertainty on this dark correction is approximately 0.01 W/m^2, or less than 10 ppm of the solar signal.

3.2. Servo Loop Gain is Monitored on Orbit by Analyzing the Response to a Step Function Input

The primary ESR is thermally balanced against a reference ESR maintained at constant temperature by an AC bridge circuit operated at 100 Hz. A DSP maintains the ESR thermal balance by applying heater power to the primary cavity. The gain of this servo loop at the 0.01 Hz shutter fundamental affects the measured signal when responding to non-equilibrium conditions. The DSP prevents such conditions by applying a feedforward signal to the ESR anticipating power changes as the shutter transitions between open and closed.

The servo gain is calibrated by the DSP monthly to a relative accuracy of 4.5×10^{-4} (see Figure 9). We further set the feedforward value within 1% of the anticipated ESR power level after a shutter transition, reducing the sensitivity of the TSI measurement to uncertainty in gain. For TIM gains, with the real part of $G \approx 60$, the gain dependent term is thus a $<10^{-3}$ relative correction causing relative uncertainty in the final TSI ~ 0.1 ppm.

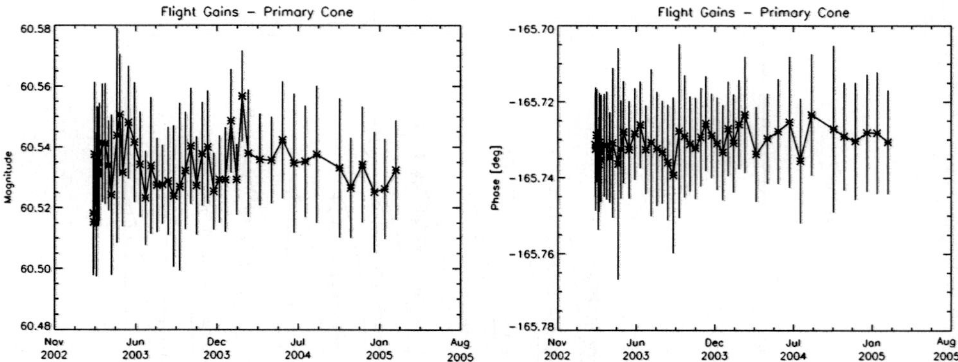

Figure 9. On-orbit servo gain calibrations. Gain of the primary cavity is measured at least monthly by analyzing the response to a DSP-provided step function. Both gain magnitude (*left*) and phase (*right*) are very constant with time. Note the small vertical scales on the plots. Uncertainties are indicated by *vertical bars*.

3.3. Degradation is Monitored via Duty Cycling ESRs and Reflectance-Tracking Photodiodes

Simultaneous, pair-wise inter-comparisons of the four cavities used to measure TSI allows monitoring for long-term changes in cavity absorption, with the exposure-dependent degradation expected to be lower in the lesser-used cavities. A photodiode monitors the light reflected from each cavity, and has gain set such that it is very sensitive to small changes in cavity reflectance. A ground program to monitor unused witness ESR cavities tracks reflectance changes of the cavity due to time alone.

The primary ESR is compared against the other, lesser-used ESRs regularly. Measurements with the primary ESR are acquired nearly continually. Simultaneous measurements of the Sun with both the primary and secondary ESR are done 1% of the time, or for one SORCE orbit per week. The secondary and tertiary ESRs acquire simultaneous measurements 0.5% of the time, and the fourth ESR is used simultaneously with the primary 0.2% of the time. Exposure effects can thus be tracked and corrected by a degradation factor that starts at unity at launch.

The primary ESR shows a slight decrease in its sensitivity that is likely due to an increase over the first 2 years of on-orbit operations in the cavity reflectance by <90 ppm (see Figure 10). This increase is due to solar brightening of the nickel phosphorus (NiP) black cavity interiors. Compared to the traditional space-borne TSI-monitoring instruments (Fröhlich *et al.,* 1997; Willson and Hudson, 1991), this is a very small change of only 0.11 ppm/day and indicates the TIM's NiP black cavity interior (described by Kopp and Lawrence, 2005) is more robust than the black paints used in other instruments. These sensitivity changes in the TIM are tracked to <10 ppm/year.

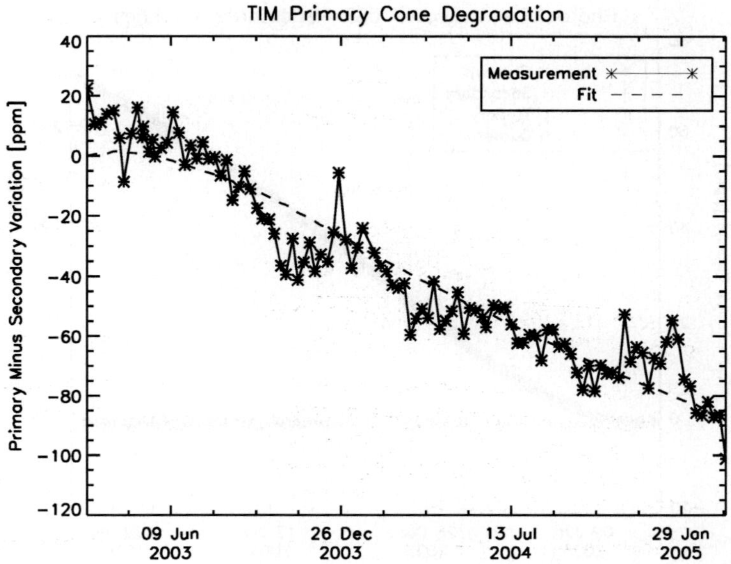

Figure 10. ESR degradation. The primary ESR is 87 ppm less sensitive than just after launch due to solar exposure related degradation through March 2005. A fit (*dashed*), based on photodiode signals, smoothes the measurements (*asterisks*).

Through March 2005, the four TIM ESRs acquired 235, 3.9, 0.98, and 0.47 days of cumulative solar exposure. At the degradation rate observed for the primary ESR and the current duty cycling rates, the other ESRs will not show any signs of solar exposure dependent degradation by the end of the SORCE's 5-year mission.

The photodiodes monitoring the reflection from each of the four cavities show results consistent with the inter-cavity comparisons. All photodiodes show similar radiation damage (independent of solar exposure) that lowers their sensitivity. After correction for this nearly exponential decrease in sensitivity, the photodiodes indicate no changes in the lesser-used ESRs, and a ~72 ppm increase in reflectance of the primary ESR cavity (see Figure 11). The off-axis view angle of the photodiodes into the cavities and/or the different spectral responses of the photodiodes from the ESRs likely account for the difference between the 72 ppm photodiode-determined increase in the primary ESR reflectance and that determined by the inter-cavity comparisons; the photodiode signals are useful mainly for indicating relative cavity reflectance changes.

3.4. Spacecraft Position is Known to High Accuracy

Spacecraft position and velocity knowledge correct measured irradiances to a constant solar distance of 1 AU. The measured irradiance varies inversely as the square

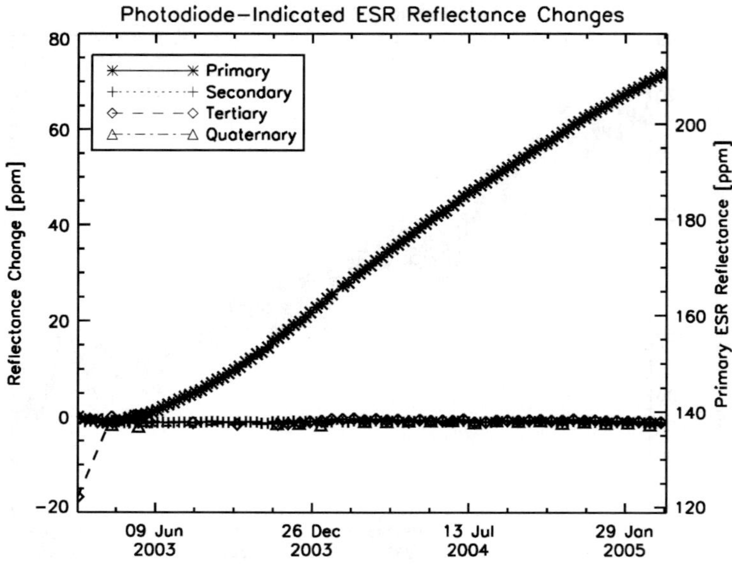

Figure 11. Photodiode signals. The TIM photodiodes monitoring ESR cavity reflectances show similar changes to the ESR comparisons, namely a brightening of the primary ESR interior with solar exposure. The primary ESR's reflectance, as indicated by the photodiode changes, is shown on the right-hand vertical axis. The three lesser-used cavities show almost no change in ESR brightness with time after correction for nearly equal radiation-induced changes in the photodiode sensitivity. The tertiary cavity has an errant initial measurement.

of the distance D (in AU) between the instrument and the Sun and directly with the radial velocity v toward the Sun. The distance and velocity corrections relate the measured value I_{meas} to that reported at a constant 1 AU by

$$I_{1\text{AU}} = \frac{I_{\text{meas}}}{(1 + 2v/c)1/D^2}, \tag{4}$$

where c is the speed of light. The factor of 2 in the velocity correction includes the effects of both a Doppler shift and a change in photon collection rate.

Contributions to the Sun-to-spacecraft distance and velocity come from the spacecraft's orbit around the Earth and the Earth's orbit around the Sun. These are precisely known from North American Aerospace Defense Command (NORAD) tracking data and NASA/JPL ephemerides respectively.

SORCE data processing involving the Earth–Sun distance and velocity uses the JPL VSOP87 ephemeris described by Bretagnon and Francou (1988). The VSOP87D solution, which gives heliocentric positions in spherical coordinates reckoned to the mean ecliptic and equinox for any desired date, is applied. The computed ephemeris positions agree with those in the *Astronomical Almanac* to ± 1 in the least significant digit tabulated.

TABLE I

JPL ephemeris and NORAD uncertainties.

	σD (km)	σv (m s^{-1})
JPL ephemeris VSOP87	3.7	0.04
NORAD TLE/SGP4	0.5	1.0
Total (RSS)	3.7	1.0

Knowledge of the spacecraft's position and velocity relative to the Earth are derived from propagations of the NORAD two-line-element (TLE) sets using the SGP4 propagation model. These TLEs, used to track thousands of Earth-orbiting objects, are reported regularly courtesy of NORAD. At an orbit of 640 km, the SORCE spacecraft's orbital parameters are fairly stable.

Using the full VSOP87 precision and NORAD TLE updates about every 18 h, the approximate distance and velocity uncertainties from both are shown in Table I. The JPL ephemeris is the dominant source of uncertainty in the spacecraft's distance to the Sun, and NORAD is the dominant uncertainty in radial velocity, although neither is significant; position and velocity uncertainties contribute to TIM accuracy uncertainties by less than 0.05 and 0.01 ppm respectively.

3.5. Sensitivity to Pointing is Low

Although several FOV maps have been done on-orbit for the benefit of other SORCE instruments, these pointing characterizations are not sensitive at the <50 ppm level needed to discern TIM pointing sensitivity from normal TSI fluctuations caused by solar oscillations. This is not unexpected, since the pre-launch analysis described in Section 2.6 indicated little sensitivity to pointing. Currently, no pointing correction is applied to the TIM data.

4. Relative Standard Uncertainties as Flown

Table II summarizes post-launch estimates of the combined standard uncertainty, σ. These estimates include the ground-based calibrations of several instrument components and the on-orbit calibrations of others, and are assembled from calibrations at CU/LASP and/or NIST or from analyses. The dominant uncertainty is the knowledge of non-linearity in the power applied to the flight ESRs, as these are only measured using ground-based TIM instruments and are adjusted to be consistent with flight-unit test measurements.

The quadrature sum of the flight uncertainties gives a total standard uncertainty slightly in excess of 200 ppm.

TABLE II

TIM uncertainty budget summary as flown (uncertainties are 1 standard deviation).

Factors/corrections	Size (ppm)	σ (ppm)
Distance, f_{AU}	33 537	0.1
Velocity, f_{Dopp}	57	0.7
Shutter waveform, S	100	1
Aperture, A	1 000 000	55
Reflectance, $1 - \alpha$	200	54
Servo gain, G	16 000	0
Standard, V^2	1 000 000	7
Non-linearity	1000	∼186
Standard, R	1 000 000	17
Equivalence, $\mathbf{Z}_H/\mathbf{Z}_R$	7	22
Dark signal	2700	10
Scattered light	100	25
Repeatability (noise)		1.5
Total (RSS)		∼205

5. Summary

The ground- and flight-based TIM calibrations are much as designed with the exception of the non-linearity in applied power. This has been corrected using precise measurements of two ground-based TIM units, although it limits the combined standard uncertainty of the TIM to ∼205 ppm.

Comparisons of the TIM to other TSI radiometers are described by Kopp, Lawrence, and Rottman (2005).

Acknowledgements

This research was supported by NASA contract NAS5-97045. Generous contributions from NIST enabled the small uncertainties achieved. LASP engineers and managers are to credit for the instrument assembly and calibration facilities.

References

Bretagnon, P. and Francou, G.: 1988, *Astron. Astrophys.* **202**, 309.

Fowler, J., Saunders, R., and Parr, A.: 2000, *Metrologia* **37**, 621.

Hanssen, L. M., Khromchenko, V., Prokhorov, A., and Mekhontsev, S.: 2003, in *Proceedings of the 2003 CALCON Program*, Logan, UT.

Fröhlich, C., Crommelynck, D. A., Wehrli, C., Anklin, M., Dewitte, S., Fichot, A., Finsterle, W., Jimenez, A., Chevalier, A., and Roth, H.: 1997, *Solar Phys.* **175**, 267.
Kopp, G. and Lawrence, G.: 2005, *Solar Phys.*, this volume.
Kopp, G., Lawrence, G., and Rottman, G.: 2005, *Solar Phys.*, this volume.
Lawrence, G. M., Rottman, G., Harder, J., and Wood, T.: 2000, *Metrologia* **37**, 407.
Lean, J.: 2000, *Geophys. Res. Lett.* **27**, 16.
Press, W., Teutolsky, S., Vetterling, W., and Flannery, B.: 1993, *Numerical Recipes: The Art of Scientific Computing*, Cambridge University Press, Cambridge.
Shirley, E.: 1998, *Appl. Opt.* **37**, 28, 6581.
Shirley, E.: 2000, *NIST Report of Modeling 0000207602*.
Spreadbury, P. J.: 1991, *IEEE Trans. Instrum. Meas.* **40**, 343.
Willson, R. C. and Hudson, H.: 1991, *Nature* **351**, 42.

Solar Physics (2005) 230: 129–139

THE TOTAL IRRADIANCE MONITOR (TIM): SCIENCE RESULTS

GREG KOPP, GEORGE LAWRENCE and GARY ROTTMAN
Laboratory for Atmospheric and Space Physics, University of Colorado, Boulder, CO 80309, U.S.A.
(e-mails: kopp@lasp.colorado.edu; lawrence@lasp.colorado.edu; rottman@lasp.colorado.edu)

(Received 7 February 2005; accepted 5 May 2005)

Abstract. The solar observations from the Total Irradiance Monitor (TIM) are discussed since the SOlar Radiation and Climate Experiment (SORCE) launch in January 2003. The TIM measurements clearly show the background disk-integrated solar oscillations of generally less than 50 parts per million (ppm) amplitude over the ∼2 ppm instrument noise level. The total solar irradiance (TSI) from the TIM is about 1361 W/m^2, or 4–5 W/m^2 lower than that measured by other current TSI instruments. This difference is not considered an instrument or calibration error. Comparisons with other instruments show excellent agreement of solar variability on a relative scale. The TIM observed the Sun during the extreme activity period extending from late October to early November 2003. During this period, the instrument recorded both the largest short-term decrease in the 25-year TSI record and also the first definitive detection of a solar flare in TSI, from which an integrated energy of roughly $(6 \pm 3) \times 10^{32}$ ergs from the 28 October 2003 X17 flare is estimated. The TIM has also recorded two planets transiting the Sun, although only the Venus transit on 8 June 2004 was definitive.

1. Introduction

Attempts to measure the total solar irradiance (TSI) began in earnest in the 1830s, with independent ground-based measurements by Claude Pouillet and John Herschel. Their results were low by nearly a factor of 2 because of absorption by the Earth's atmosphere. Even balloon-borne measurements in the 1900s lacked the instrumental accuracy to detect the ∼0.1% short-term changes due to solar activity, let alone the similar level of TSI variations over an 11-year solar cycle. It was not until multi-year measurements from space were available that changes in the TSI were accurately measured and the misconception of a "solar constant" changed. This space-borne TSI record has been uninterrupted since 1978, thanks to overlapping measurements from different missions (see Figure 1). While instrument offsets are large, each instrument has high precision and is able to detect small changes in the TSI caused by variability in solar activity. Increases of 0.1% in TSI during times of high solar activity over the 11-year solar cycle are unambiguous. Short-term changes of ∼0.3% appear directly attributable to variations in solar magnetic activity (Fligge, Solanki, and Unruh, 2000). However, from the two solar minima observed to date, possible secular changes in the TSI are at low enough level to be difficult to discern (Willson and Mordvinov, 2003; Fröhlich, 2002). Determination of such long-term solar variability will rely on longer-duration measurements of high accuracy, for which continued space-borne TSI measurements are essential.

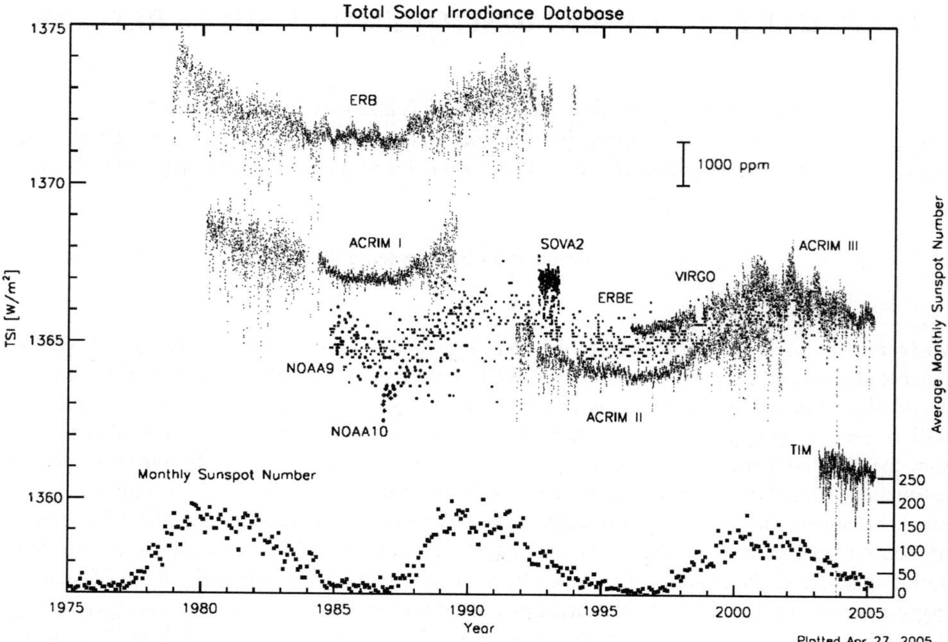

Figure 1. The space-borne TSI database. Space-borne measurements of the TSI show $\sim 0.1\%$ variations with solar activity on short and long time scales. The TIM values are lower than those reported by other instruments. Early instruments showed even larger offsets, and these may be indicative of current capabilities of space-borne radiometry. (These data are available from the following sources: the space-borne TSI plot is available via *spot.colorado.edu/~koppg/TSI.* The solar irradiance data from SORCE are available via *lasp.colorado.edu/sorce/tsi_data.html.* The solar irradiance data from VIRGO are courtesy of the VIRGO team via *www.pmodwrc.ch/.* The solar irradiance data from ACRIM are courtesy of Dr. Richard Willson via *www.acrim.com.* The solar irradiance data from ERBE are courtesy of Robert B. Lee III. The monthly sunspot data and the NOAA and ERB spacecraft TSI data are courtesy of *www.ngdc.noaa.gov/stp/SOLAR/solar.html.*)

Very simplistically, short-term decreases in the TSI can be largely attributed to dark sunspots, while values above normal are generally due to bright faculae. The actual Sun displays a much greater continuum of activity however; example solar atmosphere models by Fontenla *et al.* (1999) include eight components describing solar activity, including average supergranules, faint supergranules, average network, bright network, average plage, bright plage, sunspot umbrae, and sunspot penumbrae. It is the spatial and spectral integral of the continuum of solar activity type, each with some center-to-limb variation, that contributes to the TSI measurement.

The TSI measured by the Total Irradiance Monitor (TIM) is plotted in Figure 2. These data have been corrected for instrument degradation, background thermal emission, and instrument position and velocity. On time scales of a few minutes, the TIM measures small fluctuations in the TSI (see Figure 3) having typical amplitudes

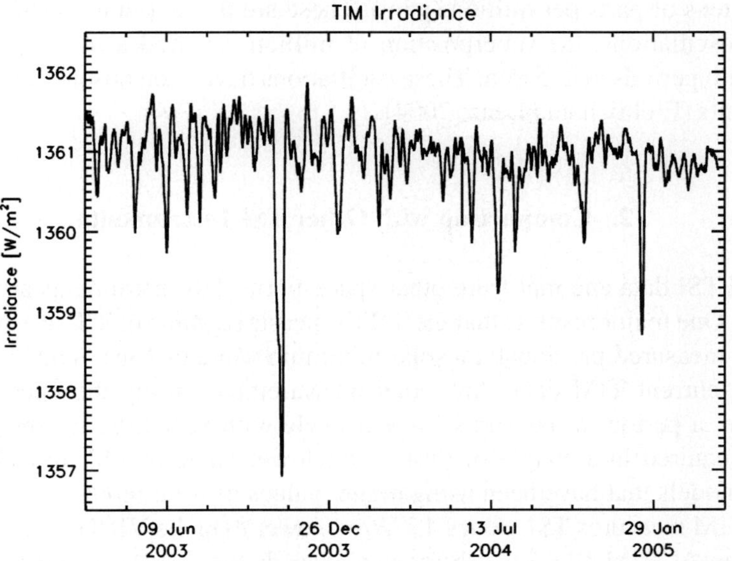

Figure 2. TIM TSI record. This plot shows the TIM Version 4 data of total solar irradiance since launch. The gradual decrease with time is from the Sun approaching solar minimum. The short-term decreases are due to the passage of sunspots.

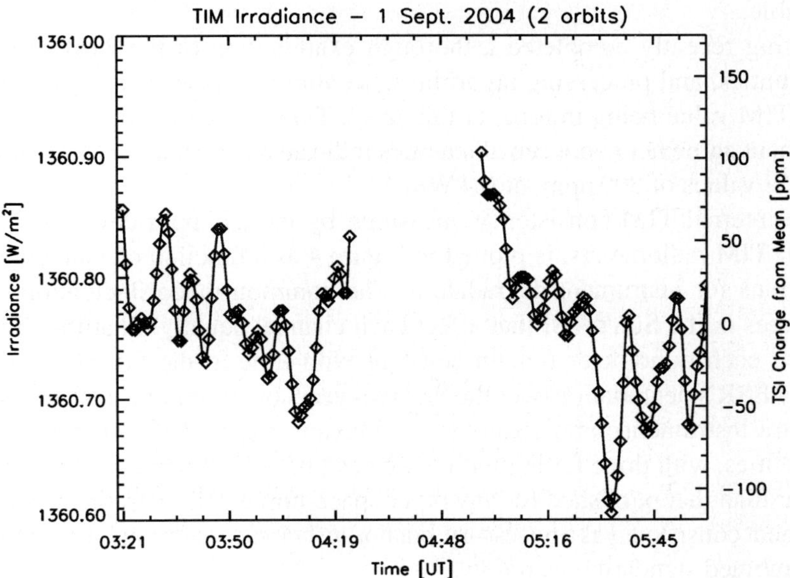

Figure 3. Solar oscillations in TSI. Changes of several tens of ppm in the TSI time series are due to globally averaged solar oscillations. The gap in these data is when the SORCE spacecraft was in the Earth's shadow.

of a few tens of parts per million (ppm). These are the disk-integrated signals due to solar oscillations, the superposition of millions of modes of trapped acoustic waves with periods near 5 min. These oscillations have been observed by other TSI instruments (Fröhlich and Lean, 2004).

2. Comparison with Other TSI Instruments

The TIM TSI data and that from other space-borne TSI instruments are plotted in Figure 1. One major result is that the TIM's measured value of TSI at 1 AU is lower than that measured previously; a solar minimum value of 1361 W/m^2 is estimated from the current TIM data. An actual measurement during the upcoming solar minimum, a period in the Sun's 11-year cycle with very little magnetic activity, will be acquired in a couple of years. This lower value of TSI may affect Earth climate models that have been using higher values of solar input.

The TIM measures TSI values 4.7 W/m^2 lower than the VIRGO[1] and 5.1 W/m^2 lower than ACRIM III.[2] This difference exceeds the $\sim$0.1% stated uncertainties on each instrument, and continues to be a source of discussion. Differences between the various data sets are solely instrumental and will only be resolved by careful and detailed analyses of each instrument's uncertainty budget. Unfortunately documentation and test samples from the earliest missions may no longer be accessible.

Having recently completed a thorough examination of the TIM calibrations, uncertainties, and processing algorithms, we find no causes that could contribute to the TIM value being in error at this level. The TIM calibrations and the intra-instrument agreement between all cavities indicate a 1-σ standard deviation on the TIM TSI values of 300 ppm, or 0.4 W/m^2.

The internal TIM consistency, measured by the 270 ppm deviations between the four TIM radiometers, is plotted in Figure 4 as a function of time without any corrections for instrument degradation. The common-mode fluctuations are due to changes in the Sun's TSI that affect each channel equally. That the differences between each radiometer remain constant with time at the 90 ppm level of the primary ESR's degradation over the first two years on orbit indicates good stability. This intra-instrument comparison shows a maximum spread of 640 ppm between all four cavities, with three falling within a spread of only 290 ppm. This consistency is better than that published for any other space-borne TSI instrument. Such intra-instrument consistency is a necessary, but not sufficient, condition for demonstrating low combined standard uncertainty.

[1]Unpublished Version 6 data from the VIRGO experiment on the cooperative ESA/NASA Mission SOHO is courtesy of the VIRGO team through PMOD/WRC, Davos, Switzerland.
[2]ACRIM data are courtesy of Dr. Richard Willson via *www.acrim.com*.

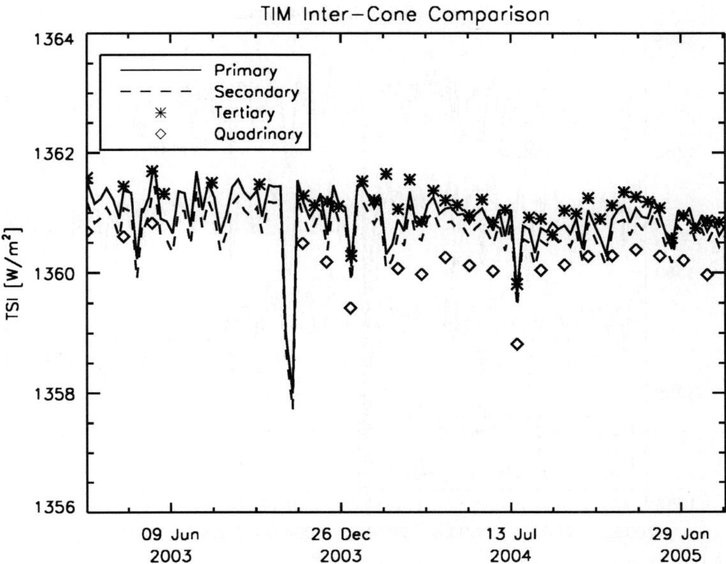

Figure 4. TIM intra-instrument comparison. The four TIM radiometers measure within a standard deviation of 270 ppm of the same value of TSI and show only a 90 ppm change with time due to tracked degradation of the primary ESR. The fluctuations in this plot common to multiple ESRs are due to normal solar variability. Degradation corrections have not been applied to these values, so this plot also indicates good fundamental instrument stability.

On a relative scale, offsetting the data to account for differences between the three currently flying TSI instruments acquiring daily data, the TIM, VIRGO, and ACRIM III instruments agree very well (see Figure 5).

3. Solar Activity during the October to November 2003 Time Frame

The TIM made the first definitive measurement of a solar flare in TSI. This occurred during a two-week period of extraordinarily high solar activity near the end of October 2003. During this time, the passage of two extremely large sunspot groups across the disk caused the largest short-term fluctuation of TSI in the 25-year record of observations (see Figure 6).

Over this two-week period the Sun released several flares, including the largest and the fourth largest X-ray flares ever recorded. These occurred respectively at 4 November 19:44 UT and 28 October 11:10 UT based on NOAA Geostationary Operational Environmental Satellite (GOES) soft X-ray measurements.

Despite the notoriously high energies of flares, they are small compared to the entire energy output of the Sun, and thus are nearly imperceptible in TSI. Indeed, in 25 years of space-based irradiance monitoring, no previous solar flare had been definitively measured in TSI, although studies by Hudson and Willson

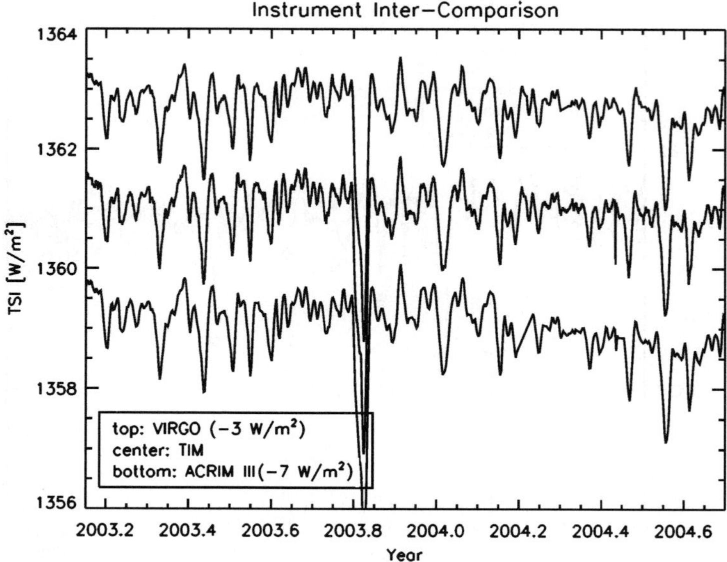

Figure 5. TSI comparison. The three TSI instruments currently providing daily measurements show good agreement on a relative scale. The spike in early June 2004 (2004.44) is from the transit of Venus across the solar disk and was not observed by the VIRGO instrument.

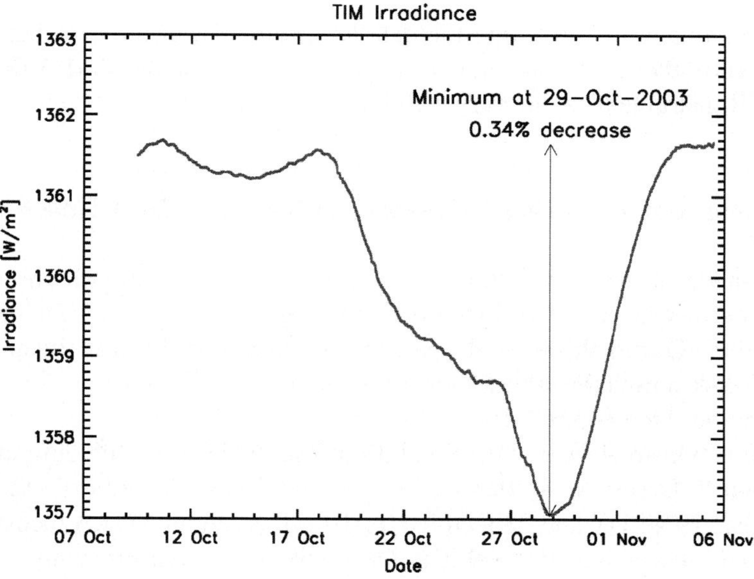

Figure 6. TSI decrease in October 2003. Two extraordinarily large sunspot groups caused the greatest short-term change in TSI on record.

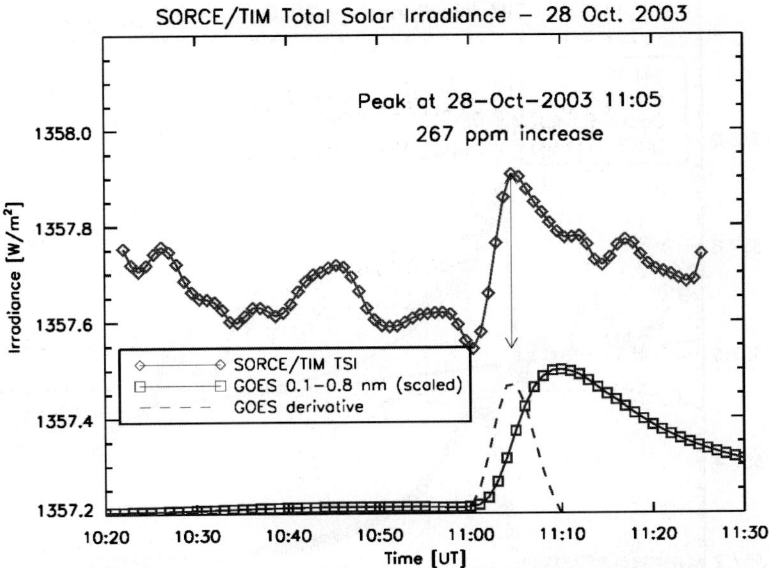

Figure 7. 28 October 2003 X17 flare. The TIM detected this large flare in TSI. The TSI peak is roughly coincident with the estimated hard X-ray peak. The TIM data are reported at a 50-s cadence using the phase sensitive detection algorithm described by Kopp and Lawrence (2005). The GOES data are reported at a 1-min cadence, although measurements are acquired every 3 s.

(1983) established flare energy levels needed for detection in TSI with an early ACRIM instrument.

The 28 October flare was large enough and well enough centered on the solar disk to cause a measurable change in the TSI. This flare came from NOAA's Solar Region Number 10486, a large sunspot group with surrounding intense magnetic activity. The NOAA ranking for this flare was X17.

The soft X-ray signature of the 28 October X17 flare as measured by the NOAA/GOES is shown in Figure 7 with a peak at 11:10 UT. The SORCE/TIM data show a sudden increase of 267 ppm peaking at 11:05 UT; this abrupt increase is the flare's signature in the TSI. The gradual decay in TSI after the flare's peak is characteristic of flares, although the complete relaxation is interrupted at 11:26 UT when the SORCE spacecraft entered the Earth's shadow. While some portion of the TSI increase may be due to normal solar variations, these are generally at the 20 ppm level and are much more gradual, so do not account for the large and abrupt peak observed.

A flare's hard X-ray emission roughly coincides with the derivative of the soft X-ray (GOES) emission, as described by Neupert (1968). As seen in Figure 7, the TSI measurements of this X17 flare coincide with this derivative, both leading the soft X-ray peak by 5 min. This indicates that portions of the visible and UV included in TSI also respond to the sudden initiation of the flare.

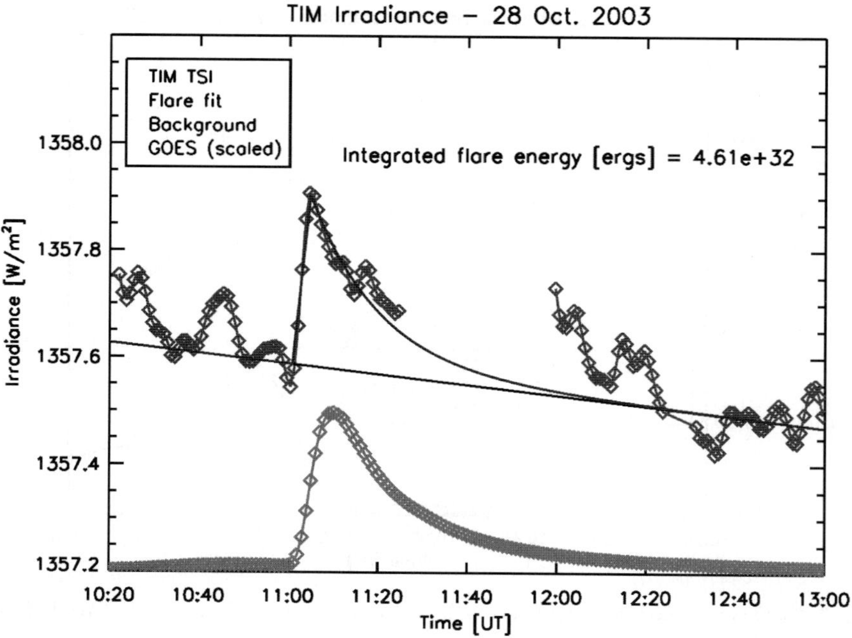

Figure 8. Total flare energy. Since the TSI measurement is sensitive to the entire solar spectrum, the TIM measurement of the X17 flare can be integrated over time to give the total flare energy.

The TIM measures radiant power across the entire solar spectrum from X-ray to far infrared wavelengths. The measurement of a flare in TSI thus provides knowledge of the total flare energy. From the TIM measurements, this X17 flare had an estimated total energy of approximately 5×10^{32} ergs (see Figure 8). A lower energy limit of 3×10^{32} and an upper limit of 9×10^{32} ergs are estimated based on possible background signals due to solar oscillations.

Combined with solar spectral irradiance measurements, we have estimated the spectral distribution of the flare energy to determine the amount released at EUV and at longer wavelengths. Preliminary estimates of the flare energy at wavelengths shorter than 100 nm, based on solar EUV measurements from other SORCE instruments and from a similar instrument on NASA's TIMED mission, only account for roughly 20% of this energy, meaning the majority of the flare's energy was emitted at wavelengths longer than 200 nm (Woods *et al.*, 2004).

The 4 November flare, although larger in X-ray emission with a NOAA ranking of X28, was only faintly detectable in TSI because its position was at or just beyond the solar limb. The correlation of the TIM TSI measurements with disk-integrated GONG intensities by Leibacher *et al.* (2004) shows a sudden increase in the TIM TSI that is likely due to this flare.

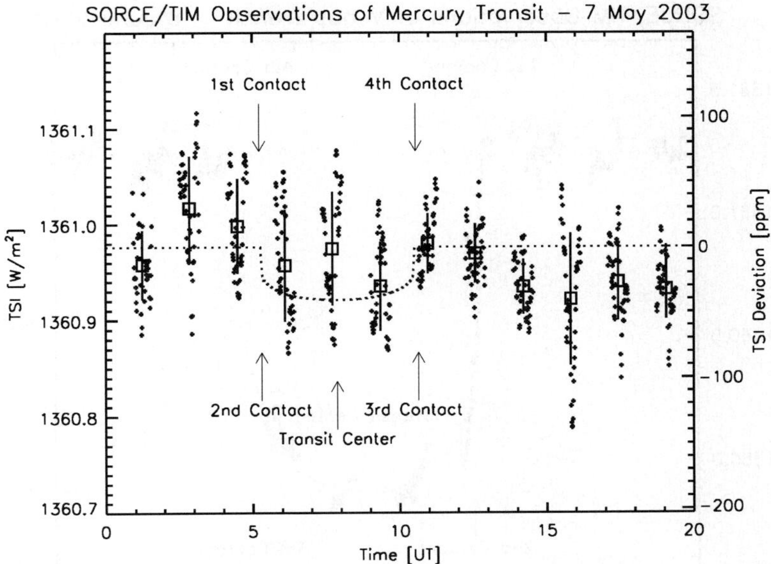

Figure 9. TIM observations of Mercury transit. The TIM observations do show roughly the expected level of decrease in TSI during the times of the Mercury transit (*smooth curve*), but these are indistinguishable from the effects of normal solar fluctuations. High-cadence data are given as *dots* and orbital averages by *squares* with their associated standard deviation error bars.

4. Planetary Transits Observed

The TIM observed during the transits across the solar disk of both Mercury (7 May 2004 from 5:13 to 10:32 UT) and Venus (8 June 2004 from 5:13 to 11:36 UT). Accounting for solar limb darkening for the off-center transits, expected decreases in the TSI for these transits are 41 and 1005 ppm, respectively.

The Mercury transit (see Figure 9) was not detectable in TIM TSI measurements. While there may be a decrease of the appropriate magnitude over the three orbits during which this transit occurred, normal solar fluctuations during prior and subsequent orbits show similar levels of variability, making any detection here ambiguous.

The Venus transit was much larger and was easily detected. The TIM observed ingress (first to second contact) for this transit, although egress occurred when the TIM was in the Earth's shadow and could not view the Sun. The observed decrease in the TSI during transit closely matches the expected decrease using the solar limb darkening profiles of Hestroffer and Magnan (1998) (see Figure 10). While there is little of scientific value from TIM observations of this transit, it does confirm that Hestroffer and Magnan's limb darkening estimates are correct at the $<5\%$ accuracy level limited here by the noise due to solar fluctuations.

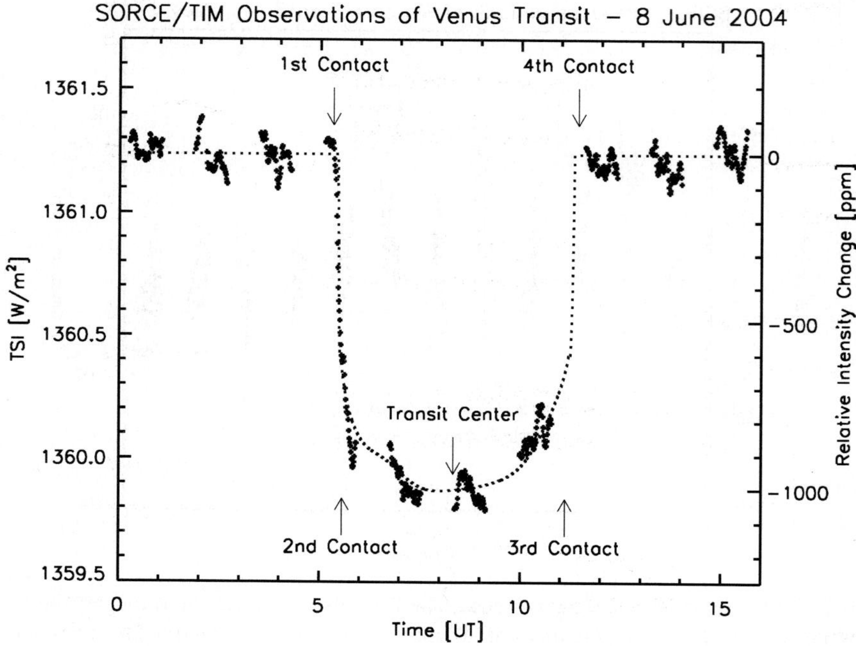

Figure 10. TIM observations of Venus transit. The TIM observations ('+' *signs*) of the Venus transit match the expected TSI decrease (*smooth curve*) to the limits of the background disk-integrated solar oscillations.

5. Summary

The SORCE/TIM measures a lower value of total solar irradiance than previous space-borne irradiance measurements. We do not attribute this discrepancy to any known calibration error in the instrument. Based on the TIM data, we predict the TSI input at the top of the Earth's atmosphere at a distance of 1 AU from the Sun during the upcoming solar minimum to be 1361 W/m^2.

We also report the first measurement of a solar flare in TSI, from which we estimate the total energy of this X17 flare to be 5×10^{32} ergs.

The TIM observed planetary transits of both Mercury and Venus. Measured TSI values closely match the expected decreases in intensity.

Acknowledgement

This research was supported by NASA contract NAS5-97045.

References

Fligge, M., Solanki, S. K., and Unruh, Y. C.: 2000, *Astron. Astrophys.* **353**, 308.
Fontenla, J., White, O. R., Fox, P., Avrett, E., and Kurucz., R.: 1999, *Astrophys. J.* **518**, 480.

Fröhlich, C. and Pap, J.: 2002, *Adv. Space Res.* **29**, 12, 1879.
Fröhlich, C. and Lean, J.: 2004, *Astron. Astrophys. Rev.* **12**, 273.
Hestroffer, D. and Magnan, C.: 1998, *Astron. Astrophys.* **333**, 338.
Hudson, H. S. and Willson, R. C.: 1983, *Solar Phys.* **86**, 123.
Kopp, G. and Lawrence, G.: 2005, *Solar Phys.*, this volume.
Leibacher, J., Harvey, J., GONG Team, Kopp, G., and Hudson, H.: 2004, AAS/SPD Meeting, Denver, Colorado.
Neupert, W. M.: 1968, *Astrophys. J.* **153**, L59.
Willson, R. C. and Mordvinov, A. V.: 2003, *Geophys. Res. Lett.* **30**, 5, 1199.
Woods, T. N., Eparvier, F. G., Fontenla, J., Harder, J., Kopp, G., McClintock, W. E., Rottman, G., Smiley, B., and Snow, M.: 2004, *Geophys. Res. Lett.* **31**, doi: 10.1029/2004GL019571, L10802.

Solar Physics (2005) 230: 141–167

THE SPECTRAL IRRADIANCE MONITOR: SCIENTIFIC REQUIREMENTS, INSTRUMENT DESIGN, AND OPERATION MODES

JERALD HARDER, GEORGE LAWRENCE, JUAN FONTENLA,
GARY ROTTMAN and THOMAS WOODS
Laboratory for Atmospheric and Space Physics, University of Colorado, Boulder, U.S.A.
(e-mail: jerald.harder@lasp.colorado.edu)

(Received 20 January 2005; accepted 30 March 2005)

Abstract. The Spectral Irradiance Monitor (SIM) is a dual Fèry prism spectrometer that employs 5 detectors per spectrometer channel to cover the wavelength range from 200 to 2700 nm. This instrument is used to monitor solar spectral variability throughout this wavelength region. Two identical, mirror-image, channels are used for redundancy and in-flight measurement of prism degradation. The primary detector for this instrument is an electrical substitution radiometer (ESR) designed to measure power levels ~1000 times smaller than other radiometers used to measure TSI. The four complementary focal plane photodiodes are used in a fast-scan mode to acquire the solar spectrum, and the ESR calibrates their radiant sensitivity. Wavelength control is achieved by using a closed loop servo system that employs a linear charge coupled device (CCD) in the focal plane. This achieves 0.67 arcsec control of the prism rotation angle; this is equivalent to a wavelength positioning error of $\delta\lambda/\lambda = 150$ parts per million (ppm). This paper will describe the scientific measurement requirements used for instrument design and implementation, instrument performance, and the in-flight instrument operation modes.

1. Introduction

This paper describes the Spectral Irradiance Monitor (SIM) on the Solar Irradiance and Climate Experiment (SORCE). SORCE was launched on 25 January 2003, and the four instruments on the satellite are designed to study spectral and total solar irradiance with very high accuracy over a broad wavelength range. SIM was developed to replace and extend the UARS SOLSTICE N-channel spectrometer (Rottman, Woods, and Sparn, 1993) that was used to study solar variability in the 280–420 nm region. The SIM instrument covers a much wider spectral range (200–2700 nm) and uses an electrical substitution radiometer (ESR) as its primary detector. A Fèry prism is used in place of a grating, allowing for full wavelength range with only one optical element, thereby simplifying the spectrometer design and operation. The preliminary design and operation of the instrument was documented in two manuscripts prior to this comprehensive study (Harder *et al.*, 2000a; Harder *et al.*, 2000b).

Starting in the mid 1980s the majority of the extra-atmospheric measurements of solar spectral irradiance have concentrated on the more variable ultraviolet spectrum (London *et al.*, 1993) because of its importance to stratospheric ozone trends and mid-latitude stratospheric circulation (Labitzske and Van Loon, 1988; Hood, 1999). These studies show that ultraviolet variability ranges from about 10% at 200 nm to about 0.1% at 300 nm over the length of a solar cycle. In the visible and near-infrared (NIR) however, there is a paucity of spectral irradiance data and the secular trends have been estimated from measurements of the total solar irradiance (TSI) performed by satellite-borne active cavity radiometers that have an exquisitely high absolute accuracy and precision (Kopp, Lawrence, and Rottman, 2003; Willson, 1988), but cannot provide the needed wavelength dependent information about spectral variability. This wavelength information is important to both solar and Earth atmospheric physics: in solar physics it is used in the interpretation of flux variability of emerging solar surface features, such as sunspots and plage (Solanki and Unruh, 1998), and in the Earth sciences it is needed to understand the strongly wavelength dependent radiation absorption processes in the atmosphere and oceans that absorb this variable solar flux (Reid, 1999; Kiehl and Trenberth, 1997). The importance of wavelength dependent information for the Earth's atmosphere is demonstrated in Figure 1 showing a MODTRAN-generated top of the atmosphere irradiance calculation at a resolution of 1 nm (Anderson *et al.*, 1999). It then shows the penetration of this radiation to the Earth's surface under equatorial, sea level, and cloud free conditions and indicates the dominant O_2, O_3, H_2O, and CO_2 atmospheric absorptions. Finally the surface radiation is propagated into the first 10 meters of ocean water (Curcio and Petty, 1951; Smith and Baker, 1978) and suggests the importance of long wavelength radiation and its variability on ocean circulation

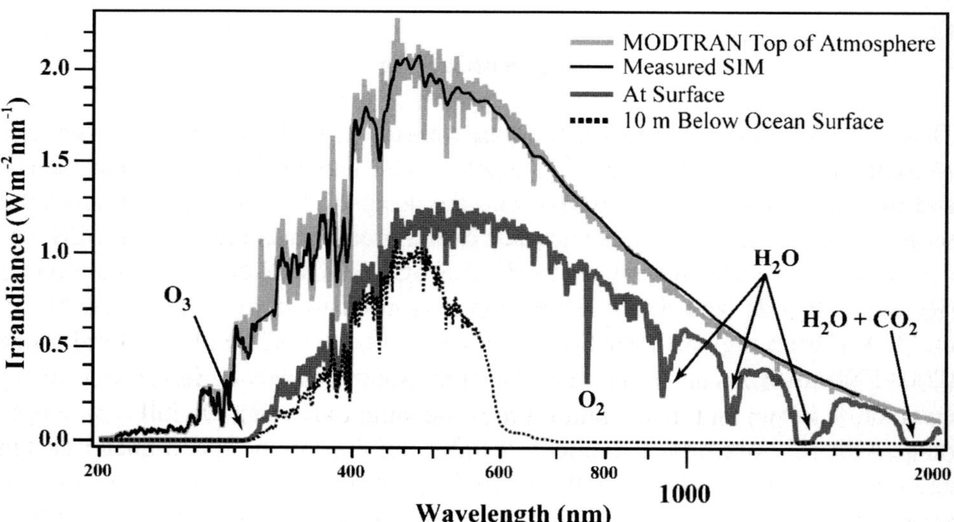

Figure 1. Penetration of the top of the atmosphere radiation into the Earth's atmosphere and oceans. The dominant molecular absorbers are noted in the graph. The graph also shows the 1 nm resolution MODTRAN top of the atmosphere solar spectrum overplotted with the measured SIM spectrum covering the wavelength range of 208 to 1604 nm.

processes (Reid, 1991; White *et al.*, 1997). Also shown in this figure is the measured SIM irradiance in the 208–1604 nm range covered by the SIM photodiode detectors discussed in Section 2.3.1.

The most frequently used modern information on the solar spectrum in the visible and NIR arises from two primary sources: the Kitt Peak solar atlas of the solar spectrum (Kurucz, 1991) in conjunction with the Neckel and Labs radiometric calibration (Neckel and Labs, 1984), and the SOLSPEC spectra (Thuillier *et al.*, 2003) acquired during the ATLAS and EURECA missions. The Kurucz spectrum is measured at a resolution of 0.055 cm^{-1} (0.0014 nm at 400 nm), so individual solar Fraunhofer lines are resolved, but do not possess an absolute calibration through the strongly scattering and absorbing Earth atmosphere. The space-based SOLSPEC spectrum, measured at a resolution of 1.0 nm in the 200–870 nm region and 20 nm in the 850–2500 nm region, has an excellent absolute laboratory calibration and does not require atmospheric corrections, but like the Labs and Neckel spectrum, cannot address solar variability because the time series is limited to only a few days during the ATLAS campaigns. Therefore, the central purpose of the SIM instrument is to measure the daily solar spectral variability in the visible and NIR and maintain an accurate absolute calibration for both solar and Earth science studies.

2. Realization of Instrument Requirements for Spectral Variability Studies

2.1. High-level instrument requirements

Instrument requirements for a spectral radiometer were derived from an analysis of the TSI record and theoretical estimates of variability in the visible and IR over solar cycle length records (Solanki and Unruh, 1998; Lean, 1991). The TSI record bounds the magnitude of the spectral variability: typically, solar rotation modulation is on the order of 0.2% with peak-to-peak differences of 0.1% over the course of the 11-year solar cycle. Theoretical studies on the wavelength dependence of variability produced by the emergence of active regions predict a much smaller response in the visible/IR regions with the relative variability in the 200–300 nm range from 2% to 0.5%, but less than 0.5% in the 300–1000 nm region. Thus the requirement to measure solar variability in the visible and IR spectral regions is that the instrument must have a combined precision and measurement drift stability of about 100 ppm and this condition must be valid over the 200–2000 nm region; an order of magnitude of wavelength coverage representing ~95% of the TSI. The accuracy is based on the current standards used to measure radiant power and the needs of the atmospheric and solar physics communities; at the present time this is about 1–2%. As solar and Earth atmospheric models become more sophisticated, and as calibration standards and methods improve, accuracy will become a more stringent requirement in future studies of solar variability. The resolution required for a spectral radiometer is a function of wavelength. The resolution and spectral sampling must be high enough in the 200 to 400 nm region to allow meaningful instrument intercomparison with currently existing instruments such as SOLSTICE, SUSIM, SOLSPEC (Rottman, Woods, and Sparn, 1993; Vanhoosier *et al.*, 1981; Thuillier *et al.*, 2003, respectively). If the resolution is too low, the spectrum becomes biased due to the rapidly changing nature of the UV spectrum, and the measurement of the important Mg II index becomes more difficult (Viereck and Puga, 1999). Therefore, in the ultraviolet the resolution must be about 1 nm. In the visible and NIR (400–2700 nm), lower spectral resolution is acceptable due to the fact that density of solar Fraunhofer lines become smaller and the effects of spectral smoothing by low resolution instruments are not as important as they are in the UV. Rottman *et al.* (2005) demonstrate that a resolution of 1–35 nm is adequate to study solar variability in this spectral region if the instrument has high enough radiometric precision. For a spectral radiometer the sources of accuracy and precision can be partitioned into three categories that must be addressed in the instrument design: (1) radiometric accuracy and precision, (2) wavelength accuracy, and (3) the maintenance of long-term instrument calibration that can be degraded by the space environment.

These needs are addressed in the SIM instrument in the following ways. (1) The radiometric accuracy scale is based on the electrical substitution principle with unit level calibrations of spectrometer components to account for light loss processes prior to the detection of radiant power. (2) The wavelength standard is based on the solar spectrum itself, and high precision of the wavelength drive is needed for spectrum-to-spectrum reproducibility without distortion. (3) SIM is designed to have two mirror-image spectrometers built into the same case that can be used for side-by-side comparisons and can be optically coupled to permit inter-instrument calibration.

2.2. Principle of Design for the SIM Instrument

The functionality needed to meet the requirements presented in the previous section can be represented in a block diagram form that describes three independent optical paths through the SIM instrument along with their associated electromechanical mechanisms. This block diagram is illustrated in Figure 2. Figure 3 shows two orthogonal cross-sectional views through one of the spectrometers (SIM A), and Figure 4 shows the focal plane assembly in greater detail to accompany Figure 3. The next three subsections refer to these three figures.

2.2.1. *The Spectrometer Path*

The spectrometer path provides the calibrated irradiance measurement. After passing through the front end baffle set, the light beam encounters a retractable light filter called the 'hard radiation trap' (HRT) which is a window made from Suprasil 300 (Hereaus Amersil, Inc.), the same material used to construct the prism, and acts as a light filter to absorb radiation of wavelengths less than 160 nm to reduce exposure

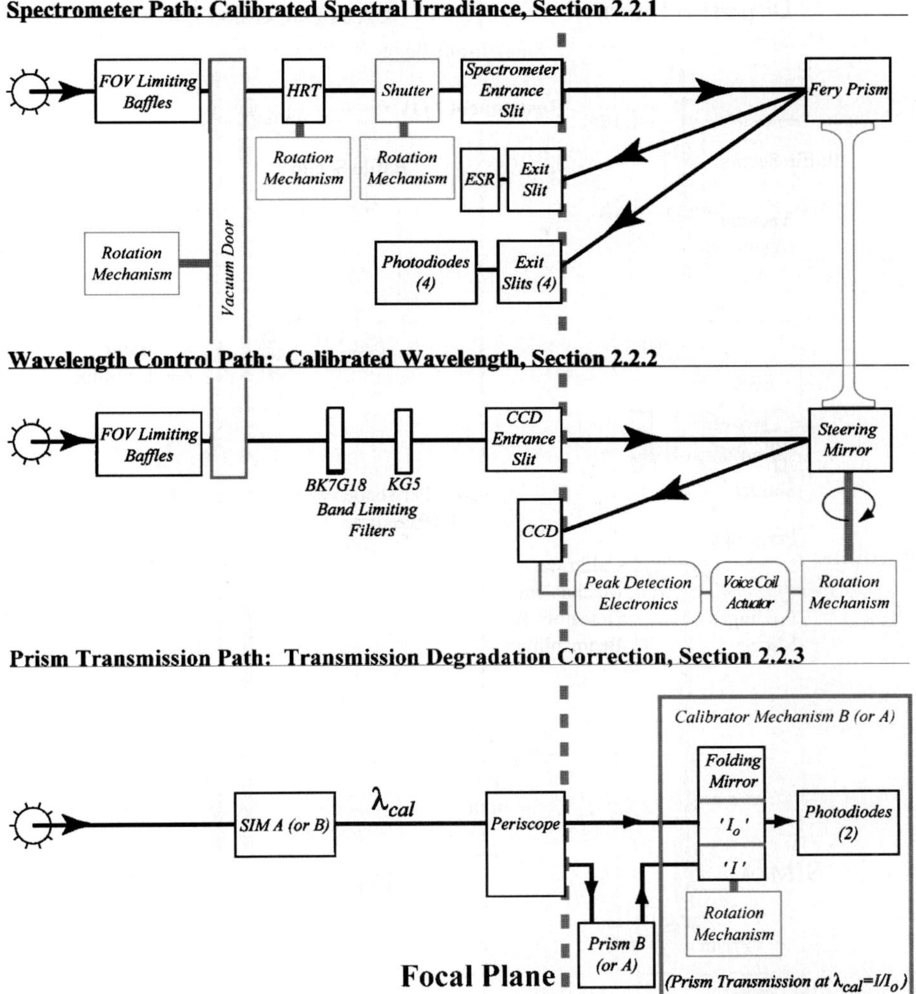

Figure 2. Block diagram of the SIM instrument. It shows three independent optical paths for irradiance measurement (spectrometer path), wavelength control, and prism transmission. The subsections in the text are labeled in the block diagram. Electrical/mechanical components are shown in gray and optical components and pathways are shown in black. Note that for the prism transmission measurement path, one of the blocks represents an entire channel of SIM.

of the prism to UV radiation known to degrade the transmission of the prism glass. The HRT is rotated out of the light path for absolute spectral measurements but is placed in the beam about 80% of the time for measurements that monitor relative solar changes. This window is tilted at 2 degrees so surface reflections will not propagate through the spectrometer, and it is wedged at 1/2° to prevent etalon effects (channeling) from distorting the spectrum. The light beam then encounters a light shutter directly in front of the spectrometer entrance slit. This shutter operates

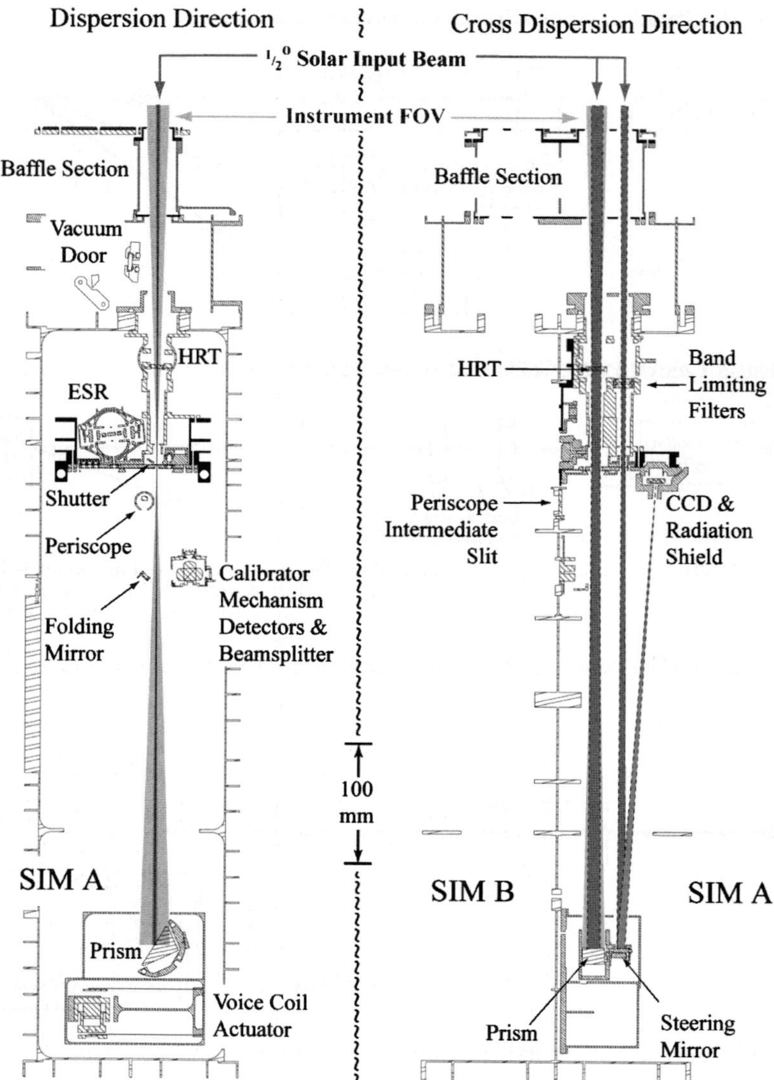

Figure 3. Two orthogonal cross-sectional views of the SIM instrument identifying critical mechanisms represented in the Figure 2 block diagram. Solar and instrument fields of view are shown as shades of gray.

synchronously with the operation of the ESR for phase sensitive detection measurements and is under control of the instrument's digital signal processor (DSP); the shutter completely blocks light from reaching the detectors in less than 7 ms. The light then enters the spectrometer chamber through the entrance slit (7×0.3 mm^2) with a calibrated area and slit width and impinges on the Fèry prism. The prism has a concave front surface and a convex aluminized back surface, so only a single

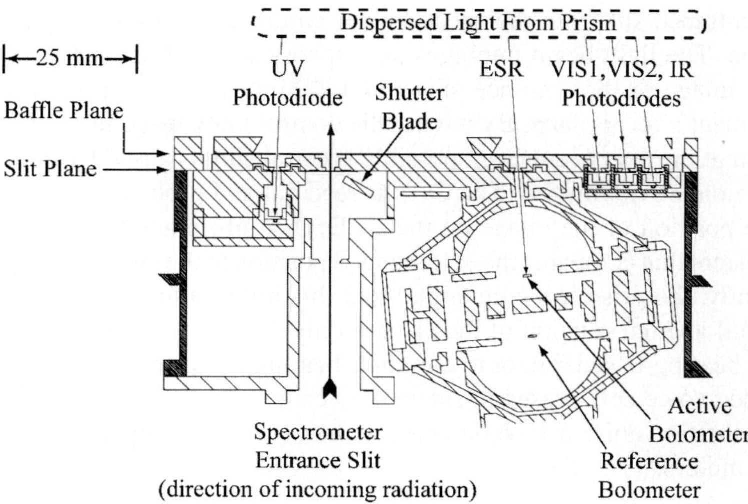

Figure 4. Detail of focal plane assembly to accompany Figure 3. It shows the locations of the five irradiance detectors, the entrance slit and light shutter. The motor to rotate the shutter is below the plane of the diagram, and the CCD (75 mm length) is above the plane.

optical element is needed to disperse and focus the incoming light beam on the focal plane that contains five irradiance detectors, the ESR plus four photodiodes.

For broad wavelength coverage applications a prism spectrometer has a distinct advantage over an equivalent grating spectrometer. The prism produces only a single spectral order whereas the grating would produce overlapping spectral orders that would need to be removed. Likewise, the well-polished prism produces far less scattered light than an equivalent grating. The resolution of the instrument is determined from the spectrometer design geometry and the glass wavelength dispersion (i.e., the first derivative of the index of refraction). The spectrometer's resolving power ($\lambda/\Delta\lambda$) is a strong function of wavelength and varies from 378 at 250 nm, to a minimum of 37 at 1260 nm, and increases slowly in the infrared to 142 at 2700 nm. Section 2.3.2 will describe the optical and spectroscopic properties of this instrument.

2.2.2. *The Wavelength Control Path*

The wavelength control path provides the prism rotation angle which is then combined with other measured instrument parameters to give the wavelength that reaches each of the focal plane detectors (see Section 2.3.3). Very precise wavelength control is needed for radiometric measurements of the solar spectrum; otherwise, small unknown wavelength shifts will translate into signal changes that could be falsely interpreted as solar variability. Sunlight enters the instrument through the CCD entrance slit (3 mm tall × 0.1 mm wide) after a series of field-of-view limiting baffles and a set of wavelength limiting glass filters. Note that this is a separate entrance slit from the spectrometer entrance slit discussed in the previous section. The light beam impinges on a spherical mirror (steering mirror) that makes an image of the entrance slit on a 12000 element linear array located in the instrument's focal plane. By having the control knowledge in the focal plane, rather than at the axis of rotation, possible spectral shifts caused by thermal drifts and distortion are minimized. The CCD is read out and peak detection electronics locate the position of the image on the CCD; this information is used in a servo control system that compares the commanded position to the current image position (equivalently, the prism rotation angle) and this information is updated every 25 ms. The full angular rotation of the drive is only $\pm 2.5^\circ$, so a flexure is used in the place of a bearing-based rotator to eliminate bearing noise and backlash. Likewise, a suspended voice coil is used to rotate the flexure since very fine motion control and low actuation noise can be obtained with this device compared to stepper and DC drive motors.

2.2.3. *The Prism Transmission Measurement Path*

Degradation in the optical system is expected, and two independent on-orbit methods are used to measure prism transmission degradation – the single largest source of long-term uncertainty in the irradiance measurement. The first method simultaneously compares the responsivity of the two SIM channels where one channel is exposed to sunlight less that 1/5th of the time of the primary channel. The second method is to define an optical path that uses one spectrometer to deliver monochromatic light to the other so a direct measure of the prism transmission can be obtained. This third optical path (see Figure 2) is symmetric, so SIM A can be used to calibrate Prism B and vice-versa. The two instruments are coupled with a periscope that consists of two concave spherical mirrors mounted at 45° with an intermediate slit located in the wall that separates the two instruments. The periscope is positioned in the instrument so an image of the entrance slit is formed at this intermediate slit with imaging properties nearly identical to the spectrometer's exit slits. The folding mirror mechanism permits the measurement of the light intensity before and after it passes through the prism and the ratio of these two intensities gives the prism transmission. This on-board prism calibration system is designed as a relative transmission measurement for tracking long-term changes. Because the periscope mirrors affect the polarization, the absolute transmission is measured with a ground calibration system that does not use the periscope system. The in-flight transmission measurement details are discussed in Section 2.3.4.

2.2.4. *Instrument Vacuum/Pressure Enclosure*

The only common component in the SIM A and SIM B spectrometers is the vacuum/pressure vessel that has separate vacuum doors for each of the two spectrometers, in case one of the doors failed to open after launch. Designing the instrument housing in this manner has a number of advantages. (1) The instrument is pumped out and degassed to remove volatile organics and water vapor thereby protecting the optical components from surface contamination. (2) The enclosure is designed to hold a pressure of 1.2 atmospheres. During spacecraft/launch vehicle integration and pre-flight environmental testing the instrument is backfilled with high purity argon to maintain cleanliness. Likewise, the instrument was launched with an Ar backfill. (3) The vacuum doors are equipped with fused silica windows so the optical paths can be stimulated for pre-flight integration and test, and first light check-out after launch. (4) The instrument case can be evacuated to a pressure $\sim 10^{-8}$ atmospheres for laboratory testing of the ESR detectors. (5) After launch and 4 weeks of spacecraft outgassing, the instrument was opened to space vacuum through a bleed valve. After final evacuation the vacuum doors were opened.

2.2.5. *Generic Channel Interface (GCI) Instrument Control*

SIM uses the same GCI unit described by McClintock, Rottman, and Woods (2005) but it is reconfigured for specific SIM activities. The GCI consists of three boards. A digital signal processor (DSP) board, based on the Temic TSC21020F processor, performs the most important tasks for the instrument. It performs the bridge excitation and readout of the ESR (see Section 3.3), multiplexing and readout of the photodiodes, shutter actuation, and prism drive control (see Section 3.2). A multifunction board has H-bridge circuitry to control bi-stable mechanisms, proportional controllers for instrument heater control, and conditioned low-voltage power. An interface board accepts command and control instructions from the flight computer, performs housekeeping monitoring, and packetizes and transmits housekeeping and DSP science data back to the flight computer.

2.3. Details of the SIM Instrument

Section 2.2 gave an overview of the SIM instrument design that meets the requirements to measure solar spectral variability, and this section will provide details on the design and performance of these subsystems.

2.3.1. *ESR and Photodiode Detectors*

The known solar irradiance spectrum, the area and width entrance slit, and the dispersion of the spectrometer can be used to estimate the power collected by the instruments detectors. The greatest power observed will be about 40 μW near 800 nm, and the signal is down about a factor of 470 in the ultraviolet (260 nm) and a factor of 65 at infrared (2700 nm). To make measurements with a precision of 100 ppm the noise equivalent power must be on the order 4 nanowatts. This value sets the design goal for the ESR. In order to keep the thermal mass of the detector small, a $1.5 \times 10\,\text{mm}^2$ bolometer is used in the place of the cone geometry typically used for TSI measurements and the bolometer is surrounded by an optical quality sphere to increase the inherent blackness of the bolometer and to thermally isolate it from its surroundings. Figures 3 and 4 show the ESR location in the instrument case and the focal plane assembly. Thermal detectors have slow response times, and therefore are not suitable detectors for spectral scanning purposes. Therefore, complementary photodiode detectors are also used to allow a fast scan mode, but the radiant sensitivity of the photodiodes are calibrated against the ESR routinely in-flight.

The primary innovation of the SIM ESR over other radiometers described in Hengstberger's comprehensive book (1989) is the use of phase sensitive detection. There are, in fact, two phase locked loops used in the detection of radiant power. The first is a high-frequency (50 Hz) loop implemented in hardware that controls the excitation of a precision AC resistor/thermistor bridge, and the second is a low-frequency (0.01 Hz) light chopper operating at the minimum of the detector's noise spectrum and the phase sensitive detection is handled in data processing (Lawrence *et al.*, 2000).

The fundamental underlying principle of electrical substitution radiometry is to establish a constant thermal environment on two independent black surfaces (bolometers) using Joule heating from resistors in intimate contact with the surfaces so that the difference in temperature between the surfaces approaches 0. When one of these surfaces (the active bolometer) is exposed to radiant energy, a measurable temperature difference is generated between the active and reference bolometers. Joule heating on the active bolometer must then be reduced until its temperature equals that of the reference bolometer ($\Delta T \simeq 0$). When this condition is met, the electrical power removed equals the radiant power incident on the active surface to within the magnitude of the noise associated with the temperature measurement.

Figure 5 shows the block diagram implementation of the ESR. Very small temperature differences are measured with a Wheatstone Bridge circuit using matched spinel thermistors in two arms of the bridge excited by a 50 Hz cosine wave

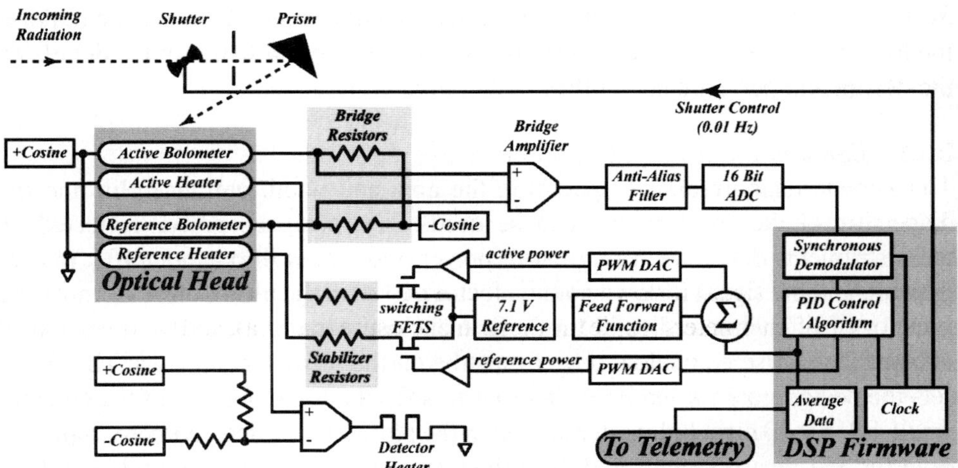

Figure 5. Block diagram of the SIM ESR circuitry. The system employs two phase locked loops: shutter operates at 0.01 Hz, and the thermistor bridge in the optical head is driven by a 50 Hz cosine wave.

generated by the DSP firmware. A precision 7.1 V voltage source (LTZ 1000, Linear Technologies Inc.) in conjunction with the heater resistors produces an equilibrium temperature value of 32 °C on both the active and reference bolometers. Switching MOSFET (Metal Oxide Semi-Conductor, Field Effect Transistor) transistors modulate the power to the heaters with a 50% duty cycle square wave. Electrical substitution on the active bolometer is achieved by pulse width modulating the heater waveform when radiant energy impinges on the detector; in this way the power delivered to the bolometer is proportional to the duty cycle. Control of the active bolometer duty cycle occurs in DSP firmware where the conditioned and digitized bridge temperature error signal is filtered by a PID (proportional/integral/differential) algorithm. The PID algorithm controls the closed-loop response of the servo system. A pulse width modulated digital-to-analog converter (PWM DAC) converts the 16-bit digital output from the loop filter to the power applied to the replacement heater. The output of the loop filter is scaled and clipped to a unipolar code ranging from 0 to M – 1, where $M = 64\,000$. This code counts the width of the pulse on the gate of the MOSFET switch that pulses the voltage source to the heaters. The replacement heaters on the SIM bolometers are not precision resistors and vary by about 100 ppm. Therefore series stabilizer resistors of equal value are included in the circuit. The nominal power per data number for the SIM ESR is:

$$A = V_{\text{ref}}^2 \frac{R_H}{(R_H + R_s)^2} \frac{1}{M} \cong 2 \times 10^{-9} (\text{Watts/DN}) \quad \text{where:} \quad \begin{aligned} V_{\text{ref}} &= 7.1\,V \\ R_H &= 100\,k\Omega \\ R_s &\approx R_H \\ M &= 64\,000 \end{aligned} \quad . \tag{1}$$

The reference thermistor also provides the signal for a proportional heater (rather than a PWM heater) to maintain a constant temperature environment for the two bolometers. This heater circuit is also controlled by the DSP using a similar PID algorithm to the one used to balance the bridge. As is discussed in greater detail in Harder *et al.* (2005), a feed-forward function is supplied to add a predetermined power step to the control loop. This allows an end-to-end measurement of the servo gain thereby giving a measure of stability and degradation of the ESR electronics.

Figure 6 shows the mechanical/optical/thermal implementation of the SIM ESR with a cross-sectional rendering of the detector showing the location of the active and reference bolometers inside the thermal enclosure (panel a), and two microscope images showing the construction of the detector (panels b and c). In Figure 6, the detector is shown in the spectrometer's cross-dispersion direction (tall dimension of the exit slit) while Figure 4 shows a cross-section in the dispersion direction. The bolometer is a 1.5 × 10 × 0.03 mm CVD (chemically vapor deposited) diamond substrate that detects radiation on the front surface and a replacement power resistor is located on the back surface. The thermal enclosure places the two bolometers in a uniform thermal environment, but optically isolates them so that light entering the

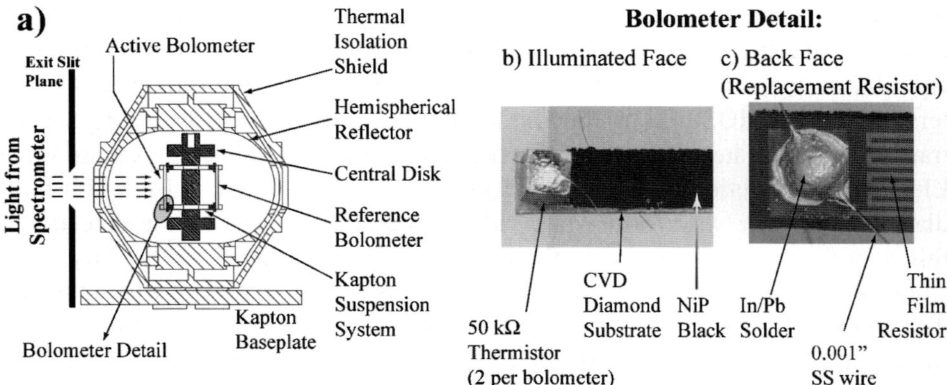

Figure 6. Panel (a) shows a cross-sectional view of the ESR detector and identifies critical components. Panel (b) shows a microscope photograph of the front face of the bolometer with one of the spinel thermistor soldered onto the diamond substrate along with the NiP surface. There are two series thermistors on the front face at opposite ends. Panel (c) shows the back face of the bolometer with the stainless steel wire soldered onto a gold pad. The thin film resistor is evident in this photograph as well.

detector cavity directly illuminates the active bolometer, but cannot propagate to the reference bolometer. The front surface of each bolometer has two 50 k Ω spinel thermistors in series soldered at opposite ends of the strip and a layer of nickel phosphorous black (NiP) is deposited in the center portion of the bolometer for radiation absorption (Johnson, 1980). The back surface of the bolometer has a thin film resistor photo-etched onto the diamond substrate and covers the same area as the NiP black layer on the front side of the detector.

The custom-made photodiodes used for SIM are $10 \times 2\,\text{mm}^2$ for the three silicon photodiodes (International Radiation Devices Inc., Torrence, CA) and $8 \times 2\,\text{mm}^2$ for the InGaAs photodiode (Hamamatsu Corporation, Bridgewater, NJ). The Vis1 and UV silicon photodiodes have *n*-on-*p* construction with a nitride passivated SiO_2 layer to stabilize their radiant sensitivity in the ultraviolet. The Vis2 photodiode is constructed similarly, but with p-on-n geometry. Figure 4 shows the location of these detectors in the instrument's focal plane. The detectors are located 2 mm behind the exit slit for the VIS1, VIS2, and IR photodiodes, but the UV diode is placed 10 mm behind the exit slit to improve the rejection of scattered light. A baffle is placed 2 mm in front of the exit slits to limit the field-of-view of the detectors to the solid angle subtended by the prism. The photo current for each detector is converted to a voltage by a precision transimpedance amplifier, and these voltages are multiplexed and digitized by the same 16-bit bipolar analog-to-digital converter.

2.3.2. *Optical and Spectroscopic Properties*

The nominal properties of the SIM spectrometer and detectors are summarized in Table I. The Fèry prism spectrometer is analogous to the Rowland Circle concave grating spectrometer (Warren, Hackwell, and Gutierrez, 1997). This spectrometer

TABLE I
SIM optical properties.

Parameter	Value				
Spectrometer					
Wavelength coverage	200–2700 nm				
Focal length	400 mm				
F-number					
Solar	115				
Spectrometer	16 (Dispersion), 22 (Cross dispersion)				
Prism glass	Suprasil 300				
Prism figure					
Front surface radius	421.48 mm				
Back surface radius	441.27 mm				
Central thickness	12.30 mm				
Wedge angle	34.49 mm				
Projected aperture	25 mm				
Prism height	18 mm				
Measured focal length	403.15				
Surface roughness	0.75 Å (RMS)				
Focal plane aberrations	5 μm				
Entrance slit dimensions					
Slit width	0.3 mm				
Slit length	7.0 mm				
Field-of-view					
Dispersion	2.8°				
Cross dispersion direction	1.7°				
Detectors	UV	ESR	Vis1	Vis2	IR
Range (nm)	200–308	255–2700	310–1000	360–1000	994–1655
Material	*n-p* silicon		*n-p* silicon	*p-n* silicon	InGaAs
Detector size (mm)	2 × 10	1.5 × 10	2 × 10	2 × 10	2 × 8
Exit slit width (mm)	0.34	0.30	0.30	0.30	0.30
Nominal location (mm)	−10	35	50	55	60

design is limited to high *f*-number applications because of significant image degradation due to coma and astigmatism that occur even at moderate numerical apertures. However, this is not an issue for the 1/2° solar geometry where the aberrations are only about 5 microns in the focal plane. The radii of curvature for the front and back surfaces of the prism are optimized so that the minimum RMS (root mean

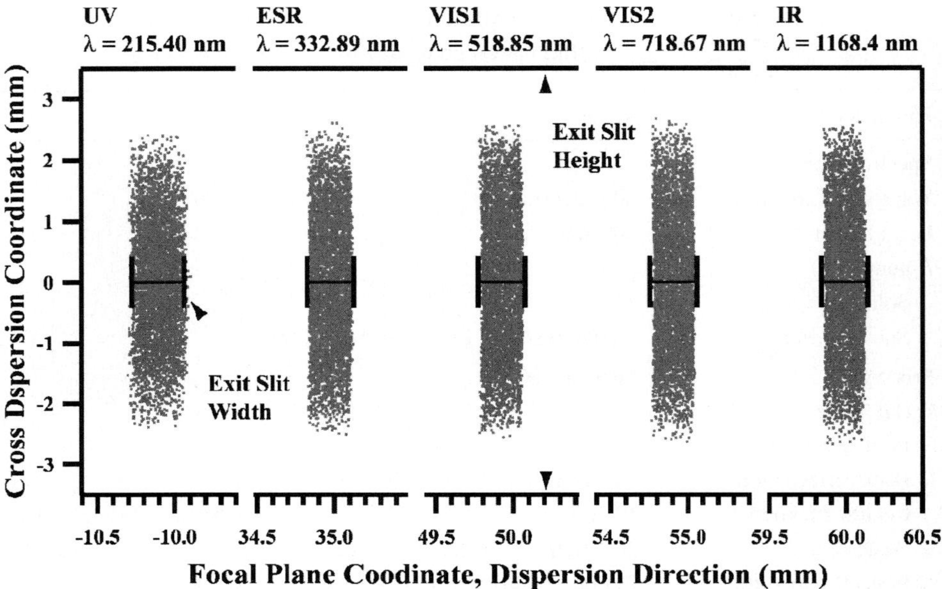

Figure 7. Spot diagrams for the five focal plane detectors at a fixed prism rotation angle of 59.5°. The figure shows the range of the slit width and height in focal plane coordinates and the center wavelength of light that goes through the appropriate exit slit.

square) spot size is attained at the location of the ESR exit slit (ZEMAX, Focus Software Inc., San Diego, CA).

Figure 7 shows a focal plane spot diagram at each detector's exit slit at a fixed prism rotation angle of 59.5° (generated by IRT, Parsec Technology Inc., Boulder, CO). At this prism rotation angle, the prism disperses white light and each detector observes a different wavelength simultaneously. The ray trace analysis assumes the origin is at the center of the entrance slit and uses the actual measured positions of the slits and prism; this figure shows that the UV photodiode detector is on the opposite side of the entrance slit from the other four detectors. This figure shows the effects of coma and spherical aberration on images; they have about 5 μm of curvature due to coma and the images are about 5 mm tall because the instrument focal plane is at the horizontal focus (focus in the dispersion direction) of the tilted spherical mirror back surface of the prism. The vertical focus (focus in the cross dispersion direction) is located behind the entrance slit at a distance of $\sim 2d$ where d is the distance from the face of the prism to the horizontal focus; ~400 mm for SIM. The exact location of the tangential focus is a function of the prism refractive index, therefore, the height and width of the image at the exit slit is wavelength dependent as well. For the ESR, VIS1, VIS2, and IR exit slits the image is demagnified in the dispersion direction, so the instrument function (the convolution of the entrance and exit slit at fixed wavelength) is trapezoidal in form. The image distortion is the worst at the location of the UV photodiode, and the image is magnified so a

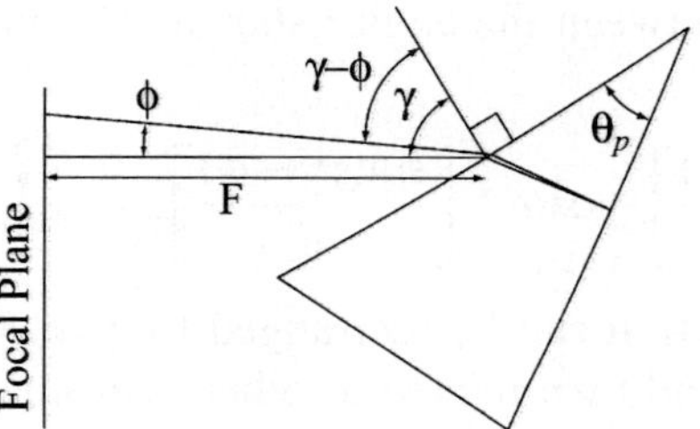

Figure 8. Shows the geometry of the Fèry prism spectrometer that is described in Equation (3).

wider exit slit (340 μm instead of 300 μm) is used to partially compensate for the significantly wider spot width. In all cases the height of the image at the exit slit is smaller than the entrance slit, but the exit slits are made to be 7 mm tall to match the entrance slit and so light is not vignetted with off-axis spacecraft pointing in the cross-dispersion direction. These wavelength dependent aberrations have a small, but significant, effect on the radiometric performance of the instrument. Because the instrument cannot be focused equally well at all wavelengths and all focal plane positions simultaneously, the exits slits are slightly over-filled and the peak of the trapezoidal instrument function does not reach a value of 1.0 for most wavelengths; a value of 1.0 corresponds to the case where all the photons at the entrance slit pass through the exit slit (assuming a perfectly transmissive prism). The worst case is a 1.4% deficit near the instrument's resolution minimum at 1.2 μm for the ESR detector. Ray tracing with measured instrument parameters can accurately model these radiometric deficits that are then included in the final radiance calculation.

The instrument dispersion model is based on prism geometry derived from Snell's law and applies to a prism in Littrow configuration (James and Sternberg, 1969). Figure 8 shows the geometry of the spectrometer: the variables are the prism incidence angle with respect to the first surface normal vector, γ, the deviation angle, ϕ, the index of refraction of fused silica, *n*, the prism wedge angle, θ_p, and the instrument focal length, F, produced by the prism's spherical surfaces. The index of refraction of fused silica is given from the three-term (six coefficient) Sellmeier Equation measured by Malitson (1965) at 20 °C and is valid to 10 ppm:

$$n_{20}(\lambda) = \sqrt{1 + \sum_{j=1}^{3} \frac{\lambda^2 K_j}{\lambda^2 - L_j^2}} \quad \begin{array}{cc} \mathrm{K_j} & \mathrm{L_j} \\ 0.6961663 & 0.0684043 \\ 0.4079426 & 0.1162414 \\ 0.8974794 & 9.896161 \end{array} . \tag{2}$$

This paper also gives the temperature dependence of the index of refraction; the data from Figure 3 of his paper was digitized and fit to 1 ppm with a 5th order polynomial as a function of wavenumber in inverse microns and is incorporated in SIM data processing. The $\mathrm{d}n(\lambda)/\mathrm{d}T$ dependence is significant so $n(T)$ is calculated at each prism rotation step with prism temperatures recorded in instrument telemetry.

The basic relation between the angles shown in Figure 8 and the index of refraction is given by

$$2\theta_p = \sin^{-1}\left[\frac{\sin(\gamma)}{n}\right] + \sin^{-1}\left[\frac{\sin(\gamma - \phi)}{n}\right]. \tag{3}$$

In general, this equation can be rearranged to give each of the variables as a function of the others and then related to other instrument subsystem parameters and the validity of this equation can be checked against ray tracing. This topic is discussed in detail in Harder *et al.* (2005).

The irradiance detectors are located at focal plane positions to maximize their wavelength coverage and remain within the ±2.5° rotation range of the prism drive mechanism; see Figure 4 for the relative locations of the detectors in the focal plane. Because the detectors are at different locations in the focal plane, the prism rotation angle reported by the CCD position encoder system has a different wavelength value at each detector position. Figure 9 shows the mapping of wavelength into prism rotation angle for each of the focal plane detectors; the extent of the curves for each detector also reflects the ranges over which the data are valid. This figure shows the position and wavelength coverage of the periscope that was described in Section 2.2.2. Like the other focal plane detectors, the extent of the curve indicates the valid operating range for the prism calibration system. The curvature seen in these traces is caused by the non-linearity of fused silica's dispersion (dn/dλ). When a vertical line is drawn through Figure 9 at a fixed prism angle, the intersection of the detector curves with this line indicates the wavelengths that each detector simultaneously observes; if this vertical line does not intersect one of the detector traces then that detector will not respond at that rotation angle. Likewise, a horizontal cut of constant wavelength indicates the needed prism rotation angle for a requested wavelength at each detector, and if this horizontal line intersects two or more detector curves, then

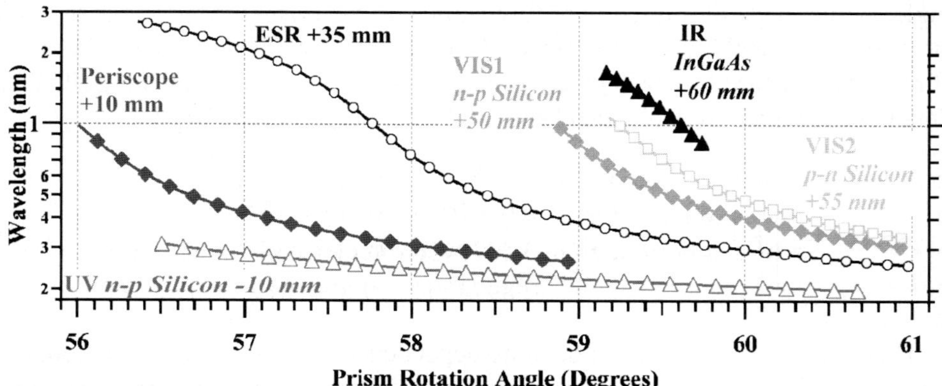

Figure 9. Wavelength is plotted as a function of prism rotation angle for each of the five focal plane detectors and the periscope that couples the two instruments together for prism transmission measurements. This figure also gives the valid operating wavelength ranges for each detector.

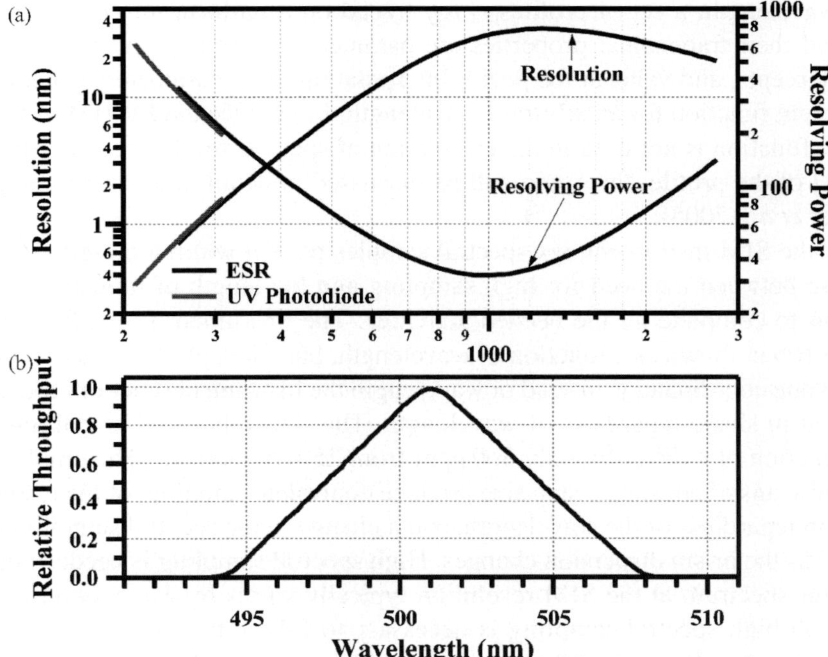

Figure 10. Panel (a) shows the FWHM resolution and the resolving power ($\lambda/\Delta\lambda$) of the SIM instrument for two detectors, ESR, and UV. The resolution function is slightly different at each detector position in the instrument focal plane, so the two traces do not overlap. Panel (b) shows the instrument function at 501.02 nm with a half-width of 7.06 nm. The throughput of the instrument at the peak is 0.996 because of image overfilling of the exit slit.

there is wavelength overlap between these detectors. This is particularly important since the ESR is used to calibrate the photodiode detectors and this indicates the range over which the calibration will be effective for each detector.

The spectral resolution and resolving power ($\lambda/\Delta\lambda$) of the instrument are given in Figure 10a for the ESR and the UV photodiode detector to show the full wavelength range. Refraction geometry produces a slightly different resolution function for each detector, but data processing accounts for these differences. The resolution, as defined here, is given by the full-width-half-maximum (FWHM) of the asymmetric trapezoidal instrument function. The asymmetry is caused by non-uniform $dn/d\lambda$ over the span of a slit width along with spectrometer aberrations. The resolving power is greatest in the ultraviolet, decreases to a minimum near the minimum deviation angle of the prism, and slowly increases again in the infrared.

Figure 10b shows a particular instrument function profile for a peak value of 501.02 nm, a FWHM profile width of 7.06 nm, and the peak throughput is 0.996 instead of 1.0 because of slit overfilling. Ray tracing the spectrometer with the known instrument geometric parameters and the fused silica refractive index generates this instrument function. Because the instrument profile changes smoothly with wavelength, a set of profiles is ray traced on a uniform index of refraction grid and their trapezoidal properties are parameterized (rising and falling slopes and intercepts, and value at the peak). Interpolating these parameters generates an instrument function for an arbitrary wavelength. In practice the FWHM of the resolution function is not used in the calculation of spectral irradiance, but rather the integral of the profile; this topic will be discussed in detail in a companion paper (Harder *et al.*, 2005).

For the SIM instrument, six spectral samples per slit width are used as a compromise between the need for high sampling and the length of time available on an orbit to complete all the needed activities. The instrument function shown in Figure 10b is shown as a function of wavelength, but when plotted as a function of focal plane coordinate, y, instead of wavelength the instrument functions are nearly identical in shape regardless of wavelength. The integrals of relative throughput as a function of y differ by only 300 ppm from 250 to 2500 nm. This implies that spectral scans of constant y step size produce equivalent sampling of the resolution function regardless of the wavelength, and a change in the spectral sampling is not needed as the prism dispersion changes. High spectral sampling is needed because the solar spectrum at the SIM resolution typically varies by 4% over an exit slit width, so high spectral sampling is necessary to follow the light curve over the course of a scan. Because of the temperature dependence of the index of refraction, the wavelengths reported by the instrument are not consistent from scan to scan, so spectra must be interpolated onto a uniform wavelength grid for spectral comparisons. The high spectral sampling used here is needed to maintain radiometric accuracy during this interpolation process.

2.3.3. *Prism Drive Operation and Block Diagram*

The wavelength control path discussed in Section 2.2.2 achieves very high precision by using a linear 12000 element CCD to measure the prism rotation angle in the focal plane rather than at the point of rotation as is done by most angular rotation encoders. The CCD, prism turntable, and the voice coil actuator form an electro-mechanical closed-loop system to control the incident angle of the prism (the variable γ in Figure 8).

Figure 11a shows an electro-mechanical block diagram for the closed-loop operation of the prism drive. The principle of operation is based on the comparison of two counters: one counter measures in near real-time the center location of the light spot on the CCD by measuring the length of time needed to reach that position at a constant video sweep rate, and the other counter measures the length of time needed to reach a predetermined target position on the CCD; the time difference between these two counters gives the servo error. A single 2.5 MHz clock inputs the position down counter and sets the CCD video readout rate after it is divided by 5; this effectively subdivides a CCD pixel by a factor of 5 and increases the spatial resolution of the measurement. The SIM DSP asynchronously loads the target position into the position down counter terminal count register, and a synchronizing

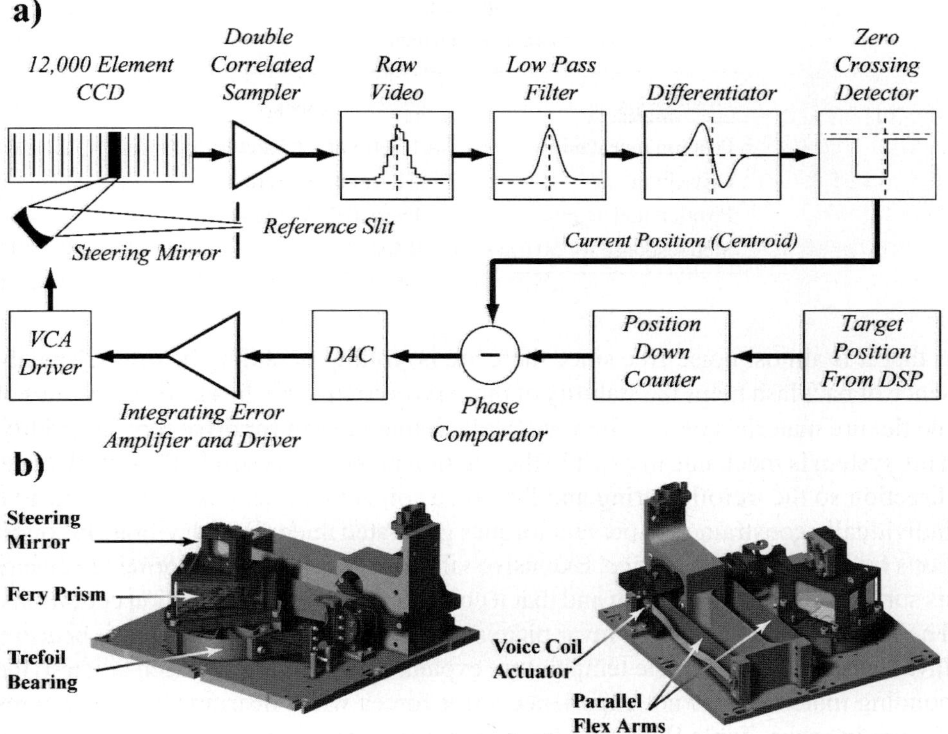

Figure 11. Panel (a) Electro-mechanical block diagram of the SIM prism drive system. The diagram shows the process of comparing a down counter with the video sweep of the CCD. Panel (b) is two views of the prism drive identifying important attributes to the mechanical design of the drive.

pulse enables the down counter and initiates the sweep of the CCD. The location of the image spot on the CCD is detected and conditioned by double correlated sampler electronics, low-pass filtered, differentiated, and this smoothed waveform is fed into a zero crossing detector so the length of time needed to reach zero crossing can be obtained. The phase comparator determines the difference in time, whether positive or negative, and creates a pulse that is fed to the bi-polar digital-to-analog converter (DAC) and then an integrating power amplifier drives the voice coil actuator (VCA) that rotates the prism/steering mirror to move the light spot on the CCD. The location of the image spot relative to the target position is updated every 25 ms. The time constant of the integrating amplifier determines the slew rate of the mirror.

For the prism drive electronics to work the rotation mechanism must provide very smooth and low noise motion. This prism drive is shown in Figure 11b. Drive power is furnished through a linear voice coil actuator suspended by parallel flex arms with ~6 mm of motion (BEI Sensors and Systems Inc., model LA13-12-000A). The voice coil is coupled to a radial flexure, a custom trefoil bearing, which supports the prism and steering mirror. The most important attribute of this flexure-based system

TABLE II

Prism drive attributes.

Parameter	Value
Position repeatability	1.3 μm or 0.65 arcsec
Drive jitter	0.3 μm or 0.16 arcsec
Position update rate	Every 0.025 s (40 Hz)
Step response for 50 μm step	0.3 s

is that it is almost noise-free since there are no rolling or sliding bearings. The absence of backlash helps the stability of the servo system, and the elastic properties of the flexure materials permit the very fine pointing needed for drive reproducibility. This system is mechanically stiff in the rotation plane, but is soft in the out-of-plane direction so the trefoil bearing and the voice coil actuator sub-assemblies must be individually constrained to prevent torques generated under launch vibration conditions from damaging the drive. Extensive vibration testing was performed to ensure its survivability during launch and that it could not 'jam' under vibration conditions. The prism was bonded to an invar plate and then mounted onto the trefoil bearing. Invar was used to match the temperature expansion coefficient of fused silica so the bonding material does not experience shear forces when instrument temperatures change in space. Table II lists the performance specification achieved for the drive.

2.3.4. *In-Flight Prism Degradation Measurements*

Two in-flight calibration operation modes track long-term degradation of the instrument. The first is a direct prism transmission calibration that uses the calibration path introduced in Section 2.2.2 and depicted in Figure 2, and the second is a direct SIM A/SIM B spectrometer comparison. Prism transmission is a first order term in the SIM radiometric measurement equation (Harder *et al.*, 2000a; Harder *et al.*, 2005), and changes from the ground-based prism transmission measurement must be tracked throughout the mission. This radiation effect is significant for SIM, but the rate and wavelength dependence of in-flight prism transmission degradation is well-characterized and presented in detail by Harder *et al.* (2005). The mechanism for the observed prism degradation is not well known, but is most likely due to hard radiation and/or energetic particle modification of the fused silica surface of the prism itself or a thin film ($< 5 \times 10^{-10}$ m) of organic material contaminating that surface. The findings of Havey, Mustico, and Vallimont (1992) from the LDEF Experiment (Long Duration Exposure Facility) are suggestive of the effects observed in SIM but do not give evidence of the degradation mechanism.

The transmission of the SIM prisms was measured in the laboratory on ground witness prisms manufactured from the same bole of Suprasil 300 as the flight units that were polished and coated simultaneously (Harder *et al.*, 2005). Ray trace analysis shows that light rays propagate almost normal to the mirrored surface of the prism so the mirror reflectivity is nearly independent of prism incident angle (see Figure 8). The transmission of the prism can be decomposed into three contributions: (1) Fresnel surface losses (both vacuum-to-glass and glass-to-vacuum), (2) light attenuation in the bulk of the glass, and (3) reflective losses on the aluminized back surface. Therefore, the transmission can then be calculated for each detector by combining the reflective/bulk losses with the Fresnel reflection loss to give the total transmission.

For the in-flight transmission re-calibration, only the relative time dependence of the transmission as a function of wavelength is needed so corrections for systematic light losses are not accounted for by the in-flight method described here. To obtain a true transmission measurement (as was done for the laboratory method), it is necessary to account for non-symmetric light losses in the I and I_o measurement paths, polarization effects, and the spatial dependence of radiant sensitivity of the detector.

A cross-sectional view of the periscope is shown in Figure 12a depicting the chief ray propagation of light through the periscope. It couples the two instruments and

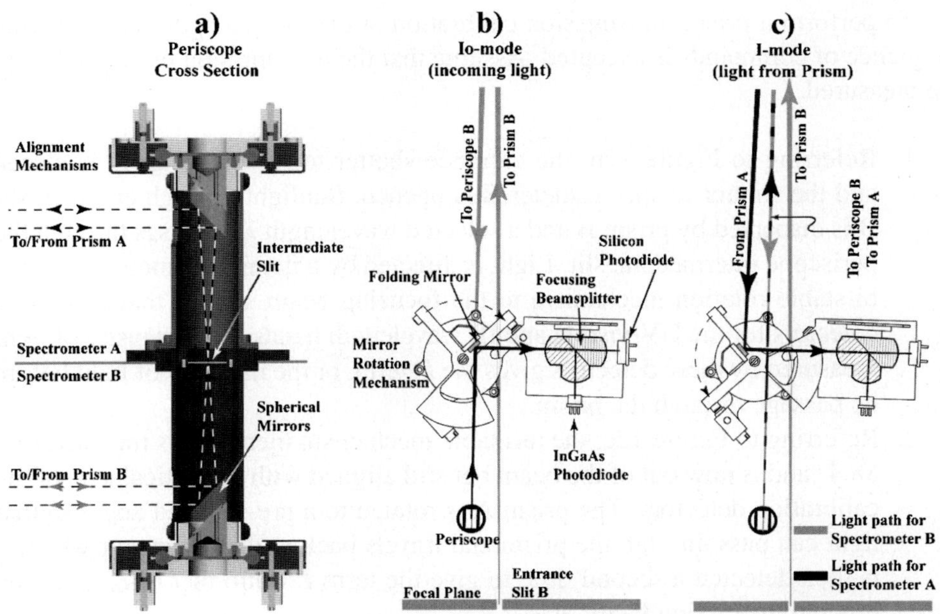

Figure 12. Panel (a) shows a cross-sectional view of the periscope and the propagation of light through it. Panels (b) and (c) show the opto-mechanical system to perform the prism transmission calibration. In this figure SIM B is calibrating the prism transmission of SIM A. The visual perspective is looking down on spectrometer A with its rays colored black. SIM B (mirror-image of SIM A) is directly underneath SIM A and hidden in this view, but the light rays of SIM B are colored in gray and projected onto the page to show their optical path. Monochromatic light is selected by SIM B and directed through the periscope to spectrometer A. Panels (b) and (c) show the light paths to give the I_o and I modes along with the orientation of the mirror rotation mechanism, respectively.

consists of two 45° spherical mirrors (1038 mm radius) separated by an intermediate slit that matches the spectrometer's entrance and exit slits (0.3×7 mm); the spherical mirrors are on-center and concave. There are two optical requirements for the periscope: (1) the periscope images prism A onto prism B (and visa-versa), and (2) the periscope images the entrance slit onto the intermediate slit just as if it were another focal plane exit slit. Figure 9 includes the wavelength range and prism rotation angle used for the calibration system.

The prism transmission relies on a mirror rotation mechanism that moves a flat mirror into one of two fixed locations. Its action is shown in Figures 12b and 12c. The design and location of the calibrator mechanism in the instrument case is set to ensure that in the event of the rotation mechanism failure no light paths to the focal detectors can be vignetted. The focusing beamsplitter is designed to have a 10 mm focal length to keep it compact, and images the folding mirror onto the two detectors. This optical element has to be able to image the diverging beam from the periscope onto essentially the same area of the detector as the converging beam returned from the prism. The beamsplitter coating is a non-polarizing broadband coating that has a flat ~35% transmission/reflection throughout the 300 to 1000 nm region.

To perform a prism transmission calibration at one wavelength, the following sequence of commands is executed. Assume that the transmission of prism A is to be measured.

1. Referring to Figure 12b, the entrance shutter to spectrometer A is closed and the shutter to spectrometer B is opened. Sunlight through entrance slit B is dispersed by prism B and a selected wavelength λ_{cal} passes through the periscope intermediate slit. Light is directed by a flat mirror mounted on the bi-stable rotation mechanism to the focusing beam splitter that feeds two detectors for the UV/visible and IR wavelength bands. The intensity of light measured by these detectors gives the I_o term, or the intensity of light before its passage through the prism.
2. Referring to Figure 12c, the turntable mechanism then rotates the mirror by 88.4° and is now out of the beam but still aligned with the optical axis of the calibration detectors. The prism A is rotated to a pre-selected angle so that light can pass through the prism and travels back to the flat mirror where it is then detected a second time to give the term I. Ratio of I to I_o gives the prism transmission factor at λ_{cal}.

Because spectrometer channels A and B are identical and mirror-image each other, the prism B transmission can be calibrated by an analogous procedure. The hard radiation trap is located in the optical train to minimize unnecessary exposure during the measurement. This procedure is repeated at 44 different wavelengths in the 300 to 1000 nm region on a weekly basis, and the data are used in the SIM exposure model described by Harder *et al.* (2005).

The second in-flight calibration mode for this instrument is to compare the two spectrometers. Spectrometer B is exposed to ionizing radiation on a much lower duty cycle so its optics will not degrade at the same rate as SIM A, which is used for daily measurements. The two spectrometers are operated simultaneously in the absolute irradiance mode so the instrument comparison excludes possible spacecraft pointing, solar, and thermal differences that would affect the comparison if the measurements were conducted in a serial fashion. The spectral irradiance for the A and B channels are calculated independently using the prism transmission procedure for each channel independently. The resultant irradiance measurements from spectrometers A and B can then be used to validate the prism transmission measurement and to identify other sources of degradation in SIM.

2.4. Flight Instrument Operation Modes

A number of instrument operation, housekeeping, and calibration modes have been developed for SIM to maintain measurement accuracy, account for the limited wavelength ranges of the detectors, and monitor changes in instrument responsivity. These operation modes are tabulated in Table III along with a description of the action and the frequency that these operation modes are executed. In this table, cadence is the number of events that occur within a specified number of days. The operating modes for SIM A and SIM B are identical except that SIM B is directly exposed about 18% of the time of SIM A. When either channel of SIM is operated, the HRT is usually in the optical train to reduce prism exposure.

The ESR is an inherently slow detector so the photodiodes are used to track the orbit-to-orbit variability of the Sun and the ESR is used to calibrate the radiant sensitivity of the diodes. During a nominal 58-min sunlit portion of an orbit, the typical plan is to perform a 24-min photodiode scan along with part of an ESR table sequence. This is repeated over nine orbits until all 63-table positions are completed, and the process is then repeated. The table wavelength positions are selected to occur at maxima and minima in the solar spectrum so the irradiance measurement is less sensitive to possible wavelength shifts over the span of a measurement. Both the ESR table and the photodiode scans are performed with the HRT both in and out of the light beam. In this way solar variability can be followed without excessive and unnecessary exposure to the prism. During the course of the day, only three orbits have the HRT out of the beam; two for 24-min scans, and one for the ESR infrared scan. The ESR infrared scan is used to acquire daily spectra at wavelengths longer than the 1.7 μm cutoff of the InGaAs photodiode. The ESR is used in a phase sensitive mode with the shutter operating at 0.05 Hz and two shutter cycles per prism step. The spectrum is measured with two samples per slit width, not quite meeting the Nyquist criteria, but permitting the spectrum to be acquired in a single orbit. The ESR full scan provides the best measurement of the solar spectrum, but requires 240 sunlit minutes to complete, so solar variability likely occurs during

TABLE III

SIM irradiance and calibration modes.

Mode name	Cadence (number/days)	Action
Irradiance measurement modes		
24-min photodiode scan	9-13/1	Measure solar irradiance with the 4 photodiodes over their full operation range; 6 samples/slit width sampling; includes measurement of photodiode dark signal; 2 scans/day are without HRT.
ESR table sequence	12/7 HRT in 1/7 HRT out	63 selected wavelengths; 0.01 Hz shutter frequency; 2 shutter cycles per wavelength step.
ESR infrared scan	1/1	Spectral scan with ESR as primary detector in the 1200–2700 nm range; 2 samples per slit width; 0.05 Hz shutter frequency; 2 shutter cycles per prism step.
ESR full scan	1/90	Complete scan of solar spectrum with ESR as primary detector; 0.05 Hz shutter frequency; 2 shutter cycles per prism step; 3 samples/slit width; 15 orbits to complete scan.
Calibration/housekeeping modes		
Fixed wavelength	1/1	Study thermal stability of instrument and detector noise. Performed with shutter frequency of 0.05 Hz.
Prism transmission	1/7	44 discrete wavelengths; see Section 2.3.4 for action; measure the transmission of both prisms.
ESR gain	2/7	Measure closed loop gain on ESR at 0.05 and 0.01 Hz.
Cruciform scan	1/7	Check alignment of instrument to spacecraft. Prism rotation angle held at a fixed wavelength by open-loop prism drive control; cycles through 6 different positions.
CCD image dump	1/7	Read out CCD video to track intensity of image spot and evaluate progress of CCD radiation damage.

(*Continued on next page*)

TABLE III
(*Continued*)

Mode name	Cadence (number/days)	Action
A/B comparison	1/30	Perform both photodiode scans and table sequence without HRT on both instruments simultaneously.
Field-of-view (FOV) map	1/180	Measure solar spectra on a 5×5 spatial grid with a 0.125° spacing. Spacecraft offset pointing is used to generate the map. Measures the spatial extent of radiation damage on the prisms; performed with HRT in the beam.

this time frame. Therefore, the ESR full scan is done in conjunction with a 24-min scan and spread over 15 orbits to complete. In this way, the comparison of the ESR full scan with multiple photodiode scans determine the degree of solar variability over the course of the day.

The weekly calibration/housekeeping modes are spread uniformly to eliminate gaps in the coverage of solar variability. The prism transmission calibration is performed at 39 wavelengths and requires 300 s to complete an individual wavelength, so six orbits are occupied to complete the experiment for SIM A and B.

The cruciform scan and the FOV map are spectrometer operations performed in conjunction with planned spacecraft offset pointing maneuvers. The SORCE spacecraft has excellent pointing accuracy of better than 1 arcmin, pointing knowledge of 10 arcsec, and jitter of 5 arcsec/s, so instrument pointing corrections are not normally required. However, effects of prism degradation have a gradient across the 3×11 mm light spot on the face of the prism so monitoring the changes in transmission across this illuminated spot is important. For the cruciform scans, the spacecraft slews across the instrument's field-of-view in the dispersion and cross dispersion directions at a rate of 1.5 arcmin/s for a range of $\pm 4°$. With pointing offsets this large, the steering mirror that illuminates the CCD will lose signal to the point where it cannot control, so open-loop fixed wavelength positioning is used for this mode of operation. These cruciform scans supply information about the relative alignment of the instrument with respect to the spacecraft attitude control reference. They supply information about the spatial extent of transmission degradation on the face of the prism, and they also give information about scattered light in the instrument. The FOV maps are a 5×5 mm, 7.5 arcsec mapping of the prism response and are performed with closed-loop control. The full wavelength range of the instrument is measured at each of these 25 positions over the span of about 4 days with frequent measurements of the reference center position to remove bias from solar variability. The FOV maps provide the best information about the degradation gradient on the face of the prism.

3. Conclusions

This paper describes the SORCE SIM instrument. It is a prism spectrometer operating over the spectral range of 200 to 2700 nm. It employs an electrical substitution radiometer as the primary detector, a CCD as an absolute encoder of prism rotation angle measured in the focal plane of the detectors, and is capable of providing self-calibration of prism degradation by using one spectrometer to deliver monochromatic radiation to the other. The instrument uses phase sensitive detection to determine the irradiance, but also employs low noise photodiodes to acquire the spectrum in a rapid scan mode.

Acknowledgements

The authors gratefully acknowledge the entire staff at LASP, both students and professionals, who contributed to the success of this new instrument throughout concept, design, fabrication, integration, test, and launch phases of the program, as well as the current data processing staff. This research was supported by NASA contract NAS5-97045.

References

Anderson, G. P. *et al.*: 1999, *SPIE Proc.* **3866**, 2.
Curcio, J. A. and Petty, C.: 1951, *J. Opt. Soc. Am.* **41**, 302.
Harder, J. W., Lawrence, G. M., Rottman, G., and Woods, T. N.: 2000a, *Metrologia* **37**, 415.
Harder, J. W., Lawrence, G., Rottman, G., and Woods, T.: 2000b, *SPIE Proc.* **4135**, 204.
Harder, J. W., Fontenla, J., Lawrence, G., Woods, T., and Rottman, G.: 2005, *Solar Phys.*, this volume.
Havey, K., Mustico, A., and Vallimont, J.: 1992, *SPIE Proc.* **1761**, 2.
Hengstberger, F.: 1989, *Absolute Radiometry*, Academic Press Inc., San Diego, California.
Hood, L. L.: 1999, *J. Atmos. Sol. Terr. Phys.* **61**, 45.
James, J. F. and Sternberg, R. S.: 1969, *The Design of Optical Spectrometers*, Chapman and Hall LTD, London, p. 41.
Johnson, C. E.: 1980, *Metal Finishing*, 21.
Kiehl, J. T. and Trenberth, K.: 1997, *Bull. Am. Met. Soc.* **78**, 197.
Kopp, G., Lawrence, G. M., and Rottman, G.: 2003, *SPIE Proc.* 5171.
Kurucz, R. L.: 1991, in A. N. Cox, W. C. Livingston, and M. S. Matthews (eds.), *Solar Interior and Atmosphere*, University of Arizona Press, Tucson, Arizona.
Labitzske, K. and Van Loon, H.: 1988, *J. Atmos. Sol. Terr. Phys.* **50**, 197.
Lawrence, G. M., Rottman, G., Harder, J., and Wood, T.: 2000, *Metrologia* **37**, 415.
Lean, J. L.: 1991, *Rev. Geophys.* **29**, 505.

London, J., Rottman, G., Woods, T., and Wu, F.: 1993, *Geophys. Res. Lett.* **20**, 1315.
McClintock, W., Rottman, G., and Woods, T.: 2005, *Solar Phys.*, this volume.
Malitson, I. H.: 1965, *J. Opt. Soc. Am.* **55**, 1205.
Neckel, H. and Labs, D.: 1984, *Solar Phys.* **90**, 205.
Reid, G. C.: 1991, *J. Geophys. Res.* **96**, 2835.
Reid, G. C.: 1999, *J. Atmos. Sol Terr. Phys.* **61**, 3.
Rottman, G. J., Woods, T., and Sparn, T.: 1993, *J. Geophys. Res.* **98**, 10667.
Rottman, G., Harder, J., Fontenla, J., Woods, T., White, O., and Lawrence, G.: 2005, *Solar Phys.*, this volume.
Smith, R. C. and Baker, K.: 1978, *Limnol. Oceangr.* **23**, 260.
Solanki, S. K. and Unruh, Y.: 1998, *Astron. Astrophys.* **329**, 747.
Thuillier, G., Hersé, M., Labs, D., Foujols, T., Peetermans, W., Gillotay, D., Simon, P., and Mandel, H.: 2003, *Solar Phys.* **214**, 1.
Vanhoosier, M. E., Bartoe, J.-D. F., Brueckner, G. E., Prinz, D. K., and Cook, J. W.: 1981, *Solar Phys.* **74**, 521.
Viereck, R. and Puga, L.: 1999, *J. Geophys. Res.* **104**, 9995.
Warren, D. A., Hackwell, J., and Gutierrez, D.: 1997, *Opt. Eng.* 3**6**, 1174.
White, W. R., Lean, J., Cayan, D., and Dettinger, M.: 1997, *J. Geophys. Res.* **102**, 3255.
Willson, R. C.: 1988, *Space Sci. Rev.* **38**, 203.

Solar Physics (2005) 230: 169–204

THE SPECTRAL IRRADIANCE MONITOR: MEASUREMENT EQUATIONS AND CALIBRATION

JERALD W. HARDER, JUAN FONTENLA, GEORGE LAWRENCE, THOMAS WOODS and GARY ROTTMAN
Laboratory for Atmospheric and Space Physics, University of Colorado, Boulder, Colorado 80309, U.S.A.
(e-mail: jerald.harder@lasp.colorado.edu)

(Received 6 April 2005; accepted 28 July 2005)

Abstract. The Spectral Irradiance Monitor (SIM) is a satellite-borne spectrometer aboard the Solar Radiation and Climate Experiment (SORCE) that measures solar irradiance between 200 and 2700 nm. This instrument employs a Fèry prism as a dispersing element, an electrical substitution radiometer (ESR) as the primary detector, and four additional photodiode detectors for spectral scanning. Assembling unit level calibrations of critical components and expressing the sensitivity in terms of interrelated measurement equations supplies the instrument's radiant response. The calibration and analysis of the spectrometer's dispersive and transmissive properties, light aperture metrology, and detector characteristics provide the basis for these measurement equations. The values of critical calibration parameters, such as prism and detector response degradation, are re-measured throughout the mission to correct the ground-based calibration.

1. Introduction

The Spectral Irradiance Monitor (SIM) is a satellite-borne spectrometer aboard the Solar Radiation and Climate Experiment (SORCE) that measures solar spectral irradiance between 200 and 2700 nm. This paper is a companion paper to Harder *et al.* (2005) that appears in this same issue of *Solar Physics*. That paper describes the overall instrument requirements, the hardware implementation, and the measurement modes needed to acquire the scientific data. This current paper emphasizes the calibration methods, in-flight corrections, and the mathematical operations (measurement equations) that are needed to convert instrument hardware signals measured in engineering units into SI units (International System of Units) of spectral irradiance with units of $\mathrm{W\,m^{-3}}$, or equivalently $\mathrm{W\,m^{-2}\,nm^{-1}}$ (Parr, 1996). Section 5 of this paper give the status of the calibration, and the corrections that have been included up to this point in time that are not covered by the measurement equations discussed in this paper.

Briefly summarizing Harder *et al.* (2005), SIM implements a number of unique design characteristics to provide: (1) broad wavelength operation, (2) multiple focal plane detectors, (3) a very high precision wavelength drive, and (4) in-flight monitoring of instrument response degradation. The instrument uses a low light scattering Fèry prism as the dispersing element that has high optical throughput within the 210–2700 nm region with a variable resolving power ($\lambda/\Delta\lambda$) ranging from 400 at 250 nm to a minimum of 33 at 1200 nm. The prism has excellent imaging properties, so multiple detectors can be used to detect incoming light in the instrument's focal plane. The primary detector is an electrical substitution radiometer (ESR). The ESR is a thermal detector that measures light from the spectrometer using phase sensitive detection to dramatically reduce the effect of thermal drift and detector noise. The input light beam to the spectrometer is chopped at 0.01 Hz by a shutter and only signal variations at the fundamental frequency are used to determine radiant power. The ESR detector calibrates the radiant sensitivity of four photodiode detectors during flight. The most important in-flight irradiance correction factor is prism transmission degradation, so the instrument is designed as two back-to-back, mirror image spectrometers that are coupled with a periscope. This provides both direct measurement of prism transmission and end-to-end comparisons by the two independent instruments. The two spectrometer configuration also provides instrument redundancy to ensure the continuity of the data record if the working spectrometer should fail.

The operation of the SIM radiometer is schematically represented in Figure 1. Solar radiation, E_λ (units of $\mathrm{Wm^{-2}}$), is incident on the instrument's rectangular entrance slit of area A (units of $\mathrm{m^2}$); it is the limiting aperture that defines the total radiant power entering the spectrometer. The light is then dispersed by the prism and imaged on an exit slit. The prism's geometry, orientation, and index of refraction, along with the entrance and exit slit widths, determine the selected wavelength (λ_s) and spectral bandpass ($\Delta\lambda$) that is transmitted through the exit slit and impinges on the detectors. This slit function convolution is effectively a low-pass filter of the spectrally complex solar spectrum over the wavelength band $\Delta\lambda$. A photometric detector, either the ESR or a photodiode, measures the incident power, P_D, within the spectral bandpass. The measured spectral irradiance, $\mathcal{E}_\lambda(\lambda_s)$ (units of $\mathrm{Wm^{-2}\,nm^{-1}}$), is then derived from three components: the determination

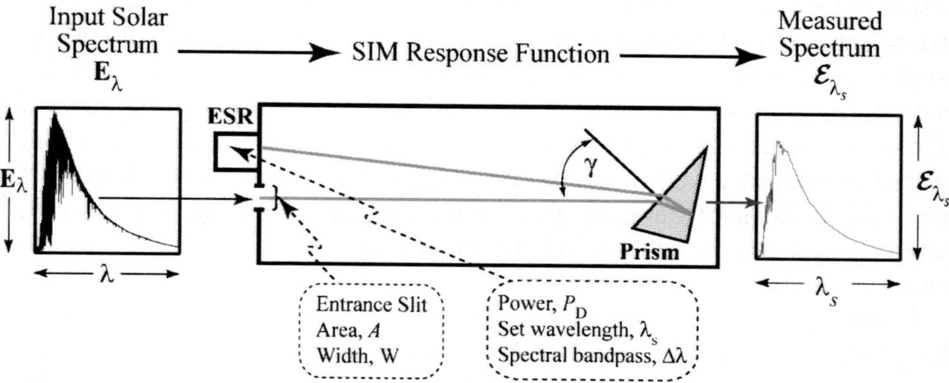

Figure 1. The figure schematically shows the spectrometer's response function and the detection of radiant power by a detector (ESR) in the instrument's focal plane.

of the instrument bandpass by the process of slit function convolution, calibration of the slit area, and the determination of radiant power by the detector:

$$\mathcal{E}_{\lambda}(\lambda_s) = \frac{P_D(\lambda_s)}{A\Delta\lambda} \quad (\mathrm{Wm^{-2}\,nm^{-1}}). \tag{1}$$

This simplified picture neglects numerous important corrections, like the orbital parameters solar distance and the Doppler effect, and wavelength-dependent corrections like prism aberrations and transmission, detector efficiency and temperature effects, diffraction, and time-dependent degradation processes. The term $A\Delta\lambda$ is, in reality, an integral over these wavelength-dependent contributions.

The block diagram in Figure 2 shows the unit level calibrations needed to measure spectral irradiance and their associated measurement equations. The calibration parameter table lists the methods used to derive their value and marks (*) the calibrations that require in-flight modification. In addition, the rounded rectangles show where these in-flight corrections are inserted into the measurement process.

The wavelength calculation equations convert prism encoder positions into wavelength information. This set of equations gives the relationships between target wavelength (λ_s), the charge coupled device (CCD) encoder reading (C), and the spectral focal plane coordinate, (y_s). Section 2 describes the dispersion geometry, and the detailed transformations between these variables are presented in Appendix A. In Section 3, a number of instrument characteristics, calibrations, and in-flight corrections are combined to give the spectral instrument profile, $\mathbf{S}(y_s)$, describing the wavelength-dependent radiometric response of the instrument. In addition, the instrument function, $S'(\lambda_s, \lambda)$, described in Section 3.1 gives the function needed to convolve other higher resolution data to the resolution of SIM. This is important for comparing other tabulated or modeled solar spectra against the measured SIM irradiance, it is also used for interpreting laboratory spectra from atomic lamps and other wavelength standards such as Schott Glass BG20 filters. Section 4 describes the methods used to determine the radiant power detected by the SIM focal plane detectors. The ESR is the absolute detector for this instrument, and phase sensitive power detection is used to minimize the effects of $1/f$ noise inherent in thermal detectors. Four photodiode detectors complement the ESR to produce low-noise, fast response spectral scans. This operational mode produces the most useful information about the time series of solar spectral variability (see Rottman *et al.*, 2005). However, the ESR continually recalibrates the in-flight photodiode detectors (Section 4.2).

2. Prism Dispersion

The set of equations needed to define the focal plane coordinate system starts with the dispersion geometry of a prism in Littrow configuration and is derived from

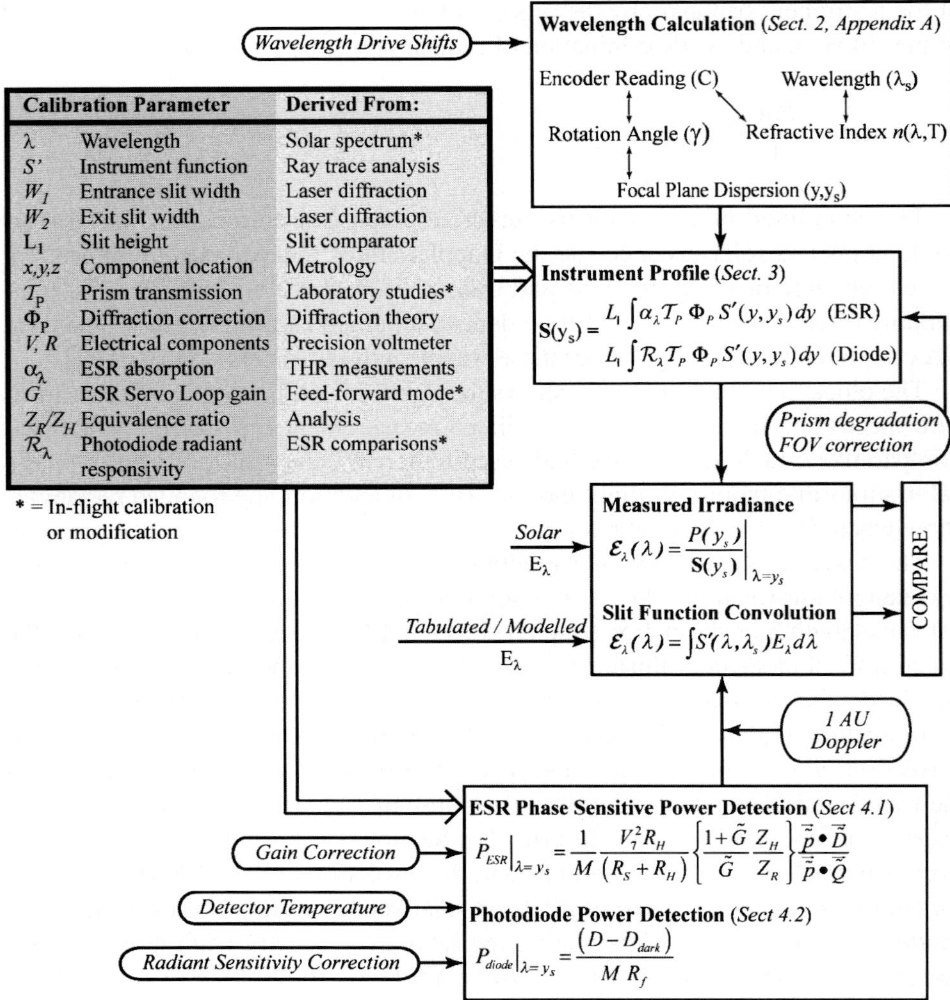

Figure 2. The SIM measurement equations used to calculate solar spectral irradiance in block diagram form. Processes shown in *square blocks* are equations and actions specific to the SIM instrument and the sections discussing these equations are noted along with the title of the block. Calibration quantities needed for these equations are identified and listed in the *gray box*. In-flight correction factors are shown as *rounded rectangles* and where they are inserted in the measurement process.

Snell's law for a plane surface prism:

$$2\theta_{\mathrm{P}} = \sin^{-1}\left(\frac{\sin(\gamma)}{n}\right) + \sin^{-1}\left(\frac{\sin(\gamma - \phi)}{n}\right). \tag{2}$$

The dispersion geometry and the definitions of the symbols in Equation (2) are shown in Figure 3. Equation (2) and Figure 3 in this paper are the same as Equation

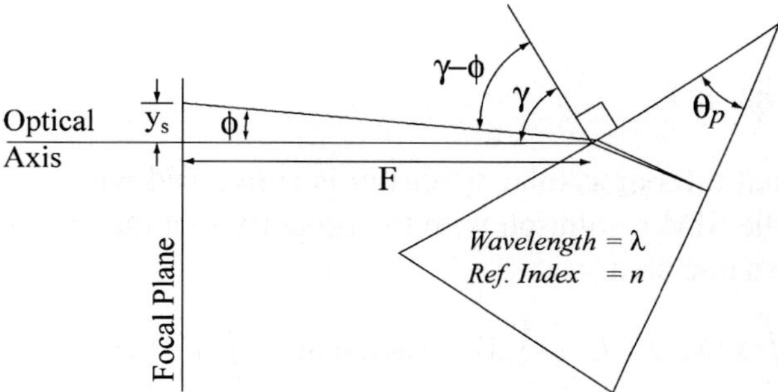

Figure 3. The dispersion geometry of the Fèry prism and the definitions of the variables described in Equation (2) and Appendix A.

(3) and Figure 8 of Harder *et al.* (2005). In this first paper, they are used to describe the dispersion properties of the instrument, and here they are the starting point for developing the analytical equations for the wavelength calculation.

In Figure 3, γ is the prism rotation angle derived from the CCD encoder position C; the refraction angle, ϕ, or its equivalent focal plane coordinate y is found from the prism rotation angle and wavelength of incident light; the index of refraction n is a unique surrogate for wavelength, λ, and is derived from the rotation angle and geometry of the prism. The analytical equations associated with these transformations are presented in four separate sections in Appendix A, and these analytical results are verified through ray tracing. While there are several equations to relate the wavelength scale to the prism angles and the focal plane CCD encoder system, it is important to note that the SIM wavelength scale is a deterministic process. Slight offsets and stretches during flight are corrected against a SIM-measured reference spectrum that is then used to adjust all other spectra. This single reference spectrum is then calibrated by comparison with a higher resolution solar spectrum convolved with the SIM instrument function. At the present time, the Thuillier *et al.* (2003) Composite 1 spectrum is used for this process. It should be noted that since 21 April 2004, no measured spectra have been shifted relative to this reference. The shift-stretch algorithm is used to correct data prior to that date, which have been affected by an operational problem related to the action of the prism drive system; see Section 5 for further discussion.

3. Instrument Function Convolution

The theoretical SIM instrument function $s(\lambda_s, \lambda)$ can be found by convolving identical entrance and exit slit rectangular functions of width W in the focal plane coordinate system to yield a triangular function with an area of W. Division by the focal plane dispersion $(\partial y/\partial n)(\partial n/\partial \lambda)$ converts it into the equivalent wavelength coordinates. The instrument function acts as a convolution kernel with unit area:

$$\int s(\lambda_s, \lambda)\, d\lambda = 1. \tag{3}$$

When an external reference solar spectrum is convolved with $s(\lambda_s, \lambda)$ a spectrum is produced at the SIM resolution with the property that the integrals of $\mathbf{E}_\lambda(\lambda)$ and $\mathcal{E}_\lambda(\lambda_s)$ have the same area:

$$\mathcal{E}_\lambda(\lambda_s) = \int s(\lambda_s, \lambda)\, E_\lambda(\lambda)\, d\lambda \quad \text{such that} \quad \int E_\lambda(\lambda)\, d\lambda = \int \mathcal{E}_\lambda(\lambda_s)\, d\lambda_s. \tag{4}$$

In practice, the function $s(\lambda_s, \lambda)$ presented in Equation (3) is inadequate to describe the actual instrument function. Additional wavelength-dependent corrections must be included within the integral that influence the overall shape of the spectral response function. The response function, represented by $\mathbf{S}(y_s)$ in Figure 2, includes corrections for spectrometer aberrations (S'), prism transmission ($\mathcal{T}_P$), diffraction loss (Φ_P), and ESR and photodiode responsivities R_λ and α_λ. All of these correction factors are smooth functions of wavelength with the exception of α_λ for the photodiode detectors that vary rapidly in the vicinity of the long wavelength cut-off, and the silicon photodiodes have additional structure in the UV part of the spectrum. The correction factors are described in the next four sections except for α_λ, which is discussed in the context of the instrument detectors in Section 4.2.

3.1. $S'(\lambda_s, \lambda)$: Instrument Function Convolution from Ray Tracing

A ray trace model based on the measured geometry of the prism and locations of the exit slits relative to the entrance slit is used to generate the instrument function, $S'(\lambda_s, \lambda)$, to account for optical aberrations and vignetting. Section 2.3.2 of Harder *et al.* (2005) describes the instrument's optical properties in detail but can be summarized here: (1) the predominant aberrations are image magnification and coma, (2) imperfect prism focusing causes some light loss at the exit slit, and (3) prism glass dispersion is a non-linear function of wavelength so the ideal trapezoidal instrument function is slightly asymmetric in wavelength space. All of these optical processes are modeled by ray tracing the spectrometer using both ZEMAX (Focus Software Inc., Bellevue, WA) and IRT (Parsec Technologies, Boulder, CO) software packages. For each of the five SIM detectors, the ray trace computes instrument functions on an index of refraction grid, giving roughly constant spacing in dispersion. The optimizer in the ray trace software determines the rotation angle for each wavelength (λ_s) so that the chief ray hits the center point of the detector exit slit. The rotation angle is then fixed and separate ray traces are performed for ± 100 wavelengths incrementally offset from λ_s. A total of 40 000 randomly distributed rays from an object forming a 0.5° beam pass through the entrance slit and are propagated through the optical system to the focal plane where a mask

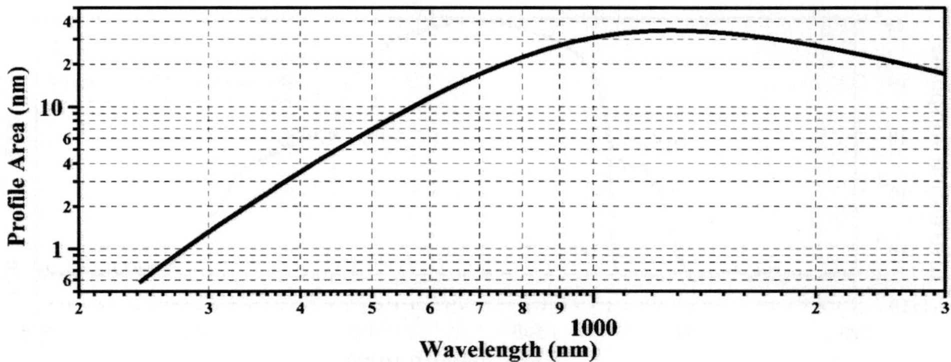

Figure 4. The profile integral over its usable wavelength band as a function of wavelength. The graph is shown on a log–log scale for the ESR detector.

with the dimensions of the exit slit is used to count the number of rays that pass it. As the wavelength shifts away from λ_s by $\delta\lambda$, a smaller fraction of rays pass through the mask. For each λ_s and $\delta\lambda$, the ratio of the number of rays collected at the exit slit to those passing through the entrance slit provides an estimate of the spectrograph's efficiency, $S'(\lambda_s, \lambda)$. One hundred separate instrument functions are generated over each detector's usable wavelength range. The profile integral (or equivalent bandwidth) is representative of the spectral bandpass, and is shown in Figure 4 versus wavelength for the ESR detector over its usable wavelength range.

This process generates the proper instrument function for processing spectral irradiances. The reciprocal process that fixes the wavelength and rotates the prism gives a nearly identical result that can be compared to an experiment in which an intense laser line is scanned so the character of the instrument can be seen in the far wings of the instrument function. Figure 5 compares the ray traced SIM slit profile with the scan of a 543.5 nm HeNe laser as measured by the Vis1 photodiode. Other scans of this kind were performed at other wavelengths using discrete laser lines and with a mercury electroless discharge lamp (EDL) and for each detector, but the data shown in Figure 5 displays the best SNR attained for this kind of measurement. The measured out-of-band stray light contribution is about 50 parts per million of the main signal, so does not contribute to the shape of the profile shown in Figure 5. The measured and ray traced profiles qualitatively agree in the core of the trapezoidal function and deviations occur at the 0.5 – 0.01% level. These differences arise from a combination of diffraction and edge scatter generated by the exit slits and baffles. The ratio of the areas of the laboratory calibration to the ray trace profile in this case is 1.0045, indicating that ray traced instrument function adequately describes the instrument function with the exception of small corrections needed for diffraction. Saunders and Shumaker (1986) used a prism and a grating in a double spectrometer configuration to perform a similar experiment, and their findings are in accord with the results for SIM. However, a number of improvements in the comparison must be made before this kind of experiment is deemed a calibration: in particular,

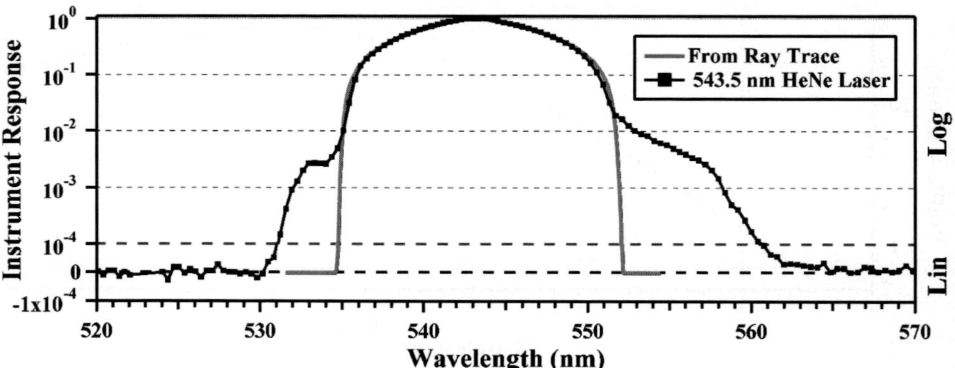

Figure 5. A comparison of instrument function ray tracing with the measurement of a 543.5 nm HeNe laser for the Vis1 photodiode. The graph is a combined log–linear plot with 4 orders of magnitude of response on a log scale and a linear scale below 10^{-4} to show the noise level of the measurement and the zero points of the ray traced trapezoid. The instrument responds to additional sources of light scattering and diffraction in the far wings of the trapezoidal instrument function.

the optical performance of the Fèry prism is strongly f/number-dependent so the input laser beam must fill the spectrometer in the same manner as the Sun (i.e., 0.5° beam) and the emergent flux from the light source must be known so light losses are properly accounted for in the calibration. With the development of the NIST SIRCUS facility (National Institute of Science and Technology, Spectral Irradiance and Radiance Calibrations with Uniform Sources; Brown, Eppaldauer, and Lykke, 2000) calibrations meeting these requirements will be possible for future missions.

3.2. *A*, *W*, *L*: Measurement of Slit Dimensions

An accurate measurement of the entrance slit area is needed to establish the radiant flux into the instrument, and the width of the entrance slit is needed for the determination of the instrument's bandpass and diffraction correction (see Section 3.4). The requirement was set so that its accuracy was commensurate with the accuracy of ESR power measurements, the ability to measure the instrument profile, and the magnitude of the diffraction correction that varies by an order of magnitude over the wavelength span of the instrument. The level of accuracy required for the slit dimension measurements is on the order of 100–500 ppm):

$$\left(\frac{\delta W_1}{W_1}\right)^2 + \left(\frac{\delta L_1}{L_1}\right)^2 = \left(\frac{\delta A_1}{A_1}\right)^2 \approx (5 \times 10^{-4})^2. \tag{5}$$

Thus, the 0.3 mm slit width must be determined to about $\pm 0.1\ \mu$m accuracy, and the 7 mm tall slit to about $\pm 2.5\ \mu$m. This same specification applies to the exit slits as well.

The slits used for SIM are copper/nickel bimetal etched slits (manufactured by Buckbee-Mears, Inc., Minneapolis, MN) that have a width tolerance of $\pm 8\ \mu$m, and a parallelism of $\pm 4\ \mu$m over the slit's 7 mm length. The bimetal etching process creates an edge with only a few microns in thickness thereby minimizing additional light scattering. Analysis of the slits by scanning electron microscopy indicates there is about 0.3 μm root mean square (RMS) roughness along the slit edge and corner-rounding is about 3 μm in radius. Because of this, slit calibrations are performed to measure the area and the effective width over its entire length. The slit length is then inferred from these two measurements. The width is measured by laser diffraction, and area is measured by comparing the light flux through the slit relative to the light flux through a known, calibrated aperture. These two methods are described in Appendix B.1 and B.2, respectively.

The practice of high-accuracy measurements of long, narrow spectrometer slits is not as advanced as the circular aperture area measurements used for TSI studies (Fowler, Saunders, and Parr, 2000), so the absolute calibration of the two methods described in Appendix B rely on known standard widths and areas. The slit width standard (Photo Sciences Inc., Torrance, CA) is a chrome-on-glass slit measured with a Nikon 2i metrological microscope with a quoted uncertainty of $\pm 0.5\ \mu$m traceable to NIST standards, and the precision of the diffraction calibration method is $\pm 0.03\ \mu$m based on multiple measurements of the slits. The flux comparator system for slit area measurements is based on a standard, 0.5 cm^2, NIST calibrated circular aperture with an area known to $\pm 3 \times 10^{-5}$ mm^2. The apparatus used to measure the area has a precision of $\pm 2 \times 10^{-5}$ mm^2. The uncertainties in the area and width measurements are comparable to the requirement limits of Equation (5), but the standard apertures are retained as a ground witness for recalibration so future improvements in the measurement methods will translate into a refined value for the flight slits.

Since the area and width of the instrument's entrance slit is a function of temperature, appropriate thermal corrections are applied to the flight data. Since the slits are fabricated from a bimetal material, the temperature coefficient of expansion has to be calculated for the bimetal combination (Gere and Timoshenko, 1990) and has a numerical value of 1.58×10^{-5} K^{-1}. The temperature of the slit is monitored in-flight with a thermistor bonded to the nearby UV photodiode.

3.3. $\mathcal{T}_P$: PRISM TRANSMISSION AND DEGRADATION

The calibration of prism transmission and monitoring the degradation of this transmission represent two of the most important activities for the SIM calibration. Prism transmission measurements are discussed in Section 3.3.1 and Appendix C, and the prism degradation model is explained in Section 3.3.2. The measurement of prism transmission is a very difficult and time consuming process, so the actual calibration was performed on two ground witness prisms rather than the actual flight units.

However, these ground witnesses were made from the same boule of Suprasil 300, and were manufactured and aluminized simultaneously with the flight prisms.

3.3.1. *Prism Transmission Measurements*

The transmission of the SIM prism results from a number of sources that are a function of incidence angle and wavelength, and the transmission must be calculated for each of the instrument's detectors since the geometry is different for each of them. Fresnel reflections on the vacuum–glass and glass–vacuum interfaces of the prism cause a loss in transmission that is a function of incidence angle and the index of refraction of the glass for a given wavelength; the intensity of the reflection is also a function of the incoming light beam polarization. Furthermore, this effect is enhanced because the prism rotation angle, γ, is $59^\circ \pm 2.5^\circ$, which is near the Brewster angle for fused silica. Light intensity losses in the bulk of the Si_xO_y glass matrix are significant only in the ultraviolet (UV) for $\lambda < 300$ nm, and in the infrared (IR) for $\lambda > 2700$ nm. Suprasil 300 fused silica glass (Hereaus Amersil Inc., Duluth, GA) is a 'dry glass' with a very low OH content ([OH] < 1 ppm), so the broad and deep hydroxyl absorption features are suppressed in the transmission spectrum (Humbach *et al.*, 1996). This glass is made by a chemical vapor deposit process, so the trace metals that give rise to color centers are present only at the part per billion level, and cannot contribute to the bulk absorption over the effective 24 mm path through the prism. The reflectivity of the aluminized rear-surface of the prism is a function of wavelength, particularly in the 700–900 nm region. Inspection of Figure 3 shows that the light path through the prism is nearly normal to the back surface regardless of rotation angle and wavelength, so there is no polarization effect in the aluminum reflectivity. Laboratory measurements of prism transmission suggest that the reflectivity of the second-surface aluminum mirror on the prism is different from the reflectivity of bare aluminum. Therefore, the combined absorption of the bulk glass and reflectivity of the surface glass–aluminum layer is explicitly measured.

Figure 6 shows the effective mirror reflectivity as measured by the method described in Appendix C. This figure also shows the Fresnel horizontal and vertical two-reflection contributions for the geometry corresponding to the ESR detector; the Fresnel contributions are different for the other detector locations in SIM but can be computed from Equations (C.2) and (C.3) presented in Appendix C. The prism transmission for unpolarized light is the product of an angular-dependent portion arising from the average of the two Fresnel reflections and the measured mirror reflectivity that includes bulk losses in the UV and IR. This formulation has the advantage that the transmission can be computed for all detector positions and all prism rotation angles. The attenuation for wavelengths longer than 2600 nm shown in Figure 6 is caused by absorption in the bulk of fused silica (Humbach *et al.*, 1996). The estimated photometric error from the transmission measurement in the 200–1000 nm is about 0.1%, but in the 1000–2900 nm region the error increases to about 1%.

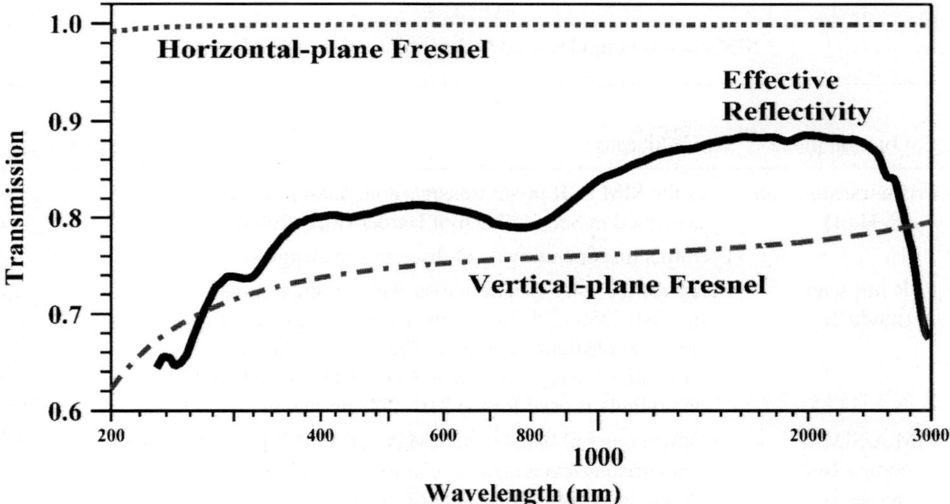

Figure 6. The measured prism transmission. The figure shows the second-surface effective reflectivity of the prism and the two-reflection Fresnel contributions in each polarization. The calibration was done for the geometry of the ESR detector. Transmission for unpolarized light is the product of the reflectivity and the average of the two Fresnel components.

3.3.2. *Prism Degradation Measurement and Model*

Exposure to the space environment causes irreversible changes to the transmissive properties of the prism that must be tracked both as a function of wavelength and time. Complete understanding of the degration characteristics is an ongoing task during the instrument lifetime. This section discusses the in-flight methods that were designed and operate to measure the degradation and the current model applied to determine and correct the degradation observed so far.

Table I summarizes the operation modes used to determine the prism degradation properties, the action of the modes, and the number of days between calibrations. The two primary modes for this purpose are the direct prism transmission calibration mode and the ESR full scan. The SIM A/SIM B comparison is not currently used in the calculation of the prism degradation, but is used as an end-to-end check to ensure that the degradation correction factors are consistent.

Without loss of generality, the prism degradation can be expressed by the equation

$$T(t, \lambda) = T_0(\lambda)\,\mathrm{e}^{-\tau(t,\lambda)}, \tag{6}$$

where τ represents this degradation in a logarithmic scale and is defined as 0 at the beginning of the mission, and T_0 is the un-degraded prism transmission. By using the in-flight measurements, the prism degradation observed until the present can be described by a model that corresponds to the following equation:

$$\tau(t, \lambda) = \kappa(\lambda)C(t). \tag{7}$$

TABLE I

SIM operation modes used to determine prism degradation.

Calibration mode	Action/purpose	Cadence (number/days)
Prism transmission (mode 1)	Uses the SIM ESR prism transmission measurement mode described in Section 2.3.4 of Harder *et al.* (2005). Performs this operation at 44 discrete wavelengths.	1/7
ESR full scan (mode 2)	Measures the solar spectrum over the operating range of the ESR (256–2700 nm) with a sampling of 3 prism steps per resolution element. The range for prism calibrations using the ESR is 300–1100 nm, and the UV photodiode is used for the 210–300 nm region.	1/90
SIM A/SIM B comparison (mode 3)	Simultaneous full scans with SIM A and SIM B provide comparisons and an end-to-end measure of the effectiveness of the degradation corrections.	1/30

With the wavelength and temporal variations accounted for by two separate functions, the absorption coefficient $\kappa(\lambda)$ and the column density $C(t)$, respectively. The function κ is obtained by comparing ESR full scans (mode 2 in Table I), at widely separated times. To date, all of the ESR full scans have been used to check the validity of Equation (7), and this relationship is applicable to within the limitations of observation noise and solar variability. The value of κ will be improved by continued analysis of ESR full scans and the SIM A/B comparisons (mode 3 of Table I). Figure 7a shows this absorption coefficient as a function of wavelength, and shows that absorption is greatest in the near UV, and drops to values indistinguishable from noise by 700 nm thereby indicating that no prism degradation is observed at these longer wavelengths.

The column value shown in Figure 7b is found from the in-flight prism transmission measurement performed on a weekly cadence. Since SIM B receives only 18% of the solar exposure of SIM A, this experiment is done symmetrically so the effects of exposure time on degradation can be assessed. These activities are performed with the hard radiation traps (HRT, see Harder *et al.*, 2005 for details) inserted in the light beam to minimize unnecessary exposure to the prisms. The transmission calibration system yields a relative change in transmission rather than an absolute value. Additional reflections in the calibrator's relay optics (prism, periscope, folding mirror, focusing beamsplitter) alter the polarization state and modify the wavelength dependence of the input light beam so the prism transmission measured through this system has additional contributions not present in the transmission function measured by the ground calibration system and described in Section 3.3.1. However, the modifications induced by the relay optics are believed to be time invariant since light flux on these elements is very small and these optical elements are common to both the I and I_0 modes.

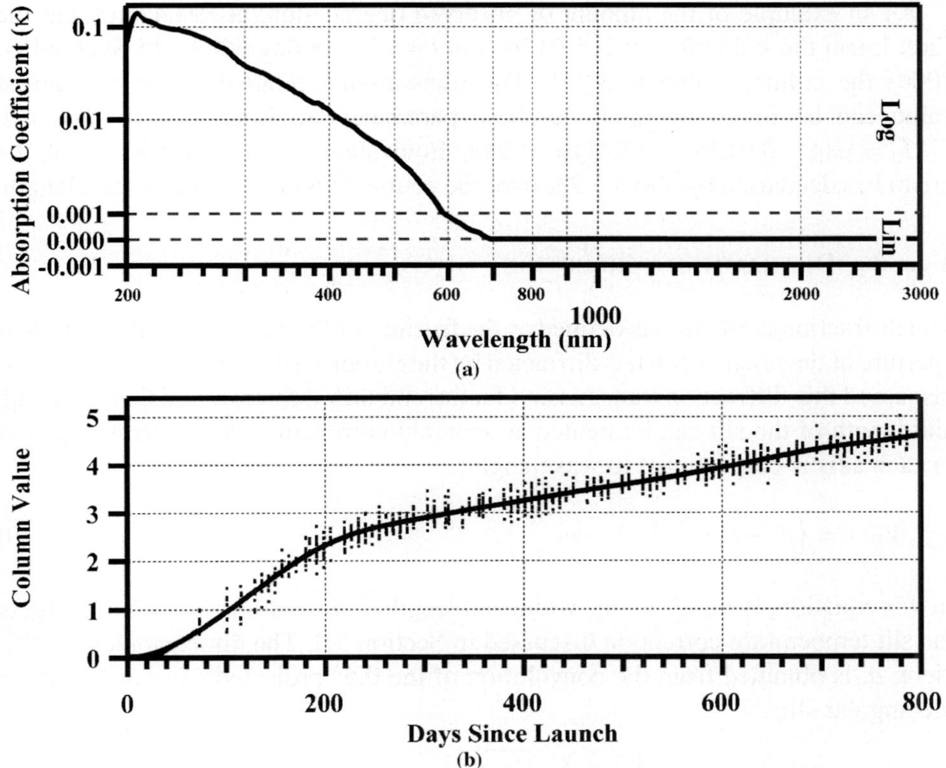

Figure 7. Panel (a) shows the time-independent absorption coefficient derived from ratios of ESR full scans and UV diode data below 300 nm. This figure is plotted as a log scale for κ value greater than 0.001, and as a linear scale below that value to indicate the function's decent to a zero value. Panel (b) shows the growth of the column layer over the course of the mission. The data is derived from prism transmission calibration experiments; the line in this panel shows the b-spline model fit to the individual data points at different wavelengths (shown as symbols). The value of prism degradation at any given time is found by applying these data to Equation (6).

The in-flight transmission measurement (mode 1 of Table I) corresponds to Equation (6), but with a different initial transmission: $\mathcal{T}_0' = \mathcal{T}_{0\,\text{prism}}' \times \mathcal{T}_{0\,\text{calibrator}}'$. The first step in the process is to iteratively find the best values of $\mathcal{T}_0'(\lambda)$ and $C(t)$ that simultaneously minimizes the differences for all wavelengths used in mode 1 of Table I:

$$C(t) = \frac{-1}{\kappa(\lambda)} \ln\left(\frac{(\mathcal{T}'(t,\lambda))}{\mathcal{T}_0'(\lambda)}\right). \tag{8}$$

Values for $C(t)$ are then found for all value of t and λ. The $C(t)$ used for the data processing is obtained from a b-spline fitting to all of these values versus time (Lawson and Hanson, 1974); it is this b-spline fit that appears in Figure 7b.

As an example of the amount of observed degradation, at 393.4 nm (near the Ca II lines) the κ coefficient is 0.0136, and on mission day 600.0 (15 September, 2004) the column value is 3.941. The transmission relative to the unexposed value can be found by applying these parameters to Equations (6) and (7): $\mathcal{T}/\mathcal{T}_0 = \exp(-0.0136 \times 3.941) = 0.948$. Equivalently, the transmission of the prism has decreased by about 5.2% over the course of 600 days at this wavelength.

3.4. Φ_P: Diffraction Correction

The diffraction correction is defined as the fraction of light that lies within the clear aperture of the prism when it is diffracted by the entrance slit. Lawrence *et al.* (1998) discussed this diffraction transmission factor, and they demonstrated that the width and length of the slit can be treated as separable problems and the fractional loss in each axis is proportional to wavelength:

$$\Phi(\lambda) = \left(1 - a_W \frac{\lambda}{W}\right)\left(1 - a_L \frac{\lambda}{L}\right). \tag{9}$$

In this equation, W and L are the width and length of the entrance slit after applying the slit temperature correction discussed in Section 3.2. The dimensionless coefficient, a, is obtained from the convolution of the 0.5° projected solar disk and the rectangular slit:

$$a_W = \frac{-\Delta W}{\lambda} = \frac{1}{\pi^2 \theta_P}\left\{\frac{2 - 2\sqrt{1-\varepsilon^2}}{\varepsilon^2}\right\}. \tag{10}$$

In this equation, θ_P is the half angle subtended by the prism and ε is the ratio of the solar angular radius to the prism half angle. An analogous equation can be written for the slit height. For the SIM geometry, these factors are tabulated in Table II.

TABLE II

Slit diffraction parameters, a_W, a_L.

	Slit width	Slit length
Prism half angle (°), θ_P	0.03124	0.02250
Solar radius (°)	0.00436	0.00436
ε	0.139661	0.193944
a, $\Delta W/\lambda$	3.2593	4.5471
Example wavelengths (nm)	$\Phi(\lambda)$	
250	0.99715	
500	0.99431	
1000	0.98862	
2500	0.97158	

4. SIM Detector Characteristics and Calibrations

4.1. ESR Operation and Calibration

The electrical, mechanical, thermal, and optical properties of the ESR detector are discussed in Section 2.3.1 of Harder *et al.* (2005). This section describes the characterization of the detector and the terms that relate to the phase sensitive detection at the shutter fundamental. Additionally, the ESR absorption factor, α_λ, introduced in Section 3 is presented here.

In the following analysis, a tilde ($\sim$) represents complex numbers and an arrow ($\rightarrow$) denotes a times series of numbers, corresponding to each data point. The detector measurement equation can then be represented by the equation

$$\tilde{P}_{\mathrm{ESR}} = \frac{1}{M}\frac{V_7^2 R_{\mathrm{H}}}{(R_{\mathrm{S}}+R_{\mathrm{H}})^2}\left\{\frac{1+\tilde{G}}{\tilde{G}}\frac{\tilde{Z}_{\mathrm{H}}}{\tilde{Z}_{\mathrm{R}}}\right\}\frac{\vec{\tilde{p}}\cdot\vec{D}}{\vec{\tilde{p}}\cdot\vec{Q}}, \tag{11}$$

where $\tilde{P}_{\mathrm{ESR}}$: detected power; M: scaling factor for the data output: 64 000 is the data number for 100% duty cycle of the pulse width modulator; V_7: value of the 7.1 V reference; $\frac{R_{\mathrm{H}}}{(R_{\mathrm{S}}+R_{\mathrm{H}})^2}$: voltage divider ratio of the series heater resistors; $\frac{1+\tilde{G}}{\tilde{G}}$: closed-loop gain from an open-loop servo gain of $\tilde{G}$; $\frac{\tilde{Z}_{\mathrm{H}}}{\tilde{Z}_{\mathrm{R}}}$: equivalence ratio; $\frac{\vec{\tilde{p}}\cdot\vec{D}}{\vec{\tilde{p}}\cdot\vec{Q}}$: projection of the data onto shutter waveform (see Section 4.1.2); $\vec{D}\equiv D_j$: time series of data numbers from the DSP; $\vec{Q}$: shutter transmission square wave, 0 or 1; $\vec{\tilde{p}}\equiv\exp(i\,2\pi f_1 t_J)\,f_1$: shutter fundamental frequency; t_J: time of each data point.

The DSP data numbers $\vec{D}$ (Harder *et al.*, 2005), are produced at a rate of 100 s^{-1} and can be decimated by factors of 1, 2, 5, 10, or 20 for telemetry. Typically, the data stream is decimated by a factor 10. The ESR data numbers are a linear function of the detected power, and conceptually the ESR power (P_{ESR}) can be written:

$$P_{\mathrm{ESR}} = A_{\mathrm{P}}\times D, \quad \text{where} \quad A_{\mathrm{P}} = V_7^2\frac{R_{\mathrm{H}}}{(R_{\mathrm{H}}+R_{\mathrm{S}})^2}\frac{1}{M}. \tag{12}$$

A light chopper then modulates P_{ESR} (Equation (12)) and converts it into the AC waveform $\tilde{P}_{\mathrm{ESR}}$.

The ratio of thermal impedances to the ESR thermistor, for radiation input and heat input, $\tilde{Z}_{\mathrm{R}}/\tilde{Z}_{\mathrm{H}}$, gives the equivalence between replacement heater power and radiant power. This equivalence ratio is determined from a model of the heat flow on the SIM bolometer as a function of frequency. The details of this model are beyond the scope of this paper, but because of the high thermal conductivity of diamond, the in-phase component of equivalence is within 10 ppm of unity. The out-of-phase component is near 3000 ppm due to mismatched delays, but this is not relevant to the determination of absorbed power.

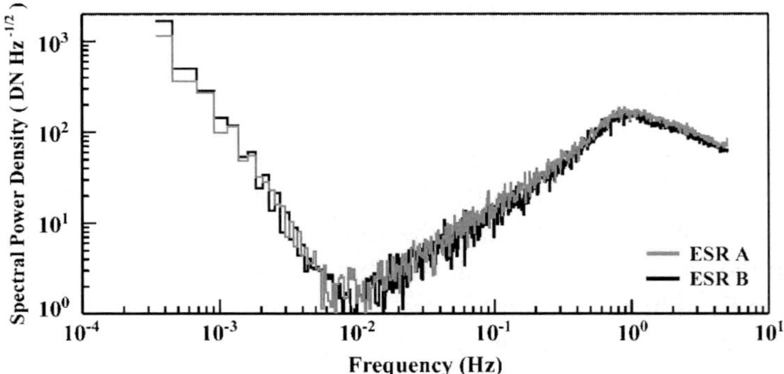

Figure 8. The noise spectral power density as a function of frequency of the two SIM ESR detectors while in flight. The minimum in the noise corresponds to the point where 1/*f*-type noise approximately equals the excess noise on the bolometer thermistors.

4.1.1. *Detector Performance and Servo Gain Recalibration*

The SIM ESR is auto-balanced by a servo-loop and the components of the loop were described in Harder *et al.* (2005). The performance of the ESR can be determined by analysis of the detector's noise spectrum. The spectrum is obtained by operating the detector in the dark for long time periods compared to shutter period, and using Fourier analysis to characterize the noise spectrum. Figure 8 shows the noise spectral power density in terms of data numbers (DN) as a function of frequency and was measured with the instrument on-orbit. The detectors show a very characteristic pattern with 1/*f* noise dominating at the lowest frequencies and with a rising excess thermistor noise power to a frequency of about 1 Hz where the servo-loop gain drops to a value near 1.0. The minimum in the noise power density at 0.01 Hz corresponds to the location where the contributions of these two noise sources cross and become comparable in magnitude. This cross over point determines the optimal shutter frequency to operate the instrument. At 0.01 Hz, the noise power is $\sim$2 DN/$\sqrt{\text{Hz}}$ and with a 200 s integration period, the noise on the measurement is $\sim$ 0.3 nW; this is the condition used for the ESR table measurements. For the ESR full scans, the instrument is operated with a 0.05 Hz shutter frequency, where the noise is a factor of 5 higher, for a 40 s dwell time per prism step giving a noise floor of $\sim$2 nW. The conditions used for this scan represent a compromise between low noise and the length of time required to complete the measurement.

The closed-loop gain is a first-order term in the ESR measurement equation (Equation (11)) and its value must be monitored throughout the flight to assure that changes in electronic component values do not change (degrade) over the course of the mission. The open-loop gain of the system can be determined in-flight by injecting a digital square wave at the shutter frequency into the servo-loop before the pulse width DAC and then measuring the system's response to this perturbation; this square wave is referred to as a feed-forward signal. The action of this feed-forward signal can be written as a control loop equation where G is the unperturbed gain, FF is the magnitude feed-forward signal, and 'out' is the output of the control loop. This equation can then be solved for the open-loop gain of the system:

$$\text{out} = \text{FF} - G \times \text{out} \Rightarrow G = \frac{\text{FF}}{\text{out}} - 1. \tag{13}$$

Figure 9 is an example of data acquisition and the gain calculation for an in-flight calibration in February of 2005. Figure 9a shows data for a 20 s feed-forward period: a 40 min time series of ESR data is collected in the feed-forward mode and about 100 cycles are co-added to reduce random noise, and the mean value is subtracted. The detector response (the variable 'out' in Equation (13)) is shown as a gray trace, and the driving feed-forward waveform (FF) is the dotted black trace. The feed-forward waveform consists of adding in a digital value of 8000 during the first half of the period, and subtracting 8000 during the second half. In this way, the system must respond to an instantaneous change of 16 000 DN at $t = 0$ and $t = 10$ s and then settle to its balanced value. If the system was perfect, the

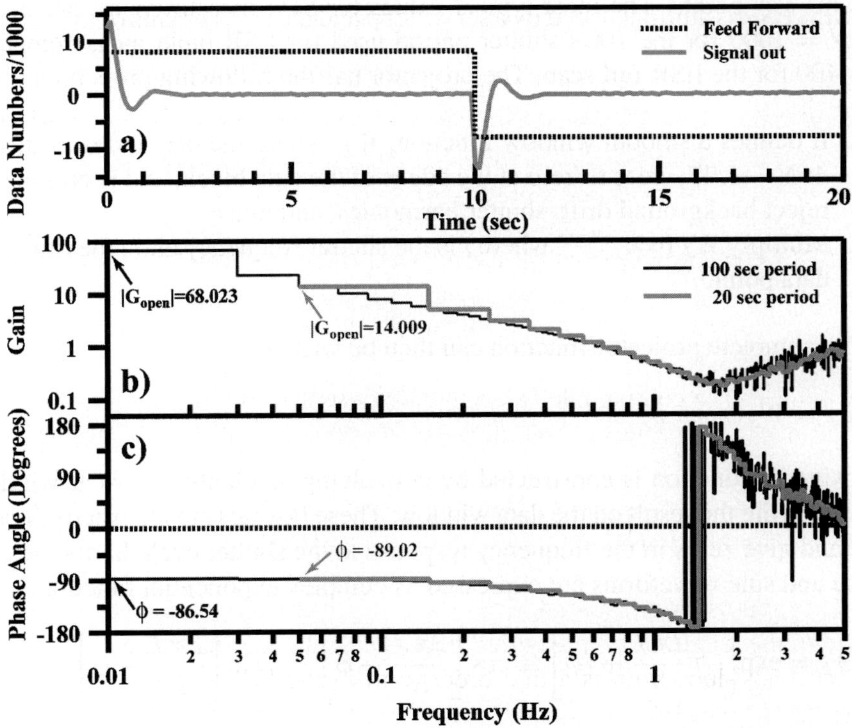

Figure 9. Gain measurements for the ESR. Panel (a) shows a time series of the SIM ESR (*gray trace*) and the feed-forward pulse (*dotted black trace*) that is driving the detector's response; the graph is for a 20 s period on the feed-forward pulse. Panels (b) and (c) show the measured gain and phase when the time series data of panel (a) is processed by Equation (13). These graphs show the results for both 100 (*black*) and 20 (*gray*) second feed-forward periods. The open-loop gain and phase at the shutter fundamental are marked on the graphs.

peak values at $t = 0$ and 10 s would be $\pm 16\,000$. The gain is found by separately performing the Fourier transforms of 'FF' and 'out' and applying Equation (13) at each frequency. The open-loop gain, G_{open}, is the modulus of G, and the phase (ϕ) is the argument and the frequency dependence of these terms are shown in Figure 9b and c, respectively. The open-loop gain attains its maximum value at the fundamental, and decreases to its minimum value at about 1.5 Hz where the phase lags by 180°. Because the incoming light from the SIM spectrometer is chopped by a shutter, only G_{open} and ϕ at the fundamental frequency are needed to calculate the closed-loop gain for the measurement equation seen in Equation (12). These values are noted in the figure for both the 100 and 20 s shutter periods. The gain and the phase have been tracked throughout the first 2 years of the SORCE mission, and they are constant and without a discernable trend to 0.1% throughout this period.

4.1.2. *Projection Operator for Phase Sensitive Detection*

The projection operator, $\vec{\tilde{p}}$, presented in Equation (11) is a discrete Fourier filter that operates on M shutter cycles and N data points per cycle; the M-cycle filter contains MN points. The most typical values for SIM measurements are $M = 2$, and $N = 1000$ for the 100 s shutter period used for ESR table measurements or $N = 400$ for the ESR full scan. The projector has the following properties:

a. It defines a smooth window function, W_J, where the data index J runs 0 to $MN - 1$. W_J goes to zero at the edges of the data block, and is optimized to reject background drift, shutter harmonics, and noise.
b. Multiply W_J by $\mathrm{e}^{-i2\pi ft}$ where f is the shutter frequency and t the time of the data point.

The complete projector function can then be written:

$$\tilde{p}_J = W_J \tilde{\Theta}_J \equiv \vec{\tilde{p}}. \tag{14}$$

The window function is constructed by convolving M identical boxcar windows, then centering the result on the data window. These boxcars are all exactly N points wide and give zeros in the frequency response at the shutter cycle harmonics. The cosine and sine waveforms are expressed as complex exponential function:

$$\tilde{\Theta}_J = \exp\left[-i\frac{2\pi J}{N} + i\phi\right] \equiv \cos\left[\frac{2\pi J}{N} - \phi\right] - i\,\sin\left[\frac{2\pi J}{N} - \phi\right]. \tag{15}$$

The phase angle ϕ is arbitrary if the same phase angle is used in the projection of the shutter waveform. Since ϕ is arbitrary, it is set to zero for simplicity. Figure 10 shows W_J, the shutter waveform, and the real and imaginary parts of $\vec{\tilde{p}}$ for the case of $M = 2$ as calculated by Equations (14) and (15).

The last step is to perform the dot product by multiplying the Jth data point and the projector element and summing all elements in the data block. This same

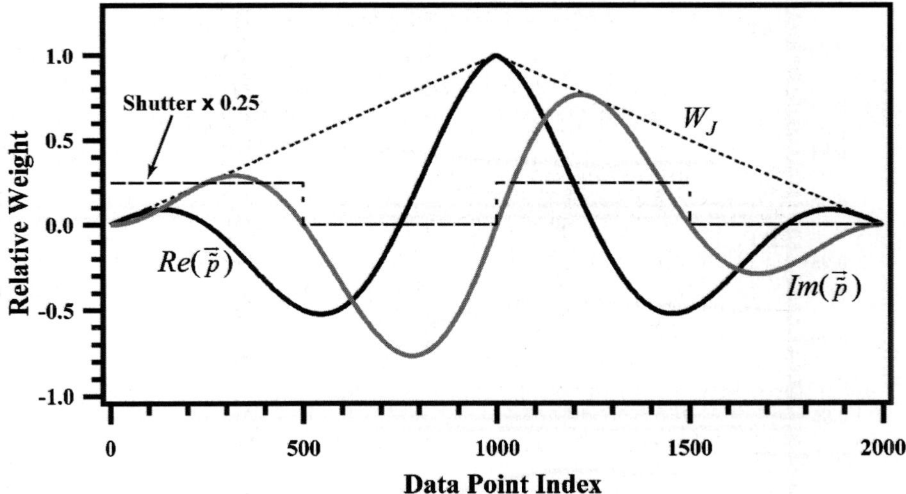

Figure 10. Waveforms of W_J, the shutter, and the real and imaginary parts of $\vec{\tilde{p}}$ for the case of $M = 2$.

process is applied to the idealized shutter wave form, Q_J, and these two quantites are ratioed to ensure proper scaling of the data numbers:

$$\frac{\vec{\tilde{p}} \cdot \vec{D}}{\vec{\tilde{p}} \cdot \vec{Q}} \equiv \frac{\sum_{J=0}^{MN-1} \tilde{p}_J D_{NK+J}}{\sum_{J=0}^{MN-1} \tilde{p}_J Q_{NK+J}}. \tag{16}$$

4.1.3. *Absorptance of Nickel Phosphorus Black*

The absorptance of the bolometer, α, results from the combined effects of the absorptance of the nickel phosphorous (NiP) and the return reflectance of the aluminized hemisphere of the ESR cavity (see Harder *et al.*, 2005, Section 2.3.1, for more discussion on the optical properties of the ESR). The value of α is wavelength-dependent and is found by summing the light absorption through the multiple absorption/reflection light path between the bolometer and its surrounding reflective hemisphere.

Assume that the intensity of light entering the ESR detector is I_o and the NiP surface of the bolometer has an absorptance ρ. On first contact with the bolometer the fraction of light absorbed is ρ. The intensity of light diffusely scattered off of the bolometer, β, is then $I_o\beta$ where $\beta = (1 - \rho)$. This light is then reflected off of the aluminum hemisphere with reflectivity, r, and re-directed to the bolometer with an intensity of βr. This light will again be absorbed, and the process is repeated until the intensity becomes diminishingly small. The overall efficiency of this process can be written as an infinite series and summed because it is a geometric progression:

$$\alpha I_0 = I_0\rho + I_0\beta r\rho + \beta^2 r^2 \rho + \cdots + I_0\beta^n r^n \rho + \cdots \quad \text{with } 0 < \beta, \quad r \leq 1,$$

$$\alpha = \rho(1 + \beta r + (\beta r)^2 + (\beta r)^3 + \cdots + (\beta r)^n + \cdots) = \rho\left(\frac{1}{1 - \beta r}\right). \tag{17}$$

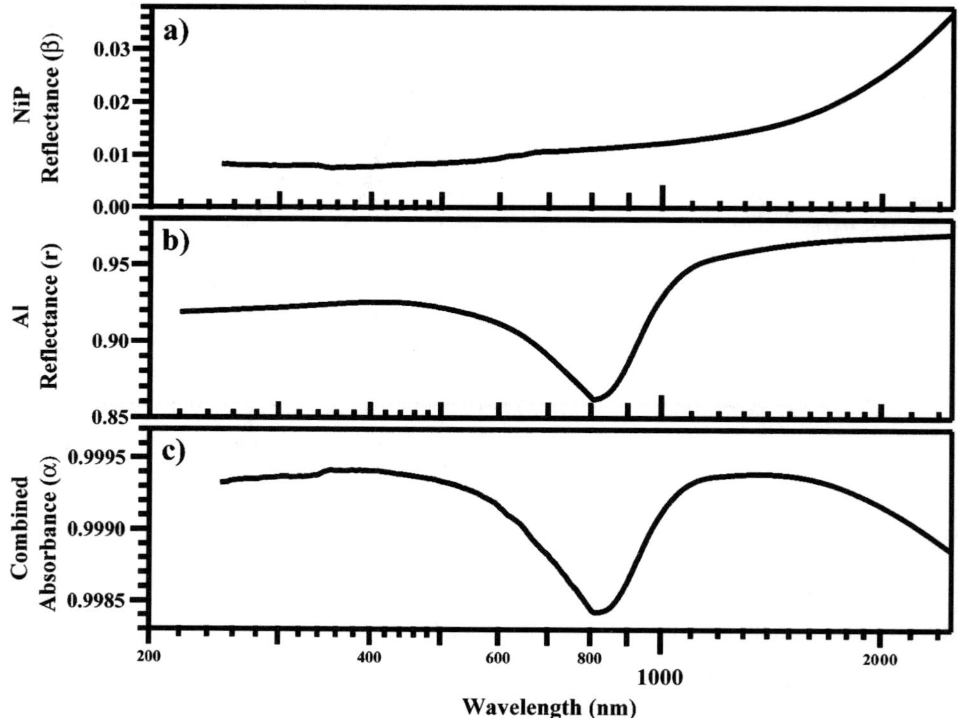

Figure 11. The reflectance of the NiP surface (β) of the bolometer; here $\beta = (1 - \rho)$, where ρ is the absorptance (panel (a)). Panel (b) is the reflectance of the hemispherical reflector surrounding the bolometer (r). Panel (c) is the combined absorbance (α) of the sphere and the bolometer as calculated from Equation (17) assuming no optical aberrations.

Figure 11 shows the wavelength dependence of ρ, r, and α. This figure shows that the hemispherical reflector significantly increases the blackness of the bolometer. In this figure, and in the derivation of Equation (17), it is assumed that the optical efficiency of the cavity is 1, and in other words, every ray reflected off of the bolometer is re-collected because of the hemispherical cavity. This assumption most likely is not true because of aberrations, particularly at longer wavelengths. Laboratory tests are needed to test this assumption. The nickel phosphorous black used for the SIM ESR bolometers was developed and produced by Custom Microwave Inc. (Longmont, CO) in conjunction with our laboratory, and a test article produced by the same production method used for the SIM bolometers was subsequently tested by Ball Aerospace Inc. (Boulder, CO). Ball Aerospace produced a report (Fleming, 1999) on these tests. This report presents measurements of the bi-directional reflection distribution function (BRDF) and total hemispherical reflectance (THR) of this material. It is assumed that optical properties of the material tested in Fleming's report are representative of the material used for the flight bolometers. This is a reasonable assumption since the electron micrographs of the test article and the black surface are comparable in structure. It is necessary to make this assumption since it was not possible to measure the flight component because the reflectance measurement requires a large target area so the light signal reflected off of the black samples is large enough to make a quality measurement. The quoted error for wavelengths greater than 800 nm is 0.5%, and about an order of magnitude less than this for the 250–800 nm range. This study demonstrated that the light reflected from the surface of the black is predominately diffuse with a small (<0.6%) specular reflectance component.

4.2. Photodiode Calibration and Degradation Correction

The radiant responsivities, $\mathcal{R}_\lambda$, (units of A W^{-1}) for each photodiode are measured on-orbit by dividing the photocurrent by the power measured by the ESR using phase sensitive detection. There are a number of small, but important, complications to this process: $\mathcal{R}_\lambda$ is not constant over a typical SIM resolution element, particularly near the red cut-off of the photodiode's response curve, whereas the ESR response is essentially flat. Therefore, the convolution over the instrument function is different for these two detectors and must be accounted for in forming the ratio. Portions of the spectrum where the photodiode's response changes rapidly cannot be accurately measured by this method. Nonetheless, the highly stable geometry, wavelength knowledge, and intensity provided by the Sun makes this measurement much more accurate and reliable than an equivalent laboratory calibration. Figure 12 shows the radiant responsivity retrieved by this method for the four photodiodes used for SIM A.

The radiant responsivity curves shown in Figure 12 define the ranges over which each detector gives reliable data. The photodiodes provide complete coverage everywhere except for a small part of the spectrum between 308 and 310 (covered by the ESR). The Vis2 and IR photodiodes overlap in the 900–1000 nm region, and the best data quality in this spectral region is from Vis2 because $\mathcal{R}_\lambda$ function is

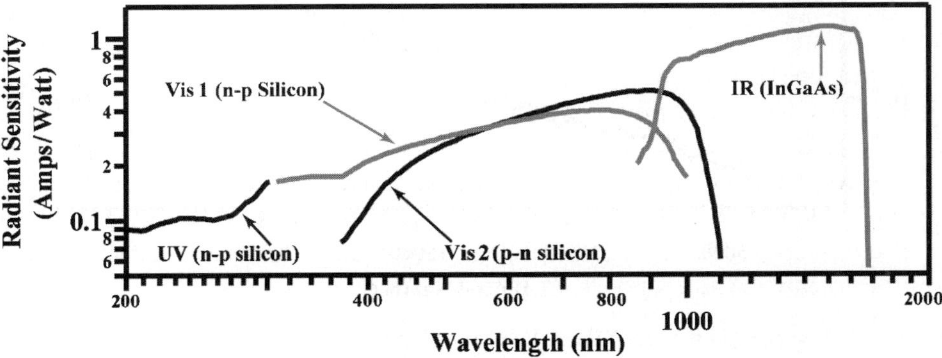

Figure 12. Radiant sensitivities for the four photodiodes used for SIM. The values are measured in-flight using the ESR to calibrate the photodiodes.

smoother. However, both silicon photodiodes have greater temperature sensitivity red-ward of the $\mathcal{R}_\lambda$ peak, so data in this regime are corrected and their usage limited because of the increased uncertainty. In summary, the operating ranges for the photodiodes that give the most reliable results are as follows: UV $= 200$–308 nm, Vis1 $= 310$–800 nm, Vis2 $= 800$–1000 nm, IR $= 1000$–1655 nm.

The photocurrent from each photodiode is measured with a radiation hardened transimpedance amplifier with an 11 Hz bandwidth, multiplexed and then converted to digital numbers with a 16-bit, bipolar, dual-slope analog-to-digital (ADC) converter. Each channel of photodiode data is sampled at 100 Hz, and then decimated at the same rate as the ESR data by the instrument's DSP. The feedback resistor for each amplifier was selected to cover most of the 2^{15} bit unipolar dynamic range of the ADC needed for each photodiode's spectral range. The converter has about 2 bits of noise per sample, so the photodiode measurements are ADC limited rather than photon noise limited. Because of this, the ultimate signal-to-noise ratio (SNR) is proportional to signal strength for a fixed integration time, and doubling the dwell time at a fixed wavelength does not improve the SNR by $\sqrt{2}$.

Since the photodiode spectral scans are used to track the orbit-to-orbit variability of the Sun (see Sections 3.2 to 3.4 of Rottman *et al.*, 2005 for more detail), their measurement precision plays an important role in interpreting solar variability. Figure 13 shows the measured photocurrent for the Vis1 photodiode as a function of CCD position and wavelength. On the right hand axis is the SNR ratio on this measurement.

The custom made photodiodes used for SIM are 10 mm $\times$ 2 mm for the three silicon photodiodes (International Radiation Devices Inc., Torrence, CA) and 8 mm $\times$ 2 mm for the InGaAs photodiode (Hamamatsu Corporation, Bridgewater,

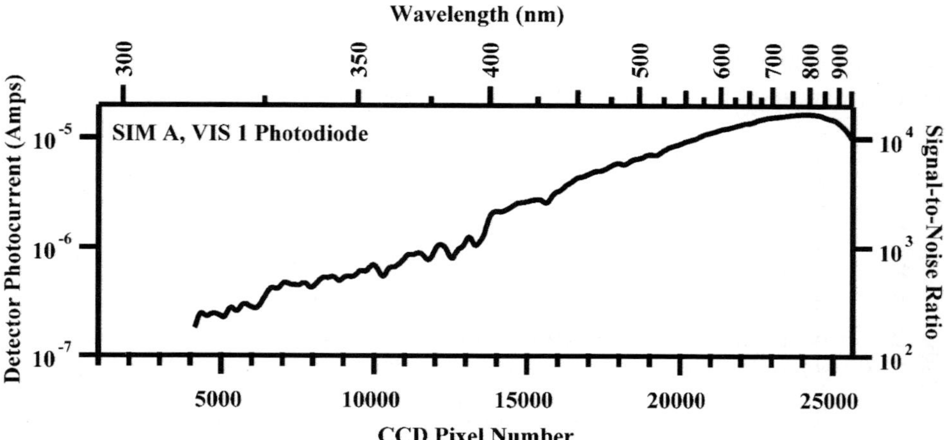

Figure 13. Raw spectral data for SIM A Vis1 photodiode. The *bottom axis* shows the CCD pixel value for the scan, and the *top axis* is the wavelength corresponding to the bottom axis. The measured detector photocurrent is shown on the *left axis*, and the approximate SNR that corresponds to the photocurrent is shown on the *right*.

NJ). The Vis1 and UV silicon photodiodes have n-on-p construction with a nitride passivated SiO_2 layer to stabilize their radiant sensitivities in the ultraviolet. The Vis2 photodiode is constructed similarly, but with p-on-n geometry. The silicon photodiodes appear to undergo a slow decrease in responsivity predominately for wavelengths longward of the responsivity peak (see Figure 12) that remains apparent in the data after prism degradation is removed. Loss of photodiode responsivity is the dominant source of instrument degradation for wavelengths greater than 650 nm where prism glass is stable (see Figure 7a). At this point in time, the radiation damage observed in the SIM photodiodes is consistent with the radiation testing described by Jorquera *et al.* (1994). They found negligible damage on the n-on-p photodiodes for the short wavelengths, and small (though not specified) damage for the long wavelengths. For the p-on-n detectors, they reported no change at 420 and 552 nm, but a 2 and 11% drop in the internal quantum efficiency at 670 and 875 nm, respectively, and for 5 MeV proton energy at a fluence of 6.0×10^8 protons cm^{-2}. The best method to correct the in-flight degradation is to match the slope of the photodiode time series to that of ESR measurements at selected wavelengths. This process is most readily done at the ESR table values. The degradation is expected to be smooth, so values between the wavelengths in the ESR table are interpolated. This method has the distinct advantage of correcting the diode degradation but not biasing the slope in the data that occurs as the intensity of the Sun decreases to the (yet unknown) solar cycle 23 minimum value over the next 3 years. The exact nature of photodiode degradation seen in-flight is still under study and, like prism degradation, will require further refinement as more data become available over the course of the SORCE mission; more detailed discussions about this degradation mechanism will be presented in subsequent publications about the SIM instrument.

5. Final Corrections and Status of SIM Solar Spectroscopy

At this juncture in time a number of additional corrections and analyses will be performed on the SIM instrument prior to assigning a final absolute calibration to the instrument. Corrections that have been implemented are:

1. The dispersion model of Section 2 and Appendix A requires slightly different parameters for each detector. In particular, the wavelength scales for each photodiode detector and the ESR can be brought into agreement with the solar spectrum of Thuillier *et al.* (2003) to within ± 0.02 nm by assigning independent wedge angles, θ_P, for each of the focal plane detectors, and by changing the effective sub-pixel size, C, on the CCD from 1.3 μm to 1.2886 μm $pixel^{-1}$ (see Equations (A.1) and (A.2)). This result was obtained by convolving the 1.3 nm-resolution Thuillier *et al.* Composite 1 spectrum with the wavelength-dependent SIM instrument function in the 300–900 nm range. The two spectra are then processed to a zero mean

differential spectrum and the θ_P and C parameters of the dispersion model are varied using Levenberg–Marquart minimization (Press *et al.*, 1992) until a minimum in the sums of squares difference between the two spectra is obtained. In using this method to find the best prism wedge angle for each detector, the worst case difference is about 0.013° out of a wedge angle of 34.497°. There is only a single prism and it is not possible to have different wedge angles, but making the correction in this parameter fixes a problem most likely caused by slightly different refraction angles produced by spherical surfaces on the prism. The dispersion equations of Appendix A do not account for curvature of the prism faces. Studies of prism refraction based on non-sequential ray tracing may lead to an improved physical understanding and correction to this problem.

2. The solar irradiance variations in the visible and near IR are on the order of 0.05–0.1%, so very small shifts in wavelength produce comparable discontinuities in the time series. The effect was clearly seen in the first 10 months of operation of SIM with the occurrence of a problem related to commanding CCD position system. This problem caused the signal on the CCD to saturate and produced a non-linear response in the drive's servo system that could not be detected in the drive housekeeping channels. Different CCD settings produced different levels of saturation and therefore a different drive response. This problem has been corrected by identifying the CCD settings that prevent CCD saturation, and the affected spectra are being corrected in ground processing with a spectral shift-stretch algorithm. The affected spectra are shifted and stretched with respect to a standard SIM spectrum without the CCD position non-linearity. The algorithm used to perform this re-mapping of the CCD position (C) converts the measured spectra into a zero-mean differential spectrum and uses a golden section search over a limited range of C values to align the spectra. This process is performed for each detector and over the full operating range of the prism drive. Prior to performing the wavelength alignment, the data are corrected for degradation so spectral slope does not bias the peak finding of the golden section search. The transfer function between the saturated and non-saturated drive positions is fitted with a third-order polynomial and then applied to the affected spectra.

The analysis of NiP absorptance, and therefore the overall efficiency of the ESR, requires further analysis and ground-based experimental verification. This study is being performed on a flight witness ESR made with the same components as the flight detectors. These measurements will impact the discussion of Section 4.1.3. Additional information on the efficiency of the detectors will be obtained through the on-orbit comparison of the two SIM spectrometers. In addition, end-to-end analysis of the prototype SIM responsivity are being planned and may provide a more definitive validation of the parameter used to derive the solar irradiances for the SIM measurements.

At the present time, and for the purpose of comparing SIM spectral irradiance time series to other measurements of solar activity, such as the Mg II core-to-wing ratio and TSI (see Rottman *et al.*, 2005), a relative calibration factor is applied to the current SIM radiometric calibration. This correction factor smoothly brings the SIM data into agreement with Thuillier *et al.* (2003) for the infrared and visible, and with UARS SOLSTICE for the 200–300 nm spectral regions. This adjustment is only significant for wavelengths longer than 600 nm and has a very smooth wavelength behavior. Thus, all spectral features present in the SIM spectra are measured and are not a consequence of these adjustments. This adjustment is complemented by a final correction through a wavelength-independent coefficient that makes the integral of SIM irradiance in the 200–1600 nm range equal to 1225 $\mathrm{Wm^{-2}}$. This value assumes the TIM TSI value of 1361 $\mathrm{Wm^{-2}}$ and estimates the irradiance in the 1.6–10 μm wavelength range at a value of 136 $\mathrm{Wm^{-1}}$ using the Fontenla *et al.* (1999) spectral synthesis calculation. This final correction is only about 1% and within the absolute accuracy of our measurements at this time.

Figure 14 represents the current spectrum with a wavelength accuracy of ±0.02 nm and an overall radiometric accuracy of about 1%. The top panel of the graph shows the irradiance spectrum, and the lower panel shows the average disk brightness temperature (Fontenla *et al.*, 1999). This plot is from 210 to 1650 nm covering the range of the SIM photodiodes. The solar spectrum in the

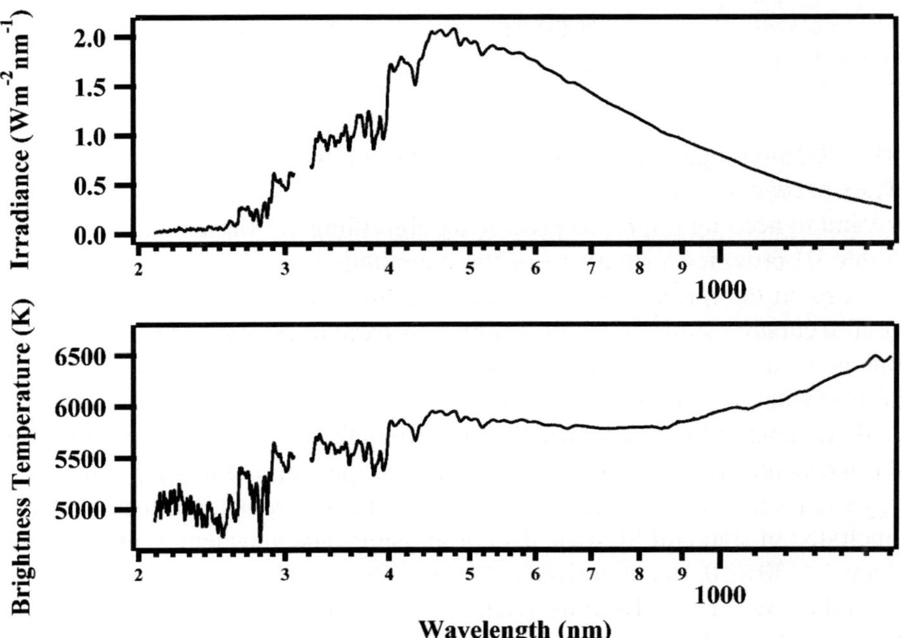

Figure 14. Current SIM solar spectrum. The *top panel* shows the spectrum in terms of irradiance, and the *lower panel* is the same data in terms of brightness temperature.

TABLE III
Current estimates of the calibration parameters for the SIM instrument.

Parameter (units)	Magnitude/range	Uncertainty
Solar distance (ppm)	+33116 to −33764	1
Doppler (ppm)	43	1
Wavelength (nm)	200–1650	$\sim$0.02 ± (150 × 10^{-6}) × λ (worst case)
Instrument function area	0.58–34.5	$\sim$0.4%
Slit parameters		
Width (μm)	300.0	0.5 ± 0.03
Area (mm^2)	2.1	3 × 10^{-5} ± 2 × 10^{-5}
Component metrology (mm)	0–400 mm	0.01
Prism transmission		
Value (%)	0.55–0.77	0.1% 200–700 nm
		$\sim$1% 1000–2700 nm
Degradation correction	$\sim$0.to 0.65	$\sim$0.1%
Diffraction correction (%)	0.3–2.2	$\sim$0.01
ESR parameters		
Standard volt (V)	7.1615 V	10 μV
ESR absorptance (%)	$\sim$99	+0 to −1 (200–700 nm)
		+0 to −10 (700–2700 nm)
Closed-loop gain	15.086	1 × 10^{-4} (0.05 Hz)
	73.205	3 × 10^{-5} (0.01 Hz)
Equivalence (ppm)	100	60

1650–2700 nm range is still under study and is not reported here due to additional analysis needed as related to incomplete pre-flight calibrations in this wavelength range and to needing improved processing algorithms for these data.

Table III provides a summary of the magnitude or range of values associated with terms in the measurement equation scheme presented in Figure 2, and the current accuracy estimates of the calibration parameters discussed in this paper. Refinements in the error estimate will occur with additional analyses and studies of ground witness components. In particular, additional work must be done on the optical efficiency of the ESR, which currently is the largest source of uncertainty in the measurement. The entry in Table III for the important solar distance and Doppler corrections are derived from the JPL Ephemeredes (Standish, 1982), and are included in standard SORCE data processing. The wavelength uncertainty of 0.02 nm $\pm (150 \times 10^6) \times \lambda$ is derived from the accuracy adjustment of the wavelength scale to the spectrum of Thuillier *et al.* (2003) and the worst case precision based on the variable resolution of the instrument. A more refined precision can be found from the ±1 subpixel CCD reproducibility of the prism drive and an analysis of Equation (A.4).

Acknowledgements

The authors would like to acknowledge the contribution of the entire LASP Engineering and Calibration Groups for their patient and thorough efforts on the calibration of this instrument. In particular, we would like to thank David Crotser, Mathew Triplett, Karl Heuerman, Anthony Canas, Miriam Adda, and Byron Smiley whose work forms the basis of this paper. This research was supported by NASA contract NAS5-97045.

Appendix A

This Appendix is associated with Section 2, and describes the equations that relate the measured prism rotation angle to refraction angle and wavelength. The terms and angles are defined in Figure 3.

A.1. Prism Incident Angle Measurement: γ from CCD Subpixel Position

The prism incident angle, γ, is found from the imaging behavior of an off-axis spherical mirror (see Section 2.3.3 of Harder *et al.*, 2005 for a description of the CCD encoder system):

$$\gamma = \gamma_z + \frac{1}{2}\tan^{-1}\left(\frac{C - C_z}{F_{\text{REF}}}\right). \tag{A.1}$$

In this equation the subscript z is the CCD subpixel count and corresponds to the condition that the centroid of the image on the CCD is aligned with the center of the spectrometer entrance slit in the dispersion plane. F_{REF} is the focal length of the spherical mirror used with the focal plane CCD. For the SIM instrument, C corresponds to 1/5th of a CCD pixel width or 1.3 μm.

A.2. Refraction Angle Calculation: ϕ from γ and n

Equation (2) in Section 2 can be solved for the angle ϕ and knowing the focal length of the prism (F), the spectral coordinate y can be obtained:

$$\phi = \gamma - \sin^{-1}\left(n \sin\left\{2\theta_{\text{P}} - \sin^{-1}\left(\frac{\sin(\gamma)}{n}\right)\right\}\right) \quad \text{and} \quad y = F\tan(\phi). \tag{A.2}$$

The index of refraction of fused silica is calculated from the three-term Sellmeier equation that is valid to about 10 ppm at 20 °C (Malitson, 1965). See also discussion of Equation (2) in Harder *et al.* (2005):

$$n_{20}(\lambda) = \sqrt{1 + \sum_{j=1}^{3} \frac{\lambda^2 K_j}{\lambda^2 - L_j}}. \tag{A.3}$$

Malitson also provides the temperature dependence dn/dT used to correct for wavelength shifts that occur as the prism temperature changes during in-flight operation.

The wavelength of an observation occurs when $y(n, \phi) = y_{\mathrm{d}}$, where y_d is the focal plane location of one of the five focal plane detectors. Thus, at a given prism incident angle, five different wavelengths are detected. Likewise, a specified wavelength can be delivered to a specific detector by changing ϕ. This topic is discussed in Section 2.3.2 of Harder *et al.* (2005) in the context of the instrument's functional capabilities.

A.3. Index of Refraction Calculation: n from γ and ϕ

Solving Equation (A.2) for n gives the index of refraction strictly from prism geometry for rotation angle:

$$n = \frac{1}{\sin(2\theta_{\mathrm{P}})} \sqrt{\sin^2(\gamma) + 2\cos(2\theta_{\mathrm{P}})\sin(\gamma)\sin(\gamma - \phi) + \sin^2(\gamma - \phi)}. \tag{A.4}$$

Numerically inverting Equation (A.3) and accounting for dn/dT determines the wavelength. Newton's method (Press *et al.*, 1992) provides an efficient numerical solution for this inversion.

A.4. Prism Incident Angle Calculation: γ from n and ϕ

Solving Equation (A.2) for γ gives the prism rotation angle:

$$\sin(\gamma) = \sqrt{\frac{\cos(\phi)\sin^2(2\theta_{\mathrm{P}}) + X\cos(2\theta_{\mathrm{P}})\sin^2(\phi) + \sin(\phi)\sin(2\theta_{\mathrm{P}})\sqrt{(\cos(2\theta_{\mathrm{P}}) + X\cos(\phi))^2 - (1 - X)^2}}{2X(\cos(2\theta_{\mathrm{P}}) + \cos(\phi))}}$$

with

$$X \equiv \frac{1}{n^2}. \tag{A.5}$$

Equation (A.1) is then solved for C and insertion of γ from Equation (A.5) gives the value of C as a function of ϕ and n.

Appendix B

Appendix B is associated with Section 3.2, and describes the experimental methods for slit width calibration (Section B.1) and slit area calibration (Section B.2).

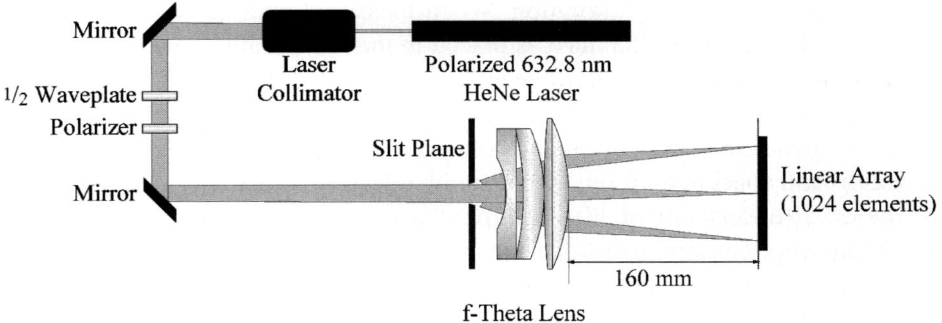

Figure B.1. Apparatus used to measure slit diffraction. An amplitude-stabilized laser beam is expanded and polarized and then impinges on the slit plane. An *f*-theta lens then images the diffraction pattern onto a linear array. Slit widths are determined from a least-squares fit of the diffraction pattern.

B.1. Slit Width Inferred from the Diffraction Pattern of a 632.8 nm Laser

The most convenient method for creating a far-field diffraction pattern at a finite distance is to employ a lens in the Fourier-transform configuration with the slit and detector placed at the conjugate focal points of the lens. It is important that the lens has low distortion and that its position in the detector focal plane linearly relates to the diffraction angle. Figure B.1 shows the apparatus used to make this slit diffraction measurement. An amplitude stabilized laser beam is collimated with a zoom expander and the quality of collimation is examined with a shear plate. The beam polarization is established with a polarizer so the polarization direction is parallel to the long-dimension of the slit. A $\frac{1}{2}$ waveplate is then inserted in the beam as a variable attenuator to keep the light level within the linear operating range of the detector. The expanded laser beam then impinges on the slit plane where both flight slits and a custom calibration target are mounted. The diffracted radiation is then passed through an *f*-theta lens (Optische Werke G. Rodenstock, München, Germany). This lens conserves the angle (θ) in focal plane instead of tan(θ), which is important since the observed diffraction depends on θ as well. The lens is optimized to work at 632.8 nm and the air wavelength is corrected to atmospheric conditions using Edlen's (1953) formulation. The lens has a distortion less than 0.1% for off-axis rays within a 25° light cone; for this experiment, only about 6° of this light cone is used for the measurement. The diffracted light is then imaged onto a 1024 element linear array with 20 μm wide $\times$ 2.5 mm tall pixels on a 25 μm pitch (Hamamatsu S3903-1024Q). The photoresponsivity non-uniformity (PRNU) is measured and removed from the data prior to analysis.

The calibration mask (Photo Sciences Inc., Torrance, CA) consists of a series of nine double slits and nine single slits etched onto a chrome-on-fused silica plate. The double slits have a 4 μm width and spacing of 200, 300, and 400 μm and are used to calibrate the apparatus. The single slits have widths of 100, 300, and 400 μm and are used to check the calibration. When measured with the apparatus shown in Figure B.1, the calibration mask is placed in the light beam perpendicular to its direction of propagation; the light beam encounters the glass substrate before the chrome layer containing the slits, and in this way its refractive properties do not disturb the measurement. After manufacturing the plate, Photo Sciences measured the widths and spacing between the slits with a Nikon 2i metrological microscope and quoted an uncertainty of $\pm 0.5\ \mu$m for measurements in the 0.1–150 mm range; their microscope calibration is traceable to NIST standards. The quoted uncertainty on this measurement is larger than the requirement for the flight slits, but the calibration mask is a ground witness, and future improvements in dimensional measurements of the mask will directly translate into a refined measurement of the flight slits. The agreement between Photo Sciences' microscope measurement and the result of this experiment are at least as good as the quoted $\pm 0.5\ \mu$m uncertainty.

The diffraction pattern from two slits separated by a distance a, with a width, W, produce an apodized cosine wave pattern that is described by the Fraunhofer theory of slit diffraction (Jenkins and White, 1976):

$$I(\theta) = 4I_o\left(\frac{\sin^2\beta}{\beta^2}\right)\cos^2\alpha, \quad \text{where} \quad \beta = \frac{\pi W}{\lambda}\sin\theta, \quad \alpha = \frac{\pi a}{\lambda}\sin\theta,$$
$$\theta = \frac{f(y)}{F}. \tag{B.1}$$

In this equation, the diffraction angle, θ, is determined from a polynomial function of position, y, along the linear diode array, and the focal length of the lens. Least-squares fitting of this pattern to the calibrated slit mask is used to remove lens distortion, and precisely measure the focal length of the f-theta lens, thereby establishing the scale of the apparatus so single slit widths can be determined when mounted in the apparatus. Figure B.2(a) shows the measured double slit pattern, the modeled output, and the fit residuals. The modeled data are found from a Levenberg–Marquardt least-squares minimization (Press *et al.*, 1992) that fits the experimental data to Equation (B.1) with additional non-linear parameters to adjust the scale, offset, slit width (held constant here), lateral shift with respect to the center of the array, and slope across the array. The fitting is performed on the noise-weighted signal. Figure B.2(a) shows a narrow light spike on the central fringe. This is caused by collimated laser light passing through the chrome plating on the calibration mask, which is then imaged into the CCD. The mask has an optical density of 5.0 (0.001% transmission), so this problem was expected and an algorithm was developed to account for its influence (see discussion below). The measurement of the flight slits is not disturbed by this problem.

Figure B.2(b) shows a similar plot for one of the slits used for the flight instrument; the graph is on a log scale to emphasize the dynamic range associated with the single slit diffraction measurement. Since the least-squares analysis is weighted by the detector noise, only the first several fringes meaningfully contribute to the

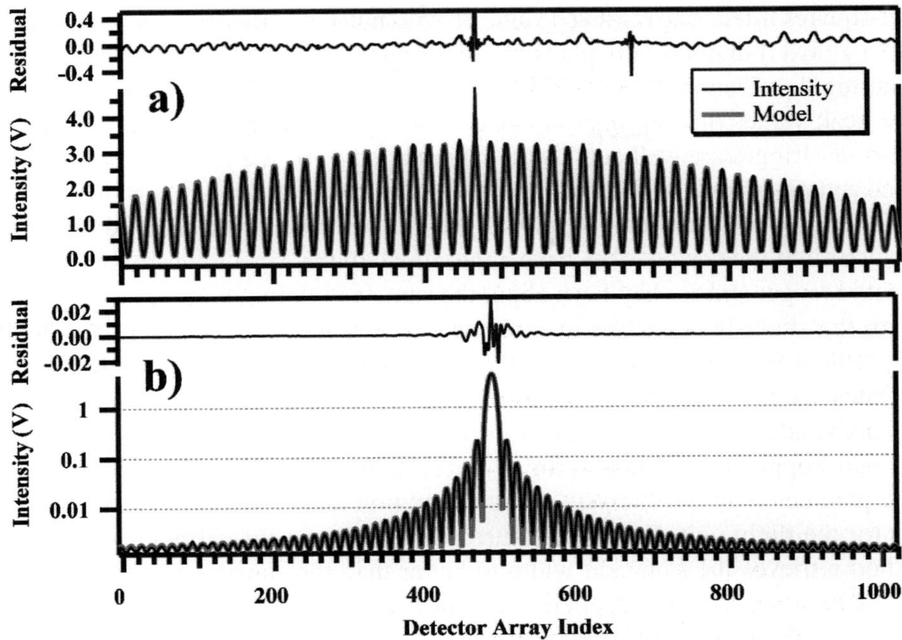

Figure B.2. Panel (a) shows the diffraction pattern generated by one of the Photo Sciences' calibrated double slits with a slit spacing of 200 μm, as well as the fitted model result and fit residuals. The central spike, caused by light transmission through the mask, is apparent in both the data plot and the residual plot. The axis range on the residual plot is set to show the fit quality in the wings of the diffraction pattern, so the central spike is off-scale. Panel (b) shows a single slit diffraction for one of the flight instrument's entrance slit. The intensity scale is logarithmic to emphasize the structure in far wings.

quality of the fit. The same least-squares method described for the double slit measurement applies to the single slit, but uses the focal length and distortion terms as fixed parameters, and fits the data using the Fraunhofer single slit diffraction law:

$$I(\theta) = 4I_o\left(\frac{\sin^2\beta}{\beta^2}\right), \quad \text{where} \quad \beta = \frac{\pi W}{\lambda}\sin\theta, \quad \theta = \frac{f(y)}{F}. \tag{B.2}$$

Prior to performing the flight slit calibrations, the validity of the double slit experiment is verified by measuring the widths of the single slits in the calibration mask. The light leak discussed in the previous paragraph is also present in the single slit measurement, but it is less pronounced due to the less intense light level used in this experiment relative to the double slit experiment. However, its influence on the least-squares fit is still significant because of the rapidly changing signal level associated with single slit diffraction and errors in the intensity of the central fringe have the greatest influence on the results of the fit. To account for this problem, the analysis method was expanded to include a step where the intensity of the central fringe was perturbed away from its measured value, and this data was then least-squares fitted. The retrieved value of χ^2 is noted and this procedure is repeated until a global (but physically plausible) minimum is found. This procedure is aided by noting that Fraunhofer diffraction theory indicates that the ±first-order fringes have peak values that are 0.046 as intense as that of the central fringe. Since the first-order fringes are well outside the region affected by the light leak, a first guess of the central fringe's true intensity is set to this value.

The light leak does not greatly affect the outcome of the double slit fitting because all of the cosine lobes have approximately the same intensity over the full detected range, and so they have about the same weighting in the least-squares fit. When the analysis described in the previous paragraph is applied to the double slit problem (with modifications to the magnitude of the first guess), the analysis indicates that the light leak makes a difference in the recovered focal length of only 12.5 ppm and a difference in the distortion term of 0.25%, but the RMS value of the residual dropped by about a factor of 2. These modified values were then used as fixed parameters in the analysis of the calibration mask's single slits and the SIM slits for the flight unit. For the measured single slits on the calibration mask, this method retrieves the same slit width to better than the stated ±5 μm uncertainty. This same procedure was applied to the measurement of the slits that do not have the light leak problem; the linearity of the detector electronics is about 0.1% of full scale and this procedure introduced corrections on the order of 0.05% thereby giving reproducibility in the slit width retrievals of better than 100 ppm.

B.2. Slit Area Inferred from the Light Flux Comparison with a Standard Aperture

The flux comparator system designed to compare SIM slit with a standard 0.5 cm^2 NIST calibrated circular aperture is shown in Figure B.3. The area ratio between the

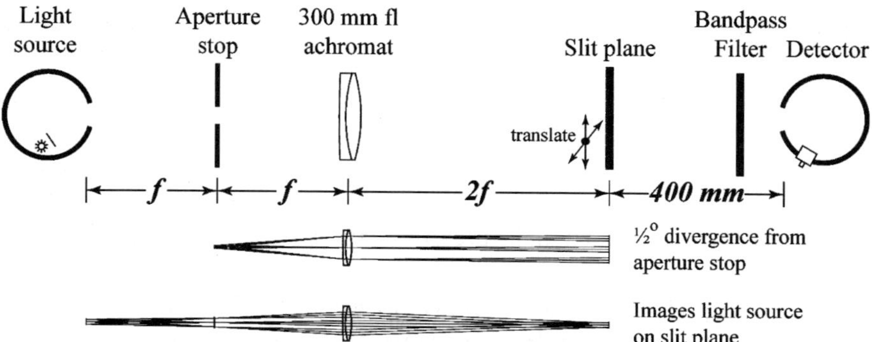

Figure B.3. The apparatus used to compare the areas of the 0.3 mm × 7 mm SIM slits to a standard 0.5 cm^2 circular aperture. The figure shows the optical arrangement based around a 300 mm focal length achromat and its associated ray traces. The slit plane is an x–y translation stage so the uniformity of the light source can be mapped in the slit plane, and the comparative measurement can be automated.

SIM slits and this standard aperture is a factor of 23.8, so the difference in intensities for uniform light will be nearly the same. This dynamic range is well within the linear operating range of photodiodes and this makes the flux comparison the most effective method to measure the area of the slits. This apparatus is similar to the one discussed by Fowler, Saunders, and Parr (2000), but has a number of attributes that make it appropriate for spectrometer slit measurements. The apparatus is based on a telecentric optical system with a 300 mm focal length achromat ($f = 300$ mm in Figure B.3) as the objective. The system's aperture stop is at a distance of f and the source and image planes are at a distance of $2f$. In this configuration, the uniform source area is imaged onto the exit plane, and the aperture stop diameter, d, is sized to produce a light beam with a 0.5° of divergence ($d = f\tan(0.5°) = 2.6$ mm). This optical arrangement ensures the calibration is performed with nearly the same geometry as the Sun, and the 1:1 imaging in the slit plane gives the optimal light intensity and radiometric accuracy. The light source illuminates a 20 cm diameter integrating sphere with a 1.27 cm circular exit port; this configuration produces a spatially uniform light beam, and the lamp power was current regulated to 0.1% to reduce intensity fluctuations. After passage through the slit plane, the light encounters a 500 ± 50 nm optical filter before being detected by a 50 mm^2 silicon photodiode located inside an integrating sphere identical to the source, and the detector's photocurrent is readout with a precision ammeter. The detector sphere has a 25 mm × 18 mm rectangular entrance port that has the same projected area as the SIM prism. This is done to match the diffracted light throughput of this apparatus to the prism. This illumination system produces a 1% total variance in irradiance at the slit plane. The variance was measured by positioning a 1 mm hole in the slit plane and recording the beam intensity. The hole is then moved in 0.5 mm steps to create a raster scan consisting of a grid of beam intensities at 194 locations covering a circular area with a diameter of 13 mm. This intensity map was then used to correct the light intensity when the standard aperture was swapped with the spectrometer slits.

The flux comparison process leads to a measurement equation; terms in this equation are written with subscripts to indicate the standard aperture (aper) and the spectrometer slits (slit):

$$A_{\text{slit}} = A_{\text{aper}} \frac{I_{\text{slit}}}{I_{\text{aper}}} \frac{J_{\text{aper}}}{J_{\text{slit}}} \frac{(1 + 2\alpha_{\text{Cu}}\{T_{\text{M}} - T_{\text{ref}}\})(1 - \beta_{\text{aper}}\lambda)}{(1 + 2\alpha_{\text{slit}}\{T_{\text{M}} - T_{\text{ref}}\})(1 - \beta_{\text{slit}}\lambda)}, \tag{B.3}$$

where $A_{\text{slit,aper}}$: geometric area, for slit and aperture; $I_{\text{slit,aper}}$: measured signal, for slit and aperture; $J_{\text{slit,aper}}$: relative intensity at aperture/slit plane, for slit and aperture; $\alpha_{\text{slit,Cu}}$: coefficient for thermal expansion, 16 and 16.5 ppm °C^{-1}, respectively; $T_{\text{M,ref}}$: temperature during measurement and reference temperature (20 °C); $\beta_{\text{slit,aper}}$: diffraction correction slope at wavelength λ (11 ppm nm^{-1} for slit, 1 ppm nm^{-1} for aperture).

This measurement is very reproducible with a precision of better than 20 ppm for the selected flight slits. Sixty slits were measured and the slits with the most precise areas and widths were then selected for the flight instrument.

Appendix C

This Appendix is associated with Section 3.3, and describes the Fresnel reflection equations and the apparatus used to deduce the bulk and mirror reflection losses of the Fèry prism.

The Fresnel reflection at the vacuum–glass interface can be characterized in terms of the usual variables (γ, λ, ϕ) presented in Figure 3. Defining two exterior and two interior angles,

$$\gamma_1 = \gamma, \quad \gamma_2 = \gamma - \phi, \quad \beta_1 = \sin^{-1}\left[\frac{\sin\gamma_1}{n}\right], \; \beta_2 = \sin^{-1}\left[\frac{\sin\gamma_2}{n}\right], \tag{C.1}$$

from the Fresnel formulas (Jenkins and White, 1976), the transmission for the horizontal and vertical polarizations of light for the two surface transits can be written:

$$\mathcal{T}_{\text{horizontal}} = \left(1 - \left\{\frac{\tan(\gamma_1 - \beta_1)}{\tan(\gamma_1 + \beta_1)}\right\}^2\right)\left(1 - \left\{\frac{\tan(\gamma_2 - \beta_2)}{\tan(\gamma_2 + \beta_2)}\right\}^2\right) \quad \text{and}$$

$$\mathcal{T}_{\text{vertical}} = \left(1 - \left\{\frac{\sin(\gamma_1 - \beta_1)}{\sin(\gamma_1 + \beta_1)}\right\}^2\right)\left(1 - \left\{\frac{\sin(\gamma_2 - \beta_2)}{\sin(\gamma_2 + \beta_2)}\right\}^2\right). \tag{C.2}$$

The dispersion plane of the prism defines horizontal polarization, and the vertical is defined by the cross-dispersion direction. The average of the two equations in Equation (C.2) would give the transmission of the prism for unpolarized light if it had no bulk losses and the rear surface mirror was a perfect reflector. The light transit from glass-to-vacuum requires some consideration since multiple internal reflections produce a source of scattered light that must be prevented from reaching the detectors. About 5% of the light intensity at this interface is directed towards the base of the prism, where most of the light will escape, but about 4% of this light is internally reflected and then escapes out the apex of the prism. Polishing the base of the prism and coating it with an index refraction matching black epoxy eliminated this problem. Most of the internally reflected light passes through the prism glass and is absorbed by the black coating without reflection (Figure C.1).

The apparatus used to measure prism transmission is schematically shown in Figure 9. A 1000 W xenon arc lamp is used for the source, an ac signal is generated by mechanically chopping the light beam, the bandpass filter is used to limit the wavelength range of light entering the monochromator removing higher orders of light, and a lock-in amplifier is used to determine the light level. The monochromator is set to produce a 0.5 nm resolution light beam for any wavelength over the

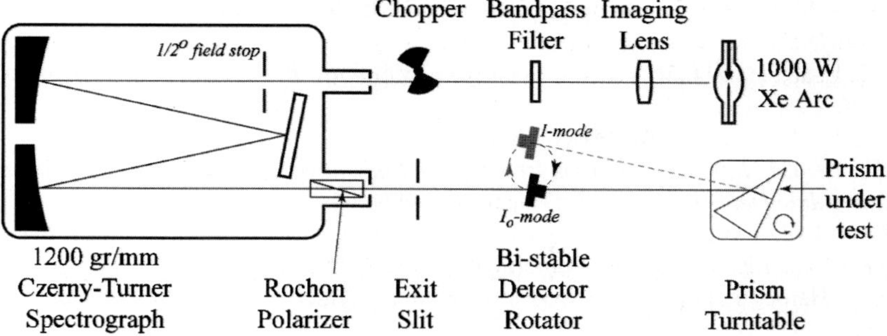

Figure C.1. The apparatus used to determine the transmission of the prism. The measurement is made by ratioing the signal between two rotation positions of the detector (one position shown in *black* and the other in *gray*).

full spectral range of the prism. The MgF_2 Rochon polarizer placed in the output beam is used in two orthogonal orientations, parallel and perpendicular to the long dimension of the exit slit. The calibration is performed in both polarizations to determine the prism transmission after computing and eliminating the contributions from Fresnel reflections. The polarized light exiting the prism is then either detected directly, or after passage through the prism depending on the position of the bi-stable detector rotator. A silicon photodiode is used for visible and UV measurements and a thermoelectrically cooled PbS cell is used for the infrared. A lock-in amplifier operating at the frequency of the light chopper measures the signal from the detectors. The prism is mounted on a precision rotation stage so precise and reproducible rotation angles are achieved. The system is set up to perform the calibration with the ESR's geometry, i.e., the refracted light beam would return to a focus at 35 mm from the entrance slit.

References

Brown, S. W., Eppeldauer, G. P., and Lykke, K. R.: 2000, *Metrologia* **37**, 579.

Edlen, B.: 1953, *J. Opt. Soc. Amer.* **43**, 339.

Fleming, J. C.: 1999, *Reflectivity and BRDF of Nickel-Phosphor Black*, Serial No. S99.41830.OPT.005, Ball Aerospace Corp., Boulder, CO.

Fontenla, J., White, O. R., Fox, P. A., Avrett, E. H., and Kurucz, R. L.: 1999, *Astrophys. J.* **518**, 480.

Fowler, J. B., Saunders, R. D., and Parr, A. C.: 2000, *Metrologia* **37**, 621.

Gere, J. M. and Timoshenko, S. P.: 1990, *Mechanics of Materials*, 3rd edn, PWS-Kent, Boston, Massachusetts, p. 76.

Harder, J. W., Lawrence, G., Fontenla, J., Rottman, G., and Woods, T.: 2005, *Solar Phys.*, this volume.

Humbach, O., Fabian, H., Grzesik, U., Haken, U., and Heitmann, W.: 1996, *J. Non-Cryst. Solids* **203**, 19.

Jenkins, F. A. and White, H. E.: 1976, *Fundamentals of Optics*, McGraw-Hill, New York.

Jorquera, C. R., Korde, R., Ford, V. G., Duval, V. G., and Bruegge, C. J.: 1994, *Geoscience and Remote Sensing Symposium: Proceedings of the IGARSS '94*, 8–12 August, Vol. 4, p. 1998.

Lawrence, G. M., Harder, J., Rottman, G., Woods, T., Richardson, J., and Mount, G.: 1998, *SPIE Proc.* **3427**, 477.

Lawson, C. L. and Hanson, R. J.: 1974, *Solving Least Squares Problems*, Prentice-Hall, Englewood Cliffs, New Jersey, p. 222.

Malitson, I. H.: 1965, *J. Opt. Soc. Amer.* **55**, 1205.

Parr, A. C.: 1996, *A National Measurement System of Radiometry, Photometry, and Pyrometry Based Upon Absolute Detectors*, NIST Technical Note 1421.

Press, W. H., Teukolsky, S. A., Vetterling, W. T., and Flannery, B. P.: 1992, *Numerical Recipes in C: The Art of Scientific Computing*, Cambridge University Press, New York.

Rottman, G., Harder, J., Fontenla, J., Woods, T., White, O., and Lawrence, G.: 2005, *Solar Phys.*, this volume.

Saunders, R. D. and Shumaker, J. B.: 1986, *Appl. Opt.* **25**, 20.

Standish, E. M.: 1982, *Astron. Astrophys.* **114**, 297.

Thuillier, G., Hersé, M., Labs, D., Foujols, T., Peetermans, W., Gillotay, D., Simon, P., and Mandel, H.: 2003, *Solar Phys.* **214**, 1.

Solar Physics (2005) 230: 205–224

THE SPECTRAL IRRADIANCE MONITOR (SIM): EARLY OBSERVATIONS

GARY ROTTMAN, JERALD HARDER, JUAN FONTENLA, THOMAS WOODS, ORAN R. WHITE and GEORGE M. LAWRENCE
Laboratory for Atmospheric and Space Physics, University of Colorado, Boulder, CO 80309, U.S.A.
(e-mail: rottman@lasp.colorado.edu; harder@lasp.colorado.edu; fontenla@lasp.colorado.edu)

(Received 29 March 2005; accepted 28 July 2005)

Abstract. This paper presents and interprets observations obtained by the Spectral Irradiance Monitor (SIM) on the Solar Radiation and Climate Experiment (SORCE) over a time period of several solar rotations during the declining phase of solar cycle 23. The time series of visible and infrared (IR) bands clearly show significant wavelength dependence of these variations. At some wavelengths the SIM measurements are qualitatively similar to the Mg II core-to-wing ratio, but in the visible and IR they show character similar to the Total Solar Irradiance (TSI) variations. Despite this overall similarity, different amplitudes, phases, and temporal features are observed at various wavelengths. The TSI can be explained as a complex sum of the various wavelength components. The SIM observations are interpreted with the aid of solar images that exhibit a mixture of solar activity features. Qualitative analysis shows how the sunspots, faculae, plage, and active network provide distinct contributions to the spectral irradiance at different wavelengths, and ultimately, how these features combine to produce the observed TSI variations. Most of the observed variability appears to be qualitatively explained by solar surface features related directly to the magnetic activity.

1. Introduction

Solar irradiance variations are likely drivers of the Earth climate system, and yet are poorly understood. Before the first reliable Total Solar Irradiance (TSI) measurements from space the solar irradiance was assumed to be a constant, at least within the ground-based observational uncertainty; hence, the term "solar constant" was commonly used. The improved space observations beginning in the 1980s showed that solar irradiance variability occurs over timescales of minutes (due to the *p*-mode oscillations), hours (corresponding to active region evolution), days (corresponding to solar rotation), and decades (due to the 11-year solar activity cycle). TSI behavior on century timescales remains speculative due to the lack of direct observations (Lean, 2000). For relevant SORCE studies see papers by Kopp (2005) for TSI, McClintock, Rottman, and Woods (2005) and Snow *et al.* (2005) for UV spectral irradiance, and Woods and Rottman (2005) for X-ray irradiance. Furthermore, this paper is the third of a series of papers concerning the SORCE SIM instrument. The first paper (Harder *et al.*, 2005a) concerns the design and operation modes of the instrument, and Harder *et al.* (2005b) describes the measurement equations and calibration of this instrument.

Existing records of variability in UV and X-ray irradiances clearly establish the strong connection between solar activity and the solar output at UV, EUV, and X-ray wavelengths. These variations strongly affect the upper layers of the Earth's atmosphere (e.g., London, 1994) but, in terms of power, the variability at UV to X-ray wavelengths alone is not sufficient to explain the magnitude of the observed TSI variations. Therefore, it is clear that variability in visible and infrared bands also contributes to the TSI variation. The amplitude of the relative variation in TSI is small ($\sim$0.1% over the 11-year solar cycle) with respect to that at UV–EUV–X-ray wavelengths, but the power variation measured in TSI is much larger than the integrated power over the UV, EUV, X-ray region. It is expected that the visible and IR irradiance variations (at wavelengths between 300 and 2000 nm) account for most of the measured TSI variations (London, 1994; Lean *et al.*, 2000). Initially, the visible and infrared irradiance variations were estimated from models that determine these variations from sunspots and faculae observed on the solar disk (e.g., Hudson *et al.*, 1982).

The first space experiments to measure the spectral irradiance in the visible and IR began with the SOLSPEC instrument onboard Atlas 1 and 2 to measure the 350–850 nm wavelength range. By combining SOLSPEC measurements on Atlas 3 with measurements from the EURECA capsule launched and recovered by the space shuttle, the wavelength range of spectral irradiance measurements was extended to 200–2400 nm (see Thuillier *et al.*, 1998, 2003). The VIRGO instrument onboard SOHO obtains a continuous record in three wavelength bands centered near 402, 500, and 862 nm (e.g., see Lanza, Rodono, and Pagano, 2004).

With the launch of SORCE in 2003, SIM began the first continuous record of the entire solar spectral irradiance from 200 to 1600 nm with sufficient precision to track the expected variations on timescales longer than a half day. Spectral irradiance in the 200–1600 nm wavelength range is measured by SIM photodiode detectors with a cadence of two observations per day. SIM also has the capability to measure solar spectral irradiance at longer wavelengths by using an Electric Substitution Radiometer (ESR) in the range from 1600 to 2700 nm.

The design and operation of the SIM instrument are described by Harder *et al.* (2005a,b). The first SIM results for the modulation of spectral irradiance by solar rotation are presented here, and these variations are considered relative to the presence of sunspots and faculae on the solar disk. Additionally, the SIM data variations over time are compared with the F10.7 flux and Mg II index time series.

The measurements of TSI by several instruments (e.g., VIRGO and ACRIM) and by SORCE's Total Irradiance Monitor (TIM) are addressed by Kopp, Lawrence, and Rottman (2005). In this study, the SIM time series are compared with the TSI obtained by TIM. UV measurements below 300 nm are obtained by SORCE's SOLSTICE (with higher spectral resolution than SIM) and discussed by Snow *et al.* (2005). SIM and SOLSTICE overlap in the wavelength range between 200 and

300 nm. Although SIM observations are used in this spectral range, the SOLSTICE observations are completely consistent.

Empirical mathematical models of the solar irradiance have been developed for TSI and UV observations using composite time series covering the last 25 years. Early regression models used a sunspot index (e.g., Hudson *et al.*, 1982), sunspot and facular indices (Chapman, Cookson, and Bobias, 1996), F10.7 radio flux (Oster, 1983), or Mg II c/w index (Viereck *et al.*, 2001) to model UV irradiance variability. Although all of these indices are related to solar activity as observed on the solar disk, the sunspot index is linked directly to TSI decreases associated with the large sunspot areas on the disk, while the other indices are related to the bright plage and enhanced network areas (e.g., see de Toma *et al.*, 2004). Correlation between these indices is expected over the solar cycle, but there is no clear physical reason why they should closely agree on shorter time scales. For instance, on days when the sunspot index is high, the TSI will decrease and the UV will increase. This is because the chromospheric lines that dominate the UV spectra will be enhanced in both plage and over the sunspots. However, on days when facular regions are present near the limb, both the TSI and the UV will increase. While existing linear regression models cannot address in detail the physical reasons behind the observed irradiance variations, they demonstrate the importance of understanding the interplay between the dark sunspots and the bright faculae and plages appearing simultaneously on the solar disk.

This paper describes the first high-precision and high-cadence observations of spectral irradiance across the entire visible and IR spectra. SIM measurements give the first picture of the solar rotation modulation effect due to spectral irradiance from 200 to 2700 nm. Comparisons of spectral irradiances at several sample wavelengths are made with TSI and with standard indices of solar activity (e.g., the Mg II index). In this paper, the observed spectral irradiance variations are qualitatively explained by the features observed on the solar disk. Detailed understanding will require sophisticated quantitative modeling that is not undertaken here.

The specific response of the Earth's atmosphere to the changes in spectral irradiance has yet to be realistically considered because of the lack of detailed observations such as those SIM is now producing. Ultimately, the irradiance variations at some wavelengths may be significant drivers for atmospheric perturbations yet to be explained.

2. SIM Spectral Irradiance and Time Series

The absolute irradiance scale of SIM is still under study (Harder *et al.*, 2005a,b). For use in this paper, the SIM irradiance scale is smoothly adjusted to fit the UARS SOLSTICE scale in the UV from 200 to 400 nm and the SOLSPEC scale (Thuillier *et al.*, 2003) at visible and IR wavelengths.

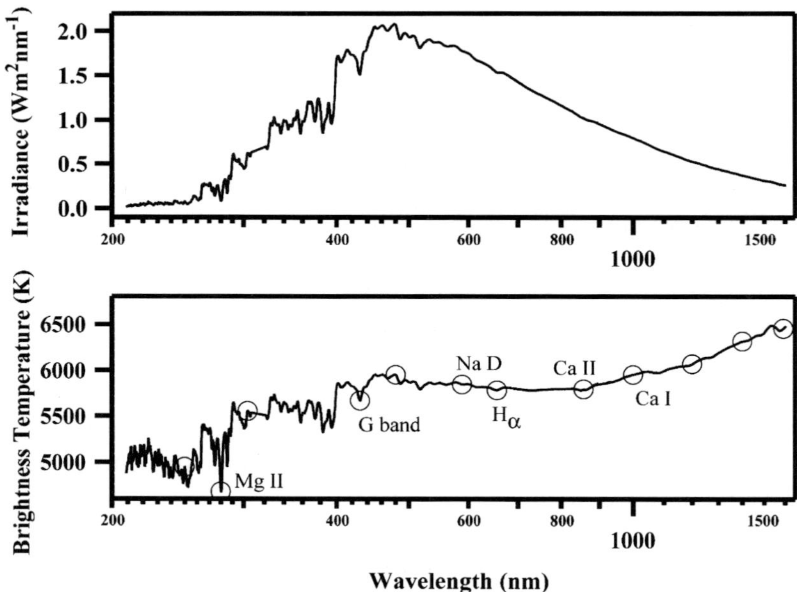

Figure 1. The irradiance (*top*) and brightness temperature (*bottom*) observed by SIM on April 21, 2004. Identification of certain spectral features are noted in the bottom panel, and the circles locate specific spectral bands with time series presented in Section 3.

The adjusted SIM spectrum for the reference day of April 21, 2004 is shown in Figure 1 in units of irradiance and brightness temperature. The wavelength-dependent brightness temperature was defined by Fontenla *et al.* (1999) as the temperature of an equivalent uniform solar disk that would produce the observed spectral irradiance at the Earth.

Despite the low resolution of SIM at visible and IR wavelengths, many spectral features are apparent in Figure 1, especially in the brightness temperature plot. These spectral features correspond to clustering of important spectral lines. It is emphasized that even the smaller features correspond to unresolved blends of well known atomic and molecular lines. The feature around 430 nm is known as the G-band and contains many molecular lines (mainly CH and C_2) and also several Fe I resonance lines. The 520 nm feature corresponds to Mg I b lines, the 656 nm feature is H alpha, and the broad feature at 855 nm is the Ca II IR triplet. While these features are well known in solar spectroscopy, other features in the SIM IR spectrum are less well known. The broad features at 1082 and 1190 nm are two clusters of lines of C, N, and O, together with the H Paschen lines and He I 1083 nm line. The broad features near 1572 are due to H Bracket series and the important CO and OH bands.

Analysis of changes in the SIM responsivity is continuing (Harder *et al.*, 2005b). Only a preliminary estimate of degradation has been applied to remove spurious trends in the SIM data. All time series are adjusted to give the same irradiance on 2 days, June 12, 2004 and August 30, 2004, when the solar disk showed few sunspots and faculae.

3. Recent Spectral Irradiance Variations

This section discusses the solar rotation modulation of the spectral irradiance between April 21, 2004 and October 1, 2004, and compares the irradiance variations with well-known measures of solar activity. In order to compare the irradiance variations at different wavelengths, each irradiance time series is plotted as a fractional difference of a daily value from the irradiance measured on June 10, 2004 using the formula

$$\delta(t) = \frac{I(t) - I(t_{\mathrm{ref}})}{I(t_{\mathrm{ref}})}, \tag{1}$$

where t_{ref} is June 10, 2004.

Since SIM measurements do not yet exist at true solar minimum in the 11-year activity cycle, this reference day was chosen because it is one of the quietest days found in the Mg II index, the F10.7, and the TSI in the study period. This is confirmed by the available images (e.g., the continuum images from SOHO MDI shown in the next section) that show the fewest sunspot groups and faculae near the limb. The Mg II index, F10.7 flux, and TSI values shown in Figure 2 are in their standard units, but their relative variations in subsequent figures use the same fractional difference normalization described in Equation (1).

3.1. Available Solar Activity Indices

The top panel of Figure 2 shows the 10.7 cm radio flux supplied by Dominion Radio Astronomy Observatory (DRAO), Penticton, Canada (K. Tapping, personal communication); the Mg II index from SBUV on NOAA16 (R. Viereck and L. Puga, *http://www.sec.noaa.gov/data/index.html*), Space Environment Center (NOAA); and the TSI measurements by TIM (Kopp, Lawrence, and Rottman, 2005). Peak values of F10.7 occur at days 2, 26, 61, 91, 116, and 142 of the study period. These peaks are generally broad and separated in time by nearly the solar rotation period (~27 days). The middle panel of Figure 2 shows the Mg II core-to-wing ratio index with a shape similar to the F10.7, but with narrower peaks and valleys and a more triangular shape. Although both the Mg II index and the F10.7 are formed mainly in the chromospheric layers, their differences can be explained by the contamination of the F10.7 by emission from extended coronal regions whose projected areas do not decrease substantially as the active regions move to the limb. For example, an active region exactly at the limb would have negligible contribution to the Mg II index, but it may have significant contribution to the F10.7 due to coronal gyrosynchrotron emission above the limb.

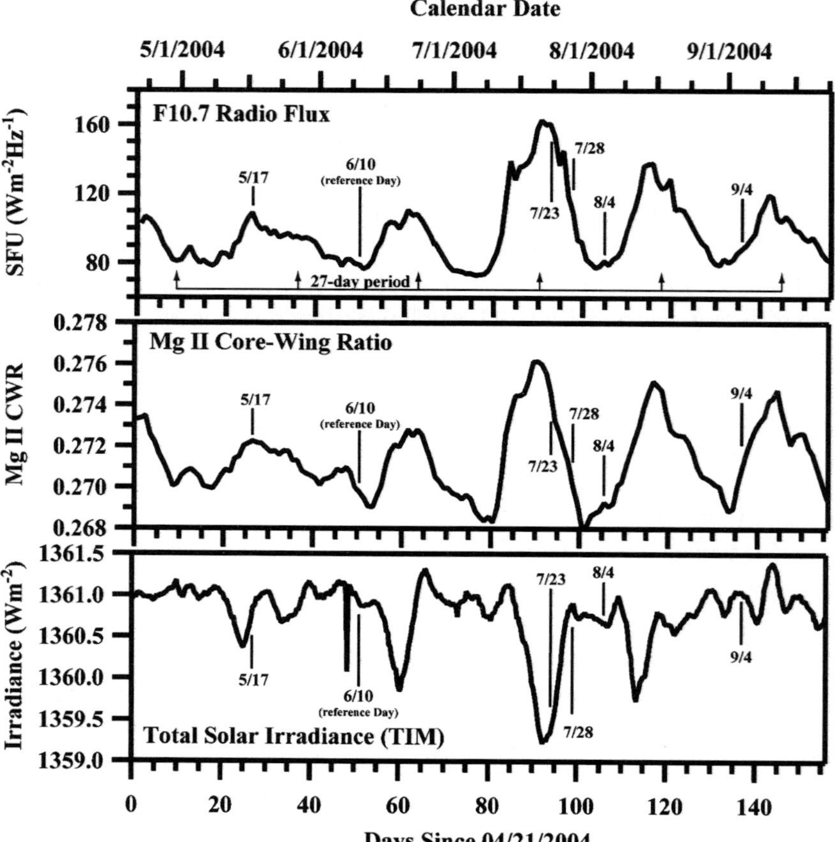

Figure 2. Time series of F10.7 radio flux (*top*), NOAA Mg II index (*middle*) and TSI, as measured by SORCE TIM (*bottom*). The arrows in the top panel correspond to the 27-day solar rotation period with respect to the maximum of the Mg II index. The dates shown in the graphs correspond to solar images discussed in Sections 4 and 5. Note the June 10, 2004 reference day corresponds to a quiescent time during this study period.

Figure 2 (bottom panel) shows the TIM TSI data. In contrast to the top two panels, the TSI displays a number of small fluctuations and four deep valleys with minima on days 24, 60, 92, and 113. The sharp downward spike on day 48 corresponds to the transit of Venus across the solar disk and not a true solar variation (Kopp, Lawrence, and Rottman, 2005). These minima in the TSI are associated with large sunspot groups near the center of the solar disk and are offset in time from the F10.7 maxima.

The recurrent peaks and valleys seen in Figure 2 suggest that a certain region on the solar surface remains active for several rotations and produces these variations as it transits the solar disk. However, this interpretation is an over simplification, and indeed the solar disk images discussed in Section 4 show a more complicated behavior.

As these wavelength-dependent variations in irradiance show, one cannot expect a "typical" rotational modulation curve during high solar activity. Sunspots, faculae, plage, and active network each make a separate contribution depending on their positions on the disk. Since each type of surface feature evolves differently, their combined effect in integrated radiation from the entire solar disk produces the complicated variations seen in the TSI time series.

Figure 2 (top panel) shows that for the June 10, 2004 reference day the F10.7 flux was very close to its minimum value, and the Mg II index (middle panel) was relatively small but not at a minimum. Indeed other minima are deeper in the Mg II index, but an examination of full disk solar images reveals that there were always some active features on the disk during this period (see Section 4). Therefore, measurements on June 10, 2004 are not of a completely quiet solar disk. As solar cycle 23 reaches its minimum, expected in ~2007, a true quiet-Sun reference will become available.

3.2. Near ultraviolet irradiance variations

Figure 3 shows the relative irradiance variations (Equation (1)) measured at several UV wavelengths throughout the 5-month study period. At 250 nm the irradiance is the integral over a pseudo-continuum weighted by the SIM resolution profile with a FWHM of 0.45 nm (see Harder *et al.*, 2005a). Thus, this measurement includes the continuum formed near the top of the photosphere combined with many deep absorption lines formed in the low chromosphere. Although variability at this wavelength is a mix of continuum and line variations, the net effect is due primarily to variability in the chromospheric lines and is similar to the Mg II index variations. Although the contribution of photospheric continuum decreases as a consequence of the presence of sunspots (the behavior shown by the TSI), the line contribution increases due to plage and dominates this SIM band. There is little variation before day 80. The broad maxima around days 90 and 120 are aligned with the Mg II index and are consistent with the large plage area seen in Ca II K images. These maxima have about 1% amplitude, but as a consequence of the large photospheric contribution, are smaller than those of the Mg II index.

The irradiance at 280 nm (0.64 nm FWHM bandwidth) includes the line cores of both Mg II h and k lines, their line wings, and a pseudo-continuum. This pseudo-continuum has the same characteristics mentioned earlier for the 250 nm band, but the line cores have a very strong upper chromospheric signature, since the emission cores (seen in high-resolution spectra) show large increases in active regions and may be enhanced over sunspots. Because of the effects of the emission cores, the variability at this wavelength is large and has a chromospheric character. Consequently, the SIM 280 nm irradiance maximizes when plage areas are largest

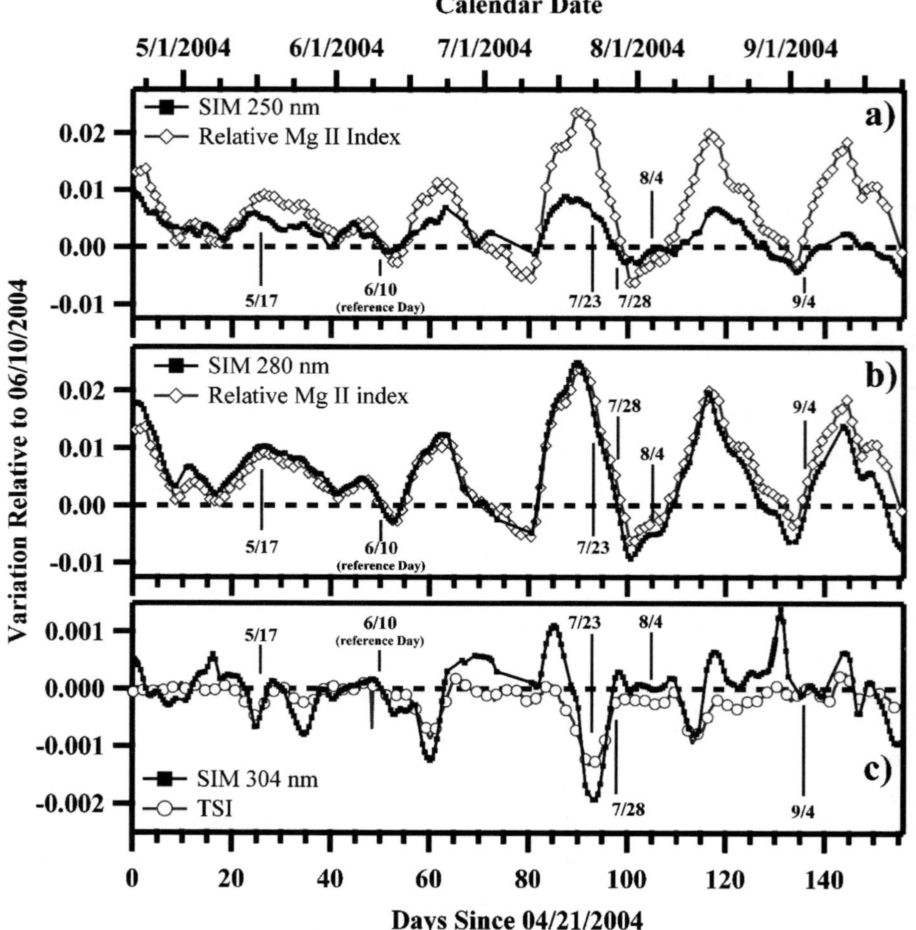

Figure 3. Relative irradiance variations measured by SIM at 250 nm (*top*), 280 nm (*middle*), and 304 nm (*bottom*) compared with the relative variations of the Mg II index (*diamonds* in top and middle panels) and the TSI (*circles* in the bottom panel).

and likely coincides with the minima of the TSI when the sunspot areas reach maximum. The SIM 280 nm irradiance tracks the SBUV Mg II index in Figure 3b because the SIM and SBUV spectral resolutions are very similar. The maximum excursion of the irradiance is about 3%. Because the SIM 280 nm irradiance and the Mg II index are both affected strongly by active network, the minimum values in this period are significantly above quiet Sun values. This is understood by examining the images on the quietest day in this period, June 6, 2004, when only very small sunspots were observed, but active network features can be seen in the Ca II K images (see Figure 7 and discussion in Section 4).

The SIM irradiance in the 304 nm band (0.97 nm FWHM bandwidth) is, in principle, also a pseudo-continuum. The continuum at this wavelength is produced slightly deeper in the photosphere, and the distribution of absorption lines is less dense than at 250 nm. As a result, the behavior is more photospheric, less chromospheric, and resembles that of the TSI but with larger variations. Figure 3c shows decreases up to almost 0.2% in the 304 nm irradiance at the times of the TSI minima, while the TSI only decreased by 0.1%. The 304 nm increases on days 83, 98, and 118 are similar but larger than the increases in TSI and occur when facular areas are near the limb as will be discussed in Section 4.

3.3. Visible Irradiance Variations

Figure 4 shows the normalized SIM irradiance variations in three spectral bands in the visible. The band centered at 430 nm wavelength (3.1 nm FWHM bandwidth)

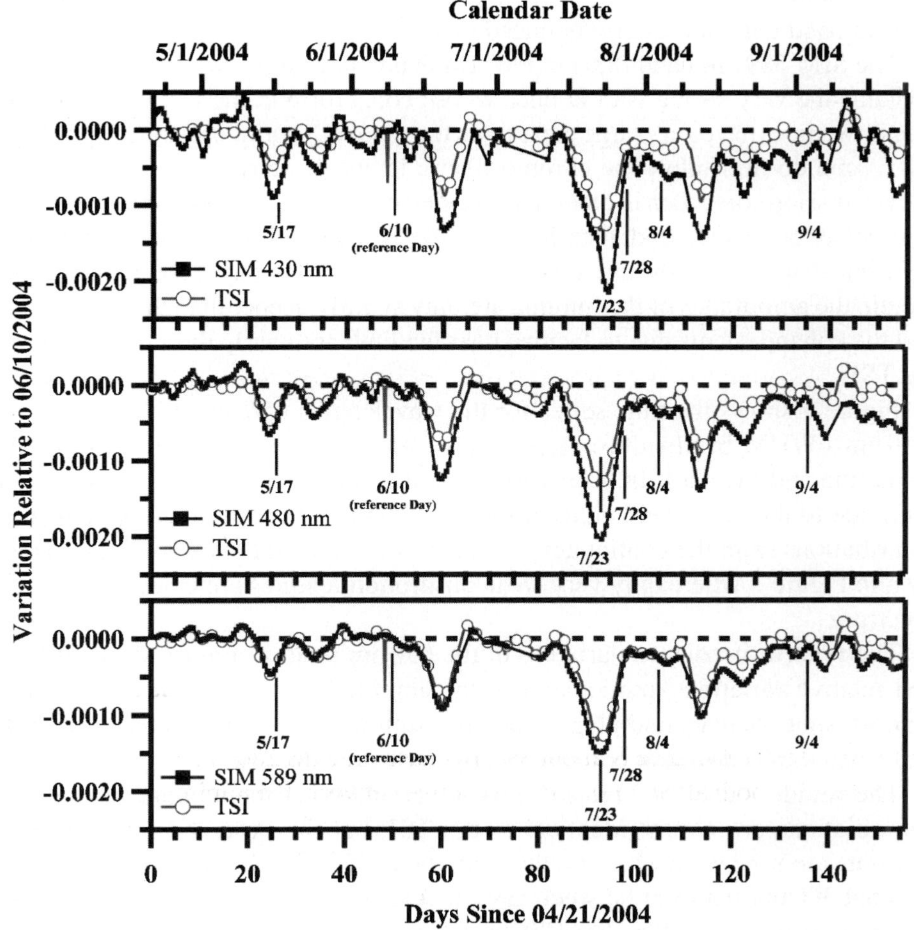

Figure 4. SIM irradiance relative variations in the blue at 430 nm (G-band), 480 nm (spectral irradiance peak), 589 nm (Na D lines) compared with the relative variations of the TSI (*circles*).

corresponds to the so-called G-band that appears as a broad absorption feature in the SIM data. In reality, it contains a large number of narrow molecular lines (mainly CH and H_2), several very deep resonance Fe I lines, and the H Balmer gamma line at ~434.25 nm. These molecular lines originate in the upper photosphere and low chromosphere, and they are deeper in sunspots where the sunspot continuum is also depressed. The behavior of this band is similar to that of the TSI except that the sunspot minima are deeper by a factor of 2 (reaching 0.2%). The irradiance increases due to faculae near the limb on days 83, 98, and 118 are also noticeable but not as conspicuous as they are at shorter wavelengths as discussed in Section 3.2 and shown in Figure 3.

The solar spectral irradiance reaches its maximum intensity near the 480 nm band (4.4 nm FWHM bandwidth) when given on a wavelength scale as in Figure 1. This spectral band is largely dominated by the continuum with lines of neutral metals. The variability in this band again has a photospheric character, and is similar to the 430 nm band but with less-pronounced increases due to faculae near the limb.

The SIM 589 nm band (8.0 nm FWHM bandwidth) has fewer narrow lines, but contains the very strong Na I D lines whose cores form in the low chromosphere. These two lines have very large departures from Local Thermodynamic Equilibrium (LTE) and do not follow the chromospheric temperature rise. These lines do not show emission cores seen in other lines formed in the mid- and upper-chromosphere. The behavior of the irradiance in this band is again very similar to that of the TSI, but displays a slower recovery than the TSI after day 120. At this wavelength, the amplitudes of the minima are only slightly deeper than the TSI minima, and the enhancements due to faculae near the limb are nearly the same as those in the TSI.

Figure 5 shows the time series for the wavelengths 656 and 857 nm (10.5 and 18.0 nm FWHM bandwidths, respectively) where the broad and deep H Balmer alpha line and the Ca II IR triplet lines are the dominant spectral features. However, due to the large bandwidth of the SIM instrument at these wavelengths, the contributions from the continuum outside of these spectral lines dominate. This is seen in Figure 2 where only very weak absorption features are evident in the SIM spectrum.

The relative irradiance variation in the 656 nm band is almost identical to the TSI relative variations, particularly in the amplitude of the variations associated with sunspot, faculae, and plage. The only significant differences occur after day 120 when the SIM data at 656 nm are slightly below the TSI.

The middle panel of Figure 5 shows variability in the 857 nm band. Again the variations are almost identical to the TSI, but the observed minima due to sunspots are now somewhat smaller than those of the TSI. Small differences (up to about 300 ppm) appear between days 20 and 30. The previous differences noted for the 656 nm band after day 120 are not present at this wavelength. Note that while the TIM TSI measurements are 6-h averages, the SIM measurements are snapshots taken twice a day, so intra-day variations are not expected to match

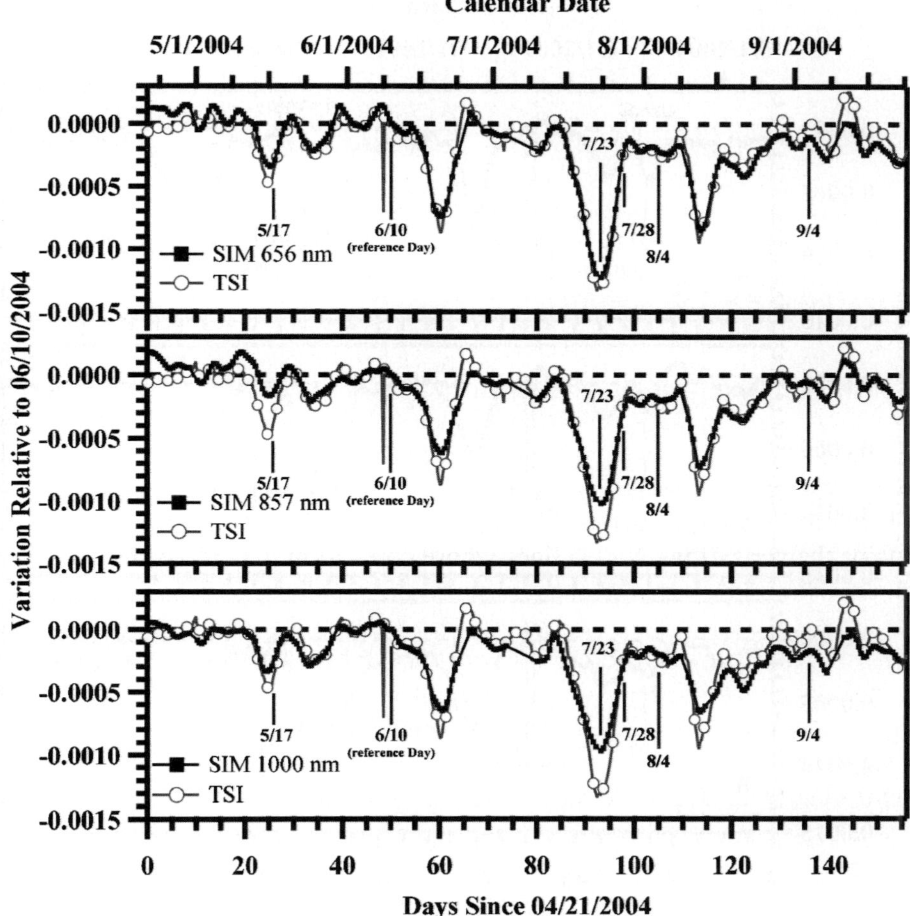

Figure 5. Relative irradiance variations measured by SIM at 656 nm (H alpha), 857 nm (Mg II IR triplet), 1000 nm (Ca I resonance line) compared with the relative variations of the TSI (*circles*).

exactly in the SIM and TIM data. In particular, higher cadence TIM data show the *p*-mode fluctuations of about 50 ppm (that cancel in the 6-h averaged data) (Kopp, Lawrence, and Rottman, 2005), yet these fluctuations appear as solar "noise" in the SIM snapshot measurements.

3.4. Near infrared irradiance variations

In the 1000 nm band (23.4 nm FWHM bandwidth), bottom panel of Figure 5, the Ca I resonance lines are the most important spectral feature. The spectral irradiance in this band again tracks the TSI very well, but displays substantially smaller amplitudes of the minima due to sunspots. It also displays much smaller amplitude

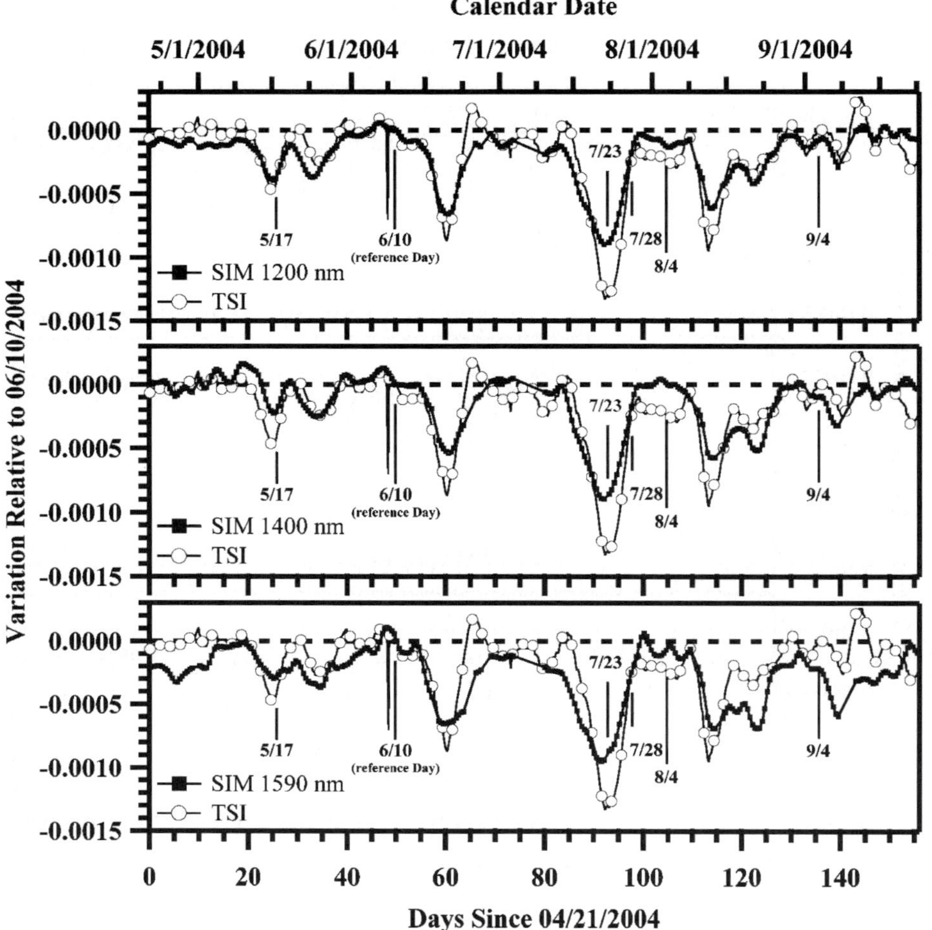

Figure 6. Relative irradiance variations measured by SIM at 1200, 1400, and 1590 nm, compared with the relative variations of the TSI (*circles*).

in the peaks on days 66, 85, 109, and 119 due to facular areas near the limb. Also, there are some small differences in the timing of these peaks with respect to visible wavelengths. At this wavelength, facular brightening at the limb barely compensates for sunspot areas. Again, as in the 656 nm band, a divergence with TSI occurs after day 120, but the difference is small (about 150 ppm).

Figure 6 shows the relative variation of the irradiance in the 1200, 1400, and 1590 mm bands. These data show the complete disappearance of the peaks due to faculae near the limb and a more gradual recovery from the valleys due to sunspots. The valleys due to sunspots are shallower than at shorter wavelengths and are substantially smaller than those of the TSI. However, more dramatic differences with the TSI are clear for days 30–34 and after day 120. These differences are attributed to plage and active network that affect the TSI and visible wavelengths, but not the IR wavelengths.

A feature similar to that described by Fontenla *et al.* (2004) appears between the two minima at days 92 and 113 and corresponds to an enhancement of the IR irradiance above the relative TSI level of about 120 ppm. This feature is not observed at wavelengths in the visible and UV. In the interval between the minima on days 60 and 93, TSI and all spectral components – UV, visible, and IR – track together quite well and do not show the differences. This may suggest that the enhancement was due to a low-contrast structure deep in the solar photosphere.

The general behavior observed in the 1400 nm band (27.0 nm FWHM bandwidth) is similar to that of the 1200 nm band (26.7 nm FWHM), but the minima are somewhat shallower. Before day 30 and after day 120, SIM data at this 1400 nm wavelength show some significant differences from the behavior of the TSI variations, not apparent in the 1000 and 1200 nm bands. It is unlikely that these differences are due to instrumental effects because there is no noticeable degradation of the instrument at these wavelengths (Harder *et al.*, 2005b).

The band at 1590 nm (25.9 nm FWHM bandpass), the longest wavelength available from the InGaAs photodiode detectors (see Harder *et al.*, 2005a), shows behavior similar to the 1400 nm band. The wavelength of this band is close to the minimum continuum absorption cross-section due to H minus; therefore, continuum radiation near this wavelength originates deep in the photosphere. However, note that within the SIM bandpass at this wavelength there are well-known molecular lines identified in SIM spectra discussed in Section 2.

4. Solar Surface Features as Drivers of the Irradiance Variations

The important features observed on the solar disk during the period of this study are qualitatively described in this section. Images obtained by the MDI and EIT instruments onboard the SOHO spacecraft, and white-light and Ca II K_3 images from the Meudon Observatory are available (see *http://umbra.nascom.nasa.gov/*). These image data are used to identify solar features and make a qualitative relation to the irradiance variations observed by SIM. More sophisticated and detailed modeling will be provided in later publications.

The study period spanned Carrington rotations number 2015 through 2020. The reference day (June 10, 2004) occurred during Carrington rotation 2017. The Carrington longitude at the disk center was about 200° for the images observed on this day.

The MDI intensity images shown in Figures 7–11 are flattened by removing the center-to-limb variation of the quiet Sun. (For details on this procedure, refer to the MDI documentation – *http://soi.stanford.edu/results/*.) The Ca II K_3 images from the Meudon Observatory have a narrow bandpass, 0.025 nm; consequently, they clearly show the quiet and active network structure.

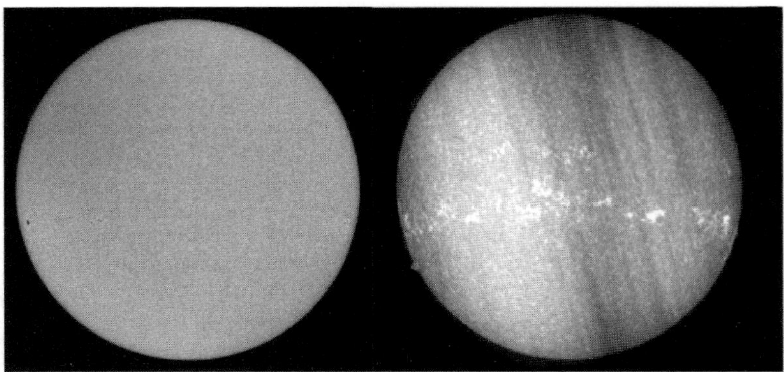

Figure 7. MDI flattened continuum (*left*) and Ca II K_3 (*right*) images taken on June 10, 2004. This is one of the quietest days of the study period and is used as the reference day. It shows three small sunspots with their associated plage, and also has additional bright regions without sunspots. (The streaks on the Ca II K_3 image are due to clouds present during the Meudon observation.)

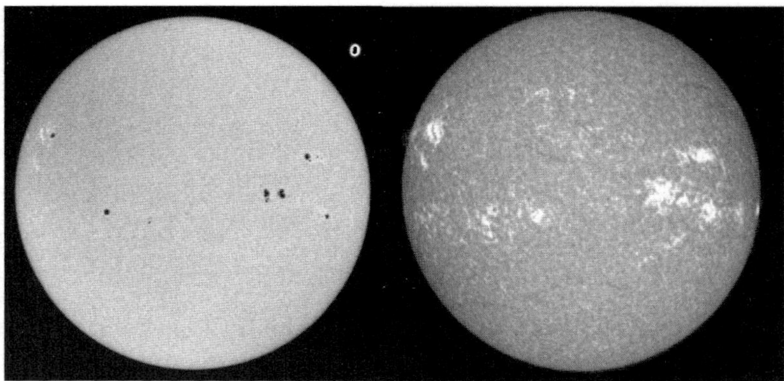

Figure 8. MDI flattened continuum (*left*) and Ca II K_3 (*right*) images taken on May 17, 2004. The rapidly evolving small sunspot groups produce a fast onset of activity, but slower decay with plage and enhanced network remaining on the disk after sunspots rotated off the visible disk.

Figure 7 shows the MDI continuum and Ca II K_3 images on the reference day with only three minor sunspot groups spanning a broad range in Carrington longitudes near the equator. As previously discussed, this relatively quiet day was not completely devoid of solar activity. Moreover, the Ca II image in Figure 7 shows the plage associated with these three sunspot groups and additional bright regions free of sunspots. Also patches of enhanced network appear at many locations across the disk. The F10.7 flux and Mg II index were close to their minimum values on this day, but are still above a true quiet Sun value. The TSI was well above the low values measured during sunspot transits across the disk, but it was below the increased values observed when large facular areas are close to the limb (see Figure 2, bottom panel).

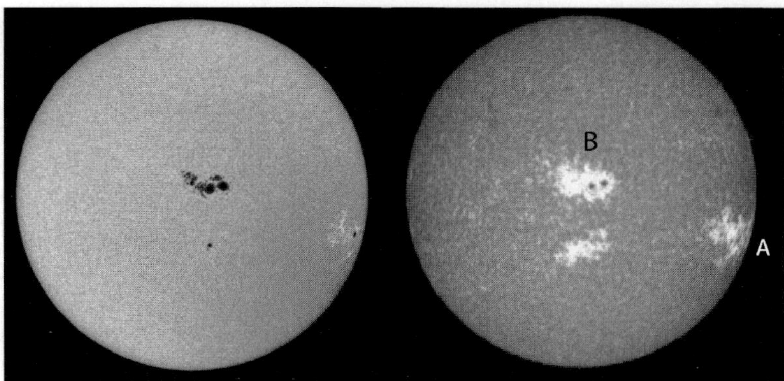

Figure 9. MDI flattened continuum (*left*) and Ca II K3 (*right*) images taken on July 23, 2004. The image shows the largest UV enhancements and lowest relative flux in the visible and TSI. It also shows the presence of two long-lived active regions A and B on the disk simultaneously (marked on the Ca II image).

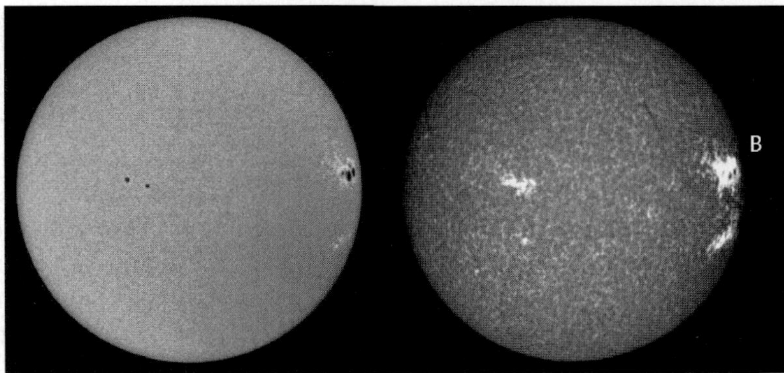

Figure 10. MDI flattened continuum (*left*) and Ca II K3 (*right*) images taken on July 28, 2004. Activity center B and a sunspot-free plage region are near the exit point on the limb. The white-light image contains two small sunspots near disk center surrounded by uniformly bright, low-contrast plage areas.

In Figure 8, the images on May 17, 2004 correspond to the first small peak in the F10.7 flux and Mg II index and the first small valley in TSI noted in Figure 2. The Carrington longitude at disk center was $\sim 158°$, and the images show many small sunspot groups between 20°N to 6°S latitude. The irradiance variations at this time correspond to the sudden appearance of many rapidly evolving small groups in Carrington rotation 2016. The sunspots of these groups disappeared in a span of a few days and did not reappear in the next rotation. The rapid rise and slow decay of the F10.7 flux, Mg II index, and SIM spectral irradiance below 300 nm are consistent with a fast onset of activity, followed by their slow decay due to the presence of plage and enhanced network after the sunspots disappeared in the full disk images. Usually, the remnants of active regions disperse and become active

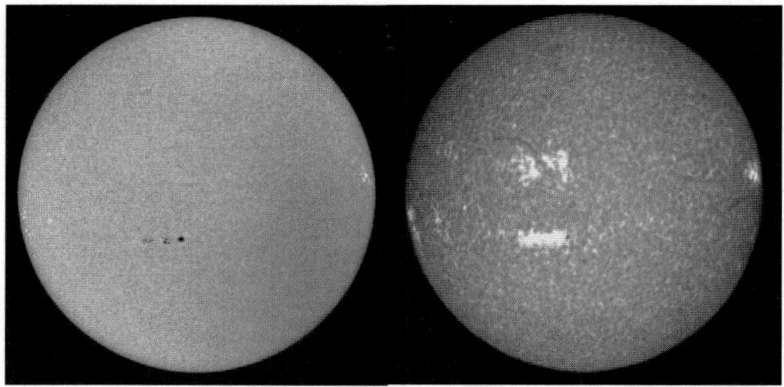

Figure 11. MDI flattened continuum (*left*) and Ca II K_3 (*right*) images taken on August 4, 2004, when the infrared bands showed enhanced irradiance not seen in the visible or TSI.

network. However, in the particular case of these small and relatively short-lived groups, only the decrease in intensity of the plage areas was observed without noticeable dispersion into the chromospheric network.

During this 5-month study period, solar activity displayed a pattern in which a few large active regions persisted on one hemisphere, while the opposite hemisphere was much less active. In particular, two widely separated and long-lasting active regions are identified – one at latitude ~10°S, 60° Carrington longitude, hereafter referred to as "active region A," and another at ~9°N, 350° Carrington longitude, hereafter referred to as "active region B." These two active regions are members of a series of active regions that appeared and decayed always near these Carrington longitudes. In fact, some residual activity was observed at these locations at all times. Since these regions are separated by about 70° in Carrington longitude, they are both present on the observable disk at some time during each solar rotation. The three strong maxima in the F10.7 flux and Mg II index, as well as the three deep valleys in the TSI, occur when both of these active regions are present on the disk. Referring back to Section 3 and Figures 2–6, the large peaks in the SIM UV spectral irradiance and the deep valleys in the visible and IR irradiance correspond to the presence of these persisting active regions on the solar disk.

The image from July 23, 2004 in Figure 9 corresponds to the deepest valley in the SIM visible irradiance (and in the TSI), and is close to the largest values of UV spectral irradiance (and also F10.7 flux and Mg II index) on 21 July. On these days, both long-lasting regions A and B were on the solar disk. At this time, region B displayed two very large sunspots and a number of smaller spots in a delta configuration sharing common large penumbrae, all spanning Carrington longitudes from 342° to 247°. Region A has smaller sunspot area; however, it is more extended in the chromosphere and corona as seen in He II 30.4 nm and the coronal lines observed by EIT at 28.4 and 17.1 nm. For the time period around the UV irradiance maximum and the visible minimum, the region B provided dominant photospheric activity, while region A showed the strongest chromospheric and coronal activity. On 23 July, the region B is close to central meridian, and region A is close to the limb.

A full quantitative study applying appropriate modeling will follow in a later publication. This investigation will include the sunspot umbrae and penumbra, as well as the effects of the bright faculae observed in the MDI image in Figure 9.

It is interesting to note that while the chromospheric UV spectral irradiance and the Mg II index shown in Figure 2 (panels a and b) decreased monotonically after 21 July, the F10.7 flux remained high and displayed a secondary maximum. Such an effect may be due to the contribution of gyrosynchrotron emissions of F10.7 in the corona above active center A.

Figure 10 shows images taken 5 days later on July 28, 2004 when activity center B was near its exit from the solar disk. Two other small spots are observed at Carrington longitudes 264° and 270°, near disk center. These minor sunspots are surrounded by uniformly bright, low-contrast areas even though they are far from the limb. It is hard to assess their importance in irradiance due to the brightness variations in the neighboring quiet areas. The most important features here are the large and bright facular areas near the west limb (right side). One of these areas is associated with region B at 352° Carrington longitude that includes large sunspots seen in the MDI image. The other feature at nearly the same Carrington longitude but latitude 20°S does not contain sunspots. At that time, both the visible spectral irradiance between ~304 and ~656 nm and the TSI, showed a rapid increase followed by a rapid decrease (see Figures 3–5). A similar case of a rapid and short-lived increase in irradiance was seen on September 9, 2004 in the visible spectral irradiance and TSI. It occurred when faculae were close to both east and west limbs. For wavelengths >1000 nm these rapid and short-lived increases are not observed.

Figure 11 shows the MDI flattened continuum and Ca II K_3 images taken on August 4, 2004. During this time, infrared bands show slightly enhanced irradiance, but the TSI and the visible spectral bands show no enhancement. Instead a small decrease from the irradiance level corresponding to the facular areas seen near the limb in Figure 10. A sunspot-free facular area appears near the east limb; however, it seems unlikely that the bright faculae compensate for the sunspots in the IR and not in the visible. A possible explanation is that the east limb sunspot-free faculae has strong IR positive contrast, while the spots on the disk have only a small negative contrast. In the MDI image of Figure 11, the enhancement around the westernmost spot may contribute to the enhanced IR irradiance. The enhancement around the sunspot in white light near disk center may not be very common, but it is similar to that reported by Rast *et al.* (2001).

5. Conclusions and Discussion

The SIM spectral irradiance observations have precision sufficient to clearly see the solar rotational modulation at all wavelengths between 200 and 1600 nm. As the

SORCE mission continues, an accurate record of solar spectral irradiance variability from the very interesting periods of activity in 2003 and 2004 to the minimum in 2007 will be available.

Studying five solar rotations in 2004 gives useful insight on how irradiance variations relate throughout the 200–1600 nm solar spectrum, and ultimately how these variations relate to surface features observed on the Sun. This study shows that at UV wavelengths below 300 nm the variations have chromospheric characteristics that are qualitatively similar to those of the Mg II index, and at visible wavelengths the variations have photospheric characteristics and are similar to those of the TSI. At infrared wavelengths, the behavior is essentially photospheric but with some important differences needing study.

Images of the solar disk show patterns of magnetic activity during the Carrington rotations 2015–2020 that qualitatively explain the irradiance variations. In particular, two active regions are identified, and their combined effects produced a clear pattern of recurrent maxima in the UV chromospheric radiation, and minima in visible and IR wavelengths. These variations correspond to minima in the TSI and maxima in the F10.7 and Mg II indices. Because these two regions are separated by $\sim 70^{\circ}$ Carrington longitude, they are simultaneously present at some times and absent at others, thereby producing a very strong rotational modulation signal.

Other features on the solar surface evolve relatively fast and produce complex spectral irradiance variations smaller in amplitude and more difficult to characterize. The sudden appearance of many active locations showing small sunspots suggests the rise and fast dissipation of a long flux rope extending in Carrington longitude over a large fraction of the Sun.

The UV chromospheric irradiance, at wavelengths shorter than 300 nm, is closely related to the persistent chromospheric magnetic heating in plage and enhanced network. The chromospheric features decay slowly, last for several rotations, and have small center-to-limb variation. On the other hand, the photospheric irradiance at wavelengths between 300 and 1000 nm is related to faculae that evolve faster. Sunspots decrease the visible and IR irradiance and their projected areas decrease toward the limb. The sunspot irradiance decreases are wavelength dependent and become shallower with increasing wavelength from the visible to IR. Individual sunspots do not last more than a few rotations, but frequently new sunspots and faculae emerge at nearby locations. Faculae increase the visible irradiance, especially when they are close to the limb. Facular contrast also decreases with increasing wavelength. Faculae decrease the IR irradiance except when they are close to the limb where faculae brightening is seen in the image. Spectral irradiance variations are most similar to the variations in the TSI at $\sim$650 nm (see Figure 5, top panel).

IR radiation is formed in the deep photosphere due to the decreasing H minus absorption cross-section that has a minimum near 1600 nm. These nearby IR wavelengths show no significant brightening due to faculae at the limb and smaller decreases (less contrast) for sunspots than is observed in the visible.

The irradiance variation time series studied here are further examples of the SIM measurements described by Fontenla *et al.* (2004) using observations earlier in 2003. Again they show the unusual phenomena that between two large sunspot decreases the IR irradiance is systematically higher than are the shorter wavelength bands and the TSI. There are no obvious features on the solar disk to account for such a difference according to solar atmospheric models (see Fontenla *et al.*, 2004).

Overall, the solar images provide a qualitative explanation for most of the behavior of the SIM irradiance variations. However, a quantitative understanding requires a detailed spectral synthesis (e.g., Fontenla *et al.*, 1999). The set of atmospheric models must include the features described in this paper, and as well as a new penumbrae model to fully account for the spectral irradiance variations. To accurately compare the synthetic spectra with the SIM data, one must take into account not only the areas of the features but also their distribution on the disk as observed in a combination of visible continuum and chromospheric emission images (e.g., the PSPT red continuum and Ca II K images). Such an analysis was used by Fontenla *et al.* (1999) and subsequently applied by S. Davis (private communication, *juan.fontenla@lasp.colorado.edu*) to understand earlier SIM observations.

For purposes of modeling Earth atmospheric response to solar input the temporal behavior and amplitude of irradiance variations at all wavelengths must be considered. For example, solar UV irradiances increase as a large sunspot group transits the solar disk, but visible and the IR irradiance decreases. When faculae are near either solar limb, the visible irradiance increases, while the UV decreases as the projected plage area decreases. Moreover, when a sunspot-free active region is present on the disk, the UV increases, while the visible irradiance shows little or no variation until the region approaches the limb. This behavior is even more complex when multiple active regions are concurrently present on the Sun.

The continuing SIM observations and data analysis, coupled with refined instrument corrections over time, will lead to better understanding of the solar irradiance variations. Ultimately, these data merged with the irradiance modeling and solar image analysis such as by Krivova *et al.* (2003), Lean *et al.* (1998), Preminger, Walton, and Chapman (2002), and Fontenla *et al.* (1999) in conjunction with analysis by S. Davis (private communication, *juan.fontenla@lasp.colorado.edu*) will be invaluable for understanding the role of the solar surface features on the solar irradiance changes. These new SIM measurements provide the first spectral irradiances in the visible and IR for input to terrestrial atmosphere studies.

Acknowledgement

This research was supported by NASA contract NAS5-97045.

References

Brueckner, G. E., Edlow, K. L., Floyd, L. E., IV, Lean, J. L., and Vanhoosier, M. E.: 1993, *J. Geophys. Res.* **98**(D6), 10,695–10,711.
Cebula, R. P. and Deland, M. T.: 1998, *Sol. Phys.* **177**, 117.
Chapman, G. A., Cookson, A. M., and Bobias, J. J.: 1996, *J. Geophys. Res.* **101**, 13541.
de Toma, G., White, O. R., Chapman, G. A., Walton, S. R., Preminger, D. G., and Cookson, A. M.: 2004, *Astrophys. J.* **609**, 1140.
Donnelly, R. F., Heath, D. F., and Lean, J. L.: 1982, *J. Geophys. Res.* **87**, 10318.
Floyd, L. E., Reiser, P. A., Crane, P. C., Herring, L. C., Prinz, D. K., and Brueckner, G. E.: 1998, *Solar Phys.* **177**, 79.
Fontenla, J., White, O. R., Fox, P. A., Avrett, E. H., and Kurucz, R. L.: 1999, *Astrophys. J.* **518**, 480.
Fontenla, J. M., Harder, J., Rottman, G., Woods, T. N., Lawrence, G. M., and Davis, S.: 2004, *Astrophys. J.* **605**, L85.
Harder, J., Lawrence, G., Fontenla, J., Rottman, G., and Woods, T.: 2005a, *Solar Phys.*, this volume.
Harder, J., Fontenla, J., Lawrence, G., Rottman, G., and Woods, T.: 2005b, *Solar Phys.*, this volume.
Hudson, H. S., Silva, S., Woodard, M., and Willson, R. C.: 1982, *Solar Phys.* **76**, 211.
Kopp, G., Lawrence, G., and Rottman, G.: 2005, *Solar Phys.*, this volume.
Krivova, N. A., Solanki, S. K., Fligge, M., and Unruh, Y. C.: 2003, *Astron. Astrophys.* **399**, L1–L4.
Lanza, A. F., Rodono, M., and Pagano, I.: 2004, *Astron. Astrophys.* **425**, 707.
Lean, J.: 2000, *Geophys. Res. Lett.* **27**, 2425.
Lean, J. L., Cook, J., Marquette, W., and Johannesson, A.: 1998, *Astrophys. J.* **492**, 390–401.
London, J.: 1994, *Adv. Space Res.* **14**, 33.
McClintock, W. E., Rottman, G., and Woods, T.: 2005, *Solar Phys.*, this volume.
Oster, L.: 1983, *J. Geophys. Res.* **88**, 9037.
Preminger, D. G., Walton, S. R., and Chapman, G. A.: 2002, *J. Geophys. Res.* **107**, 1354.
Rast, M. P., Meisner, R. W., Lites, B. W., Fox, P. A., and White, O. R.: 2001, *Astrophys. J.* **557**, 864.
Rottman, G. J., Woods, T. N., and Sparn, T. P.: 1993, *J. Geophys. Res.* **98**, 10667.
Snow, M., McClintock, W., Rottman, G., and Woods, T.: 2005, *Solar Phys.*, this volume.
Thuillier, G., Herse, M., Simon, P. C., Labs, D., Mandel, H., Gillotay, D., and Foujols, T.: 1998, *Solar Phys.* **177**, 41.
Thuillier, G., Herse, M., Labs, D., Foujols, T., Peetermans, W., Gillotay, D., Simon, P. C., and Mandel, H.: 2003, *Solar Phys.* **214**, 1.
Viereck, R., Puga, L., McMujllin, D., Judge, D., Weber, M., and Tobiska, W. K.: 2001, *Geophys. Res. Lett.* **28**, 1343.
Woods, T. N. and Rottman, G.: 2005, *Solar Phys.*, this volume.

Solar Physics (2005) 230: 225–258

SOLAR–STELLAR IRRADIANCE COMPARISON EXPERIMENT II (SOLSTICE II): INSTRUMENT CONCEPT AND DESIGN

WILLIAM E. McCLINTOCK, GARY J. ROTTMAN and THOMAS N. WOODS
Laboratory for Atmospheric and Space Physics, University of Colorado, Boulder, U.S.A.
(e-mail: mcclintock@lasp.colorado.edu)

(Received 1 January 2005; accepted 10 May 2005)

Abstract. The Solar–Stellar Irradiance Comparison Experiment II (SOLSTICE II) is one of four experiments launched aboard the Solar Radiation and Climate Experiment (SORCE) on 25 January, 2003. Its principal science objectives are to measure solar spectral irradiance from 115 to 320 nm with a spectral resolution of 1 nm, a cadence of 6 h, and an accuracy of 5% and to determine solar variability with a relative accuracy of 0.5% per year during a 5-year long nominal mission. SOLSTICE II meets these objectives using a pair of identical scanning grating monochromators that can measure both solar and stellar irradiance. Instrument radiometric responsivity was calibrated to ~3% absolute accuracy before launch using the Synchrotron Ultraviolet Radiation Facility (SURF) at the National Institute for Standards and Technology (NIST) in Gaithersburg, MD. During orbital operations, SOLSTICE II has been making daily measurements of both the Sun and an ensemble of bright, stable, main-sequence B and A stars. The stellar measurements allow the tracking of changes in instrument responsivity with a relative accuracy of 0.5% per year over the life of the mission. SOLSTICE II is an evolution of the SOLSTICE I instrument that is currently operating on the Upper Atmosphere Research Satellite (UARS). This paper reviews the basic SOLSTICE concept and describes the design, operating modes, and early performance of the SOLSTICE II instrument.

1. Introduction

The Solar–Stellar Irradiance Comparison Experiment II (SOLSTICE II) is one of four experiments launched aboard the Solar Radiation and Climate Experiment (SORCE) on 25 January 2003. SORCE is a component of the NASA Earth Observing System (EOS) dedicated to measuring solar irradiance and its variability over a nominal 5-year mission lifespan. The SOLSTICE principal science objectives are to measure solar spectral irradiance from 115 to 320 nm with a spectral resolution of 1 nm, a cadence of 6 h, and an accuracy of 5% and to determine its variability with a relative accuracy of 0.5% per year. SOLSTICE II is a follow-on to the SOLSTICE I instrument (Rottman, Woods, and Sparn, 1993), which is operating aboard NASA's Upper Atmosphere Research Satellite (UARS).

SOLSTICE II consists of a pair of identical scanning grating monochromators, referred to as SOLSTICE A and SOLSTICE B, that measure both solar and stellar irradiance using a single optical-detector chain. Each instrument covers the entire wavelength range 115–320 nm, providing both redundancy against hardware failure and simultaneous measurements for data validation. They are co-aligned on the SORCE optical bench to view the same target simultaneously but with their grating dispersion planes perpendicular. This arrangement provides a measure of the pitch-yaw offset in stellar position for both instruments (Rottman, 2005). Instrument radiometric responsivity was calibrated before launch using the Synchrotron Ultraviolet Radiation Facility (SURF) at the National Institute for Standards and Technology (NIST) in Gaithersburg, MD. During orbital operations, the SOLSTICE II instruments make daily measurements of both the Sun and an ensemble of bright, stable, main-sequence B and A stars. The stellar measurements track changes in instrument responsivity with a relative accuracy of 0.5% per year over the life of the mission.

The assumption that the average ultraviolet irradiance (100 – 400 nm) from an ensemble of main-sequence B and A stars varies by significantly less than 0.5% per year over the SORCE mission lifetime is the keystone for the success of the SOLSTICE technique. Ultraviolet irradiance from an early-type star, whose temperature is typically greater than 15 000 K, arises from blackbody emission in its stable lower atmosphere. Stellar theory predicts that the irradiance in this wavelength range from such a star is stable to better than 1% over timescales of 10 000 years (Mihalas and Binney, 1981). The same theory predicts that visible and near infrared irradiance from a late-type (solar-like) star that arises from blackbody radiation in its photosphere, should also be stable to better than 1% over timescales of 1000 years. This is consistent with recent measurements that suggest that total solar irradiance, 95% of which is emitted in the 300–2500 nm visible to near infrared wavelength range, has varied by less than 0.5% over the last two solar cycles (Kopp, Lawrence, and Rottman, 2005).

Only normal B and A main-sequence stars are used for SOLSTICE comparisons. At the beginning, 31 stars, excluding type O, rapid rotators, and magnetic stars, were selected for the SOLSTICE I experiment. After repeated observation through the first several years of the UARS mission, 18 of these, which are listed in Table I, have been selected for inclusion in the ensemble measurements for SOLSTICE II. With the exception of α Eri, which appears to be variable, most stars removed from the original ensemble are relatively faint or are located in cluttered regions of the sky. The remaining stars have been prioritized based on intensity, separation and isolation from neighboring bright stars, and irradiance stability, as established directly from the UARS observations.

Using the ensemble average rather than a single star to track in-flight instrument performance greatly improves the validity of the SOLSTICE solar variability data measured over more than a decade (Snow *et al.*, 2005a). In addition, when the SOLSTICE missions are complete, they will have established the ratio of solar irradiance to the mean flux from the stellar ensemble. Measurements of this ratio, made by future generations of observers, can be used to determine relative solar variability on timescales of tens to hundreds of years. Moreover, as radiometric measurement techniques improve, it will also be possible to reduce the current 3–5% uncertainties in the SOLSTICE solar irradiance values by recalibrating the solar observations using the solar-stellar ratios.

TABLE I

SOLSTICE calibration stars.

Star name	RA (2000)	Decl. (2000)	V magnitude	Spectral type
ε Per	3 h 57.8 min	40°0′	2.90	B0.5 III
α CMa	6 h 45.1 min	−16°43′	−1.46	A1 V
κ Vel	9 h 22.1 min	−55°1′	2.50	B2 IV–V
α Leo	10 h 8.4 min	11°58′	1.35	B7 V
δ Cen	12 h 8.4 min	−50°43′	2.60	B2 IVne
α Cru	12 h 26.6 min	−63°7′	1.35	B0.5 IV + B1 V
α Vir	13 h 25.1 min	−11°10′	0.97	B1 IV + B2 V
η UMa	13 h 47.5 min	49°15′	1.86	B3 V
ζ Cen	13 h 55.5 min	−47°17′	2.55	B2.5 IV
β Cen	14 h 3.8 min	−60°22′	0.61	B1 III
γ Lup	15 h 35.1 min	−41°10′	2.78	B2 IV
δ Sco	16 h 0.3 min	−22°37′	2.32	B0.5 IV
τ Sco	16 h 35.9 min	−28°13′	2.82	B0 V
α Lyr	18 h 36.9 min	38°47′	0.03	A0 Va
σ Sgr	18 h 55.3 min	−26°18′	2.02	B2.5 V
α Pav	20 h 25.6 min	−56°44′	1.94	B2.5 V
α Gru	22 h 8.2 min	−46°58′	1.74	B7 IV
α PsA	22 h 57.7 min	−29°37′	1.16	A3 V

2. SOLSTICE Measurement Implementation

The unique requirement for SOLSTICE is that a single optical-detector configuration must be used to measure both solar and stellar irradiance at ultraviolet wavelengths (115–320 nm). Figure 1 compares the solar irradiance to that of a typical SOLSTICE star, η UMa. The stellar irradiance is on the order of 10^4 photons cm^{-2} s^{-1} nm^{-1} and its dynamic range is only a factor of 2–4 throughout the ultraviolet. In contrast, the solar irradiance is on average a factor of 10^9 brighter than that and changes by $\sim 10^5$ from 115 to 320 nm.

The SOLSTICE instruments accommodate these large-range factors by interchanging stellar and solar entrance apertures and exit slits (2×10^5) and by increasing stellar observing times by a factor of 10^2 to 10^3.

Figure 2 shows schematic diagrams of the SOLSTICE optical system solar and stellar observing modes. In solar mode sunlight enters the instrument through a small (0.1 mm square) entrance aperture, which is placed on the optical path using a two-position mechanism, and the divergent sunlight is diffracted toward an ellipsoidal camera mirror by a plane grating. The solar entrance aperture is located at the far conjugate of the ellipse and its demagnified image is focused at the near conjugate. A small wavelength band from the dispersed image then passes

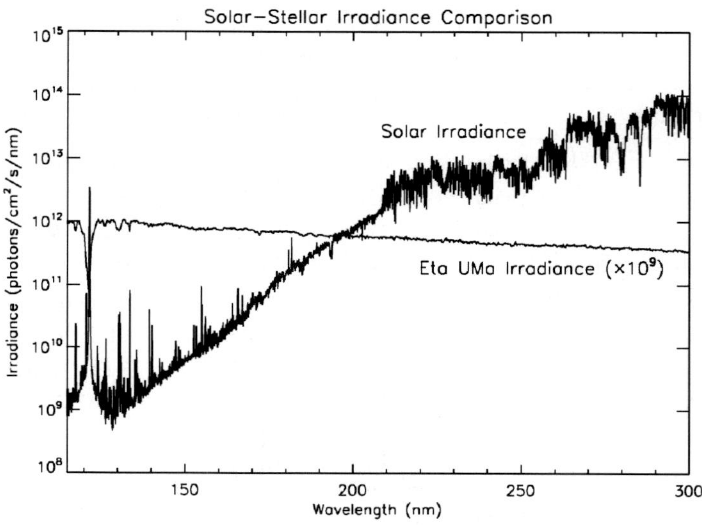

Figure 1. Comparison of solar and stellar irradiance from the bright B3 V type star η UMa. The stellar flux, which is on average 10^9 lower than mean solar value, is relatively constant. In contrast, the solar flux varies by 10^5 over the wavelength range.

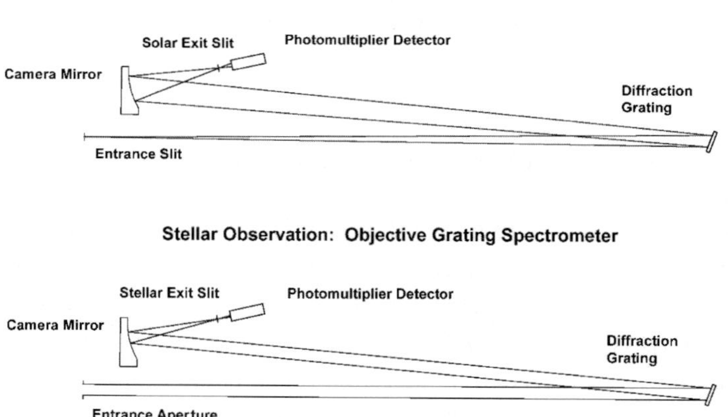

Figure 2. Schematic diagrams of the SOLSTICE optical system solar and stellar observing modes.

through a narrow exit slit located in the camera's solar focal plane and impinges upon the input window of a photomultiplier detector operating in pulse counting mode. This optical arrangement, in which a plane grating is placed in a divergent beam, is referred to as a modified Monk-Gillieson mount (Kaneko, Namioka, and Seya, 1971).

Alternately, the entrance slit mechanism selects a much larger aperture (16 mm diameter circle) for stellar observations. In this configuration the instrument operates as an objective grating spectrometer. Star light, which is already collimated when it arrives at the SOLSTICE aperture, is imaged in the camera mirror's stellar focal plane where it passes through a wide stellar exit slit before impinging upon the detector input widow. The solar and stellar focal planes are separated by a few millimeters. This difference is accommodated by using a second two-position mechanism to place either the solar or stellar exit slit at the appropriate distance from the vertex of the camera mirror. The ratio of stellar entrance aperture to solar entrance aperture is $\sim 2 \times 10^4$ and the ratio of stellar wavelength bandpass to solar wavelength bandpass is 10–20 (see Section 3.1.2). Solar and stellar spectra are recorded by rotating the grating in discrete steps and counting the number of photons detected in a fixed time interval (typically 1 s and 100–500 s, respectively) at each step.

SOLSTICE I covers the wavelength range from 115 to 430 nm using three separate optical-detector channels stacked within a single mechanical housing (Rottman, Woods, and Sparn, 1993). The designations of these channels as G (115–185 nm), F (170–320 nm), and N (280–430 nm) are historical references to the chemical composition of their detector photocathodes (Cs-I, Cs-Te and Sb-K-Cs, respectively). Figure 3 illustrates the optical-mechanical design for a single UARS SOLSTICE channel. Each channel includes a pair of fold mirrors that reduce the overall length of the instrument package by approximately a factor of two. A crossed arrangement of grating input and output beams eliminates re-entrant light from the system. All three gratings are mounted to a single shaft and the spectrum is scanned by simultaneously rotating them through 7.8° in 2048 steps. The ellipsoidal camera mirrors are designed to provide stigmatic imaging at nominal solar mode conjugate object and image distances. Astigmatism caused by placing the plane grating in a diverging beam results in a wavelength dependent shift in image location so that each channel is focused only at the middle of its spectral range. This slight de-focus causes a negligible degradation in solar spectral resolution.

The pulse counting detectors in SOLSTICE I are too sensitive for direct observations of solar irradiance at wavelengths >250 nm; therefore, the F and N channels employ interference filters, located immediately after their exit slits, to attenuate

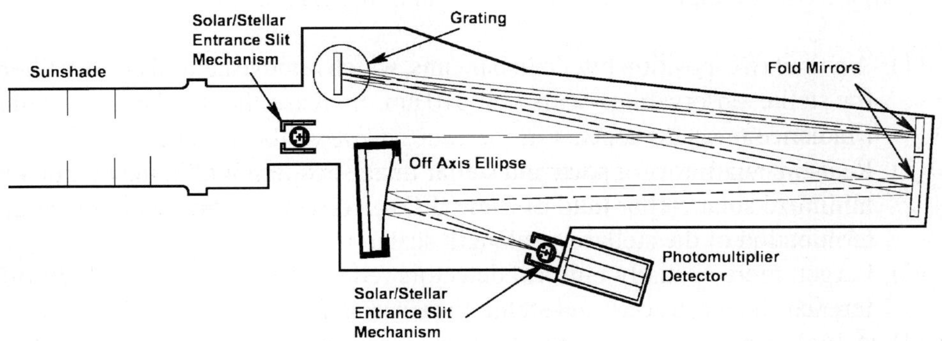

Figure 3. Optical-mechanical configuration of a single SOLSTICE I channel.

and shape the spectrum before it reaches the detector. These filters also significantly reduce the stellar signal reaching the detectors, limiting the precision of individual stellar observations at the longer wavelengths ($\lambda > 250$ nm).

3. SOLSTICE II Design Modifications

SOLSTICE I absolute sensitivity was measured using SURF II (Woods, Rottman, and Ucker, 1993). It was launched aboard the UARS on 12 September 1991 with the requirement to achieve 10% accuracy and 2% relative accuracy over the nominal 18-month UARS mission. In-flight comparison with other irradiance experiments indicates that SOLSTICE I achieves or exceeds these requirement (Woods *et al.*, 1996).

As SOLSTICE I continues to return daily solar and stellar irradiance measurements it has become clear that determination of solar-cycle UV variability for $\lambda > 210$ nm requires a relative accuracy ~0.5% per year. This tighter requirement led to the development of a second-generation SOLSTICE II instrument. It now continues the SOLSTICE I data set by measuring solar irradiance from 115 to 320 nm with a spectral resolution of 1 nm and a cadence of 6 h, with an accuracy of 5%, and with a relative accuracy of ~0.5% per year. These improvements in performance are the result of improvements in SURF (a decrease from 2% to 1% in the uncertainty of the SURF irradiance) and new features incorporated in the SOLSTICE II design.

The success of SOLSTICE I clearly demonstrates that the SOLSTICE technique works; therefore, SOLSTICE II retains the principal features of its predecessor: (1) spectral resolution and coverage, (2) optical design and layout with two channel wavelength coverage, and (3) general optical-mechanical implementation. In addition, SORCE SOLSTICE incorporates new features that increase its reliability, reduce the complexity of in-flight calibration and irradiance retrieval, and improve the accuracy of both absolute solar irradiance measurements and solar/stellar irradiance comparisons.

Major SORCE improvements to the original SOLSTICE include:

(1) A pair of two-position filter mechanisms, which remove neutral density filters for stellar observations with $\lambda > 210$ nm, increase the precision of stellar irradiance measurements.
(2) Real-time monitors of solar and stellar image position within the instrument minimize solar/stellar field-of-view (FOV) corrections and improve the determination of the stellar wavelength scale.
(3) Larger, more spatially uniform detectors reduce the magnitude of FOV differences between solar and stellar measurements.
(4) A high-precision grating position encoder increases the accuracy and repeatability of the solar and stellar wavelength scales.

(5) Redundant detector assemblies increase reliability and provide for more accurate tracking of changes in instrument sensitivity. And
(6) A composite optical bench improves the stability and reproducibility of the wavelength scale.

3.1. Optical-Mechanical Design

SOLSTICE II employs a pair of fully redundant spectrometers. Unlike SOLSTICE I, which has a single detector per optical channel, each SOLSTICE II spectrometer is equipped with a pair of Hamamatsu R-2078 photomultiplier tubes; one has a CsI photocathode to measure the 115–180 nm, far ultraviolet (FUV corresponding to the SOLSTICE I G channel) wavelength range, and the other has a CsTe photocathode to measure the 170 to 320 nm, middle ultraviolet (MUV corresponding to the SOLSTICE I F channel) wavelength range. During normal operations one spectrometer measures FUV and the other simultaneously measures MUV. Although it is used less frequently, the alternate detector in each channel is exercised on a routine basis in order to track their in-flight performance over time. In the event of a catastrophic failure in one spectrometer, SOLSTICE II will retain its full wavelength capability, but the time to acquire a complete spectrum will double.

Figure 4 illustrates the optical layout for a single spectrometer. In an arrangement, which is identical to that for SOLSTICE I, light enters the instrument through an entrance slit assembly and is reflected by a fold mirror toward the diffraction grating. A small range of wavelengths diffracted from the grating is reflected toward the elliptical camera mirror by the second fold mirror. The ellipse images

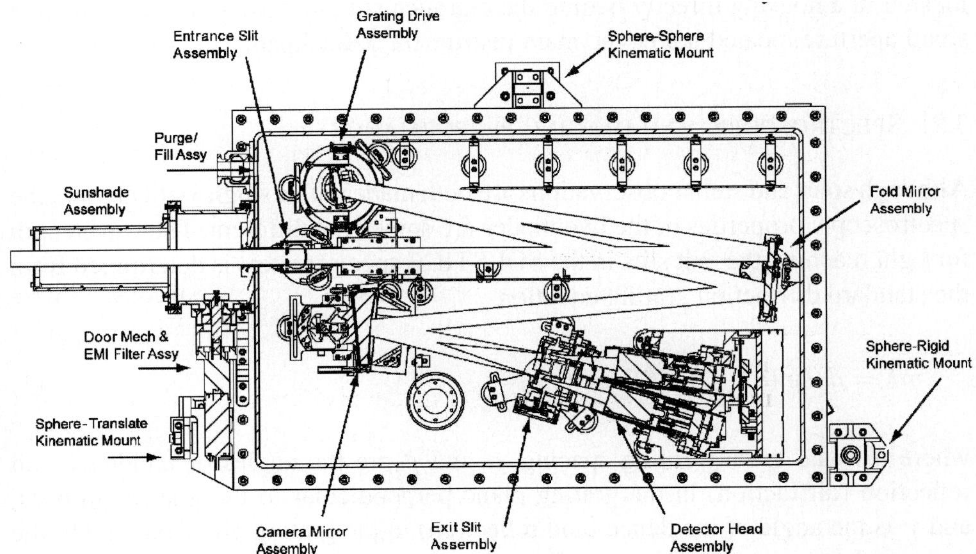

Figure 4. Optical-mechanical configuration of a single SOLSTICE II instrument.

the diffracted beam onto an exit slit assembly. SOLSTICE II optical elements are identical to those used in SOLSTICE I except for the ellipse, which was changed to increase the solar and stellar image distances by a factor of 1.5 to package side-by-side exit slit-detector assemblies. A two-position mechanism articulates the camera mirror $\pm 2^\circ$ about its vertex to illuminate either the inboard or outboard solar/stellar exit slit-photomultiplier tube. Two 10% transmitting neutral density filters, which are located between the inboard exit slits and CsTe photomultiplier tube, replace the fixed interference filter used in SOLSTICE I. These filters are mounted on two-position mechanisms and can be independently inserted in and withdrawn from the optical path. Moreover, the filters are tilted to prevent multiply reflected light from reaching the detector. The mechanisms carry both the filter and an uncoated filter substrate (window). Whenever a filter is removed, the window is automatically inserted in its place so that the illuminated area of the detector remains unchanged. An extended lightshade and a baffle assembly, located between the grating and the exit slit assembly, prevent stray light and out-of-band diffracted light from reaching the exit slit plane. Additional baffles trap light reflected from the grating into zero order, preventing it from scattering from other internal instrument surfaces. A thin-walled aluminum housing, not shown in Figure 4, completely encloses the exit slit-detector assembly so that the only light that passes through the exit slits can reach the detector input windows.

A solar position sensor (SPS) provides a direct real-time measurement of the Sun's location in the instrument field of view (FOV) and acts as a bright object sensor (BOS). In the event that the Sun comes within the FOV while the stellar entrance and exit slits are in place, a signal from the BOS will actuate the slit mechanisms, setting them to their proper solar positions. The SPS-BOS sensor is located in a housing directly behind the entrance slit and views the sky through a small aperture located below the main instrument optical path.

3.2. Spectroscopic and Imaging Performance

Although solar and stellar observations are both made with a single optical train, the spectroscopic properties of the two modes are somewhat different. The wavelength for light reaching the exit slits in the SOLSTICE spectrometers is determined from the standard diffraction grating equation

$$m\lambda = d\,(\sin(\alpha) + \sin(\beta))\,\cos(\gamma), \tag{1}$$

where d is the grating ruling spacing, α and β are the angles of incidence and reflection (diffraction) in the grating plane perpendicular to the grating grooves, and γ is the angle of incidence (and reflection) in the grating plane parallel to the grooves. It is convenient to recast Equation (1) in terms of the SOLSTICE layout geometry parameters. Assuming that $\cos(\gamma) = 1$, the solar wavelength equation becomes

$$\lambda_{\mathrm{Solar}} = 2d \sin(\theta_{\mathrm{S}}) \cos(\phi_{\mathrm{G}}/2), \tag{2}$$

where $\phi_{\mathrm{G}} = \beta - \alpha$ is the angle between the centers of the two fold mirrors, seen from the grating, and θ_{S} is the grating rotation angle measured from the bisector of ϕ_{G}. The instrument functions as an objective grating spectrometer in stellar mode; therefore, pointing offsets between a stellar target and the spectrometer optic axis give rise to first order wavelength shifts and the stellar mode wavelength equation becomes

$$\lambda_{\mathrm{Stellar}} = 2d \sin(\theta_{\mathrm{S}} + E/2) \cos(\phi_{\mathrm{G}} - E/2), \tag{3}$$

where E is the pointing offset, measured in the same direction as α.

The ellipse is designed to provide stigmatic imaging at the nominal solar mode conjugate object distance ($O = 1775.4$ mm) and image distance ($I = 200.5$ mm); however, astigmatism, caused by placing the plane grating in a diverging beam, results in a wavelength dependent shift in image location. This requires that each channel be focused near the middle of its spectral range where $I \sim 201$ mm. The image is slightly out of focus by $\Delta I \sim \pm 0.15$ mm and $\Delta I \sim \pm 0.21$ mm at the ends of the G and F scan ranges, respectively. This causes a negligible degradation in solar spectral resolution. In stellar mode the grating is illuminated by collimated light and the ellipse images all wavelengths in the stellar focal plane at $I = 180.1$ mm.

The spectral bandpasses for the two modes are given by

$$\Delta\lambda_{\mathrm{Solar}} = d \cos(\beta) \frac{\Delta w}{f_2} \frac{f_1}{D_{\mathrm{g}}}, \quad \text{and} \quad \Delta\lambda_{\mathrm{Stellar}} = d \cos(\beta) \frac{\Delta w'}{f_2'}. \tag{4}$$

β is the grating angle of diffraction, $f_1 = 1775.4$ mm and $f_2 = 201$ mm are the ellipse conjugate distanced in solar mode, and $f_2' = 180.1$ mm is the stellar focal distance, $D_{\mathrm{g}} = 920$ mm is the distance from the solar entrance aperture to the grating, and Δw and $\Delta w'$ are the solar and stellar exit slit widths, respectively. Values for the instrument spectroscopic parameters are summarized in Table II.

The SOLSTICE II spectrometers use replicas from the 3600 grooves mm^{-1} holographic master grating produced by Jobin-Yvon for the Space Telescope Imaging Spectrograph (Content *et al.*, 1996) and the exit slit widths have been increased from the UARS design by a factor of 1.5 to match the increased camera focal length. This doubles the spectral resolution in the F channel and increases the number of grating steps from 2048 to 4350 required to scan the 170–320 nm spectral range.

The ellipsoidal camera mirror in SOLSTICE II is designed to produce a stigmatic image of the solar entrance slit at its near conjugate focus. Rotating it by $\pm 2°$ about its vertex to select either the G or F detector introduces astigmatism and coma in the resulting off-axis images. Ray trace analysis was used to determine the magnitude of these defects. The calculations included a complete simulation of Fraunhofer diffraction, which expands the nominal $f/108$ solar input beam for the longer wavelengths to $\sim f/45$ as it passes through the 0.1 mm square solar

TABLE II
SOLSTICE II spectroscopic parameters.

Parameter	FUV channel	MUV channel
Wavelength range	115–180 nm	170–320 nm
Grating ruling density	3600 grooves mm^{-1}	3600 grooves mm^{-1}
Solar entrance slit	0.1 mm × 0.1 mm	0.1 mm × 0.1 mm
Solar exit slit	0.0375 × 6 mm	0.0375 × 6 mm
Solar bandpass	0.1 nm	0.09 nm
Stellar entrance slit	16 mm diam.	16 mm diam.
Stellar exit slit	0.75 mm × 6 mm	1.5 mm × 6 mm
Stellar bandpass	1.1 nm	2.2 nm
Detector photocathode	Cesium iodide (CsI)	Cesium telluride (CsTe)

entrance aperture. This widening of the Sun's geometrical image is accommodated by orienting the square entrance slit 45° to the dispersion plane and designing all the optical elements (except for the first mirror which is already oversized to accept the stellar beam) to have square apertures. In this way the extreme rays of the diffraction pattern are aligned to the diagonals of the grating and mirrors. Ellipse imaging quality degrades more rapidly toward outboard angles and the best balance is achieved when the mirror is rotated 2.5° inboard and 1.5° outboard. Ray trace simulations of instrument spectral line profiles for both the inboard ($\Theta = 2.5°$) and outboard ($\Theta = -1.5°$) demonstrate that the instrument spectral profile is very nearly triangular (inboard) or trapezoidal (outboard) with a full width at half maximum (FWHM) equal to the nominal slit width and that chromatic aberration arising from the Monk-Gillieson configuration causes a negligible change in SOLSTICE II spectral resolution (McClintock, Rottman, and Woods, 2000).

3.3. Instrument Optical Bench and Housing

The SOLSTICE II mechanical assembly, shown in Figure 5, consists of an optical bench and cover that provide a vacuum enclosure for the optical components. Both the bench and cover are composite structures that were fabricated by Program Composites Incorporated using a pair of graphite-epoxy face sheets bonded to a titanium honeycomb core. Face sheet materials were selected to provide a coefficient of thermal expansion (CTE) in the plane of the bench that is for all practical purposes equal to zero. This approach resulted in a rigid structure that reduced requirements for instrument thermal control and provided a stable mounting surface for the optical and detector elements. The face sheets were processed to minimize expansion from water vapor absorption by baking them and laminating them with thin aluminum films before bonding them to the honeycomb core. Four U-shaped

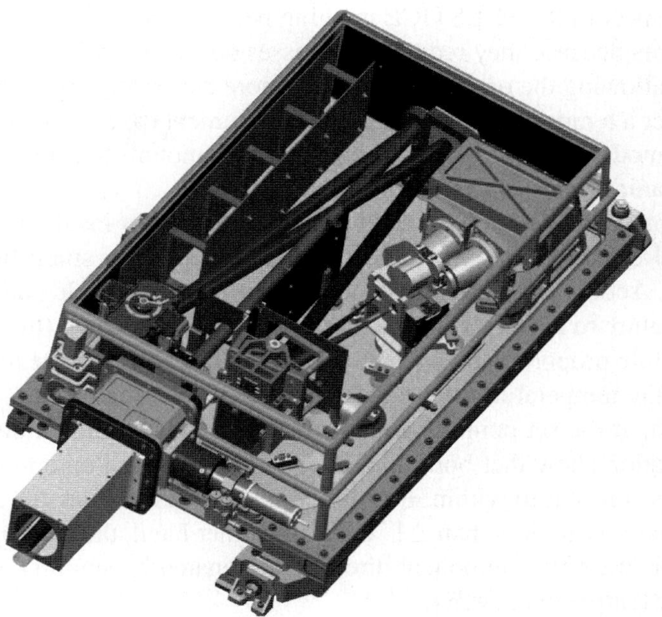

Figure 5. SOLSTICE instrument assembly.

closeouts, designed to match the mechanical characteristics of the face sheets, surround the core. They protect the core from damage and strengthen the bench at the cover attachment locations. Vents in the top of the cover and the bottom of the optical bench and small punctures in the honeycomb walls allow the pressure of interior volumes of the bench and cover to equilibrate with the external environment.

Interferometric tests performed on the final assemblies in both ambient conditions and vacuum indicate that the benches are dimensionally stable, eliminating any requirement for tight on-orbit temperature control to maintain instrument focus. On the other hand, there is evidence of a slight twisting when the benches are differentially heated to produce a front-to-back temperature gradient. The magnitude of the distortion is on the order of 1–2 arcsec for each centigrade degree of front-to-back temperature difference. These values are essentially unchanged when the bench and cover were tested as a unit. One proposed cause for this twisting is the mismatch between the bench CTE perpendicular to its plane, which was designed to be nearly equal to that of aluminum, and that of the closeouts, which are nearly zero.

Titanium inserts, bonded to the top face sheet, provide attach points for aluminum mounts, which hold the optical components and detectors. The attachment points for these mounts are flexures that accommodate the bench-to-aluminum mismatch in CTEs over the large temperature range over which the instrument must survive during ground test and qualification (−40 °C to +50 °C). Their orientation and placement on the bench reduce residual twist and defocus below the detection limit with respect to the SOLSTICE imaging performance. Flexures are preferred to bolted joints because they reduce the stresses on the insert bond lines and minimize creep allowing the optical elements to more accurately return to their initial locations after a temperature excursion. The instrument case is also attached to the observatory module structure using three kinematic mounts to avoid thermal stress in that mechanical interface as well.

Instrument thermal control is achieved using multi-layer insulation (MLI) blankets, thermal coatings on selected external surfaces, and a small heater located on the cover. Acting alone, the MLI and thermal coatings would cause the instrument temperature to fall below the design operating temperature (the operating set point). A simple proportional control system supplies power to the heater in order to maintain the temperature of a single thermistor, mounted in the center of the optical bench, at the set point. Data returned from the instrument during the first year of operation show that both optical bench and internal component temperatures remains constant to within $\pm 0.5\,^{\circ}$C on time scales of hours or days and that seasonal variations are less than $\pm 1\,^{\circ}$C. On the other hand, the instrument vacuum door, which is the only component directly illuminated by sunlight, shows a $2\,^{\circ}$C orbit-to-orbit temperature swing.

Two temperature set points, $17\,^{\circ}$C and $31\,^{\circ}$C, are defined for SOLSTICE. During the first years of the SORCE mission the instrument is controlled at the lower set point because the thermal coatings and MLI are efficient at removing heat from the instrument. As blankets and coatings age, they may become less efficient in which case the heater power may lower to zero. At that time, switching to the higher set point will still maintain a stable temperature environment.

An o-ring machined into the flange of the instrument cover creates a vacuum seal between it and the bench. The use of a vacuum housing is required for contamination control, allowing the instrument to be pressurized to 1.2 atmospheres with ultra-pure argon during ground testing and launch. A hand-actuated purge-fill valve, mounted to the cover at the front of the instrument, provides access for evacuating and back-filling the instrument without opening the main door, which could expose the entrance slit edges to damage caused by differential pressure. After launch and during vacuum test, a remotely operated valve mounted on the back of the cover is used to evacuate the instrument.

Light enters the instrument through a door equipped with an o-ring seal. This arrangement allows the instrument to be transported to the SURF facility under pressure and attached to the calibration beam line before being evacuated with the vent valve; thus, the ultra-clean SURF beamline is exposed to only the instrument interior. After calibration the door is closed before removing the instrument from the beamline. On-orbit, the door is actuated once at the beginning of instrument observations and not reclosed.

The SOLSTICE II is designed to minimize contamination of the optical surfaces. This begins with the use of a vacuum enclosure to isolate the instrument interior from the external environment. Careful attention is also given to the internal components.

Both the bench and cover that comprise the interior of vacuum housing are covered with an aluminum film to eliminate large reservoirs of water vapor inside the instrument as well as to cover large areas of potential organic materials with clean metallic surfaces. Paint is strictly forbidden. When necessary, baffle and housing surfaces are blackened using a nickel-plating – deep etch process developed at the Goddard Space Flight Center. The electrical harness is made from a copper-plated kapton-based rigid circuit rather than from bundles of Teflon coated stranded wire. Large printed circuit boards are all mounted outside the vacuum enclosure. To this end, the detector housing itself is a stand-alone vacuum housing with o-ring seals that expose only the photomultiplier tube windows to the instrument interior. A titanium vacuum feed-through bonded into the optical bench provides electrical cable access and venting for the interior of the detector head. A similar arrangement also isolates the SPS printed circuit boards from the optical cavity. In most cases, it is possible to reduce the electrical connections to the various internal mechanisms to short cable runs of 4–8 wires each. Although both SOLSTICE II instruments exhibited an unexpected loss of responsivity during ground test (up to $\sim$15% but only in $\pm$20 nm wavelength range centered near 200 nm), they show only a slight ($\sim$5% per year) additional loss over their entire operating range since launch (McClintock, Snow, and Woods, 2005; Snow *et al.*, 2005a).

3.4. Solar Position Sensor

The largest in-flight uncertainty in SOLSTICE I measurements arises from uncertainties in instrument pointing and corrections for field-of-view (FOV) nonuniformity (Woods, Rottman, and Ucker, 1993, their Table III). SOLSTICE II minimizes the uncertainty in pointing correction by directly measuring solar and stellar positions rather than relying upon values derived from spacecraft attitude sensors as done for SOLSTICE I.

TABLE III

Grating drive performance requirements.

Parameter	Value
Angular range	50°
Science step size	13.5 arcsec
Grating step size	0.5 arcsec
Accuracy	$\pm$1 arcsec
Repeatability	$\pm$1 arcsec
Settling time	To within 0.25 arcsec in 0.05 s
Jitter amplitude	0.25 arcsec for $f < 2.5$ Hz
	0.40 arcsec for $f > 10$ Hz
Slew rate	2.5° s^{-1}

An SPS, which is mounted in a housing that attaches to the optical bench directly behind the solar/stellar aperture mechanism, measures solar position in the instrument FOV. It consists of 0.9 mm square aperture 17 mm in front of a quadrant-diode sensor. In this configuration, the aperture casts a nearly square pinhole image of the Sun onto the diodes, which has an approximate width of 1.1 mm. Each quadrant is a separate diode with its own amplifier, which converts diode current into voltage, followed by a voltage to frequency converter. Output frequencies are fed into counters that produce a digital signal from each circuit. Count values are read during each science integration period and telemetered as part of a science data packet (see Section 3.8).

If the diodes are labeled 1, 2, 3, and 4 clockwise from the upper left corner, then

$$\theta_{\mathrm{SPS}} = K_\theta \left[\frac{(C_1 + C_4 - C_2 - C_3)}{(C_1 + C_2 + C_3 + C_4)} - X_0 \right] = K_\theta [X - X_0], \quad \text{and}$$
$$\phi_{\mathrm{SPS}} = K_\phi \left[\frac{(C_1 + C_2 - C_3 - C_4)}{(C_1 + C_2 + C_3 + C_4)} - Y_0 \right] = K_\phi [Y - Y_0]. \tag{5}$$

θ and ϕ are the position angles of Sun center relative to the instrument optic axis that are computed from the normalized outputs X and Y, which vary linearly with angle over the central 0.5° of the SPS FOV. The SPS aperture-to-diode spacing was chosen so that K_θ and K_ϕ are approximately equal to 1°. Accurate values for these constants as well as the offsets between the SPS and instrument optic axes (X_0 and Y_0) were determined during ground calibration at SURF. In flight cruciform alignment scans using the spacecraft pointing system to raster-scan the Sun across the instrument FOV were performed to measure small SPS-to-instrument shifts caused by launch (McClintock, Snow, and Woods, 2005). Additional cruciform scans are used to check that quadrant electrical offset and amplifier gains have remained stable since launch. Both ground and flight experiments indicate that solar position values calculated from Equation (5) are accurate to within 1 arcmin over the central 0.5° of the SPS FOV.

In stellar mode SOLSTICE is an objective grating spectrometer, and the wavelength scale depends directly on star position in the entrance aperture. Field-of-view variations and wavelength scale offsets are both determined by scanning the grating through zero order at the beginning and end of each stellar observation. The position calculated from the zero order centriod is a direct measure for the value of '*E*' in Equation (3). Placing one SOLSTICE instrument perpendicular to the other (Rottman, 2005) provides a measure of the stellar offset both parallel and perpendicular to each instrument's plane of dispersion.

Because SOLSTICE is essentially a pinhole camera, the 0.1 mm square entrance aperture forms a 4.75 mm diameter image of the Sun on the first mirror. This image produces an average irradiance on the mirror that is 1800 times smaller than the irradiance that would result from viewing the Sun through the stellar entrance aperture. SOLSTICE I experienced significant degradation in responsivity from a single-day exposure with the stellar aperture (Woods *et al.*, 1998). Therefore, the SPS was designed to also serve as a bright object sensor (BOS) that is used to prevent the instrument from viewing the Sun when the stellar entrance aperture is in place. Output voltages from the four quadrants are added in the SPS electronics to produce an SPS sum signal that is fed directly to the instrument generic channel interface (GCI) electronics board (see Figure 10). If the sum value exceeds a preset limit while the stellar aperture is in place, a control circuit actuates the entrance mechanism to switch to the solar aperture. The trigger level for the BOS, which was accurately determined before launch, can be modified by ground command to account for changes in SPS sensitivity during the mission.

3.5. Grating Drive Assembly

Because SOLSTICE is a monochromator, the wavelength of light passing through the exit slit and impinging upon the detector is a function of the diffraction grating rotation angle (Equation (1)) and spectra are acquired by stepping the grating through a series of discrete, usually equally spaced, angles and recording the spectrum with one of the detectors. The step size for standard solar irradiance measurements is 0.00375° (13.5 arcsec), which changes the wavelength at the exit slit by approximately 1/3 of the nominal spectral resolution (~0.1 nm). Experience from the SOLSTICE I indicates that each grating position from step to step should be both accurate and repeatable to 1 arcsec of angle. These position requirements and considerations related to data collection efficiency led to the grating drive performance requirements summarized in Table III.

The grating drive assembly is shown in Figure 6. It consists of a cylindrical housing, which supports a single axis rotation stage, actuated by an Aeroflex 2T10Y10 brushless direct current torque motor, and a high-resolution optical rotary encoder. A pair of ABEC-7 (Angular Bearing Engineers Committee) bearings, one at either end of the shaft, supports the shaft, which is pre-loaded from the top by a segmented diaphragm. This arrangement effectively removes the side play and maintains near constant contact pressure between the balls and races of the bearings over the operating temperature of the mechanism (−40 °C to +50 °C). The grating is mounted in a housing, which is an integral part of the shaft, using soft-side pads and spring contacts that press the grating face against three hard locating pads. A steel dowel pin, located just below the cell, secures the motor rotor to the shaft. The encoder disk is bonded to a split hub, which is pinned to the shaft below the top bearing. Two encoder read heads, mounted to the housing on opposite sides of the assembly, sense the rotational motion of the encoder disk.

The encoder was designed and developed by MicroE, Inc. of Natick, MA and repackaged for spaceflight application by LASP. Conventional optical encoders employ a pair of grids to produce and detect geometrical shadows. For high-resolution applications, the detailed shape of the intensity pattern produced by the two grids

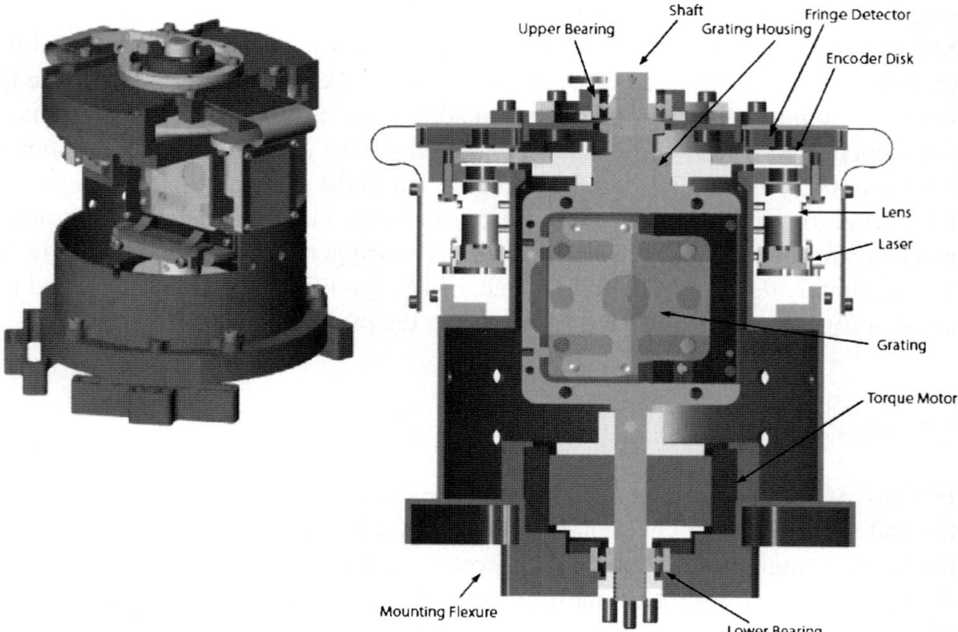

Figure 6. Grating drive assembly. The *upper left* figure is an isometric drawing of the grating drive assembly. Internal components are visible in the central cross-section view shown in the larger *right panel.*

is critically dependent upon their spacing and usually places very tight mechanical tolerances on the mechanical structure. In contrast, the MicroE device uses diffractive optics to generate a pair of overlapping coherent beams that produce a sinusoidal fringe pattern, which is detected by an array of photodiodes.

As long as the detector remains within the overlap region of the two beams, the period and shape of the fringe pattern is independent of its position; therefore, this approach leads to significantly relaxed mechanical tolerances and results in a simple system that has sub-arcsec resolution (Horwitz, 1996).

Figure 7 shows a schematic diagram of the SOLSTICE II encoder. An aspheric, plano-convex lens collimates the output from a diode laser ($\lambda = 920$ nm) to form a coherent beam, approximately 5 mm in diameter. As it passes through the diffractive optic wavefront compensator (WFC), the beam is divided into two equal parts traveling at angles α and $-\alpha$ with respect to the original direction. These two beams impinge upon a diffraction grating where they are diffracted into 1st order and – 1st orders of the grating, corresponding to angles $-\beta$ and $+\beta$, respectively. After diffraction they travel a distance (*d*) to the diode array where they produce a sinusoidal interference pattern that has spatial period, $P_{\rm N}$, where

$$P_{\rm N} = \lambda / \sin(2\beta). \tag{6}$$

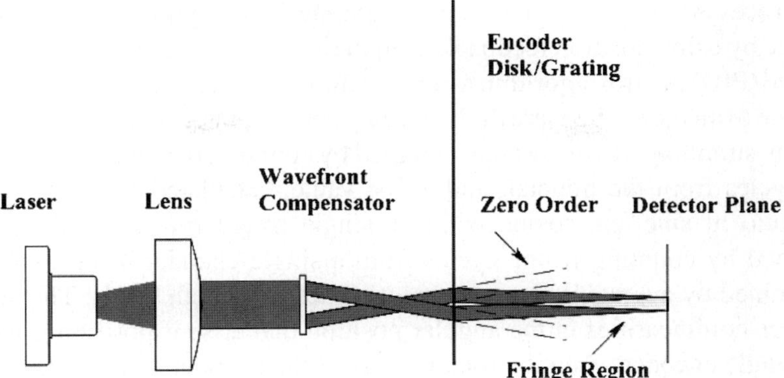

Figure 7. Schematic diagram of the MicroE encoder. The encoder disk/grating diffracts two impinging beams toward the detector plane where they interfere producing sinusoidal fringes, which are detected by a diode array.

The fringe period is adjusted by choice of α and β so that one complete cycle of width P covers four photodiodes. As the grating moves relative phase shifts in the ± 1 first order beams cause the fringe pattern on the detector array to move in the same direction producing nearly sinusoidal outputs from the four diodes that are sequentially shifted in phase by $\pi/2$.

In the SOLSTICE II encoder, the two laser beams impinging upon the grating have angles $\alpha = \pm 10.067°$ so that they are diffracted into angles $\pm\beta = 0.527°$ producing fringes having spatial period, $P_{\mathrm{N}} = 50\ \mu$m. The diode array has 32 elements spaced on 12.5 μm centers with the outputs from every fourth diode (e.g., outputs from diodes 1, 5, 9, ...) electrically connected within the detector package. As the grating moves, the array produces four phased sinusoidal signals, which are averages of eight fringe cycles. The encoder grating was manufactured by photoetching a pattern of 41 000 equally spaced radial groves, 0.5 mm long, into an annulus of a fused silica disk at a distance of 32.63 mm from the center. This produces grating with nonparallel grooves and a mean spatial period of 5 μm. The WFC is designed to compensate for nonparallel grooves so that the fringes produced at the detector are straight and parallel to the array axis. Rotating the grating through an angle of 15.8 arcsec causes the fringe pattern at the detector to undergo a phase change 2π.

The encoder disk has a second annular track, located at a radius of 31.76 mm. A binary zone lens, etched into this track, focuses light from a third aperture in the WFC into a 20 μm wide image that impinges upon a pair of diode chips, displaced by 0.85 mm with respect to the main diode array. Output from these two diodes is used to produce an index pulse that provides an absolute angular fiducial for the grating drive controller.

An application specific integrated circuit (ASIC) processes the raw output from the detector module and produces sine and cosine signals, both scaled to 2 V peak-to-peak, and a nominal 3.5 V index pulse. These are fed to Temic TSC21020F digital signal processor (DSP), which controls grating drive motion to a resolution of 0.5 arcsec by using position feedback from the encoder and a proportional-integral-derivative (PID) control algorithm to actuate the torque motor. Analog values of sine and cosine from the encoder are digitized and used to compute grating position. This is done by summing a coarse value, obtained by counting the number of 15.8 arcsec fringe cycles from the fiducial, and a fine value, calculated from the arctangent of the ratio of sine and cosine within a single fringe cycle. Coarse position is determined by counting fringe cycles from a single encoder head. Fine position is determined by averaging the arctangent values from both heads. This technique minimizes nonlinearities in the angular position that arise if position is calculated from a single encoder when the rotation axis of the encoder disk is displaced from the center of the radial grating pattern. After the DSP determines the current grating position, it compares that value with a target value and computes an error signal that is proportional to the difference. The PID algorithm processes this signal to produce a digital drive signal, which is converted to voltage in a digital-to-analog converter (DAC) and sent to the torque motor. No control voltage is sent to the motor if the difference between target position and current position is less than ± 0.5 arcsec.

The grating drive has two operational modes – slew and science step. Slew is used to rapidly move the grating over large angle at a rate of 2.5° per second. In science step mode, which is used during data acquisition, the DSP rotates the grating through a small angular displacement. This is followed by an integration time, during which the photomultiplier detector counts photons (see Section 3.10).

McClintock, Snow, and Woods (2005) characterized grating drive performance during ground test using multiple emission lines from a mercury discharge lamp. At constant temperature the relative angular position, measured from the fudicial, was accurate to $\sim$1 arcsec over the entire wavelength range ($\sim$40° of grating rotation). On the other hand, the absolute angular position, which was determined by locating the fiducial (during grating drive initialization or whenever a 'find index' command is executed), was much less accurate. Changes in fiducial location up to ± 35 arcsec were occasionally observed after the grating drive was power cycled. It is not clear if these shifts were caused by hardware, software, or both. The SOLSTICE wavelength scale is also temperature dependent. Measurements of reference line positions made during thermal vacuum testing showed that the wavelength scale shifts at a rate of 0.004 nm per centigrade degree over the temperature range of 20–30 °C. These performance characteristics have also been verified in flight.

3.6. Exit Slit Filter

The detectors in SOLSTICE I are too sensitive for direct observation of solar irradiance for wavelengths longward of 250 nm; therefore, the F and N channels use fixed interference filters, located behind their respective exit slits, to attenuate and shape the spectrum before it reaches the detectors. The F channel filter has a peak transmission of ~35% at a wavelength of 185 nm, which results in a reduction approaching 99% at 320 nm. This approach has two disadvantages. First, it also reduces the stellar flux arriving at the detectors by an identical amount. Second, the multiple thin layers in an interference filter diffuse into one another over time resulting in wavelength dependent changes in transmission (Woods *et al.*, 1996).

As an improvement SOLSTICE II uses a neutral density filter, which is constructed from a fused silica substrate coated with a thin chromium film, to attenuate the solar flux reaching the detector by a factor of 10 for wavelengths longer than 220 nm. A bi-stable mechanism is used to insert either the filter or a bare fused silica substrate into the light path between the exit slit and the detector. Inserting the window maintains the optical path length through the system without significantly reducing the intensity at the detector. The MUV detectors can accurately record the solar spectrum over the entire SOLSTICE II wavelength range; nonetheless, the filter is used for routine solar observations to prolong the life of the photomultiplier tube. A second, independent, filter-window pair is located between the first assembly and the detector for redundancy. Both filter transmissions (relative to their respective window transmissions) are measured daily (weekly since the beginning of the second year of operations) during a single spacecraft orbit by scanning the spectrum first with the filter in the beam and then with the window in the beam. No change in transmission has been observed for either filter, and for either instrument, during the first 18 months of the mission.

3.7. Detector Assembly

Light diverges as it passes through the exit slit-filter assembly until it impinges upon one of two (depending on ellipse orientation) photomultiplier detectors. Hamamatsu R2078s, which were developed for oil well logging applications, were selected for SOLSTICE II. They replace the EMR Photoelectric 510 series detectors used in SOLSTICE I, which have been discontinued. The SOLSTICE II detectors operate in pulse counting mode with grounded photocathodes and output pulses capacitively coupled into Amptek A-121 pulse-amplifier-discriminators. A-121 PADs are high sensitivity ($\sim 5 \times 10^4$ e^-), hybrid devices with a pulse pair resolution of ~ 60 ns.

Although the R2078s' borosilicate glass envelopes make them less rugged than 510 detectors, their larger active area significantly increase the instrument FOV uniformity (Drake *et al.*, 2000, 2003). Their multiplication sections have a 10-stage circular-cage configuration with Be-Cu dynodes, which produce stable, saturated pulse height distributions. Multiplier modal gain is photocathode dependent. Typical gain values for the SOLSTICE II detectors, which operate with a multiplier potential of ~1850 V, are $G \sim 2 \times 10^6$ electrons for CsTe and $G \sim 1 \times 10^6$ electrons for CsI.

The R2078 detectors were subjected to a comprehensive characterization and qualification program because they had not previously been used for a long-duration space-flight application. Characterization measurements in all tubes included quantum efficiency, photocathode spatial uniformity, and pulse height distribution, measured as a function of accumulated counts ranging from an initial value of 10^8 to a final value of 10^{10}. Two life test tubes were exposed to accumulated total count doses of 7×10^{12} (CsTe) and 5×10^{11} (CsI) over a period of 6 months. Although the modal gains declined by 27% for CsTe and 57% for the CsI, there was no detectable loss in responsivity with amplifier thresholds set at flight values (2×10^5 e^- for CsTe and 1×10^5 e^- for CsI), and no detectable change in spatial uniformity. The total test doses are equivalent to 2.5 and 8.3 years of nominal on-orbit operations for the SOLSTICE II primary detectors, which are accumulating counts at rates of $\sim 2.8 \times 10^{12}$ per year for the MUV channel (CsTe) and 6.1×10^{10} per year for the FUV channel (CsI). Because the alternate detector rates are lower ($\sim 3.2 \times 10^{10}$ per year and 2.8×10^{10} for MUV and FUV, respectively), SOLSTICE II should easily complete the nominal mission without significant change in detector characteristics.

Figure 8 shows a cross section drawing of a single detector assembly. The 25 mm diameter $\times$ 54 mm long tube is registered inside a Delrin® sleeve by a pair of o-rings. Twelve feed-through pins, which provide the electrical connections to the dynodes and photocathode, exit the detector at the rear. The front o-ring serves to both mechanically locate the detector and provide a vacuum seal between the detector input window and the instrument housing. Everything to the left of the front o-ring resides in the instrument optical cavity, while everything to the right is part of the detector housing. A 25 mm diameter circuit board, which provides the resistor divider chain for the high voltage, is soldered directly to dynode pins. The entire area around the output pins and divider circuit is encapsulated to provide

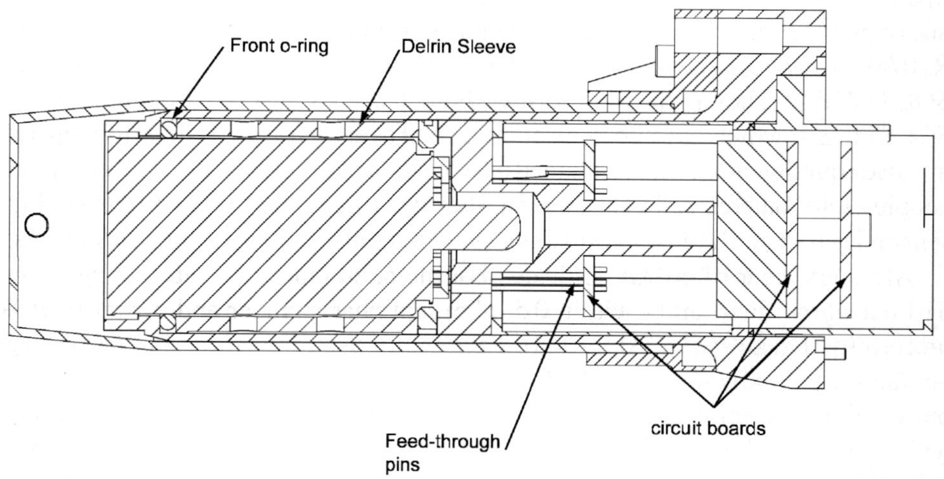

Figure 8. SOLSTICE II photomultiplier tube assembly cross-section view.

mechanical rigidity and protection from high-voltage arcing. Two additional 25 mm diameter boards carry an A-121 amplifier and its output-pulse driver electronics. A μ-metal housing with a 15 mm diameter aperture, located 10 mm in front of the input window, surrounds the entire photomultiplier tube to protect it from magnetic fields. Laboratory tests show that this shield geometry is adequate to eliminate magnetic field effects for strengths at the detector of up to 15 gauss, which is a factor of 5 greater than ambient fields produced by torque motors, located in the exit slit housing ($\sim$2.5 gauss at the tube) and the Earth's field (0.5 gauss).

Two effects cause the SOLSTICE II detector outputs to exhibit a small, but nonnegligible dependence on temperature. The smaller of the two arises from variations in the high-voltage power supply output and divider string resistance, which cause the multiplier gain to change with temperature. For the SOLSTICE II detectors, the measured values for the relative change in output is $\sim 5 \times 10^{-5}\,^{\circ}\mathrm{C}^{-1}$. The larger temperature sensitivity arises from the fact that the quantum efficiencies of all photocathodes are temperature dependent. Temperature coefficients depend upon both photocathode material and the wavelength of the incoming light. The relative changes for the SOLSTICE II detectors can be as large as $0.008\,^{\circ}\mathrm{C}^{-1}$ for CsTe at the longest observed wavelengths. Controlling the detector head temperature limits the magnitude of these two effects to less than 0.5%. Precise measurement from a pair of thermistors, one attached to the phtotmultiplier tube envelope near the input window and another embedded in the encapsulation material near the rear of the tube body, are then used to normalize detector output to the single standard temperature adopted for the instrument ground calibration performed at SURF.

The complete instrument detector head is shown in Figure 9. It consists of a vacuum enclosure to which the two photomultiplier tube assemblies mount using o-ring seals. All of the electronic components required to support the detectors are contained within this housing. These include a pair of high-voltage power supplies, a power conditioning board, and a heater, which is used to maintain the detector head temperature stability to $\pm 1\,^{\circ}\mathrm{C}$, independent of the main instrument housing thermal control system. All these components are isolated from the optical cavity by the vacuum enclosure. A vacuum feed through, located in the optical bench, provides a vent to the exterior environment of the instrument as well as a pass through for the detector head electrical harness.

The detector head thermal control system has a pair of set points, 19.5 and 33.0 °C, which are only 2 °C warmer than the main instrument set points. Data from the first year of operations show that the photomultiplier tube temperatures have remained constant to within 0.5 °C except near summer and winter solstice when they increase by $\sim$1 °C and then return to the set point on a 4 week time scale.

3.8. Generic Channel Interface (GCI)

Instrument control and interface electronics are located in a GCI box that is mechanically separate from the main instrument housing. Figure 10 shows the system

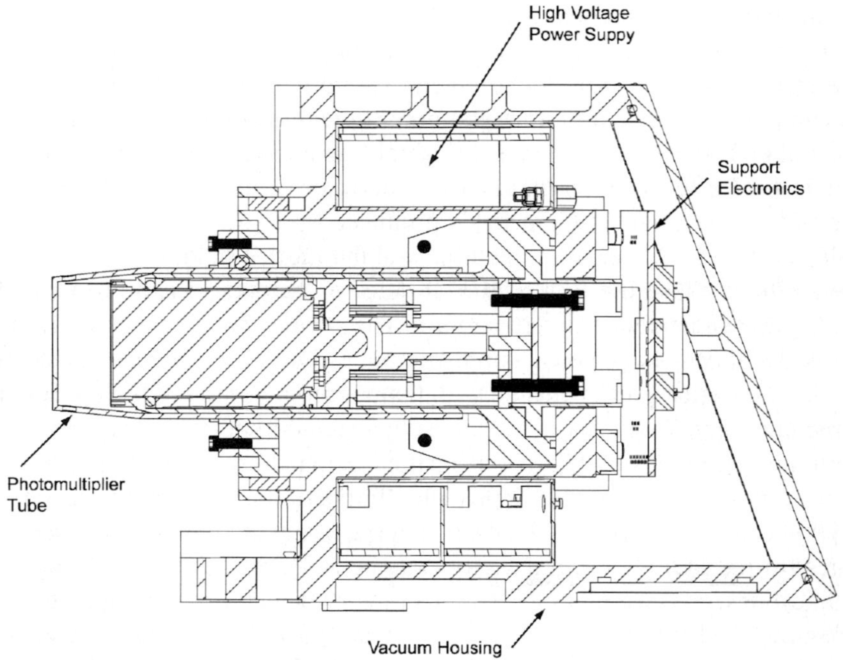

Figure 9. Section view of the detector head. It shows the placement of the high-voltage power supplies and support electronics within the detector vacuum housing.

electrical block diagram that summarizes the instrument external electrical interfaces and the interconnections among the three GCI boards and the instrument internal components.

The GCI board provides the power and data interfaces between the instrument and spacecraft. Switched +28 V power from the spacecraft is filtered and passed to a pair of Interpoint, Inc. DC-to-DC converters to produce conditioned ±15, +5, and +3 VDC for the instrument components. Command and telemetry links are provided by a field programmable gate array (FPGA), which communicates with the SORCE instrument microprocessor unit (MU) using a pair of universal asynchronous receiver transmitters (UARTs). The FPGA parses commands, interfaces with the grating drive DSP, controls instrument functions, and receives and packetizes both housekeeping and science data for transmission to the MU. Housekeeping data include voltage monitors from the power supplies (both high voltage and low voltage), temperature data from eight thermistors located on the optical bench, cover, detectors, and selected mechanisms, and pressure data from a single transducer mounted on the rear of the instrument cover. The FPGA samples and digitizes these analog voltages sequentially using a 32-channel multiplexer that feeds a 16-bit analog-to-digital converter (ADC). During normal operations the FPGA generates a housekeeping packet once per minute. The FPGA also assembles science packets from data taken at each grating step. These packets contain grating

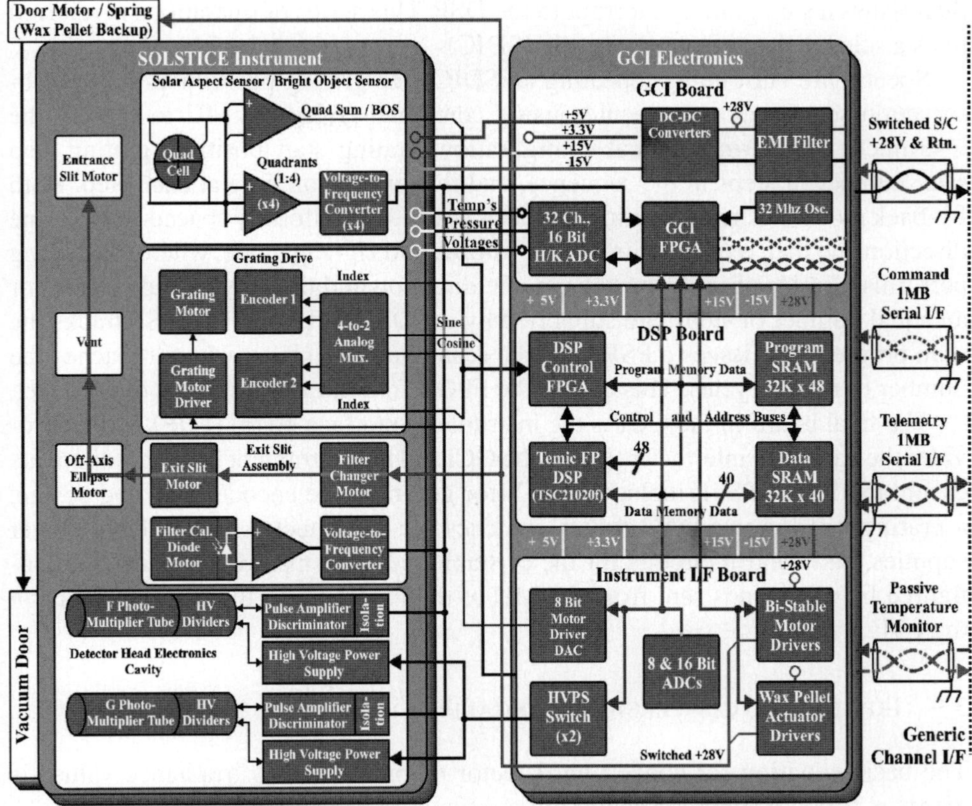

Figure 10. SOLSTICE II electrical block diagram. It shows the instrument components and the GCI, which contains three electronics boards, referred to as the GCI board, the DSP board, and the instrument interface (I I/F) board.

encoder positions, count values from both photomultiplier tubes, and SPS/BOS solar position data.

The DSP board contains a Temic TSC21020F floating point DSP, program and data static random access memory (SRAM), and an FPGA that provides DSP to GCI and handshaking and I/O. Grating drive commands issued by the MU are parsed by the FPGA and sent to the DSP, which responds with a command received valid (CRV) for valid commands. The DSP executes configuration commands (e.g., find index slew to position, and set grating step size) promptly and returns a position status report (PSR) to the MU.

During science data acquisition the DSP controls grating drive position using timing interrupts generated by the main FPGA on the GCI board. Once the grating is configured, DSP sends a PSR, which contains the absolute grating drive position, and issues an initiate integration command to the FPGA. The FPGA performs a clear counter, integrate, and read counter sequence for the PMTs and the SPS. It then issues a step grating interrupt to the DSP. This series of operations is referred to as a science data integration cycle (SDIC).

Spectra are built up by repeating an SDIC, one grating position at a time. Observation sequences are executed using commands from the MU to specify the instrument optical-mechanical configuration, grating start position, grating step size, number of steps in the sequence, and the integration time at each step. Both fly-back mode, where the grating performs a series of uniformly spaced steps in one direction and then returns to its initial position, and zig-zag mode, where the grating performs a series of uniform steps in one direction and then reverses direction for an equal number of steps, are supported by the DSP software. The DSP tracks the scan progress and issues a PSR with the scan complete bit set when it reaches the number of repeat cycles. This causes the FPGA to terminate a science data packet.

The final board in the GCI is the instrument interface board (I I/F), which provides the hardware interface between the GCI and the instrument for all components except the BOS/SPS. It includes ADCs for grating drive encoder sine and cosine, a grating drive motor DAC driver, switches for the detector high-voltage power supplies, and control circuits for the bi-stable mechanisms. Mechanisms are configured by commands sent from the MU to the FPGA using the driver circuits on the I I/F.

3.9. Irradiance Conversion and Calibration

The basic equation for converting detector output counts to irradiance values is identical for both solar and stellar observations:

$$E_{\mathrm{AU}}(\lambda) = \frac{C(\lambda, \tau, \mathrm{Dc}, \mathrm{Sl})}{R_{\mathrm{C}}(\lambda, T, \Omega)\mathrm{FOV}(\lambda, \Omega, \theta, \phi) A_{\mathrm{Entrance}} \Delta\lambda\, T_{\mathrm{Filter}}(\lambda) \mathrm{DEG}(t, \lambda, \Omega, \theta, \phi) f_{\mathrm{AU}}}, \tag{7}$$

where

$$C(\lambda, \tau, \mathrm{Dc}, \mathrm{Sl}) = \frac{S(\lambda)N(\tau) - \mathrm{Dc} - \mathrm{Sl}(\lambda) - \mathrm{St}}{\Delta t}. \tag{8}$$

$C(\lambda, \tau, \mathrm{Dc}, \mathrm{Sl})$ is instrument count rate at wavelength λ, which is computed by correcting the observed detector signal counts ($S(\lambda)$) for nonlinearity by applying a correction for detector electronic dead time ($N(\tau)$), for dark count (Dc), for scattered light ($\mathrm{Sl}(\lambda)$) and for stray light (St), and then dividing by the instrument integration period (Δt). $R_{\mathrm{C}}(\lambda, T, \Omega)$ is the pre-flight ($t = 0$) instrument quantum-efficiency-transmission function at the center of the field of view, which depends on wavelength, detector temperature (T), and the angular size of the target (Ω). A_{Entrance} is the area of the entrance aperture, and $\Delta\lambda$ is the spectral bandpass. $\mathrm{FOV}(\lambda, \Omega, \theta, \phi)$ and $\mathrm{DEG}(t, \lambda, \Omega, \theta, \phi)$ are factors that correct for instrument sensitivity variations, which are a function of target viewing direction (θ, ϕ), and time-dependent degradation, which is caused by extended on-orbit solar exposure, respectively. T_{Filter} is the transmission of the neutral density filter that attenuates the solar photon flux impinging upon the MUV detector. f_{AU} is a factor, which is calculated from earth orbital elements, that normalizes the solar irradiance values to a mean solar distance of 1 AU.

Pre-launch values for the various terms in Equations (7) and (8) were determined during instrument calibration and characterization. SURF III calibrations provided the primary data for establishing values for the product $R_C(\lambda)\Delta\lambda\, A_{\text{Entrance}}$ and for measuring $N(\tau)$, T_{Filter}, and $\text{FOV}(\lambda, \Omega, \theta, \phi)$. Detector temperature gain and dark count were measured during detector head characterizations using the calibration and test equipment (CTE) at the Laboratory for Atmospheric and Space Physics (LASP) at the University of Colorado (Drake *et al.*, 2000). Grating scatter measurements were performed at the Goddard Space Flight Center in Greenbelt, MD using the Space Telescope Imaging Spectrograph facility (Content *et al.*, 1996). These results were combined with instrument level measurements at SURF and LASP to determine values for $Sl(\lambda)$. Parameters appearing in Equations (7) and (8) that are likely to change during on-orbit operations, particularly $\text{DEG}(\lambda, \Omega, \theta, \phi)$ and Dc, are also routinely measured in flight. Routine experiments are also performed to search for changes in T_{Filter} and detector temperature gain, but none have been detected.

The application of Equations (7) and (8) to the three SOLSTICE II primary measurements and the associated error budgets are discussed in detail by McClintock, Snow, and Woods (2005) and Snow *et al.* (2005a).

4. SOLSTICE II Operations and Performance

4.1. Observing Strategy and Operations

One of the principal measurement goals for SOLSTICE II is to determine the solar ultraviolet irradiance with an accuracy of 5%. The strategy adopted to achieve this goal was to perform an accurate pre-flight calibration using SURF and to transfer that calibration to a subset of the SOLSTICE ensemble of stars before initiating solar observations. These first stellar observations provide a fiducial for the instrument ground calibration and minimize the possibility that initial solar observations will result in an undetected degradation in instrument sensitivity. Nighttime stellar observations are interleaved with daytime solar observations to provide a nearly continuous monitor of instrument in-flight sensitivity in order to meet goals for tracking solar irradiance variations with a relative accuracy of 0.5% per year. Additional in-flight solar and stellar experiments are interspersed with the nominal orbit-to-orbit observations in order to track instrument performance parameters as described below.

SOLSTICE II observations are built up from a single generalized detector integration-grating step sequence referred to as a science data integration cycle (SDIC), which is described in Section 3.8. SDICs are assembled into a number of standard observations as follows:

1. Nominal solar science scans: during nominal solar scans SOLSTICE A measures the MUV wavelength range (170–320 nm), while SOLSTICE B measures the FUV wavelength range (115–190 nm) using 0.5 and 1.0 s integration times, respectively. These scans, which are performed on approximately two out of every three spacecraft orbits, require approximately 50 min to setup and execute. A series of mini scans, one centered on HI, $\lambda = 121.6$ nm and the other centered on Mg II, $\lambda = 279.6$ nm, complete the observing on the daylight part of the orbit. These are used to study short-term variations in solar chromospheric emission.
2. A–B solar comparison experiments: once per week both instruments perform simultaneous FUV scans followed by simultaneous MUV scans. These provide a direct comparison of the relative degradation functions, $f_{\text{Degradation}}(\lambda, \theta, \phi)$, for the two instruments.
3. MUV filter calibrations: a filter transmission calibration is executed during a single orbit once per week (once per day during the first year of operations). It consist of a sequence of 4 MUV quick-scans (0.1 s integration time) with filter 1 in, filter 2 out; filter 1 out, filter 2 in; filter 1 in, filter 2 in; and filter 1 out, filter 2 out.
4. Solar 'T' scans and FOV maps: 'T' scans are cruciform maneuvers, covering $\pm 1^\circ$ in pitch and yaw, performed by the spacecraft once per week, that are used to map instrument response at each of four fixed wavelengths for both the FUV and MUV detectors. These measurements provide two quick-look cuts through $f_{\text{Degradation}}(\lambda, \theta, \phi)$. Field-of-view maps are complete spectral scans performed as the instrument boresight is rastered in angle over a 5×5 grid of points about Sun center with a step-to-step spacing of 5 arcmin. These maps provide a more complete picture of $f_{\text{Degradation}}(\lambda, \theta, \phi)$ than the T scans.
5. Stellar fixed wavelength observations: stellar fixed wavelength observations are the keystone for the SOLSTICE technique. Forty wavelengths throughout the entire wavelength range are used to track instrument sensitivity changes. Repeated 1 s integrations at fixed grating position are accumulated in order that the total counts recorded is on the order of 10^4, providing $\sim$1% statistical uncertainty in the measurement. All stellar observations are made during the nighttime. In most cases several wavelengths from a single star are measured during each observing session. An expert ground scheduling system selects the stars and wavelengths to be observed for any specific orbit based on the stellar priority listed in Table I and the time interval since the star was last observed at that wavelength. SOLSTICE A and SOLSTICE B simultaneously

observe a given star and a given wavelength, alternating days between the FUV detector and the MUV detector.

6. Stellar scans: Stellar spectral scans are performed on the brightest stars using a 0.28 nm grating step size (eight times the solar step size).
7. Stellar zero order measurements: In stellar mode, SOLSTICE is an objective grating instrument; therefore, offsets in instrument pointing lead to offsets in the wavelength scale. These potential offsets are measured at the beginning and end of each stellar observation by measuring the wavelength position of the stellar zero order image.
8. Stellar 'T' scans: Stellar 'T' scans serve the same purpose as solar 'T' scans, providing two cuts through $f_{\mathrm{Degradation}}(\lambda, \theta, \phi)$. These measurements provide a check on the solar 'T' scan results.

4.2. Data Products

The SORCE ground data system produces three data product levels (Level 1 through Level 3) for SOLSTICE (Pankratz *et al.*, 2005):

1. Level 1A and 1B are unprocessed instrument data at full resolution, which have been time-referenced and sorted by experiment type, and have all the ancillary information required for their conversion into physical units.
2. Level 2 represents the lowest level of scientifically useful data, where all calibration parameters are applied to the data, providing irradiances at full time and instrument spectral resolution. Both solar and stellar irradiances for the various standard observations, described in the previous paragraph, are produced at this level.
3. Level 3A are processed instrument data, time averaged and spectrally resampled onto a 1nm spaced wavelength scale. They include any time-dependent corrections that are not based on empirical models. Level 3B data are similar to Level 3A except that they include any additional time-dependent corrections that require the application of empirical models. This is the standard data product that is available to the general public on the project World Wide Web (WWW) site and distributed to a NASA Distributed Active Archive Center. Currently it consists of a single solar spectrum per calendar day, which is produced from the average of the Level 2 spectra that is resampled to the 1 nm wavelength scale. Future reprocessings of the Level 2 data will produce additional spectra with a 6-h cadence. Daily averages of the full resolution solar irradiances and both stellar fixed wavelength irradiance as well as stellar wavelength scans for the brightest stars are also produced at Level 3 and stored in the project database. Additional data products are being developed for public release. These include a high-resolution Mg II index (Snow *et al.*, 2005b) and daily averages of selected solar emission line strengths. These are further described by Pankratz *et al.* (2005).

4.3. Spectral Resolution, Coverage, and Reproducibility

SORCE was launched on 25 January 2003 and instruments were slowly checked out and commissioned during the following month. Routine solar observations began on 6 March, after a 1-week stellar observing campaign. These early solar observations verified that the instruments exceeded their design expectations for wavelength scale stability, spectral resolution, radiometric stability and sensitivity, and temperature stability.

Examples of Level 1 (detector counts versus grating position or wavelength) solar data are shown in Figure 11. FUV spectra are typically acquired using a 2265-step scan covering 115–190 nm with 1-s long integrations. MUV spectra cover 150–320 nm in 4975 steps using 0.5-s long integrations. The MUV scan shown in Figure 11 was acquired with both exit slit filters removed. During normal operations the grating is halted near $\lambda = 212$ nm and a filter is inserted in the light path. The grating then slews approximately 25 steps toward shorter wavelengths before continuing the scan, providing a small overlap region and reducing the count rate by a factor of 10 for $\lambda > 211.8$ nm.

Level 1 data are calibrated using algorithms that are based on Equations (7) and (8) resulting in 0.1 nm resolution solar irradiance spectra. Spectra collected during each 6-h interval are averaged and binned to 1 nm spectral intervals to produce the main SOLSTICE II Level 3 data product that has a radiometric precision better than 1% over the entire wavelength range.

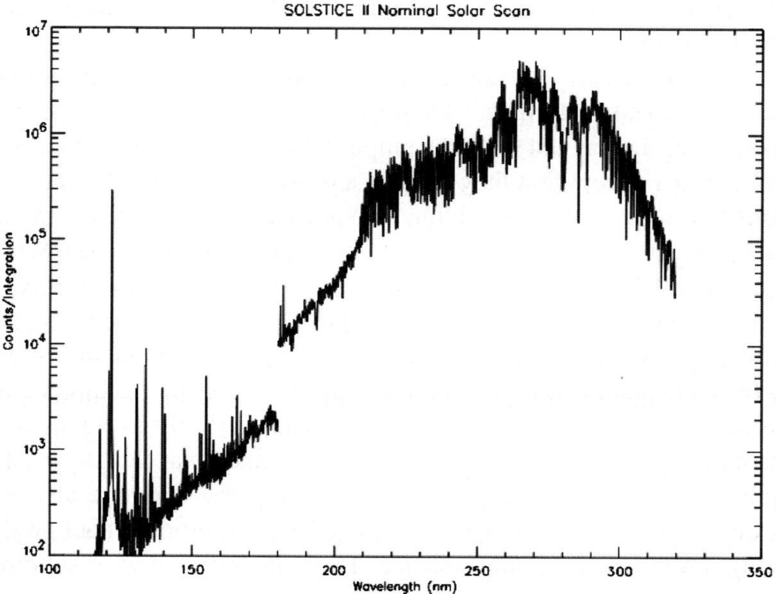

Figure 11. Nominal Level 1 solar data.

On an orbit-to-orbit basis the relative accuracy and stability of the wavelength scale is approximately ±0.001 and ±0.002 nm across the entire FUV and MUV range, respectively. The absolute position of the wavelength scale, which is much less accurate, can drift by ±0.03 nm in 24 h. The former error is likely caused by changes in the grating drive encoder performance, while the latter results from a combination of optical bench-case distortion and grating drive encoder performance. On occasion, offset errors as large as ±0.5 nm can be encountered if the grating drive is commanded to reset.

These errors appear to be associated with the algorithm that is used to locate the fiducial. It is not known if they result from hardware performance, software design, or a combination of both. In any event, both relative and absolute errors in the solar wavelengths are reduced to less than ±0.001 nm in the processed data (about 1% of the instrument bandpass) by fitting positions of emission lines in the observed spectrum to laboratory wavelength values using Equation (2) with

$$\theta_S = \theta_0 + N\Delta, \tag{9}$$

where θ_0 and Δ are the fitted zero order position and grating drive step size (nominally 0.5 arcsec), respectively, and N is the grating step number. Wavelength errors in stellar spectra are corrected by using $\theta_S + E/2$ (Equation (3)), which is measured directly from the zero order scans that proceed and follow each star observation, and the values of Θ_0 and Δ are taken from the most recent solar scan. This approach is adequate because the required accuracy for the stars is an order of magnitude less than for the Sun.

Wavelength scale registration and spectral resolution are illustrated in Figure 12, which shows a portion of a composite MUV spectrum near the Mg II h and k resonance lines. The spectrum was constructed by combining 16 nominal MUV solar science scans, obtained over 48 h beginning on 22 May 2003. First, values for θ_0 and Δ were determined for each scan in order to assign a specific wavelength to each count value. This produced 4967 pairs of wavelength-count values for each scan. Although the grating step numbers have the same values for each observation, grating encoder errors, case distortions, and Doppler shifts resulting from spacecraft orbital motion, cause the values for θ_0 and Δ to be slightly different for each scan. Next, the ensemble of pairs for all 16 spectra was sorted by increasing wavelength to produce a highly oversampled spectrum with an average wavelength spacing of 0.002 nm, which is a factor of 15 smaller than the 0.03 nm spacing for a single nominal scan. (If there were no differences in θ_0 and Δ, the resulting composite spectrum would have 16-fold degeneracy at each wavelength position.) The solid line in Figure 12 is a 0.01 nm resolution high-altitude balloon spectrum (Anderson and Hall, 1989), which has been convolved with a 0.07 nm wide triangular slit function and divided by a factor or 1.2 to match the solar absolute irradiance values. (The Anderson and Hall absolute irradiances near 280 nm are ~20% larger than currently accepted values while those from SOLSTICE II agree with them to within ~3% (Woods *et al.*, 1996; McClintock, Snow, and Woods, 2005).) The agreement

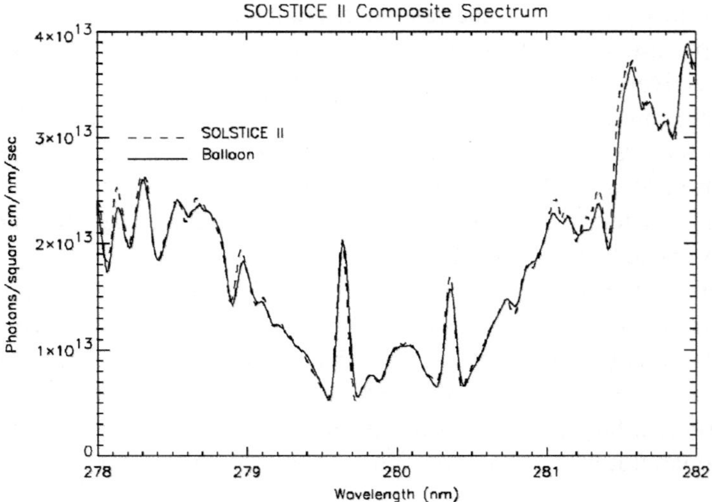

Figure 12. Composite SOLSTICE II solar irradiance spectrum (*dashed curve*) constructed from 16 individual nominal scans acquired during a 48-h period. The *bold curve* is a 0.01 nm resolution spectrum obtained from a research balloon (Anderson and Hall, 1989) that has been convolved with a 0.07 nm wide triangular slit function to match the SOLSTICE II resolution.

between the two measurements indicates that 0.1 nm resolution spectra from the SOLSTICE II instruments are reproducible at the 0.5% level over periods of days.

4.4. Radiometric Performance

McClintock, Snow, and Woods (2005) and Snow *et al.* (2005a) discuss the SOLSTICE II radiometric calibration and in-flight performance in detail. Their conclusions are summarized here.

Both SOLSTICE A and SOLSTICE B were calibrated at SURF III in the fall of 2001 before delivery to the spacecraft for flight integration and test. SOLSTICE A was returned to SURF in September 2002 for a recalibration. Those measurements showed that the instrument sensitivity was unchanged over most of the wavelength range. The exception was a ±20 nm wide region centered near 200 nm that showed a significant loss in responsivity (up to ∼15%). The SORCE project has adopted the results from September 2002, which are shown in Figure 13, as the official SOLSTICE A responsivity used for computing solar irradiance. Stellar responsivities for the FUV and MUV wavelengths are approximately 2.5×10^5 and 5.7×10^5 times larger and include small differences arising from Ω_{Sun} and Ω_{Star} (McClintock, Snow, and Woods, 2005). SOLSTICE B was not recalibrated before launch. A calibration was transferred to it during the first week of on-orbit observations by simultaneously observing the Sun with A and B and comparing their spectra. Those measurements showed that SOLSTICE B had also experienced a sensitivity loss, similar to that of SOLSTICE A, since its initial calibration.

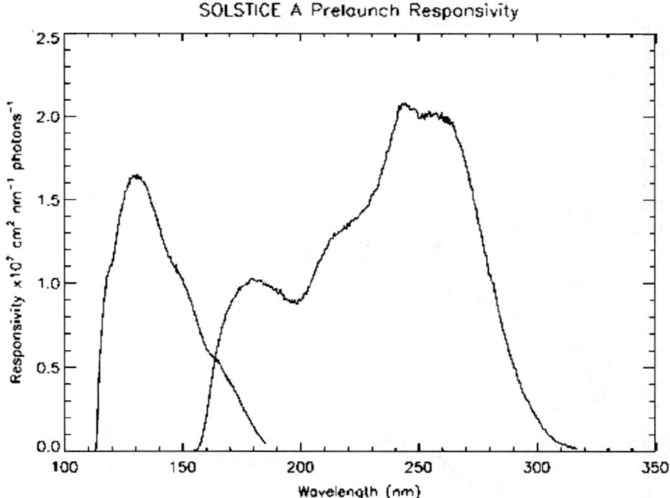

Figure 13. SOLSTICE A flight solar responsivity measured at SURF in September 2002.

Although electromagnetic theory of synchrotron radiation and the knowledge of the SURF geometry indicate that the irradiance at the center of the beamline used for the SOLSTICE calibration is known to 0.2% (Arp *et al.*, 2000), McClintock, Snow, and Woods (2005) conclude that the overall uncertainty in transferring those values to SOLSTICE II was ~0.75%. When this uncertainty was combined with uncertainties in determining the instrument wavelength scale for the relatively featureless SURF spectrum and errors in averaging the instrument response over the angular size of the sun and stars (Ω_{Sun} and Ω_{Star}), the overall uncertainty in instrument responsivity in the fall of 2001 was ~2%.

Launch schedule constraints limited the time available for the 2002 recalibration resulting in larger uncertainty in the values of Ω_{Sun} and Ω_{Star} relative to those obtained in 2001. This increased the uncertainties in the SOLSTICE A flight responsivities, which are shown for both solar and stellar observations in the top panel of Figure 14. They vary from ~1.5% to 6%, depending on wavelength. Stellar uncertainty values are slightly larger than solar values because they contain an additional small term that accounts for measurement of the stellar entrance aperture area (McClintock, Snow, and Woods, 2005).

The bottom panel in Figure 14 shows the uncertainties in observed solar and stellar irradiance. Uncertainties in irradiance include both error estimates in responsivity (top panel) and error estimates in the solar and stellar corrected count rates (Equation (8)). McClintock, Snow, and Woods (2005) show that the largest uncertainty in corrected MUV solar counts arises from the detector nonlinearity correction and is ~1.5% near 265 nm. Uncertainties in corrected FUV solar counts are less than 0.75% except near the bright hydrogen Lyman α emission line (121.5 nm) where uncertainty in the grating scattered light correction is ~2.5%. Although

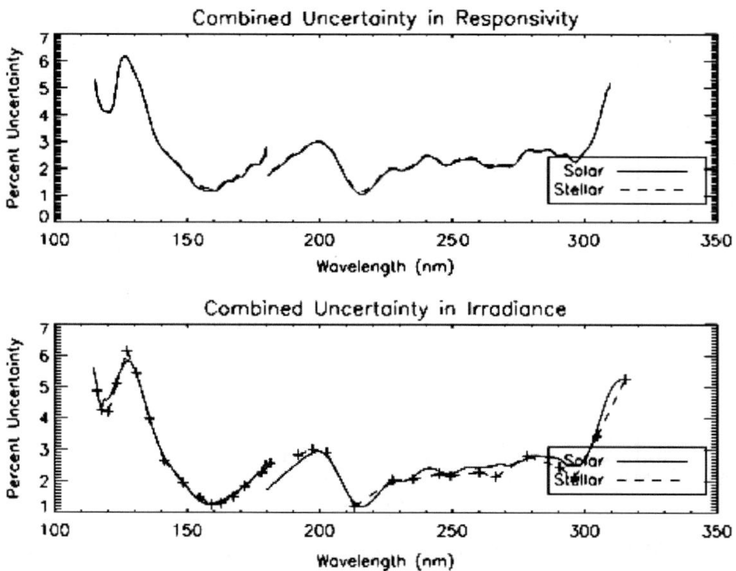

Figure 14. The *top panel* is the uncertainty in the SOLSTICE A flight responsivity for solar (*solid curve*) and stellar (*dashed curve*) observations. These are combined with uncertainties in corrected solar and stellar count rates to produce the overall uncertainties in solar and stellar irradiance, which are shown in the *lower panel*. The symbols (+) mark the 31 standard wavelengths that are used for the stellar measurements.

continuous spectral scans are acquired for only the brightest stars, all program stars are routinely observed at 40 fixed wavelengths for tracking instrument degradation. Snow *et al.* (2005b) show that uncertainties for individual stellar count rates for these 40 wavelengths vary from ~0.5% to ~1.2%, depending on target brightness.

For both the Sun and stars the count rate uncertainties are small compared to the responsivity uncertainties, which largely determine the radiometric accuracy of the SOLSTICE experiment at the beginning of mission. McClintock, Snow, and Woods (2005) show that the solar-to-stellar irradiance ratio is determined to better than 1.5% over most of the wavelength range because errors in determining the absolute responsivity do not affect the ratio measurement.

McClintock, Snow, and Woods (2005) have validated the accuracy of the SORCE SOLSTICE calibration at the beginning of the mission by comparing both the solar and stellar irradiance values with UARS SOLSTICE measurements. They show that solar values obtained by the two instruments on 3 April 2003 agree to within 5%. Stellar irradiances measured with UARS are ~10% larger than those obtained from SORCE. This discrepancy is most likely an error in the solar-stellar entrance slit area or in the solar-stellar spectral bandpass ratio. McClintock, Snow, and Woods (2005) suggest that the techniques used to establish the values for SORCE are inherently more accurate than those used for UARS. The discrepancy is being investigated.

Snow *et al.* (2005a) have analyzed the stellar and solar degradation experienced by SOLSTICE during the first 2 years of SORCE operations. They conclude that the stellar degradation function is determined to $\sim$1% during this time and that the solar function in the FUV is known with the same accuracy. On the other hand, the solar function near 200 nm is only determined to $\sim$2% during the first 2 years. They argue that recent modifications to flight calibration procedures will improve the accuracy across the entire SOLSTICE spectral range to $\sim$0.5% per year in the future.

5. Conclusion

The SOLSTICE II instruments, which were launched on 25 January 2003 on the SORCE spacecraft, have been operating successfully for over 2 years. This paper describes the design changes that enable SOLSTICE II to extend and improve the measurements of solar and stellar irradiance initiated in 1991 by SOLSTICE I, which is currently operating aboard the UARS. A new instrument case design, improved grating drive and detector assembly, and moveable exit slit filters have all contributed to a general improvement in instrument capability. Observations obtained during the first 24 months of operation demonstrate that SOLSTICE II is meeting its measurement objectives for spectral resolution, wavelength accuracy, and radiometric precision and reproducibility. The $\sim$5% agreement between SOLSTICE I solar irradiance measurements and those of SOLSTICE II validates the accuracy of both experiments providing an accurate record of solar ultraviolet irradiance variability that spans approximately 14 years.

Acknowledgements

More than a dozen engineers, technicians, and instrument makers at LASP contributed to the design, fabrication, and test of the SOLSTICE instruments. Their expertise, dedication, and hard work transformed the SOLSTICE concept into a world-class scientific instrument. This research was supported by NASA contract NAS5-97045.

References

Anderson, G. P. and Hall, L. A.: 1989, *J. Geophys. Res.* **94**, 6435.

Content, D. A., Boucarut, R. A., Bowler, C. W., Madison, T. J., Wright, G. A., Lindler, D. J. *et al.*: 1996, *SPIE Proceedings* **2807**, 267.

Drake, V. A., McClintock, W. E., Kohnert, R. A., Woods, T. N., and Rottman, G. J.: 2000, *SPIE Proceedings* **4135**, 402.

Drake, V. A., McClintock, W. E., Woods, T. N., and Rottman, G. J.: 2003, *SPIE Proceedings* **4796**, 107.

Horwitz, B.: 1996, *Laser Focus World* **32**, 10.

Kaneko, T., Namioka, T., and Seya, M.: 1971, *Appl. Opt.* **10**, 367.

Kopp, G., Lawrence, G., and Rottman, G.: 2005, *Solar Phys.*, this volume.

McClintock, W. E., Rottman, G. J., and Woods, T. N.: 2000, *SPIE Proceedings* **4135**, 24.

McClintock, W. E., Snow, M., and Woods, T. N.: 2005, *Solar Phys.*, this volume.

Mihalas, D. and Binney, J.: 1981, *Galactic Astronomy Structure and Kinematics*, W. H. Freeman, New York, p. 135.

Pankratz, C. K., Knapp, B. G., Reukauf, R. A., Fontenla, J., Dorey, M. A., Connelly, L. M. *et al.*: 2005, *Solar Phys.*, this volume.

Rottman, G.: 2005, *Solar Phys.*, this volume.

Rottman, G. J., Woods, T. N., and Sparn, T. P.: 1993, *J. Geophys. Res.* **98**, 10667.

Snow, M., McClintock, W. E., Rottman, G., and Woods, T. N.: 2005a, *Solar Phys.*, this volume.

Snow, M., McClintock, W. E., Woods, T. N., White, O. R., Harder, J. W., and Rottman, G.: 2005b, *Solar Phys.*, this volume.

Woods, T. N., Rottman, G. J., and Ucker, G. J.: 1993, *J. Geophys. Res.* **98**, 10678.

Woods, T. N., Prinz, D. K., London, J., Rottman, G. J., Crane, P. C., Cebula, R. P. *et al.*: 1996, *J. Geophys. Res.* **101**, 9541.

Woods, T. N., Rottman, G. J., Russell, C., and Knapp, B.: 1998, *Metrologia* **35**, 619.

Solar Physics (2005) 230: 259–294

SOLAR–STELLAR IRRADIANCE COMPARISON EXPERIMENT II (SOLSTICE II): PRE-LAUNCH AND ON-ORBIT CALIBRATIONS

WILLIAM E. McCLINTOCK, MARTIN SNOW and THOMAS N. WOODS
Laboratory for Atmospheric and Space Physics, University of Colorado, Boulder, CO, U.S.A.
(e-mail: mcclintock@lasp.colorado.edu, snow@lasp.colorado.edu)

(Received 21 April 2005; accepted 11 July 2005)

Abstract. The Solar–Stellar Irradiance Comparison Experiment II (SOLSTICE II), aboard the Solar Radiation and Climate Experiment (SORCE) spacecraft, consists of a pair of identical scanning grating monochromators, which have the capability to observe both solar spectral irradiance and stellar spectral irradiance using a single optical system. The SOLSTICE science objectives are to measure solar spectral irradiance from 115 to 320 nm with a spectral resolution of 1 nm, a cadence of 6 h, and an accuracy of 5%, to determine its variability with a long-term relative accuracy of 0.5% per year during a 5-year nominal mission, and to determine the ratio of solar irradiance to that of an ensemble of bright B and A stars to an accuracy of 2%. Those objectives are met by calibrating instrument radiometric sensitivity before launch using the Synchrotron Ultraviolet Radiation Facility at the National Institute for Standards and Technology in Gaithersburg, Maryland. During orbital operations irradiance measurements from an ensemble of bright, stable, main-sequence B and A stars are used to track instrument sensitivity. SORCE was launched on 25 January 2003. After spacecraft and instrument check out, SOLSTICE II first observed a series of three stars to establish an on-orbit performance baseline. Since 6 March 2003, both instruments have been making daily measurements of both the Sun and stars. This paper describes the pre-flight and in-flight calibration and characterization measurements that are required to achieve the SOLSTICE science objectives and compares early SOLSTICE II measurements of both solar and stellar irradiance with those obtained by SOLSTICE I on the Upper Atmosphere Research Satellite.

1. Introduction

The Solar Stellar Irradiance Comparison Experiment II (SOLSTICE II) is one of four instruments launched aboard the Solar Radiation and Climate Experiment (SORCE) on 25 January 2003. SORCE is a component of the NASA Earth Observing System (EOS) dedicated to measuring solar irradiance and its variability over a nominal 5-year mission lifespan. The SOLSTICE principal science objectives are to measure solar spectral irradiance from 115 to 320 nm with a spectral resolution of 1 nm, a cadence of 6 h, and an accuracy of 5% and to determine its variability with a long-term relative accuracy of 0.5% per year. SOLSTICE II is a follow-on to the SOLSTICE I instrument (Rottman, Woods, and Sparn, 1993) that is operating aboard the Upper Atmosphere Research Satellite (UARS). SORCE uses a pair of identical scanning grating monochromators, referred to as SOLSTICE A and SOLSTICE B, that measure both solar and stellar irradiance using a single optical-detector chain. Each instrument covers the entire wavelength range 115 to 320 nm, providing both redundancy against hardware failure and simultaneous measurements for data validation. During orbital operations, the SOLSTICE II instruments make daily measurements of both the Sun and an ensemble of bright, stable, main-sequence B and A stars. The stellar measurements track changes in instrument sensitivity with a long-term relative accuracy of 0.5% per year over the life of the mission (Snow *et al.*, 2005).

This paper describes the pre-flight and in-flight calibration and characterization measurements that are required to achieve the SOLSTICE science objectives. These requirements are based on an error analysis of the radiometry equation that defines the conversion from instrument telemetry output to irradiance. The Synchrotron Ultraviolet Radiation Facility III (SURF III) at the National Institute for Standards and Technology (NIST) in Gaithersburg, Maryland, provided the data for determining SOLSTICE radiometric sensitivity. SURF measurements were augmented by unit level tests that characterized detector, grating, and grating drive performance. All of these measurements were combined to produce the pre-flight instrument radiometric sensitivity including an estimate for its uncertainty. After launch additional calibrations provided *in situ* measurements for those parameters in the radiometry equation that could not be determined during ground test (e.g., in-flight detector dark counts). Comparison of early SOLSTICE II solar and stellar irradiances to those measured by SOLSTICE I provide important data for validating the measurement accuracy of both experiments.

2. Instrument Description

McClintock, Rottman, and Woods (2005) give a complete description of the SOLSTICE II design. SOLSTICE II employs a pair of fully redundant spectrometers. Unlike SOLSTICE I, which has a single detector per optical channel, each SOLSTICE II spectrometer is equipped with a pair of Hamamatsu R-2078 photomultiplier tubes. One channel has a CsI photocathode to measure the 115 to 180 nm wavelength range, known as the far ultraviolet (FUV). The other channel has a CsTe photocathode to measure the 170 to 320 nm wavelength range, known as the middle ultraviolet (MUV). On the SOLSTICE I instrument, the FUV channel was known as the G channel, and the MUV was the F channel.

During normal operations one spectrometer measures FUV and the other simultaneously measures MUV. Although it is used somewhat less frequently, the alternate detector in each spectrometer is exercised on a routine basis in order to track its in-flight performance over time. In the event of a catastrophic failure in either spectrometer, SOLSTICE II would retain its full wavelength capability, but the time required to acquire a complete spectrum would double. Table I summarizes the SOLSTICE II spectroscopic parameters.

Figure 1 illustrates the optical layout for a single spectrometer. In this arrangement, which is identical to SOLSTICE I, light enters the instrument through a door

TABLE I

SOLSTICE II spectroscopic parameters.

Parameter	FUV channel	MUV channel
Wavelength range	115–180 nm	170–320 nm
Grating ruling density	3600 grooves mm^{-1}	3600 grooves mm^{-1}
Solar entrance slit	0.1 mm × 0.1 mm	0.1 mm × 0.1 mm
Solar exit slit	0.0375 × 6 mm	0.0375 × 6 mm
Solar bandpass	0.1 nm	0.09 nm
Stellar entrance slit	16 mm diam.	16 mm diam.
Stellar exit slit	0.75 mm × 6 mm	1.5 mm × 6 mm
Stellar bandpass	1.1 nm	2.2 nm
Detector photocathode	Cesium iodide (CsI)	Cesium telluride (CsTe)

followed by an entrance slit assembly and is reflected by a fold mirror toward the diffraction grating. A small range of wavelengths leaving the grating is reflected toward an elliptical camera mirror by the second fold mirror. The ellipse images the diffracted beam onto an exit slit assembly. SOLSTICE II optical elements are identical to those used in SOLSTICE I except for the ellipse, which was re-designed to increase the solar and stellar image distances by a factor of 1.5 to package side-by-side exit slit-detector assemblies. A two-position mechanism articulates the camera mirror $\pm 2^{\circ}$ about its vertex to illuminate either the inboard or outboard solar/stellar

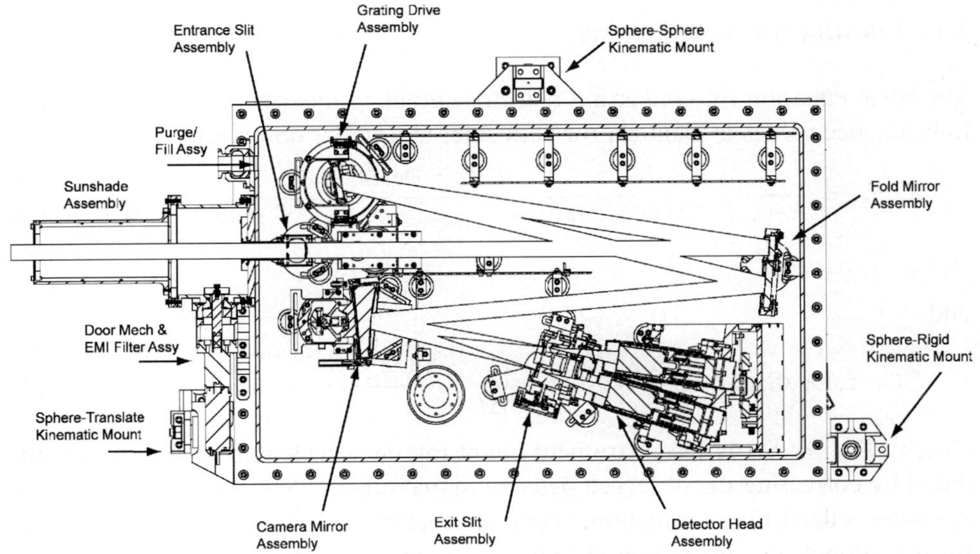

Figure 1. Optical-mechanical configuration of a single SOLSTICE II spectrometer.

exit slit-photomultiplier tube. Two 10% transmitting neutral density filters, which are located between the inboard exit slits and CsTe photomultiplier tube and tilted to prevent multiply reflected light from reaching the detector, replace the interference filter used in SOLSTICE I (Rottman, Woods, and Sparn, 1993). These filters are mounted on their own two-position mechanisms, which can be independently inserted and withdrawn from the optical path by instrument command. Each mechanism carries both the filter and an uncoated filter substrate (window). Whenever a filter is removed, the window is inserted in its place so that the illuminated area of the detector remains unchanged. A sunshade, mounted near the door, and a baffle assembly, located between the grating and the exit slit assembly, prevent stray light and out-of-band diffracted light from reaching the exit slit plane. Additional baffles trap light reflected (the zero-order light) from the grating, preventing it from scattering from other internal surfaces. A thin-walled aluminum housing, not shown in Figure 1, completely encloses the exit slit-detector assembly so that the only light that passes through the exit slits can reach the detector input windows.

A solar position sensor (SPS) provides a direct measurement of the Sun's location in the instrument field of view and acts as a bright object sensor (BOS). In the event that the Sun comes within the field of view while the stellar entrance and exit slits are in place, a signal from the BOS will exceed a preset threshold and actuate the slit mechanisms, setting them to their solar positions. The SPS-BOS sensor is located in a housing directly behind the entrance slit and views the sky through a small aperture located below the main instrument optical path.

3. Irradiance Conversion and Calibration Requirements

3.1. Irradiance Algorithms

The basic equation for converting detector output counts (the telemetered signal) to irradiance values is identical for both solar and stellar observations:

$$E_{\mathrm{AU}}(\lambda) = \frac{C(\lambda, \tau, \mathrm{Dc}, \mathrm{Sl}, \mathrm{St})}{R_C(\lambda, T, \Omega)\, \mathrm{FOV}(\lambda, \Omega, \theta, \phi)\, A_{\mathrm{Entrance}}\, \Delta\lambda_{\mathrm{BP}}\, T_{\mathrm{Filter}}(\lambda)\, \mathrm{DEG}(t, \lambda, \Omega, \theta, \phi)\, f_{\mathrm{AU}}} \tag{1}$$

and

$$C(\lambda, \tau, \mathrm{Dc}, \mathrm{Sl}, \mathrm{St}) = \frac{S(\lambda)\, N(\tau) - \mathrm{D_C} - \mathrm{Sl}(\lambda) - \mathrm{St}}{\Delta t}. \tag{2}$$

$C(\lambda, \tau, \mathrm{Dc}, \mathrm{Sl}, \mathrm{St})$ is the instrument count rate at wavelength λ, which is computed by correcting the observed detector signal counts, $S(\lambda)$, for nonlinearity by applying a dead time correction, $N(\tau)$, appropriate for electronics dead time, τ, for dark count, Dc, for scattered light, $\mathrm{Sl}(\lambda)$, and for stray light, St, and then dividing by the instrument integration period, Δt. $R_C(\lambda, T, \Omega)$, which depends on both wavelength and detector temperature, T, is the pre-flight ($t = 0$) instrument quantum-efficiency-transmission function at the center of the field-of-view, averaged over the angular size of the target, Ω:

$$R_C(\lambda, T, \Omega) = \int_{\Omega(0,0)_{\text{Target}}} r(0, \lambda, T, \theta', \phi')\, d\Omega' \Big/ \int_{\Omega(0,0)_{\text{Target}}} d\Omega'. \quad (3)$$

A_{Entrance} is the area of the entrance aperture, and $\Delta\lambda_{\text{BP}}$ is the spectral bandpass. $\text{FOV}(\lambda, \Omega, \theta, \phi)$ and $\text{DEG}(t, \lambda, \Omega, \theta, \phi)$ are factors that correct for instrument sensitivity variations, which are a function of target viewing direction (θ, ϕ), and time-dependent degradation, caused by extended on-orbit solar exposure. Where

$$\text{FOV}(\lambda, \Omega, \theta, \phi) = \int_{\Omega(\theta,\phi)} r(0, \lambda, T, \theta', \phi')\, d\Omega' \Big/ \int_{\Omega(0,0)} r(0, \lambda, T, \theta', \phi')\, d\Omega' \quad (4)$$

and

$$\text{DEG}(t, \lambda, \Omega, \theta, \phi) = \int_{\Omega(\theta,\phi)} r(t, \lambda, T, \theta', \phi')\, d\Omega' \Big/ \int_{\Omega(\theta,\phi)} r(0, \lambda, T, \theta', \phi')\, d\Omega'. \quad (5)$$

$r(0, \lambda, T, \theta, \phi)$ and $r(t, \lambda, T, \theta, \phi)$ are the pre-flight and time-dependent, in-flight response functions for angle (θ, ϕ), relative to the instrument optic axis and Ω is the target (sun or star) angular extent. T_{Filter} is the transmission of the neutral density filter(s) that attenuate the solar photon flux impinging upon the MUV detector. f_{AU} is a factor, which is calculated from Earth orbital elements, that normalizes the solar irradiance values to a mean solar distance of 1 AU. In Equation (1) the viewing direction and time dependence of the instrument responsivity are explicitly shown as a pair of independent terms as a bookkeeping convenience:

$$R(t, \lambda, T, \Omega, \theta, \phi) = R_C(\lambda, T, \Omega)\, \text{FOV}(\lambda, \Omega, \theta, \phi)\, \text{DEG}(t, \lambda, \Omega, \theta, \phi). \quad (6)$$

Calibrations at SURF III, which is a source irradiance standard, establish the primary values of $\text{FOV}(\lambda, \Omega, \theta, \phi)$ and of the products $R_C(\lambda, T, \Omega)\, \Delta\lambda_{\text{BP}}\, A_{\text{Entrance}}$ and $R_C(\lambda, T, \Omega)\, \Delta\lambda_{\text{BP}}\, A_{\text{Entrance}}\, T_{\text{Filter}}(\lambda)$ used in Equation (1) for both solar and stellar irradiances. When a source standard is used to determine the instrument responsivity with the solar entrance and exit apertures in place, the irradiance equation becomes

$$E_{\text{AU}}(\lambda) = \frac{C(\lambda, \tau, \text{Dc}, \text{Sl}, \text{St})}{R'_{\text{Sun}}(\lambda, T_0) G(\lambda, T_0, T) \text{FOV}_{\text{Sun}}(\lambda, \theta, \phi) \text{DEG}_{\text{Sun}}(t, \lambda, \theta, \phi) f_{\text{AU}}} \quad (7)$$

and

$$R'_{\text{Sun}} = R_C(\lambda, T_0, \Omega_{\text{Std}}) A_{\text{Entrance}}\, \Delta\lambda\, \Gamma(\lambda, \Omega_{\text{Sun,Std}}) = \frac{C_{\text{Std}}(\lambda, \tau, \text{Dc}, \text{Sl}, \text{St})}{E_{\text{Std}}(\lambda)} \Gamma(\lambda, \Omega_{\text{Sun,Std}}). \quad (8)$$

$R_C(\lambda, T_0, \Omega_{\rm Std})\, A_{\rm Entrance}\Delta\lambda$ is the instrument response, measured at temperature T_0, in counts per second to the irradiance source standard with angular size Ω, which is aligned to the instrument optic axis $(\theta, \phi) = (0, 0)$, and whose output, $E_{\rm Std}(\lambda)$, is specified in units of radiant power per unit area per unit wavelength interval. $G(\lambda, T_0, T)$ is a detector gain correction that accounts for changes in detector sensitivity as a function of the difference between SURF calibration temperature, T_0, and flight temperature, T (see Section 4.2). $C_{\rm Std}(\lambda, \tau, {\rm D_C}, {\rm Sl})$ is the instrument count rate corrected for nonlinearity, dark count, and scattered light. $\Gamma(\lambda, \Omega)$ is a geometrical correction factor that accounts for differences in the instrument response at its field-of-view center to sources with different angular sizes:

$$\Gamma(\lambda, \Omega_{\rm Sun,Std}) = \frac{\int_{\Omega{\rm Sun}(0,0)} r(0, \lambda, T, \theta', \phi')\, {\rm d}\Omega' / \int_{\Omega{\rm Sun}(0,0)} {\rm d}\Omega'}{\int_{\Omega{\rm Std}(0,0)} r(0, \lambda, T, \theta', \phi')\, {\rm d}\Omega' / \int_{\Omega{\rm Std}(0,0)} {\rm d}\Omega'}. \quad (9)$$

Stellar irradiances are also calculated using Equation (7), but $R_C(\lambda, T_0, \Omega_{\rm Std})$ is measured with the stellar entrance and exit apertures in place, $f_{\rm AU}$ set equal to 1, and FOV and Γ are calculated using the instrument stellar field of view instead of the solar field of view. Using SURF measurements to determine the parameters in Equations (7) through (9) establishes the solar and stellar irradiance at the beginning of mission when DEG $= 1$. On-orbit calibrations are used to determine both $\rm DEG_{Sun}$ and $\rm DEG_{Star}$ after launch.

3.2. Uncertainty Analysis and Calibration Requirements

The requirements for the three primary SOLSTICE II measurements (determine solar irradiance with 5% combined standard uncertainty at the beginning of life, determine the relative solar and stellar irradiances with 2% accuracy at the beginning of mission, and track solar variability with a long-term relative accuracy of 0.5% per year) set limits on the combined uncertainty in Equations (1), (2), (7), and (8). These total uncertainties are used to set calibration requirements for the various terms in the equations.

3.2.1. *Measure Solar Irradiance with 5% Accuracy*

Equations (10) and (11) are used to calculate the solar irradiance based on calibrations performed at SURF. If the explicit dependences on wavelength, temperature, and field angle are suppressed for clarity, these equations can be rewritten as

$$E_{\rm AU} = \frac{C_{\rm Sun}}{R'_{\rm Sun}\, G\, {\rm FOV_{Sun}}\, {\rm DEG_{Sun}}\, f_{\rm AU}}, \quad (10)$$

$$R'_{\rm Sun} = \frac{C_{\rm Std}}{E_{\rm Std}} \Gamma(\Omega_{\rm Sun,Std}), \quad (11)$$

$$C_{\rm Sun} = (S_{\rm Sun}N - {\rm Dc_{Sun}} - {\rm Sl_{Sun}} - {\rm St_{Sun}})/\Delta t_{\rm Sun}, \quad (12)$$

and

$$C_{\text{Std}} = (S_{\text{Std}}N - \text{Dc}_{\text{Std}} - \text{Sl}_{\text{Std}} - \text{St}_{\text{Std}})/\Delta t_{\text{Std}}. \tag{13}$$

The relative uncertainty in irradiance given by

$$\sigma_E^2 = \sigma_{C_{\text{Sun}}}^2 + \sigma_{R'_{\text{Sun}}}^2 + \sigma_G^2 + \sigma_{\text{FOV}_{\text{Sun}}}^2 + \sigma_{\text{DEG}_{\text{Sun}}}^2 + \sigma_{f_{\text{AU}}}^2, \tag{14}$$

$$\sigma_C^2 = \frac{S^2N^2(\sigma_S^2 + \sigma_N^2) + \text{Dc}^2\sigma_{\text{Dc}}^2 + \text{Sl}^2\sigma_{\text{Sl}}^2 + \text{St}^2\sigma_{\text{St}}^2}{(S\,N - \text{Dc} - \text{Sl} - \text{St})^2} + \sigma_{\Delta t}^2 + \sigma_{\Delta\lambda}^2, \tag{15}$$

and

$$\sigma_{R'_{\text{Sun}}}^2 = \sigma_{C_{\text{Std}}}^2 + \sigma_{E_{\text{Std}}}^2 + \sigma_{\Gamma(\Omega)}^2. \tag{16}$$

Equations (10) through (13) provide the basis for determining the SOLSTICE calibration requirements and establishing budgets for the various uncertainty terms required to measure the solar irradiance to 5%. Errors for the various terms are assumed to be independent and are combined in quadrature in Equations (14) through (16), where σ for a quantity represents the relative uncertainty in that quantity, i.e., σ_x = (uncertainty in X)/(magnitude of X). Contributions for uncertainty in dark count and scattered light scale as $\sigma_{\text{Dc}}^2 \frac{\text{Dc}^2}{(S\,N-\text{Dc}-\text{Sl}-\text{St})^2}$, $\sigma_{\text{Sl}}^2 \frac{\text{Sl}^2}{(S\,N-\text{Dc}-\text{Sl}-\text{St})^2}$, and $\sigma_{\text{St}}^2 \frac{\text{St}^2}{(S\,N-\text{Dc}-\text{Sl}-\text{St})^2}$. The other terms scale nearly 1:1.

σ_C^2 contains an explicit term for wavelength scale errors ($\sigma_{\Delta\lambda}^2$) that occur in the observed detector signal, $S(\lambda)$, when the wavelength being observed differs from the expected value during observations and ground calibrations (see Sections 4.2 and 5.2). Contributions from the two terms $\sigma_{\Delta t}^2$ and $\sigma_{f_{\text{AU}}}^2$ are negligible for the SOLSTICE error analysis. Since Δt_{Sun} and Δt_{Std} are derived from the same instrument clock, there is negligible uncertainty in their ratio. This applies to stellar measurements as well. Corrections for solar distance (f_{AU}), which can be as large as $\pm 3.3\%$, are accurate to better than $10^{-4}\%$ (Kopp, Heuerman, and Lawrence, 2005).

During instrument definition, budgets were developed for the individual terms in Equations (14) through (16). These budgets were used to define instrument performance (e.g., wavelength scale accuracy) and to determine calibration requirements. Table II summarizes the adopted budgets and the achieved accuracy for each of the terms. Columns two and three show the average budget numbers and columns four and five show the wavelength dependent range of values achieved in the final design and calibration. These results are discussed in detail in Sections 4 and 5.

3.2.2. *Determine the Solar and Stellar Irradiance Ratio with 2% Accuracy*

Ratios of Equation (10) also form the basis for comparing solar and stellar irradiance immediately after launch:

TABLE II

Uncertainty contributions to solar absolute irradiance measurements.

Parameter	Symbol	FUV budget	MUV budget	FUV achieved	MUV achieved
Solar signal	$\sigma_{S\mathrm{Sun}}$	0.5	0.1	0.2–0.7	0.01–0.05
Nonlinearity	σ_N	1.0	1.1	0.01–0.1	0.1–1.5
Dark count	σ_{Dc}	0.5	0.5	0.005–0.01	0.01–0.02
Scattered light	σ_{Sl}	0.5	0.5	0.1–2.8	0.01–0.1
Stray light	σ_{St}	0.5	0.5	0.05–0.1	0.01–0.02
Wavelength scale	σ_λ	0.5	0.5	0.1–0.5	0.1–0.5
Corrected solar signal	$\sigma_{C\mathrm{Sun}}$	1.5	1.5	0.3–2.8	0.3–1.5
SURF signal	$\sigma_{S\mathrm{Std}}$	0.5	0.5	0.5	0.5
Nonlinearity	σ_N	0.5	0.5	0.01	0.01
Dark count	σ_{Dc}	0.5	0.5	0.1	0.1
Scattered light	σ_{Sl}	0.5	0.5	1.0	1.0
Stray light	σ_{St}	0.5	0.5	0.1	0.1
Wavelength scale	σ_λ	0.5	0.5	0.5–1.5	0.5–1.8
Corrected SURF signal	$\sigma_{C\mathrm{Std}}$	1.2	1.2	1.87	1.87
SURF irradiance	$\sigma_{E\mathrm{Std}}$	2.0	2.0	0.75	0.75
Geometric correction	$\sigma_{G(\Omega)}$	3.0	3.0	1.0–6.0	1.0–3.0
Pre-flight response	σ_{Rc}	3.8	3.8	1.5–6.0	1.5–5.0
Corrected solar signal	$\sigma_{C\mathrm{Sun}}$	1.5	1.5	0.3–2.8	0.3–1.5
Pre-flight response	σ_{Rc}	3.8	3.8	1.5–6.0	1.5–5.0
Detector gain	σ_G	0.5	0.5	0.1	0.1
Pointing error	σ_{FOV}	1.0	1.0	0.0	0.0
Degradation	σ_{Deg}	0.0	0.0	0.0	0.0
Correction to 1AU	$\sigma_{f\mathrm{AU}}$	0.01	0.01	0.0	0.0
Total error		4.2	4.2	1.2–6.0	1.2–5.0

$$\frac{E_{\mathrm{AU}}}{E_{\mathrm{Star}}} = \frac{C_{\mathrm{Sun}}}{C_{\mathrm{Star}}} \frac{R'_{C_{\mathrm{Star}}} G_{\mathrm{Star}} \mathrm{FOV}_{\mathrm{Star}}}{R'_{C_{\mathrm{Sun}}} G_{\mathrm{Sun}} \mathrm{FOV}_{\mathrm{Sun}} f_{\mathrm{AU}}} \tag{17}$$

$$= \frac{C_{\mathrm{Sun}}}{C_{\mathrm{Star}}} \frac{A_{\mathrm{Star}} \Delta\lambda_{\mathrm{Star}} G_{\mathrm{Star}}}{A_{\mathrm{Sun}} \Delta\lambda_{\mathrm{Sun}} G_{\mathrm{Sun}} f_{\mathrm{AU}}} \Gamma(\lambda, \Omega_{\mathrm{Star,Sun}}) \tag{18}$$

$$= \frac{C_{\mathrm{Sun}}}{C_{\mathrm{Star}}} \frac{G_{\mathrm{Star}}}{G_{\mathrm{Sun}} f_{\mathrm{AU}}} R_{\mathrm{A\text{-}BP}} \Gamma(\lambda, \Omega_{\mathrm{Star,Sun}}), \tag{19}$$

$$C_{\mathrm{Sun}} = (S_{\mathrm{Sun}} N - \mathrm{Dc}_{\mathrm{Sun}} - \mathrm{Sl}_{\mathrm{Sun}} - \mathrm{St}_{\mathrm{Sun}})/\Delta t_{\mathrm{Sun}},$$

(repeat of Equation 12)

$$C_{\mathrm{Star}} = (S_{\mathrm{Star}} N - \mathrm{Dc}_{\mathrm{Star}} - \mathrm{Sl}_{\mathrm{Star}} - \mathrm{St}_{\mathrm{Star}})/\Delta t_{\mathrm{Star}}, \tag{20}$$

$$\Gamma(\lambda, \Omega_{\mathrm{Star,Sun}}) = \frac{\int_{\Omega\mathrm{Star}(0,0)} r(0, \lambda, T, \theta', \phi')\, \mathrm{d}\Omega' / \int_{\Omega\mathrm{Star}(0,0)} \mathrm{d}\Omega'}{\int_{\Omega\mathrm{Sun}(0,0)} r(0, \lambda, T, \theta', \phi')\, \mathrm{d}\Omega' / \int_{\Omega\mathrm{Sun}(0,0)} \mathrm{d}\Omega'}, \tag{21}$$

and

$$R_{A\text{-BP}} = \frac{A_{\text{Star}}\,\Delta\lambda_{\text{BP}_{\text{Star}}}}{A_{\text{Sun}}\,\Delta\lambda_{\text{BP}_{\text{Sun}}}}. \tag{22}$$

$R_{A\text{-BP}}$ is the ratio of the product of the stellar/solar entrance slit areas and stellar/solar spectral bandpasses. It is determined directly from the SURF calibrations with the instrument in stellar and solar modes, $R_{A\text{-BP}} = C_{\text{STD}}(\text{Stellar Mode})/C_{\text{STD}}(\text{Solar Mode})$. Γ' is the ratio of the in-flight instrument stellar responsivity to the in-flight instrument solar responsivity at the beginning of mission, which is determined from the SURF beam maps and updated with the in-flight field-of-view experiments. The relative error in the irradiance ratio is

$$\sigma^2_{E/E} = \sigma^2_{C_{\text{Sun}}} + \sigma^2_{C_{\text{Star}}} + \sigma^2_{G/G} + \sigma^2_{R_{A\,\text{BP}}} + \sigma^2_{f_{\text{AU}}} + \sigma^2_{\Gamma}, \tag{23}$$

where $\sigma_{C_{\text{Sun}}}$ and $\sigma_{C_{\text{Star}}}$ are calculated from Equation (15). Table III summarizes the budgets and achieved accuracy for the uncertainty terms in Equation (23).

TABLE III

Uncertainty contributions to solar stellar irradiance comparison.

Parameter	Symbol	FUV budget	MUV budget	FUV achieved	MUV achieved
Solar signal	$\sigma_{S\,\text{Sun}}$	0.5	0.1	0.2–0.7	0.01–0.05
Nonlinearity	σ_N	1.0	1.1	0.01–0.1	0.1–1.5
Dark count	σ_{Dc}	0.5	0.5	0.005–0.01	0.01–0.02
Scattered light	σ_{Sl}	0.5	0.5	0.1–2.8	0.01–0.1
Stray light	σ_{St}	0.5	0.5	0.05–0.1	0.01–0.02
Wavelength scale	σ_λ	0.5	0.5	0.1–0.5	0.1–0.5
Corrected solar signal	$\sigma_{C\,\text{Sun}}$	1.5	1.5	0.3–2.8	0.3–1.5
Stellar signal	$\sigma_{S\,\text{Star}}$	0.5	0.5	0.2–0.7	0.01–0.05
Nonlinearity	σ_N	0.1	0.1	0.01–0.1	0.1–1.5
Dark count	σ_{Dc}	0.5	0.5	0.005–0.01	0.01–0.02
Scattered light	σ_{Sl}	0.5	0.5	0.1–2.8	0.01–0.1
Stray light	σ_{St}	0.5	0.5	0.05–0.1	0.01–0.02
Wavelength scale	σ_λ	0.5	0.5	0.1–0.5	0.1–0.5
Corrected stellar signal	$\sigma_{C\,\text{Sun}}$	1.1	1.1	0.2–0.7	0.2–2.6
Corrected solar signal	$\sigma_{C\,\text{Sun}}$	1.5	1.5	0.3–2.8	0.3–1.5
Corrected stellar signal	$\sigma_{C\,\text{Star}}$	1.1	1.1	0.2–0.7	0.2–2.6
Area bandpass ratios	$\sigma_{A\text{-BP}}$	0.5	0.5	0.5	0.5
Solar/stellar detector gain	$\sigma_{G/G}$	0.1	0.1	0.1	0.1
Correction to 1AU	σ_{AU}	0.01	0.01	0.0	0.0
Solar/stellar Γ	σ_Γ	0.5	0.5	1.0	1.0
Total error		1.9	1.9	1.2–2.5	1.2–5.0

TABLE IV

Uncertainty contributions to tracking changes in solar irradiance.

Parameter	Symbol	FUV budget	MUV budget	FUV achieved	MUV achieved
Corrected solar signal	$\sigma_{C'_{\mathrm{Sun}}}$	0.25	0.25	0.2	0.2
Corrected stellar signal	$\sigma_{C'_{\mathrm{Star}}}$	0.25	0.25	0.2	0.2
Solar/stellar detector gain	$\sigma_{G/G}$	0.1	0.1	0.1	0.1
Correction to 1AU	σ_{AU}	0.01	0.01	0.0	0.0
Change in solar/stellar Γ	$\sigma_{\Delta\Gamma}$	0.1	0.1	0.5	0.1
Total error		0.38	0.40	0.60	1.04

3.2.3. *Track Changes in Solar Irradiance with 0.5% per Year Long-Term Relative Accuracy*

Tracking changes in solar irradiance by monitoring stars requires knowledge of the changes in response functions for stars and the Sun:

$$\Delta \frac{E_{\mathrm{Sun}}}{E_{\mathrm{Star}}} = \frac{C'_{\mathrm{Sun}}}{C'_{\mathrm{Star}}} \frac{G_{\mathrm{Star}}}{G_{\mathrm{Sun}} f_{\mathrm{AU}}} \Delta\Gamma_{\mathrm{DEG}}(t, \lambda, \Omega_{\mathrm{Star,Sun}}), \tag{24}$$

$$\Delta\Gamma_{\mathrm{DEG}}(t, \lambda, \Omega_{\mathrm{Star,Sun}}) = \\ = \frac{\int_{\Omega\,\mathrm{Star}(0,0)} r(t, \lambda, T, \theta, \phi)\,\mathrm{d}\Omega \Big/ \int_{\Omega\,\mathrm{Star}(0,0)} r(0, \lambda, T, \theta, \phi)\,\mathrm{d}\Omega}{\int_{\Omega\,\mathrm{Sun}(0,0)} r(t, \lambda, T, \theta, \phi)\,\mathrm{d}\Omega \Big/ \int_{\Omega\,\mathrm{Sun}(0,0)} r(0, \lambda, T, \theta, \phi)\,\mathrm{d}\Omega}, \tag{25}$$

and

$$\sigma^2_{\Delta_{E/E}} = \sigma^2_{C'_{\mathrm{Sun}}} + \sigma^2_{C'_{\mathrm{Star}}} + \sigma^2_{G/G} + \sigma^2_{f_{\mathrm{AU}}} + \sigma^2_{\Delta\Gamma_{\mathrm{DEG}}}. \tag{26}$$

Table IV summarizes the budgets and achieved accuracy for tracking solar irradiance changes. $\sigma^2_{C'_{\mathrm{Sun}}}$ and $\sigma^2_{C'_{\mathrm{Star}}}$ are the uncertainties in solar and stellar count rates that result from random errors in detected counts (S) dark count (Dc) and wavelength scales ($\Delta\lambda$). Daily averaged solar spectra have $\sigma^2_{C'_{\mathrm{Sun}}} < 0.2\%$. Snow *et al.* (2005) show that $\sigma^2_{C'_{\mathrm{Star}}}$ is also <0.2%; therefore, the primary error in determining $\Delta E/E$ arises from uncertainty in $\Delta\Gamma_{\mathrm{DEG}}$ (see Section 5.6.3).

4. Pre-Flight Calibrations

Values for the terms in the radiometry equations were determined during instrument pre-flight calibration and characterization. SURF III calibrations provided the primary data for establishing values for the product $R_C(\lambda, T_0\Omega_{\mathrm{SURF}})\,\Delta\lambda A_{\mathrm{Entrance}}\,T_{\mathrm{Filter}}(\lambda)$, $R_C(\lambda, T_0, \Omega_{\mathrm{SURF}})\,\Delta\lambda A_{\mathrm{Entrance}}$, and for measuring $N(\tau)$ and $r(\lambda, T_0, \theta, \phi)$. $G(\lambda, T_0, T)$ and Dc were measured during detector head characterizations using the calibration and test equipment (CTE) at the Laboratory for Atmospheric and Space Physics (LASP) at the University of Colorado (Drake *et al.*, 2000). Grating scatter measurements were performed at the Goddard Space Flight Center in Greenbelt, Maryland, using the Space Telescope Imaging Spectrograph Facility (Content *et al.*, 1996). These results were combined with instrument level measurements at SURF and at LASP to determine values for $Sl(\lambda)$. The following sections describe the details of these calibrations and summarize the results.

4.1. Wavelength Scale

Precise knowledge of the wavelength scale during both ground test and on-orbit is essential for determining the accurate irradiance values. Errors in wavelength produce errors in irradiance values predominantly through changes in $C(\lambda)$ and $R'(\lambda, T_0)$ (Equations (7) and (8)). For 1 nm resolution solar measurements, $C(\lambda + \Delta\lambda)$ introduce up to $\pm 1\%$ difference from $S(\lambda)$ with a 0.01 nm shift in wavelength scale. Stellar count values are significantly less sensitive to wavelength shifts; nonetheless, the dependence of the SOLSTICE stellar wavelength scale on spacecraft pointing offsets (see Equation (28)) and hour-to-hour uncertainties in the grating drive fiducial position require a measurement of the zero order image at the beginning and end of each stellar observation (Snow *et al.*, 2005). Change with wavelength in the relatively featureless R' is an order of magnitude smaller: $\Delta R'/R' \sim 0.1\%$ for a 0.01 nm wavelength error. $G(\lambda, T_0, T)$, $\text{FOV}(\lambda, \Omega, \theta, \phi)$, and $\text{DEG}(\lambda, \Omega, \theta, \phi)$ all exhibit even weaker wavelength dependences.

Wavelength values for the observed counts, $S(\lambda)$, are computed from the standard grating equation adopted to the SOLSTICE instrument geometry (McClintock, Rottman, and Woods, 2005):

$$\lambda_{\text{Solar}} = 2d\ \sin(\theta_s)\cos(\phi_G), \tag{27}$$

$$\lambda_{\text{Stellar}} = 2d\ \sin(\theta_s + E/2)\cos((\phi_G - E)/2), \tag{28}$$

and

$$\theta_S = \theta_0 - N\Delta\theta = (N_0 - N)\Delta\theta. \tag{29}$$

In these equations $\theta_S = (\alpha + \beta)/2$ is the grating rotation angle, and ϕ_G, which is a constant ($\phi_G = 2.32°$), is the half angle between the diffracted and incident beams. E is the pointing offset between stellar targets and the instrument optic axis in the grating dispersion plane (the plane perpendicular to the grating grooves), measured in the same sense as α and β. The grating moves through its range in a series of discrete steps with an angular resolution $\Delta\theta \sim 2.42\ \mu$ radians $= 1$ grating step data number (DN). During normal operations the solar spectrum is recorded using a grating step size equal to 0.00375° (27 DN), resulting in a sampling interval equal to approximately 33% of the instrument spectral bandpass. Step sizes for stellar observations are typically 10–20 times larger.

The SOLSTICE grating drive control system employs a high precision angular encoder based on a diffractive optics design to determine grating position (McClintock, Rottman, and Woods, 2005). Dynamic step response tests of the control loop, which measure the encoder position error as a function of time, indicate that during normal solar observations (step sizes ~13.5 arcsec) the grating settles to within 0.5 arcsec of its commanded value within 0.025 s. Laboratory measurements of the step size using the technique described by Woods, Rottman, and Ucker (1993) indicate that at constant temperature, the angular position, measured from the fiducial, is accurate to ~1 arcsec over its 40° range. On the other hand, the absolute angular position, which is determined by locating the fiducial (during grating drive initialization or whenever a "find index" command is executed), is much less accurate. Changes in fiducial location up to ±35 arcsec are occasionally observed in-flight after the grating drive is power cycled. It is not clear whether these shifts are caused by hardware, software, or both.

The SOLSTICE wavelength scale is also temperature dependent. Measurements of reference line positions made during thermal vacuum testing showed that the wavelength scale shifts at a rate of 0.004 nm per Centigrade degree over the temperature range of 20°–30°C.

4.2. Detector Performance

The SOLSTICE detectors are photon-counting photomultiplier tubes (PMTs) equipped with pulse-amplifier-discriminators (PADs). Because photon detection is a random process, which is approximately described by Poisson statistics, the observed count rate, $S(\lambda)$, must be corrected for dead time, τ, in the PMT-PAD-counting circuit. In the SOLSTICE detectors, the Hamamatsu R-2078 photomultipliers, which have an approximate 5 ns pulse width, the Amptek A-121 PADs, which have an approximate 50 ns pulse width, and the counting electronics state machines, which have a 10 ns cycle time, all contribute to the system dead time. The observed count rate C_0 and the "true" count rate C for SOLSTICE are related by the standard expression for nonparalyzable counting circuits:

$$C = \frac{C_0}{1 - C_0\tau}. \tag{30}$$

Dead time was measured during calibrations at SURF by operating the instrument at a fixed wavelength while varying the storage ring electron current over approximately 3.5 orders of magnitude. Since the radiance from SURF is directly proportional to beam current, the "true" count rate is given by: $C = K$ Beam_Current. Measurements were made on each detector at 10–15 current settings as the electron beam decayed over the course of approximately 1 h. K, an arbitrary coefficient, and τ, the dead time, were determined by a nonlinear least squares fit to Equation (30). Values of τ for the four detectors ranged from 62.0 to 66.7 ns with 5–10%

formal uncertainties from the fitting routine. The resulting dead time correction for a 1 MHz observed count rate is $\sim 6.9 \pm 0.69\%$ for $\tau = 65$ ns.

Detector warm-up and recovery were characterized by shuttering light from the SURF beam to create a series of 2 min long "photon pulses" of varying intensity. In these experiments, count rates typically increased by about 0.5% for the first 5 s after first light. This was followed by a smaller increase ranging from 0.1 to 0.3% in the following 100 s. All of the detectors exhibited elevated dark counts immediately after the light was shuttered. Count rates for both CsI detectors decreased by factors of 10^5 to 10^6 immediately after the removal of light. This was followed by an additional factor of 5 decline during the following 60 s. CsTe detectors were about an order of magnitude noisier, showing an initial drop of 10^4 declining to 10^{-5} after approximately 1 min. No corrections for detector warm-up or recovery are currently included in the SOLSTICE irradiance calculations.

Photomultiplier tubes also exhibit wavelength dependent changes in sensitivity that are a function of photocathode temperature. Temperature coefficients for all of the SOLSTICE detector assemblies were measured before instrument integration and have been reported by Drake *et al.* (2003). Worst-case values for CsI and CsTe detectors, which occur at the long wavelength ends of their operating ranges (180 and 310 nm, respectively), are $\sim 0.5\%$ per degree Centigrade. Differences between detector temperatures during flight and during SURF calibration produce relative changes in $R'_C(\lambda)$ that are accounted for in Equation (1) as a gain term $G(\lambda, T_0, T)$ that multiplies the instrument responsivity derived from SURF data acquired at temperature T_0.

During ground test, dark counts from the instrument detectors were typically $\sim$1 and 5 Hz for the CsI and CsTe detectors, respectively. Flight values, which are typically 10 to 30 times larger than pre-flight values are discussed in Section 5.

4.3. Scattered and Stray Light

The observed count rate at wavelength λ in SOLSTICE is the sum of the true spectral signal, scattered light within the observed spectrum, and background arising from stray light and detector dark count (e.g., Equation (2)):

$$S(\lambda) = \int S_T(\lambda')\mathrm{GDF}(\lambda - \lambda')\,\mathrm{d}\lambda' + \mathrm{Stray} + \mathrm{D_C} \tag{31}$$

and

$$S(\lambda) = S_T(\lambda) + S_{G\,\mathrm{Scat}}(\lambda) + \mathrm{Stray} + \mathrm{D_C}. \tag{32}$$

Both stray and dark are backgrounds that are subtracted from $S(\lambda)$. On the other hand, grating scatter conserves photons, redistributing them over wavelength range that is diffracted by the grating. (Light scattered to wavelengths outside the instrument's observation wavelength range leads to a small reduction in responsivity that is properly taken into account in the normalization procedure described below.)

The term GDF in Equation (31) is the grating distribution function, which is usually represented as a Lorentzian profile plus constant background normalized so that its integral over the instrument wavelength range is equal to unity (Woods *et al.*, 1994):

$$\mathrm{GDF} = K\left[\frac{w^2}{(\lambda-\lambda_0)^2+w^2} + A_B(\lambda_0)\right] \tag{33}$$

and

$$w = \frac{1}{\sqrt{2\pi}}\frac{\lambda_0}{N}. \tag{34}$$

$2w$, the full width at half maximum of the Lorentizan, is inversely proportional to the number of coherently illuminated grating grooves, N. During unit level testing (gratings mounted in an external test fixture) using lasers, $N = d^{-1}W_{\mathrm{G}}$, where $d^{-1} = 3600$ grooves/mm is the grating groove spacing and W_{G} is the beam width input to the grating. For SURF calibrations and solar observations SOLSTICE functions as a pinhole camera and the coherence width is determined by the entrance slit diffraction pattern projected onto the grating, $N = \lambda d^{-1}/75$ (λ in nm). Thus, N varies from ~5500 grooves to ~15500 grooves over the SOLSTICE wavelength range, but $w \sim 4.5 \times 10^{-3}$ nm remains constant. K is a constant that normalizes the integral of the GDF over wavelength to 1, ensuring conservation of observed photons:

$$K^{-1} = w\left[\tan^{-1}\left(\frac{\lambda_H-\lambda_0}{w}\right) - \tan^{-1}\left(\frac{\lambda_L-\lambda_0}{w}\right)\right] + \\ +(\lambda_H-\lambda_H)A_B \approx w\pi + \Delta\lambda_{\mathrm{Norm}}A_B. \tag{35}$$

The normalization wavelength range extends from the instrument lower wavelength scan limit to the upper scan limit (λ_L to λ_H).

Kuznetsov, Content, and Boucarut (2001) measured the GDF at four wavelengths in the far ultraviolet-visible range for a variety of source-detector combinations using the Space Telescope Imaging Spectrograph Facility (Content *et al.*, 1996). Their observed GDFs were well matched to Equation (33) with values of A_B ranging from 3.7×10^{-6} at 152 nm to 1.9×10^{-7} at 266 nm to 7.2×10^{-8} at 325 nm. These values, which follow the relation $A_B(\lambda_0) = 3 \times 10^5/\lambda^5$ per nm, are consistent with results from instrument level testing and in-flight measurements described below.

In addition to scattered light, which is a consequence of illuminating a finite number of grating grooves (the Lorentzian term) and to the micro-roughness of those grooves (constant term), the SOLSTICE gratings also exhibit a pair of "ghost" spectra that appear to be artifacts of their manufacturing process. Figure 2 shows

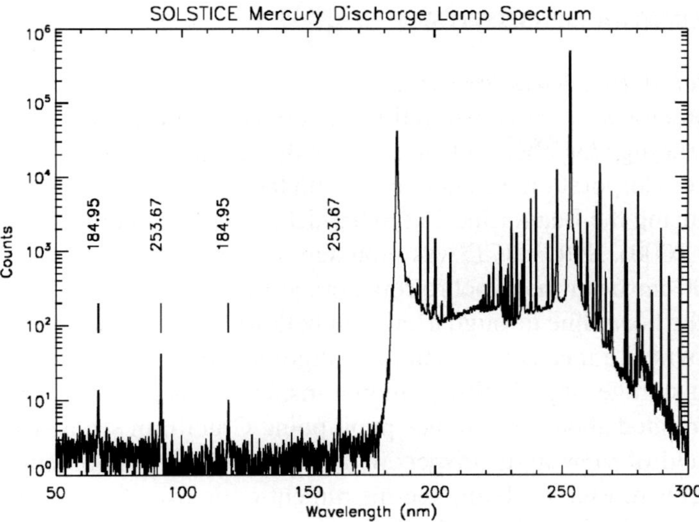

Figure 2. SOLSTICE MUV spectrum of a mercury discharge lamp showing emission lines from two "ghost" spectra.

a laboratory observation of a mercury discharge lamp acquired at ambient pressure. The complex emission spectrum, which is dominated by two lines at 184.9 and 253.7 nm, also displays four weak features at wavelengths below the air cutoff at ~180 nm. These four lines all appear in focus suggesting that they are artifacts produced by the grating rather than glints from internal instrument surfaces, which would produce diffuse, out-of-focus images. Their wavelength locations and intensities can be reproduced by assuming that the grating structure is a superposition of three different spacings with values $d^{-1} = 3600$, 2300, and 1300 grooves per mm, respectively, and that the efficiency of the latter two rulings relative to the first is 10^{-4}. This conclusion is supported by flight FUV solar observations, which exhibit features at 43.9 and 77.7 nm that arise when the bright hydrogen Lyman-α line, located at 121.57 nm, is diffracted by periodic structures on the grating having spacings of 1300 and 2300 per mm respectively. Corrections for these weak secondary spectra, which must be included for solar MUV observations in the 170 to 195 nm wavelength range, are negligible for ground calibration results.

Stray light is caused by diffuse scatter from instrument internal surfaces and is approximated by a constant background, which is proportional to the integral of $S(\lambda)$ over the entire wavelength range for which the detector is sensitive. Observations of instrument count rates for wavelengths outside the instrument sensitivity range (e.g., less than 110 nm for FUV and less than 150 nm for MUV) measure the sum of stray light plus grating scatter plus dark counts. Dark counts can be independently estimated from observations when no source is present.

4.4. Radiometric Sensitivity

4.4.1. *Initial SURF Measurements*

Initial radiometric sensitivity calibrations for both instruments were performed during August through October 2001 just before the instruments were delivered to Orbital Sciences Corporation for integration with the spacecraft. These measurements were made using the Beam Line 2 at SURF III at NIST in Gaithersburg, Maryland (Arp *et al.*, 2000). SOLSTICE was mounted inside a large vacuum chamber, located ~1700 cm from the synchrotron storage ring tangent point, and connected directly to the beam line through a vacuum bellows. External actuators attached to the tank provided a means for accurately aligning them to the orbital plane of the synchrotron storage ring. A gimbal mount inside the chamber allowed the instrument to be rotated about its entrance pupil point. Cruciform scans, which mapped the entire field-of-view in the dispersion and cross dispersion directions at a fixed wavelength, were used to define the instrument optic axes for both the FUV and MUV channels. Additional cruciform scans were used to calibrate the center of the instrument Solar Position Sensor (SPS) relative to the channel optic axes. The SPS is used in-flight to monitor the position of the Sun relative to these axes.

Because synchrotron radiation is highly polarized, initial calibrations with SOLSTICE A were preformed in two orthogonal orientations, one with the grating grooves perpendicular to the plane of the storage ring (vertical) and one with the grooves parallel to the storage ring (horizontal), and these two orthogonal measurements were averaged to provide the response to an unpolarized source. Later tests performed with SOLSTICE B showed that the response obtained from averaging two orthogonal orientations was identical to that obtained when the grating grooves are oriented at either ±45° with respect to vertical.

$R_C(\lambda, T_0\,\Omega_{\mathrm{SURF}})\Delta\,\lambda_{\mathrm{BP}} A_{\mathrm{Entrance}}\,T_{\mathrm{Filter}}(\lambda)$ and $R_C(\lambda, T_0, \Omega_{\mathrm{SURF}})\Delta\,\lambda_{\mathrm{BP}} A_{\mathrm{Entrance}}$ (see Equation (8)) were measured in solar mode using the solar flight entrance and exit slits, with and without the filter(s) in place, respectively. The SURF beam was too intense for direct measurements with the stellar entrance aperture. Instead, a 1 mm diameter pinhole was used in place of the 16 mm diameter flight aperture and the measured instrument responsivity was multiplied by the calibration aperture to flight aperture area ratio to obtain the true values. Areas for the 1 mm pinhole and the 16 mm flight apertures were measured at NIST to better than 2 parts in 10^4; therefore, the uncertainty in $R_{A\cdot\mathrm{BP}}$, which is ~0.5% is dominated by the photon noise in the SURF responsivity measurements. The stellar to solar mode area-bandpass ratios ($R_{A\cdot\mathrm{BP}}$ in Equation (22)) obtained by this procedure are summarized in Table V.

The instrument responsivities measured with the various slit and filter combinations must be multiplied by Γ and by $\mathrm{FOV}(\lambda, \theta, \phi)$ to produce the appropriate solar and stellar quantities for the irradiance calibrations (Equation (9)). These functions were computed from measurements of $\mathrm{FOV}_{\mathrm{SURF}}(\lambda, \theta, \phi)$, which were obtained by using the SURF gimbal to map instrument field-of-view in azimuth

TABLE V
Area-bandpass ratios.

Channel	Stellar aperture (mm^2)	Calibration aperture (mm^2)	Area-BP ratio
Sol A FUV	201.108 ± 0.014	0.97415 ± 0.00015	2.5053×10^5
Sol A MUV	201.108 ± 0.014	0.97415 ± 0.00015	5.7431×10^5
Sol B FUV	201.195 ± 0.014	0.97415 ± 0.00015	2.4892×10^5
Sol B MUV	201.195 ± 0.014	0.97415 ± 0.00015	5.7840×10^5

and elevation. Once the optic axes for the FUV and MUV were established from cruciform scans, spectra were collected at each point of a 5×5 coarse grid covering $\pm 0.5°$ and a 3×3 fine grid covering $\pm 0.125°$. Because the SURF beam at the storage ring tangent point is only 3 mm in diameter, spot sizes on the SOLSTICE optical elements are determined by diffraction at the instrument entrance aperture and vary from $\sim 0.06°$ for $\lambda = 115$ nm to $\sim 0.18°$ for $\lambda = 320$ nm. Thus, Ω_{SURF} is small and the $FOV_{SURF}(\lambda, \theta, \phi)$ measurements approximate the normalized responsivity $r(\lambda, T_0, \theta, \phi)/r(\lambda, T_0, 0, 0)$. Although the coarse grid somewhat under-samples FOV_{SURF}, the results are acceptable because the cruciform scan measurements demonstrated that response function is smoothly varying within $\pm 0.75°$ of the instrument optic axis. Figure 3 shows measurements of FOV_{SURF} at three wavelengths in the FUV channel for SOLSTICE B and three wavelengths in the MUV channels for SOLSTICE A, respectively.

Solar and stellar Γ values were calculated from the integrals of FOV_{SURF} (λ, θ, ϕ) over the Ω. $\Gamma(\lambda, \Omega_{Sun}, \Omega_{SURF}) = R'_{Sun}/R'_{SURF}$ and $\Gamma(\lambda, \Omega_{Sttar}\Omega_{SURF}) = R'_{Star}/R'_{SURF}$ are shown in Figure 4 as solid and dashed curves. The dotted lines show Γs for the 3×3 maps that cover the central $0.25°$ square of the field of view. Except for wavelengths greater than 250 nm in SOLSTICE A, there is a slow progression in the curves with values changing by 1% or less as Ω increases from 3×3 to solar to stellar. This behavior is consistent with the photomultiplier tube spatial non-uniformity response measured during laboratory test (Drake *et al.*, 2003) and reflected in the maps displayed in Figure 3.

Multiplying the Γ functions by the SURF responsivity curves gives the instrument on-axis solar and stellar responsivities (R'_{Sun} and R'_{Star} in Equation (8)). Results for the solar modes without MUV filters are shown in Figure 5. Values for R'_{Sun} and FOV_{Sun} (see Figure 6) include corrections that account for radiation that is diffracted past the instrument aperture stop (the diffraction grating) by the solar entrance slit. Corrections, which range from $\sim$1% at 110 nm to $\sim$4% at 320 nm, were determined using optical ray trace models. Diffraction losses are negligible for the stellar entrance aperture and for the SURF beam.

The $FOV_{Sun}(\lambda, \theta, \phi)$ and $FOV_{Star}(\lambda, \theta, \phi)$ functions are required to correct solar and stellar responsivities for pointing errors. They were obtained by integrating the $FOV_{SURF}(\lambda, \theta, \phi)$ over the appropriate solid angles. FOV_{Sun} for selected

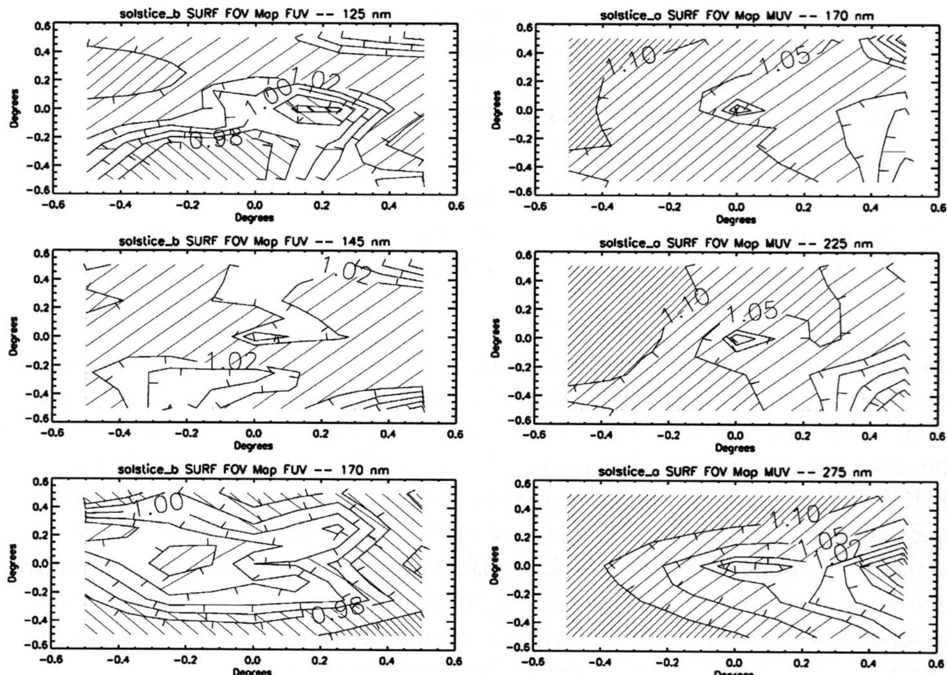

Figure 3. Measurements of $FOV_{SURF}(\lambda, \theta, \phi)$ for three wavelengths (125, 145, and 170 nm) of the SOLSTICE B FUV channel and three wavelengths (170, 225, and 275 nm) of the SOLSTICE A MUV channel are shown in the *left* and *right panels*, respectively. Since Ω_{SURF} is small, these values are nearly equal to the normalized responsivity $r(\lambda, T_0, \theta, \phi)/r(\lambda, T_0, 0, 0)$. The *shading line density* indicates departure from unity (denser lines are farther from unity). Lines rising to the right are contours greater than unity, while contours less than unity rise to the left.

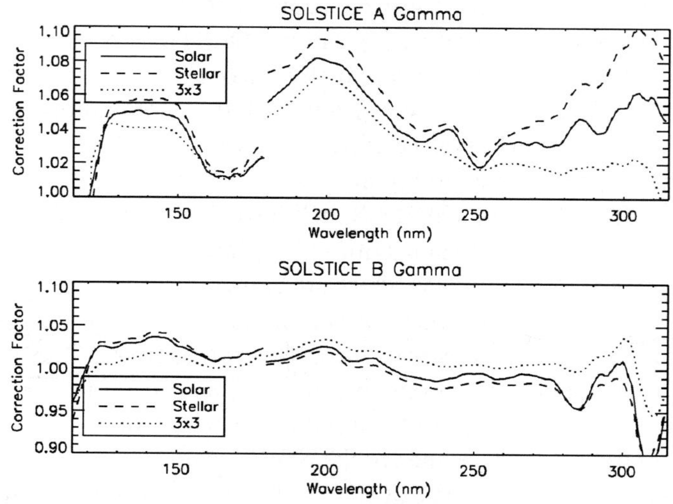

Figure 4. Γ values for SOLSTICE A and SOLSTICE B calculated from the SURF $FOV_{SURF}(\lambda, \theta, \phi)$ measurements.

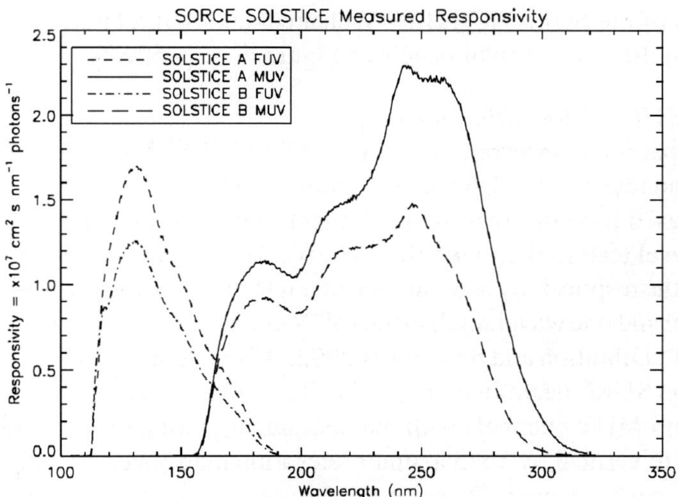

Figure 5. On axis R'_{Sun} values for the FUV and MUV channels of the two SOLSTICE instruments measured at SURF. The differences in responsivity between SOLSTICE A and SOLSTICE B arise primarily from the lower quantum efficiencies of the SOLSTICE B photomultiplier tubes.

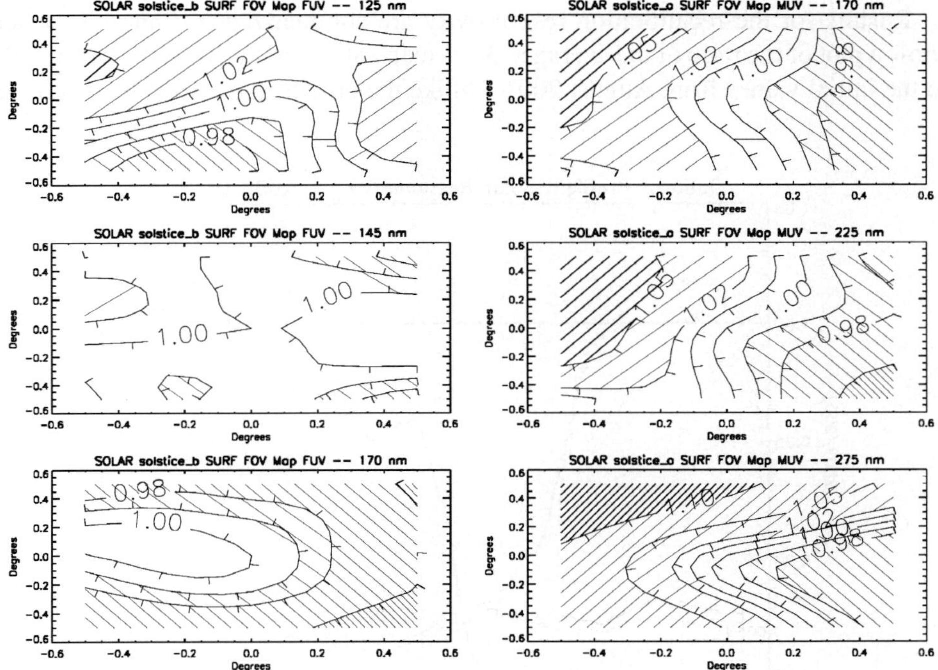

Figure 6. Measurements of $\mathrm{FOV}_{\mathrm{Sun}}(\lambda, \theta, \phi)$ for three wavelengths (125, 145, and 170 nm) of the SOLSTICE B FUV channel and three wavelengths (170, 225, and 275 nm) of the SOLSTICE A MUV channel are shown in the *left* and *right panels*, respectively. The *shading line density* indicates departure from unity (denser lines are farther from unity). Lines rising to the right are contours greater than unity, while contours less than unity rise to the left.

wavelengths of the SOLSTICE B FUV channel and SOLSTICE A MUV channel are plotted in the left and right panels of Figure 6, respectively.

4.4.2. *SOLSTICE A Recalibration*

Following spacecraft integration and test, SOLSTICE A was removed from the spacecraft and returned to SURF in September 2002 for final calibration. Cruciform scans were used to verify that the instrument FOV had not changed as a result of spacecraft level testing (i.e., vibration, shock). These scans also demonstrated that the shape of the responsivity as a function of angle measured at one wavelength in the FUV channel and one wavelength in the MUV channel had not changed between the August 2001 calibration and September 2002. After the optic axes were established relative to the SURF beam line, $R_C(\lambda, T_0, \Omega_{\mathrm{SURF}})\Delta\lambda_{\mathrm{BP}} A_{\mathrm{Entrance}}$ was measured for both FUV and MUV channels with the instrument grating grooves oriented at 45° with respect to vertical. A 3×3 azimuth-elevation map covering $\pm 0.125°$ was also obtained for each channel. Detector head temperatures for these measurements, which varied from 29.3° to 31.2°C, were measured to within ± 0.1°C and the final responsivities were corrected to a standard temperature of $20.0° \pm 0.1$°C using the temperature gain curves measured during detector head characterization (Drake *et al.*, 2003).

Results for the recalibration responsivity are summarized in Figure 7, which show a plot of the ratio of the averaged 3×3 maps obtained during September 2002 to the initial values from August 2001. Unlike the cruciform scans, which showed

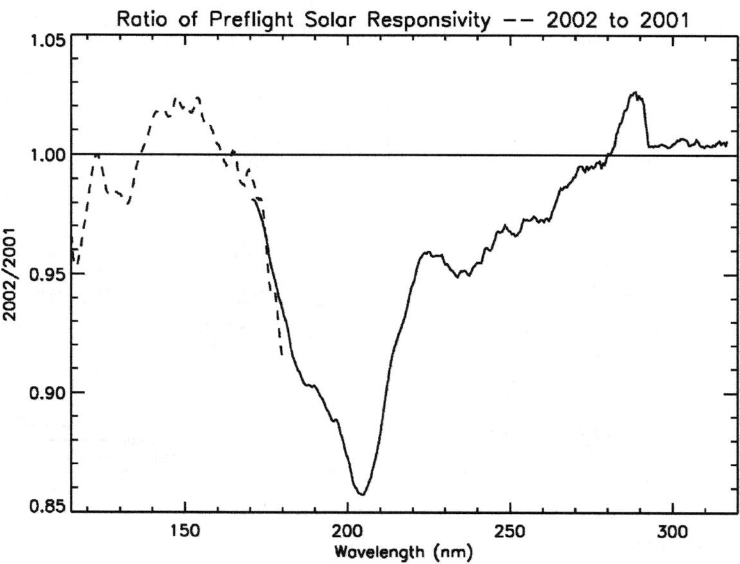

Figure 7. The ratio of SOLSTICE A solar responsivities (R'_{Sun}) measured during September 2002 and August 2001, respectively. The September results have been adopted as the official instrument calibration.

no change, there is clear evidence for a wavelength dependent loss in sensitivity between the two SURF calibrations. While the minimum near 220 nm suggests absorption by hydrocarbons, there is no corresponding decrease in FUV channel sensitivity, which would be expected to decline as typical molecular absorption cross-sections increase toward shorter wavelengths. Analysis of flight cruciform scans shows that the degradation is most likely the result of contamination on the diffraction grating (Snow *et al.*, 2005).

No 5×5 maps were measured during the 2002 recalibration. As a result, values for $\Gamma(\lambda, \Omega_{\text{Sun}}, \Omega_{\text{SURF}})$ and $\Gamma(\lambda, \Omega_{\text{Sttar}}\Omega_{\text{SURF}})$ were calculated by scaling values from the 3×3 maps $(\Gamma(\lambda, \Omega_{3\times3}, \Omega_{\text{SURF}}))$. Since there is no evidence for responsivity changes over the field-of-view, the 2001 and 2002 Γs for the two 3×3 maps were first averaged before scaling, e.g.,

$$\Gamma_{2002}(\lambda, \Omega_{\text{Sun}}, \Omega_{\text{SURF}}) = \frac{\overline{\Gamma(\lambda, \Omega_{3\times3}, \Omega_{\text{SURF}})}}{\Gamma_{2001}(\lambda, \Omega_{3\times3}, \Omega_{\text{SURF}})} \Gamma_{2001}(\lambda, \Omega_{\text{Sun}}, \Omega_{\text{SURF}}). \quad (36)$$

The final values are shown in Figure 8.

The SORCE project has adopted the results from the September 2002 calibration, summarized in Figures 7 and 8, as the as-launched values for computing flight solar and stellar irradiances for SOLSTICE A. Launch schedule constraints prevented a similar recalibration of SOLSTICE B. Instead, the SOLSTICE A calibration was transferred to SOLSTICE B during the first week of on-orbit operations by simultaneously observing the Sun with the two instruments and comparing their

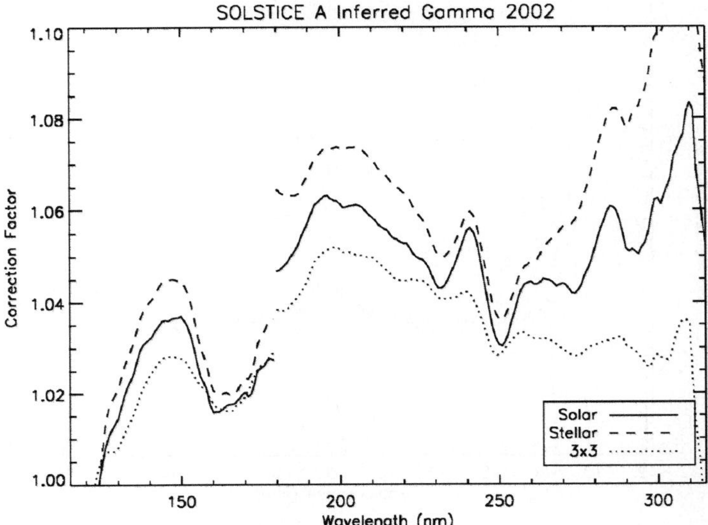

Figure 8. Γs for SOLSTICE A obtained from the 2002 recalibration. Solar and stellar values were scaled from mean of the 2001 and 2002 $\Gamma_{3\times3}$ measurements (*dotted curves* here and in Figure 4) using Equation (36).

spectra. These measurements show that SOLSTICE B had also experienced a loss of sensitivity after its initial calibration. Solar A/B comparison experiments are performed once each week in order to track relative changes in sensitivity. Error estimates in responsivity are discussed in Section 4.4.4.

4.4.3. *Unresolved SURF Issues*

In the nominal SURF III operational mode the transverse (vertical) size of the electron bunch is enlarged by adjusting the relative strengths of the horizontal and vertical betatron oscillation frequencies (Arp *et al.*, 2000). This process, which is referred to as adding "fuzz" to the beam, increases the beam lifetime by an order of magnitude. In SURF II, beam fuzz was introduced by artificially exciting the vertical betatron motion (Arp *et al.*, 2000), which introduced strong wavelength dependent oscillations in the irradiances observed with UARS SOLSTICE (Woods, Rottman, and Ucker, 1993). These artifacts were not observed during the SORCE calibrations on SURF III. On the other hand, spatial nonuniformities, which appear to be related to diffraction around apertures, were observed during beam centering experiments. Figure 9 shows profiles, obtained with the MUV channel at fixed grating position corresponding to 229 nm, made by translating the instrument aperture from edge to edge of the beam line using the vacuum tank manipulators. For $\lambda = 229$ nm, both horizontal and vertical scans exhibit approximately a 3% increase in irradiance at beam line center suggesting that the pattern is axially symmetric. Ratios of both

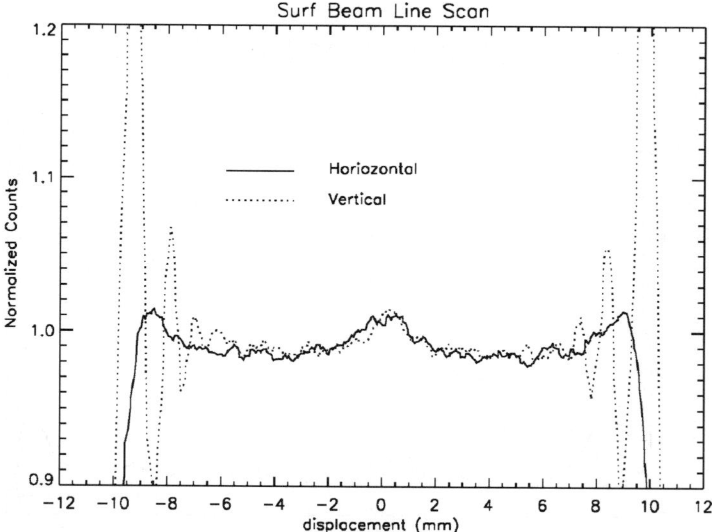

Figure 9. Horizontal and vertical scans of the SURF beam line exhibit a wavelength dependent increase in irradiance at beam center. This example was acquired with the MUV channel for $\lambda =$ 229 nm. The oscillations near the edges caused by diffraction from the aperture edge are expected, but the center bump is unexpected.

the FUV and MUV spectral scans acquired with the instrument aperture offset by 5.1 mm in the horizontal from beam center to scans acquired with the aperture at beam center showed that the increase varies from ~1% near 115 nm to ~4% at 320 nm. The cause of this nonuniformity is not well understood, but is thought to arise from diffraction around circular metal baffles located along the beam line (Furst, private communication, 2004). In that case, the irradiance calculated from the standard synchrotron equations underestimates the actual irradiance arriving at the SOLSTICE instrument entrance apertures by 1–4% across the SOLSTICE wavelength range, and therefore, the values for $E_{Std}(\lambda)$ in Equation (8) have been increased by this factor (Section 4.4.4, Equation (41)).

4.4.4. *Radiometric Calibration Uncertainty Analyses*

The uncertainty estimates for the SOLSTICE A, R', measured in September 2002 were computed from Equation (16) and are summarized in Figure 10, where the upper panel shows the contributions from the three individual terms and the lower panel shows the combined total.

Corrections to the observed signal required to compute the uncertainty in C_{Std} include nonlinearity, scattered light, and stray light. Since the count rates at SURF do not exceed 10^5 per second, the nonlinearity correction was always less than 0.5%. The scattered light correction was approximated by subtracting the observed counts from the convolution of the observed counts with the Grating Distribution Function (GDF) after applying the nonlinearity correction and subtracting stray

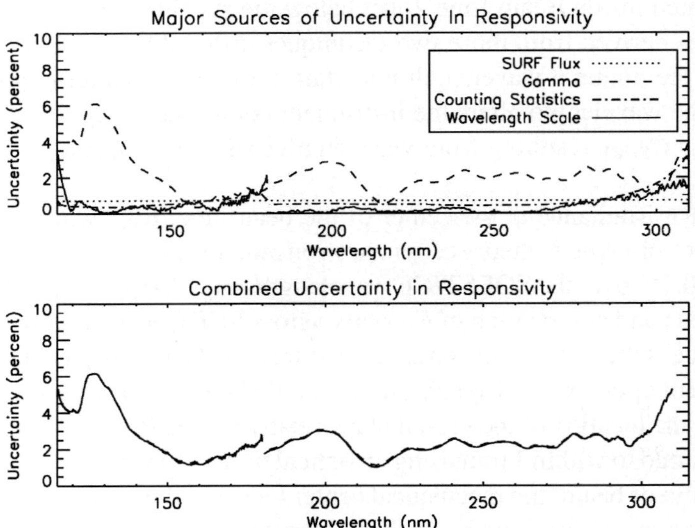

Figure 10. SOLSTICE A solar responsivity relative uncertainty, $\sigma_{R'_{Sun}}$. The largest contribution arises from $\Delta\Gamma$, which must be scaled from the 3 × 3 maps. Total uncertainties for the solar responsivity (*solid curve in the lower panel*) and stellar responsivity (*dashed curve*) are nearly identical.

light and dark counts:

$$C(\lambda)_{\text{Obs}} = C(\lambda)_{\text{True}} \otimes \text{GDF} = C(\lambda)_{\text{True}} + C(\lambda)_{\text{Scat}}, \tag{37}$$

$$C(\lambda)_{\text{Obs}} \otimes \text{GDF} = (C(\lambda)_{\text{True}} + C(\lambda)_{\text{Scat}}) \otimes \text{GDF}, \tag{38}$$

$$C(\lambda)_{\text{Obs}} \otimes \text{GDF} \approx C(\lambda)_{\text{True}} + 2C(\lambda)_{\text{Scat}}, \tag{39}$$

and then

$$C(\lambda)_{\text{True}} \approx 2C(\lambda)_{\text{Obs}} - C(\lambda)_{\text{Obs}} \otimes \text{GDF}. \tag{40}$$

This approach, which assumes that the convolution, $\otimes$, of the scattered light term with itself is negligible for the featureless SURF spectral irradiance, was validated by constructing a "zero scattered light" synthetic spectrum (C_{True}) from the instrument responsivity and the SURF flux, convolving the result with the GDF to simulate the observed spectrum (C_{Obs}), retrieving C_{True} from C_{Obs} and comparing that result with the input C_{True}. Count rates for wavelengths shorter than the detector window transmission (less than 110 nm), which have been corrected for scattered light, provide a measure of detector dark count plus stray light. $1-\sigma$ uncertainties in the SURF count rates, C_{SURF}, including counting statistics, nonlinearity, dark count, scattered light, and stray light are shown as dot-dashed lines in Figure 10.

The instrument wavelength scale was measured by scanning the grating to zero order to directly measure Θ_0 in Equation (29) and by observing an emission line source, located inside Beam Line 2 just below the synchrotron orbit plane. Wavelength scales derived from these two techniques differed by ~ 0.1 nm, which was adopted as the nominal wavelength uncertainty for SURF calibrations. A 0.1 nm change in the wavelength alters the instrument count rate by up to 1%. $1-\sigma$ uncertainties in C_{SURF} resulting from wavelength scale errors appear in Figure 10 as solid lines.

The SURF irradiance at the center of the beam line, E_{Std}, which is calculated from the electromagnetic theory of synchrotron radiation and the SURF geometry, is accurate to 0.2% over the SOLSTICE wavelength range (Arp *et al.*, 2000). Because the magnitude and polarization of E_{Std} vary across SURF beam, additional errors in the irradiance at the instrument entrance aperture arise from a displacement between the instrument optic axis and synchrotron orbital plane. Before and after every calibration run, the location of the orbital plane relative to the Beam Line 2 mechanical axis is measured to within 1 mm using an optical transit. Since SOLSTICE is placed within 0.5 mm of beam line mechanical center by using the vacuum tank actuators, calibration errors arising from beam displacement and centering are less than 0.5% and the total uncertainty in E_{Std} is $\sim 0.54\%$, independent of wavelength. The beam nonuniformity described in Section 4.4.3 also contributes to the irradiance uncertainty at the instrument entrance aperture. For the official SOLSTICE calibration, values for $E_{\text{Std}}(\lambda)$ required to calculate responsivities using Equation (8) have been increased by a wavelength dependent factor ranging from 1 to 4%:

$$E'_{\text{Std}}(\lambda) = E_{\text{Std}}(\lambda)(1.01 + 1.43 \times 10^{-4}(\lambda - 110)). \quad (41)$$

Formally, the uncertainty in this correction, which results from fitting a straight line to the ratio of two spectra is 0.5%, and the overall formal uncertainty in E_{Std} entered in Equation (16) is therefore $\sim$0.75%. A detailed calculation of the diffraction contribution to the beam line flux is beyond the scope of the work reported here.

Errors in Γ arise from misalignments, from under sampling in the instrument field-of-view measurements, $\text{FOV}_{\text{SURF}}(\lambda, \theta, \phi)$, and from extrapolating the 3×3 maps to obtain Γ_{Sun} and Γ_{Star}. If the angular dependence in $r(t, \theta, \phi)$ were unchanged between 2001 and 2002, then the difference between the 3×3 measurements in 2001 and 2002, $\Delta = \Gamma_{2001}(\lambda, \Omega_{3\times3}, \Omega_{\text{SURF}}) - \Gamma_{2002}(\lambda, \Omega_{3\times3}, \Omega_{\text{SURF}})$, can be used as the error in measuring $\Gamma_{3\times3}$ and the uncertainty in the solar and stellar values calculated using Equation (36) is $\sqrt{3}\Delta$. Δ is typically $\sim$1–2% except for wavelengths less than 140 nm where it increases to $\sim$3.5% and 120 nm. This translates to $\sim$2–3% uncertainties in Γ_{Sun} and Γ_{Star} except for the shortest FUV wavelengths where they rise to $\sim$6%.

The total uncertainty estimate for R'_{Sun} derived from the SURF calibrations (bottom panel of Figure 10) is the sum of the uncertainties in counts, which include corrections for nonlinearity, scattered light, stray light, and dark count (dotted lines), in E_{Std} (dashed lines), in wavelength scale (dot-dashed lines), and in $\Gamma(\lambda, \Omega_{\text{Sun}}, \Omega_{\text{SURF}})$, all of which are combined in quadrature to produce the grand total. The overall calibration uncertainty is dominated by the uncertainty in $\Gamma(\lambda, \Omega_{\text{Sun}}, \Omega_{\text{SURF}})$ and is approximately 2% except for wavelengths less than 140 nm where it rises to $\sim$6%. The dashed curve in the lower panel in Figure 10 shows the uncertainties in the stellar responsivity, which includes an additional term, $\sigma(R_{A\cdot\text{BP}})$. Since $\sigma(R_{A\cdot\text{BP}})$ is less than 0.5% (Section 4.4.1), it results in less than 5% fractional increase in the solar values.

5. Flight Calibrations

During flight, a number of routine calibration/characterizations are performed to complement and validate pre-flight calibrations and to track instrument performance. In addition to an accurate flight wavelength scale, measurements of scattered light, stray light, and dark counts, which are required for calculation of the corrected solar and stellar count rates (e.g., Equations (12) and (20)), can only be obtained *in situ*. Stellar irradiance measurements, which are the hallmark of the SOLSTICE investigation, solar field-of-view maps, and solar cruciform scans, are the primary tools for determining the degradation function, $\text{DEG}(t, \lambda, \Omega, \theta, \phi)$.

Finally, special solar experiments provide more accurate determinations of filter transmission, and detector nonlinearity than ground calibrations.

5.1. Wavelength Scale

The flight wavelength scale for solar observations is calculated from the Level 1 data, which consist of sequences of grating position versus detector counts. In the FUV channel, grating positions for eight emission lines, which are calculated by fitting Gaussian functions to their profiles, are least squares fit to laboratory values (Kelly and Palumbo, 1973) using Equation (27) to derive an average step size ($\Delta\Theta$) and zero order fiducial (Θ_0) for the grating drive. During nominal operation, the wavelengths for the eight emission lines, calculated from the resulting parameters, agree with the laboratory values to better than 0.01 nm.

For wavelengths greater than 200 nm the solar spectrum is dominated by closely spaced absorption lines, which are sufficiently unresolved by SOLSTICE to render precise identification of their wavelength centers ambigious. Therefore, the Si I line, located at 180.801 nm, and the Mg II *k* line, located at 279.635 nm (vacuum value), are used to uniquely calculate $\Delta\Theta$ and Θ_0 in the MUV channel. Experiments with the FUV data also confirm that a highly accurate wavelength scale can be calculated from a pair of isolated emission lines. Although the average step size and zero order fiducial calculated using only the hydrogen Lyman-α line (121.567 nm) and the Si I line (180.801 nm) produces a wavelength scale that is slightly less accurate than one derived from fitting multiple lines to Equation (27), the results are still accurate to ± 0.01 nm.

Over the course of months, the average step sizes for individual solar spectral scans agree with each other within eight parts in 10^5. On the other hand, the zero order fiducial can vary by as much as ± 35 arcsec (approximately ± 0.04 nm) from orbit to orbit if the instrument power is cycled or a grating drive "find index" command is executed.

It is not practical to determine stellar wavelength scales directly from the spectral measurements because long integration times are required for stellar observations. Instead, the grating is rotated to scan the region around zero order immediately after the spacecraft points to the star, but before a measurement sequence on that star begins. These data are used to more accurately establish the wavelength scale in the presence of both grating drive offsets and pointing errors. If the grating drive position has an offset ε, then Equation (29) becomes $\Theta_S = (N_0 - N)\Delta\Theta + \varepsilon$ and the step number for zero order, N_Z, is related to the pointing offset and grating drive offset: $e + E/2 = (N_0 - N)\Delta\Theta$ and Equation (28) becomes

$$\lambda_{\text{Stellar}} = 2d \sin((N_Z - N)\Delta\theta)\cos((N_0 - N_Z)\Delta\theta + \varepsilon + \phi_G/2). \tag{42}$$

ε is unknown; nonetheless, since the argument of the cosine term in Equation (42) is less that 3°, ignoring ε, which is typically ± 30 arcsec or less, introduces an error less than 1 part in 10^5 ($< 3 \times 10^{-3}$ nm at $\lambda = 300$ nm) in the stellar wavelength scale. This is significantly less than the 0.02 nm uncertainty introduced by the algorithm that determines the value for N_Z (Snow *et al.*, 2005). Zero-order scans are also performed after each measurement sequence in order to verify that spacecraft pointing did not drift during the observation. Additional discussion of the stellar wavelength scale calibration appears in Snow *et al.* (2005).

5.2. Dark Count, Stray and Scattered Light

Dark count, and stray and scattered light, which can only be measured *in situ*, are required for calculating the corrected solar and stellar counts (e.g., Equations (12) and (20)). Typical pre-flight dark count rates were 0.3 Hz and 5 Hz for the FUV and MUV channels respectively. In-flight nominal rates, which are typically 3–10 Hz and 50–100 Hz, respectively, are 10 to 20 times larger than the pre-flight values. The instantaneous levels appear to be correlated with spacecraft geographic location and can exceed 10^3 Hz during passage through the South Atlantic Anomaly (SAA), a region over the South Atlantic Ocean off the coast of Brazil where the lower Van Allen belt reaches down to the SORCE spacecraft altitude.

The flight levels of dark count have a negligible effect on nominal solar measurements acquired outside the SAA where the signal levels are typically 3–5 orders of magnitude greater than the dark. In this case a daily average dark level is constructed from nighttime dark sky observations and applied to the data. In contrast, real time measurements of dark count are necessary for accurately determining stellar irradiances because the stellar count rates are lower. These data are acquired before and after each stellar observation as part of the zero order scan sequence. Stellar observation sequences with significantly anomalous dark measurements are discarded. All solar and stellar observations acquired in the SAA are discarded.

Stray and scattered light for solar observations are determined from spectral scans that extend below the detector window cutoff wavelength (115 and 165 nm for the FUV and MUV, respectively) for the individual channels. FUV and MUV detector counts recorded at these wavelengths (typically 6 and 1600 Hz), which appear to be independent of grating position, are the sum of dark, stray, and scattered light. MUV detector counts also include the "ghost" spectrum contribution for $\lambda > 100$ nm. Its relative strength, which is about 5% of the total count rate below 160 nm, is computed by compressing the solar MUV wavelength scale by a factor of 1.565 and scaling a typical solar spectrum by a factor of 10^{-4} (see Section 4.3).

MUV dark counts contribute 5–10% to the total 1600 Hz, depending on geographic location. Scattered light, calculated from Equations (39) and (40) (with $w = 0.017$ nm and $A_B(\lambda) = 3 \times 10^5/\lambda^5$ nm^{-1} in Equation (33)) accounts for an additional 15% leaving the 70–75% contribution from stray light as the dominant component. In the FUV channel dark counts and scattered light (assuming $w = 0.017$ nm and $A_B(\lambda) = 3 \times 10^5/\lambda^5$ nm^{-1} in Equation (40)) each contribute

50% to the total. The relatively large stray light contribution in the MUV channel is thought to arise from its greater sensitivity to visible sunlight scattered from the instrument internal surfaces. Larger values of A_B would increase the MUV scattered light contribution, but those values would be inconsistent with the FUV requirement that scattered light contribute no more than ~3 Hz to the count rate observed at 100 nm.

5.3. Initial Alignments and Field-of-View Response

Cruciform scan experiments, performed immediately after instrument check-out by spacecraft maneuvers of $\pm 2^\circ$ about two orthogonal axes normal to the Sun line, were used to calibrate the instrument SPSs and to determine the location of the instrument optic axes relative to the spacecraft pointing control system. These data were also used to validate the SURF alignment cruciform scans and to determine offsets between $\mathrm{FOV}_{\mathrm{Sun}}(\lambda, \theta, \phi)$, calculated from the SURF field-of-view maps, and the nominal spacecraft solar boresight. The measurements for SOLSTICE A showed that except for the dispersion axis of the FUV channel, which was displaced by -0.09° from Sun center, all alignments were within 0.02°. On the other hand the dispersion axis alignments for the SOLSTICE B FUV and MUV dispersion axes were displaced by -0.36° and -0.12°, respectively, suggesting that one or more optical elements shifted during launch. Instrument SPS measurements, which are recorded as part of the SOLSTICE science data, show that nominal spacecraft control system pointing to Sun center has maintained these offsets to within $\pm 0.03^\circ$ during the entire mission.

Complete solar spectra, collected over a 5×5 point grid covering $\pm 0.125^\circ$ five weeks after the beginning of flight operations, were used to compute an initial in-flight measurement for $\mathrm{FOV}_{\mathrm{Sun}}(\lambda, \theta, \phi)$. These maps are nearly identical to those obtained during SURF calibration. Reference to Figure 6 indicates that changes in SOLSTICE A responsivity caused by launch displacements pointing errors are completely negligible for SOLSTICE A. For SOLSTICE B they produced less than a 1% change in the instrument responsivity relative to its on-axis value. Since the SOLSTICE B flight calibration is derived from the September 2002 SOLSTICE A SURF calibration and because the spacecraft points SOLSTICE A to the Sun to $\sim 0.1^\circ$, no pointing correction is applied to the irradiance values and $\mathrm{FOV}_{\mathrm{Sun}}(\lambda, \theta, \phi) = 1$.

5.4. Filter Transmissions and Nonlinearity

During routine solar observations for wavelengths greater than 200 nm, a single neutral density filter is inserted in the optical path behind the exit slit to attenuate the photon flux reaching the detector by about a factor of 10. The filter transmission ($T_{\mathrm{filter}}(\lambda)$ in Equation (1)) is measured weekly from the ratio of a spectrum with the filter in place to one with the filter removed. These two observations, which cover the entire MUV wavelength range, are acquired during a single orbit. Because the MUV solar spectrum is rapidly varying from wavelength to wavelength and count rates with the filter removed exceed 10^6 Hz at some wavelengths, these ratios provide measurements of both $T_{\mathrm{filter}}(\lambda)$ and the detector dead time τ. Values of τ derived from these measurements agree with those obtained at SURF to better than 5%. Measurements of the backup filter transmission are also obtained during the same orbit. No change in the dead time or the transmission for either filter has been detected.

5.5. Degradation

In principle, daily exposure to solar X-rays and extreme ultraviolet radiation ages the first fold mirror (M1) and, to a lesser extent, the diffraction grating, which is only exposed to the longer wavelength far ultraviolet radiation, causing a reduction in optical throughput. (Although loss of diffraction grating reflectivity caused by hydrocarbon absorption appears to be the dominant mechanism for changes in instrument transmission for both SOLSTICE A and SOLSTICE B. See Section 4.4.2 and Snow *et al.*, 2005.) In addition, the photomultiplier detectors modal pulse gains decrease with total dose (Drake *et al.*, 2003), which also leads to a decline in instrument sensitivity. Both of these effects cause the instrument responsivity to decrease over the life of the mission. The parameters in the irradiance equations that track these changes are $\mathrm{DEG}(t, \lambda, \Omega, \theta, \phi)$ and $\Delta\Gamma_{\mathrm{DEG}}(t, \lambda, \Omega_{\mathrm{Star,Sun}})$.

Nightly irradiance measurements of an ensemble of bright B and A stars provide the primary data for directly determining $\mathrm{DEG}(t, \lambda, \Omega, 0, 0)$ for the stars, which includes both optical and detector degradation. Cruciform scans, performed weekly, and solar FOV maps, performed semi-annually, are used to calculate $\Delta\Gamma_{\mathrm{DEG}}(t, \lambda, \Omega_{\mathrm{Star,Sun}})$ which is combined with the stellar $\mathrm{DEG}(t, \lambda, \Omega, 0, 0)$ to compute the solar $\mathrm{DEG}(t, \lambda, \Omega, 0, 0)$. These measurements, which are performed independently for SOLSTICE A and SOLSTICE B, are described in detail by Snow *et al.* (2005).

5.6. Irradiance Uncertainty Analyses

5.6.1. *Measure Solar Irradiance with 5% Accuracy*

The uncertainty estimates for 1 nm spectral resolution solar irradiance averaged over 6 h (typically the average of 4 independent spectra acquired on successive orbits) at the beginning of mission were computed from Equation (14) using the results for instrument responsivity, which are described in Section 4.4.4 and summarized in Figure 10, and assuming $\sigma_{\mathrm{DEG}} = 0$, $\sigma_{\mathrm{FOV}} = 0$, and $\sigma_{\mathrm{AU}} = 0$. Since the same instrument clock was used to generate the calibration and flight integration times, $\sigma_{\Delta t}$ arising from the ratio of solar count rates to SURF count rates was also 0. Error estimates for the detector gain correction, which were computed using the wavelength dependent gain coefficients measured during detector calibration (Drake *et al.*, 2003) and a 0.2°C uncertainty in the difference between SURF (Section 4.4.2) and flight detector head temperatures, are less than 0.1% and are neglected. Thus, the uncertainty in solar irradiance is dominated by uncertainties in the solar counts and the uncertainty in instrument responsivity at the beginning of mission.

Solar count rates were corrected for dead time, dark counts, scattered light, and stray light before binning to 1 nm. Uncertainties in dark count and scattered and stray light are only important in the MUV channel for $\lambda < 200$ nm. Estimates for errors arising from wavelength scale shifts were computed by shifting a nominal resolution spectrum by 0.01 nm and comparing its 1 nm binned values with binned values from an unshifted spectrum. Errors in wavelength scale, detected counts, and dark counts are random and can be made arbitrarily small by averaging a large number of spectra (e.g., producing daily averages). On the other hand, errors in the dead time and scattered light corrections are systematic because they arise from errors in the parameters used in Equations (30) and (40), respectively. These errors can only be further reduced by refining the values of those parameters.

The top panel in Figure 11 shows the major components and combined uncertainty in the corrected solar count rates, calculated from Equation (16). Uncertainties in the dead time correction dominate in the MUV for wavelengths >250 nm, while uncertainties in the scattered light correction dominate the FUV

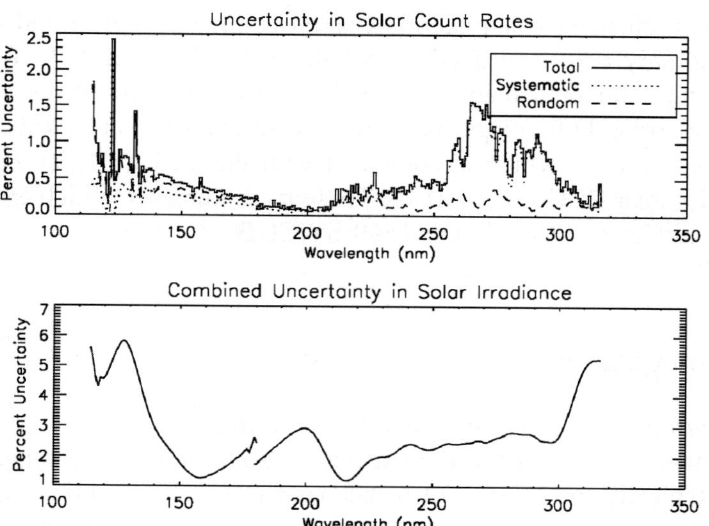

Figure 11. The *top panel* summarizes the uncertainty in solar counts (Equations (12) and (15)) for the average of four spectra that comprise a single 6-h solar observation. Count-rate uncertainties are combined with uncertainties in instrument responsivity (Figure 10 and Equation (14)) to produce the total uncertainty in solar spectral irradiance shown in the *bottom panel*.

for wavelengths <135 nm, with the largest value occurring for the correction to the hydrogen Lyman-α line at 121.567 nm. The bottom panel shows the total uncertainty in the measured 1 nm solar irradiance at the beginning of mission, calculated from Equation (15). It is dominated by the uncertainty in $\Gamma(\lambda, \Omega_{\mathrm{Sun}}, \Omega_{\mathrm{SURF}})$, which arises primarily from the extrapolation of the 3×3 maps to Ω_{Sun}.

5.6.2. *Determine the Solar and Stellar Irradiance Ratio with 2% Accuracy*

Figure 12 shows that the total uncertainty for the ratio of solar to stellar irradiance at the beginning of mission calculated from Equation (23) is less than 2% across the SOLSTICE wavelength range. It includes terms for errors in corrected solar count rates, stellar count rates, the stellar–solar area-bandpass ratio, and stellar–solar instrument responsivity ratio.

Errors in solar count rates (top panel in Figure 11) are less than 0.75% except for a narrow band of wavelengths centered at 275 nm and for Lyman-α (121.6 nm). Snow *et al.* (2005) show that error in average stellar counts for the brightest stars is 0.5% or less across the entire spectrum. The area-band pass ratio was measured at SURF by directly comparing calibrations using the flight solar entrance apertures and exit slit with calibrations using a 1 mm diameter pin-hole entrance aperture and the stellar exit slits. Later the areas of the pin-hole and flight stellar entrance aperture were measured at NIST to better than 1 part in 10^4 resulting in a ratio uncertainty, $\sigma_{\mathrm{R}} = 0.5\%$, which is dominated by the photon counting noise in the respective SURF responsivity measurements.

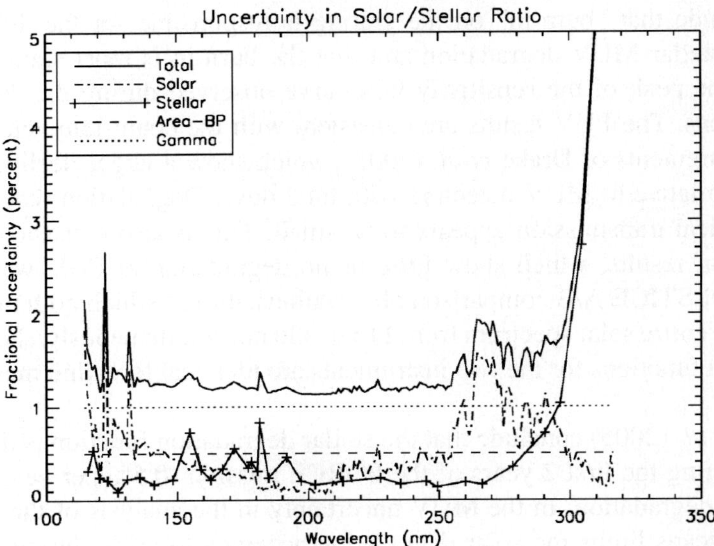

Figure 12. The solar/stellar irradiance ratio uncertainty, which is less than 2% across the SOLSTICE wavelength range, is shown as a *solid line*. Contributions from the individual components, solar, stellar counts, area-bandpass, and gamma are identified in the legend.

Uncertainties in the solar and stellar responsivity ratio, which arise from errors in determining $\Gamma(\lambda, \Omega_{\text{Sun}}, \Omega_{\text{Star}})$ (Equation (21)), are more difficult to estimate. In any event, they make only a small contribution to the overall uncertainty in the solar–stellar irradiance ratio because the correction is always small. Except for wavelengths greater than 270 nm, Γ_{Sun} and Γ_{Star} differ by 1% or less and difference rises to only ~3.5% at 300 nm. Ratios Γ_{Sun} and Γ_{Star} for the four polarization runs made with SOLSTICE B during the 2001 SURF calibration are consistent to within 1%. This is the magnitude of σ_Γ used in Equation (23). The uncertainty in photomultiplier detector temperature gain correction should also be a term in Equation (23), but its magnitude is completely negligible because the detector temperature is stable to ±0.1°C during any given spacecraft orbit.

5.6.3. *Track Changes in Solar Irradiance with 0.5% per Year Long-Term Relative Accuracy*

Snow *et al.* (2005) analyze the stellar and solar degradation functions experienced by SOLSTICE during its first 2 years of operation. They show that the instrument stellar FUV and MUV responsivities have declined by an average of 3% per year and 0.5% per year, respectively. Analysis of cruciform alignment scans, which are obtained weekly, show that the degradation in the solar FUV responsivity is tracking the stellar function. On the other hand, the MUV cruciform scans show that the degradation in the solar responsivity is up to a factor of three greater than the stellar values.

Snow *et al.* (2005) argue that the similarity of the FUV functions suggests that they are caused primarily by changes in detector modal gain. On the other hand, they conclude that "burn-in" on the grating is responsible for the difference in solar and stellar MUV degradation and that the burn-in is most significant near 220 nm (the peak of the sensitivity loss curve observed during the 2002 SURF recalibration). The FUV results are consistent with the count-rate versus modal-gain measurements of Drake *et al.* (2003), which show a larger decline for FUV detectors, relative to MUV detectors with total dose. Degradation resulting from loss of optical transmission appears to be small. This is also consistent with the recalibration results, which show little or no degradation at FUV wavelengths. Weekly SOLSTICE A/B comparison observations, during which both instruments observe the entire solar spectrum from 115 to 320 nm simultaneously, show that the degradation functions for the two instruments are identical to within measurement errors.

Snow *et al.* (2005) conclude that the stellar degradation function is determined to ~1% during the first 2 years of the SORCE mission (0.5% per year) as is the solar FUV degradation. In the MUV uncertainty in the analysis of the cruciform alignment scans limits the solar degradation accuracy to ±2% during the first 2 years. Snow *et al.* (2005) propose modification of the flight cruciform experiments and an expansion of the field-of-view maps to cover a larger field of view than the ±0.1°. It is anticipated that these new calibrations will improve the MUV accuracy to $\sim 0.5\%$ per year. The spacecraft pointing accuracy is $\sim\pm 0.03°$; no additional FOV corrections are required and the DEG(t, λ, Ω, 0, 0) adequately represent the instrument performance.

6. Validation

The accuracy of the SORCE SOLSTICE calibrations at the beginning of the mission can be validated by comparison of solar and stellar irradiance values to those obtained by UARS SOLSTICE. The Solar Extreme ultraviolet Experiment (SEE) aboard the Thermosphere Ionosphere Mesospheric Energy and Dynamics (TIMED) mission also measures solar irradiance at wavelengths less than 180 nm, but its calibration is only accurate to 15% (Eparvier *et al.*, 2001; Woods *et al.*, 2005).

Figure 13 compares a daily averaged SORCE solar observation made on 3 April 2003 to UARS. The UARS FUV spectrum was obtained from a simultaneous observation, which was processed using software recently developed at LASP. Processing software is not in place for the UARS MUV channel and that spectrum is the average of spectra obtained on three days (7–8 August 1992 and 8 April 1993) when the solar F10.7 and its 81-day average ($\langle$F10.7$\rangle$) were equal to those during 3 April 2003. Rottman *et al.* (2001) show that errors introduced by using a spectrum with similar F10.7 introduces less than 3% error for wavelengths greater than $\sim$200 nm because the solar irradiance only varies by this magnitude over

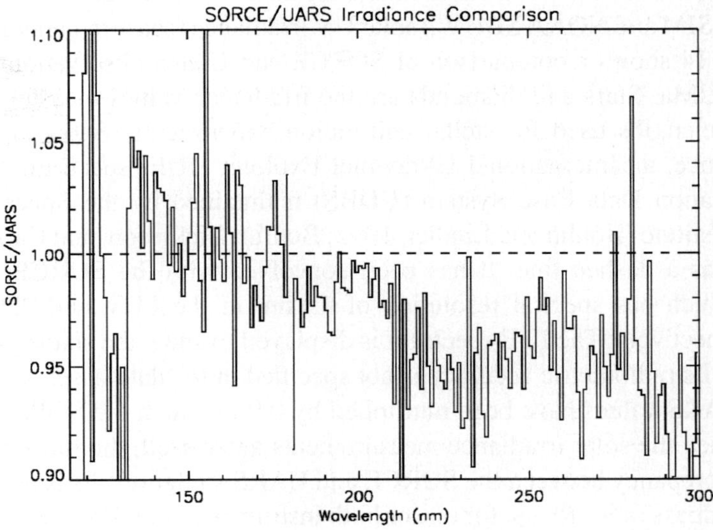

Figure 13. SORCE solar irradiance measurements agree with those of UARS to $\sim$5%. FUV values were obtained from simultaneous measurements made 3 April 2003. MUV values were obtained from spectra with equal F10.7 and $\langle$F10.7$\rangle$. The agreement is better than the stated uncertainty in the accuracy of the two experiments.

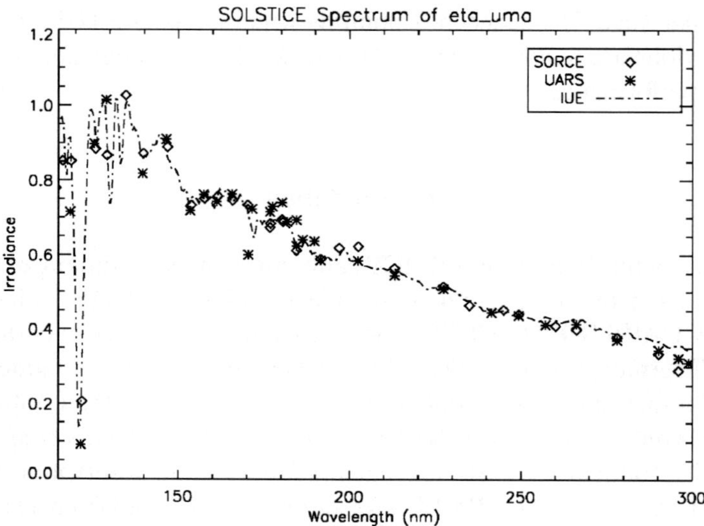

Figure 14. Diamonds and *stars* mark the irradiance values measured for η UMa for SORCE and UARS, respectively. UARS values have been multiplied by 0.9 to match the SORCE observations. An IUE spectrum is shown as a *dashed line* for reference. The scale is 10^4 photons s^{-1} cm^{-2} nm^{-1}.

the 11-year cycle. These independent observations generally agree to ~5%. This agreement, which is better than the stated uncertainties in the accuracy for the two experiments, validates both the SORCE calibration and the UARS calibration and FUV degradation function from launch until 2003. More detailed validation with UARS SUSIM and NOAA SBUV-2 solar UV irradiance measurements are planned.

Figure 14 shows a comparison of SORCE and UARS observations of the B3 V star, η UMa. Stars and diamonds are the irradiance values measured at the 40 fixed wavelengths used for stellar calibration experiments (Snow *et al.*, 2005). For reference, an International Ultraviolet Explorer (IUE) spectrum, taken from the Calibration Data Base System (CDBS) maintained by the Space Telescope Science Institute (Bohlin and Lindler, 1992; Bohlin, Dickinson, and Calzetti, 2001) is shown as a dashed line. It has been convolved with the SORCE instrument profile, which has spectral resolution of 1.1 nm in the FUV and 2.2 nm in the MUV, respectively. The IUE spectrum is displayed to show the shape of the stellar spectrum. Its radiometric accuracy is not specified in the data base.

The UARS values have been multiplied by 0.9 to match the SORCE measurements. Since the solar irradiance measurements agree well, the most likely cause for the discrepancy between the SORCE and UARS stellar results is an error in the area – bandpass ratio, $R_{A\cdot BP}$, for one or both instruments. For SORCE a direct comparison of the SURF beam and NIST measurements of the flight stellar apertures and an intermediate calibration aperture were used to obtain $R_{A\cdot BP}$ (Section 4.4.1). For UARS the entrance and exit slits were measured separately, and the bandpass was calculated using the optical parameters of the instrument to determine $R_{A\cdot BP}$

(Woods, Rottman, and Ucker, 1993). Since the slit-width measurements are inherently more uncertain than the NIST aperture measurements, the relatively simpler SORCE calibration is expected to be the more accurate of the two. This discrepancy will be investigated further.

7. Summary

The SOLSTICE II science objectives are to measure solar spectral irradiance from 115 to 320 nm with a spectral resolution of 1 nm, a cadence of 6 h, and an accuracy of 5%, to determine its variability with a long-term relative accuracy of 0.5% per year during a 5-year nominal mission, and to determine the ratio of solar irradiance to that of an ensemble of bright B and A stars to an accuracy of 2%. This paper describes the pre-flight and in-flight calibration and characterization measurements of the instrument that are required to meet those objectives. These requirements are based on an error analysis of the radiometry equation that defines the conversion from instrument telemetry output to irradiance.

The Synchrotron Ultraviolet Radiation Facility (SURF) at the National Institute for Standards and Technology (NIST) in Gaithersburg, Maryland provided the data for determining SOLSTICE radiometric sensitivity. SURF measurements were augmented by unit level tests that characterize detector, grating, and grating drive performance. All these measurements were combined to produce the pre-flight instrument radiometric sensitivity including an estimate for its uncertainty. After launch additional calibrations provided in situ measurements for those parameters in the radiometry equation that could not be determined during ground test (e.g., in-flight detector dark counts). Comparison of early SOLSTICE II solar and stellar irradiances to those measured by SOLSTICE I indicate that the experiment meets its 5% accuracy objective for determining both solar and stellar irradiances and for determining solar-to-stellar irradiance ratios to 2% at the beginning of mission. The accuracy for tracking solar irradiance changes at FUV wavelengths is $\sim$0.5% per year. In the MUV uncertainties in the ratio of the solar to stellar degradation function limit the accuracy to $\sim$1% per year during the first 2 years of operation. New calibration procedures are expected to improve the MUV accuracy to 0.5% per year for the remainder of the SORCE mission.

Acknowledgements

The authors thank Mitch Furst, Alex Farrell, Ed Hagley, Lu Deng, and Charles Clark at the National Institute for Standards and Technology who prepared and operated the SURF facility for the SOLSTICE calibrations. Ginger Drake, Frank Eparvier, Chris Pankratz, Greg Ucker, Don Woodraska, and Ann Windnagel from LASP deserve special thanks for spending many weeks at the SURF facility during data acquisition. This research was supported by NASA contract NAS5-97045. We also thank the anonymous referee for diligently reviewing this work and providing many useful comments.

References

Arp, U., Friedman, R., Furst, M. L., Makar, S., and Shaw, P. S.: 2000, *Metrologia* **37**, 357.

Bohlin, R. C. and Lindler, D.: 1992, *STSci Newsletter* **9**, 19.

Bohlin, R. C., Dickinson, M. E., and Calzetti, D.: 2001, *Astrophys. J* **122**, 2118.

Content, D. A., Boucarut, R. A., Bowler, C. W., Madison, T. J., Wright, G. A., Lindler, D. J., Hauang, L. K., Puc, B. P., Standley, C., and Norton, T. A.: 1996, *SPIE Proceedings* **2807**, 267.

Drake, V. A., McClintock, W. E., Kohnert, R. A., Woods, T. N., and Rottman, G. J.: 2000, *SPIE Proceedings* **4135**, 402.

Drake, V. A., McClintock, W. E., Woods, T. N., and Rottman, G. J.: 2003, *SPIE Proceedings* **4796**, 107.

Eparvier, F. G., Woods, T. N., Ucker, G., and Woodraska, D. L.: 2001, *SPIE Proceedings* **4498**, 91.

Kelly, R. L. and Palumbo, L. J.: 1973, *Atomic and Ionic Emission Lines below 2000 angstroms: Hydrogen through Krypton*, NRL Report **7599**, Washington, DC.

Kopp, G., Heuerman, K., and Lawrence, G.: 2005, *Solar Phys.* this volume.

Kuznetsov, I. G., Content, D. A., and Boucarut, R. A.: 2001, *SORCE/SOLSTICE UV-VUV Diffraction Gratings Scatter Characterization,* NASA GSFC Optics Branch.

McClintock, W. E., Rottman, G. J., and Woods, T. N.: 2005, *Solar Phys.* this volume.

Rottman, G. J., Woods, T. N., and Sparn, T. P.: 1993, *J. Geophys. Res.* **98**, 10667.

Rottman, G., Woods, T., Snow, M., and deToma, G.: 2001, *Adv. Space Res.* **27**(12), 1927.

Snow, M., McClintock, W. E., Rottman, G.J., and Woods, T. N.: 2005, *Solar Phys.* this volume.

Woods, T. N., Rottman, G. J., and Ucker, G. J.: 1993, *J. Geophys. Res.* **98**, 10678.

Woods, T. N., Wrigley III, R. T., Rottman, G. J., and Haring, R. E.: 1994, *Appl. Opt.* **33**, 4273.

Woods, T., Eparvier, F., Bailey, S., Chamberlin, P., Lean, J., Rottman, G., Solomon, S., Tobiska, K., and Woodraska, D.: 2005, *J. Geophys. Res.* **110**, A01312.

Solar Physics (2005) 230: 295–324

SOLAR–STELLAR IRRADIANCE COMPARISON EXPERIMENT II (SOLSTICE II): EXAMINATION OF THE SOLAR–STELLAR COMPARISON TECHNIQUE

MARTIN SNOW, WILLIAM E. McCLINTOCK, GARY ROTTMAN and THOMAS N. WOODS

Laboratory for Atmospheric and Space Physics, University of Colorado, Boulder, CO 80309, U.S.A.
(e-mail: snow@lasp.colorado.edu)

(Received 30 March 2005; accepted 3 June 2005)

Abstract. The Solar–Stellar Irradiance Comparison Experiment (SOLSTICE) measures the solar spectral irradiance from 115 to 320 nm with a resolution of 0.1 nm. The Sun and stars are both observed with the same optics and detector, changing only the apertures and integration times. Pre-launch calibration at SURF allows us to measure both with an absolute accuracy of 5%. The in-flight sensitivity degradation is measured relative to a set of stable, early-type stars. The ensemble of stars form a calibration reference standard that is stable to better than 1% over timescales of centuries. The stellar irradiances are repeatedly observed on a grid of wavelengths and our goal is to measure changes in the absolute sensitivity of the instrument at the 0.5% per year level. This paper describes the details of the observing technique and discusses the level of success in achieving design goals.

1. Introduction

The Solar–Stellar Irradiance Comparison Experiment II (SOLSTICE II) is a grating spectrograph on the Solar Radiation and Climate Experiment (SORCE) satellite designed to measure the solar irradiance from 115 to 320 nm. The design of the instrument is more fully described by McClintock, Rottman, and Woods (2005). The ground and on-orbit calibration is detailed by McClintock, Snow, and Woods (2005). The absolute sensitivity of the instrument was calibrated before launch at the NIST SURF III facility in Gaithersburg, MD (Arp *et al.*, 2000). As described by McClintock, Snow, and Woods (2005), the uncertainty in the absolute calibration is on the order of 5%. This calibration was transferred to the stars during an early-orbit observing campaign. These stars then became the in-flight irradiance reference as described later. Both the Sun and the stars are observed with the same optics and detectors, changing only apertures and integration times. The ratio of the stellar irradiance measurements to the solar measurements is then independent of instrumental degradation and depends only on the invariant ratio of the solar and stellar apertures and any differences in the illumination of the optics. The illumination difference can be significant, and is discussed in detail in Section 5. Thus, observed changes in the stellar ensemble irradiances represent changes in the instrument sensitivity, and knowledge of these changes can be used to correct the solar data. This ratio technique establishes a baseline for long term solar variability studies, since the ratio of the solar to stellar irradiance can be re-measured by future missions. This paper describes the technique for analyzing the stellar observations and field-of-view (FOV) corrections to produce the degradation correction.

A set of UV-bright, early-type stars provides the SOLSTICE irradiance reference. The full list of stars is given in Table I of McClintock, Rottman, and Woods (2005). Of the original 31 stars in the UARS catalog, only 18 remain in the SORCE catalog of stellar targets. A few were removed because they were found to be variable, but mostly they were removed because they were either too dim in the UV or else they were in the same part of the sky as a much brighter target. McClintock, Rottman, and Woods (2005) give a more detailed discussion of the SOLSTICE stellar catalog history.

These stars individually have variability less than 1% on timescales of centuries (Mihalas and Binney, 1981) and therefore the overall variability of an ensemble of such stars is negligible. The time series irradiance for each star relative to the others is used to remove any stars that do not meet a stability criteria. After the commissioning phase of the SORCE mission, but before normal solar observations began, a special stellar observing campaign was conducted. During this week-long campaign, many of the program stars were observed at a series of standard wavelengths. As of 1 March 2003, SOLSTICE had made 395 middle ultraviolet (MUV) – 180–320 nm – stellar irradiance measurements, 175 normal far ultraviolet (FUV) – 135–180 nm – measurements, and 207 FUV airglow-corrected measurements – 115–135 nm. These initial observations set the reference levels to which the subsequent solar and stellar irradiances are compared. As the instrument ages, the observed stellar irradiance will decrease relative to those initial measurements. Since there is little chance of a systematic long-term trend in the irradiances of the entire ensemble of stars, any measured changes will be due to the instrument alone.

The SOLSTICE II instrument is an evolution of the SOLSTICE I instrument on board the Upper Atmosphere Research Satellite (UARS). The design and calibration of SOLSTICE I is described by Rottman, Woods, and Sparn (1993) and Woods, Ucker, and Rottman (1993). Analysis of the in-flight degradation of the SOLSTICE I instrument is described by Woods *et al.* (1998). Throughout the rest of this paper, all references to SOLSTICE will mean SOLSTICE II unless specifically noted otherwise.

2. Individual Stellar Observations

Stellar observations are made during the eclipse portion of every orbit. An expert planning system determines which stars will be available each stellar period, and then prioritizes them based on their assigned ranks and observing history. This system ensures that each target/wavelength combination is observed on a regular basis. Brighter stars are observed more often than dimmer ones. A given star may have a high priority at some wavelengths and a lower priority at other wavelengths depending on the shape of the stellar spectrum. Stellar observations are only scheduled when the spacecraft is outside the South Atlantic Anomaly (SAA) because the enhanced count rate from the particle environment in that part of the orbit strongly contaminate the stellar signal.

2.1. Standard Stellar Observation

The standard SOLSTICE stellar observation consists of a brief scan of the spectrometer zero order to check for shifts in the wavelength scale. For the SOLSTICE optical design, wavelength shifts result from pointing offsets or from shifts in the grating drive (McClintock, Rottman, and Woods, 2005). The star is then observed at a fixed wavelength for at least 100 s. The count rate is recorded in a series of 1 s integrations separated by 50 ms as shown in Figure 1.

A χ^2 analysis indicates that the count rates are well fit by a normal distribution as would be expected if the errors are random noise due to root-N counting statistics. A Gaussian function is fit to the histogram of observed count rates. The uncertainty of the mean count rate for a single observation is therefore the Gaussian sigma divided by the square root of the number of samples. Each observation is planned so that wavelengths with lower count rates have longer dwell times keeping the product of the square root of the total counts and the square root of the number

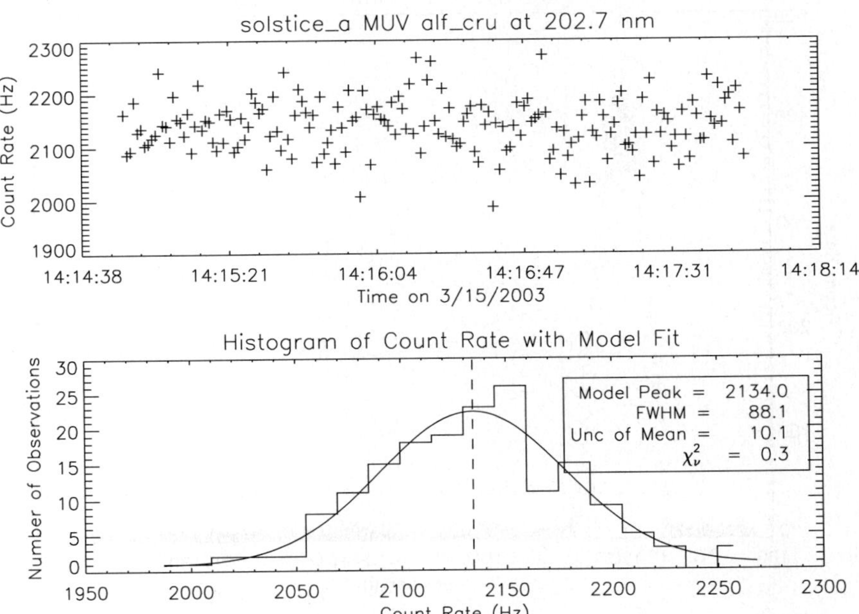

Figure 1. Typical SOLSTICE stellar observation. The *top panel* shows the raw count rates for a stellar observation of α Cru with the MUV channel of SOLSTICE A. The *bottom panel* shows the corresponding reduced observation. The histogram of count rates is modeled with a Gaussian.

of integrations below 0.01, so the random error in each observed irradiance is less than 1%.

2.2. Airglow Correction

As described by McClintock, Snow, and Woods (2005), when SOLSTICE is in stellar mode, off-axis light entering the instrument is detected at a shifted wavelength. In stellar mode, the FOV is approximately 2° and diffuse Lyman α emission from thermospheric hydrogen produces a substantial background within 10 nm of 121.6 nm as shown in Figure 2. This background is highly variable as a function of time and viewing geometry. In order to remove this background signal from the stellar observations, a special experiment observes the airglow component a few degrees away from the star immediately before and after the stellar irradiance measurement.

A subset of the standard star catalog is used for these special "companion" experiments. In addition to the observational criteria listed previously, these stars must have a nearby dark region (the "companion") that is free of stars with magnitude greater than $M_V = 8$ (Rottman, Woods, and Sparn, 1993). These dark regions are the same as observed by SOLSTICE I, listed in Table II of Rottman, Woods, and

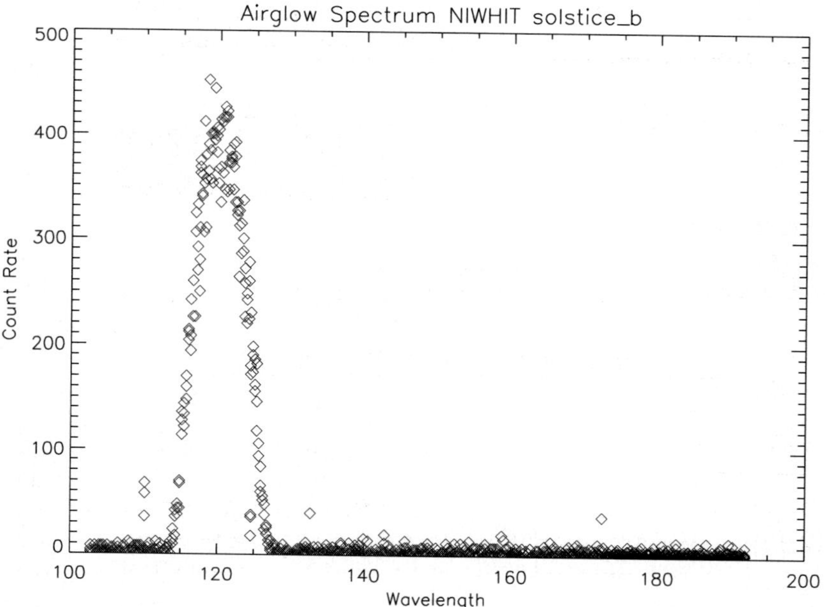

Figure 2. FUV spectral scan of stellar dark area showing contamination from airglow near Lyman α. The normal background level of a few counts per second rises due to geocoronal hydrogen emission. The value of the airglow count rate, here about 400 Hz, is dependent on observation zenith angle, solar and atmospheric conditions.

Sparn (1993). A single companion experiment consists of a 200 s observation of the dark region, slewing the spacecraft to the star, measuring the position of zero order, a 200 s observation of the desired wavelength, a second zero-order scan, slewing the spacecraft back to the dark region, and another 200 s observation of the dark region. A typical companion observation is shown in Figure 3.

The measured stellar signal is determined as follows. A second-order polynomial is fit to the three pieces of the observation (dark-star-dark) with an (additive) offset to the on-star portion as a free parameter. The calculated offset is the count rate from the star alone. To verify that the airglow-removal has not introduced a systematic

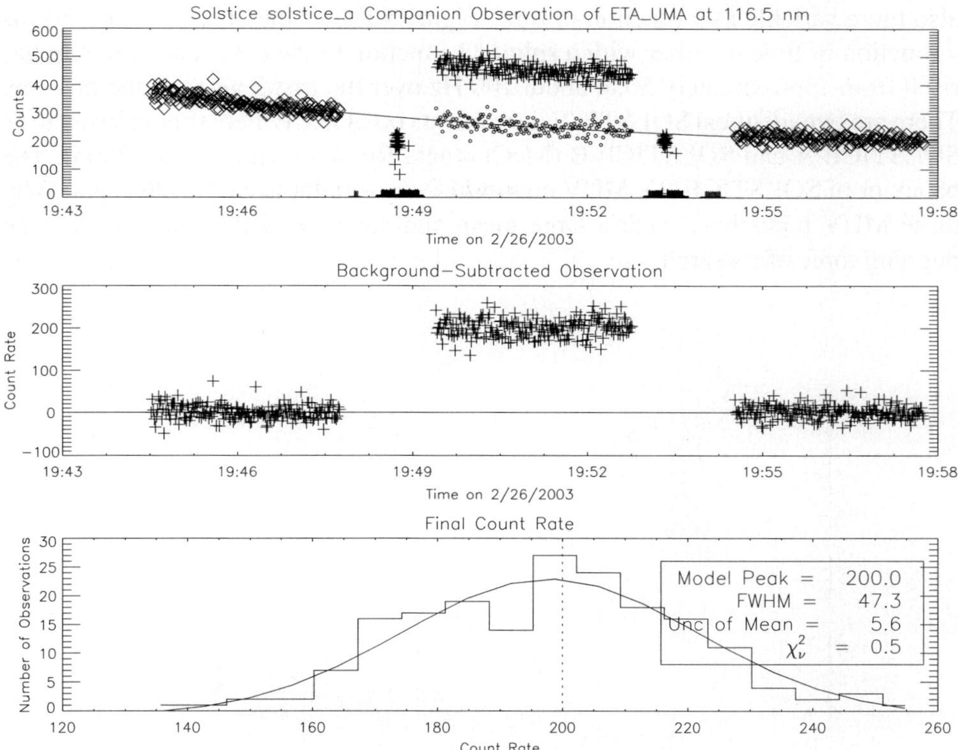

Figure 3. Typical SOLSTICE companion observation. Special care is required to correct for Lyman α airglow emission near 121.6 nm. The *top panel* shows both the raw counts of the background airglow and star and the polynomial model fit to the background and the inferred background during the stellar observation. The *diamonds* are observed background counts, the *plus symbols* indicate signal from star plus background. The *small circles* are the stellar counts with the model offset removed. The clusters of points before and after the main stellar observation are the zero-order scans. They have a shorter integration time and thus fewer counts. The *middle panel* shows the observation after the polynomial background has been removed. The *bottom panel* shows the final count rate model for the observation. The *dotted line* is the offset from the model. The histogram is the residual count rates after background subtraction (i.e., the difference between the plus signs and the circles in the top panel). The two measurements of the stellar count rate are entirely consistent.

trend in the data, the polynomial background can be subtracted from the stellar count rate, leaving only the stellar component. The background-subtracted stellar observation has noise characteristics that are very similar to the routine non-airglow stellar observations, and examination of the χ^2 indicates that the residual count rates are well-fit by a normal distribution with zero residual background.

2.3. Dark Rate

As shown in Figure 2, the dark rate in the FUV channel away from Lyman α is about 3 Hz. This value has remained very stable since launch. However, the dark rates of the MUV photomultiplier tubes are much greater than the FUV tubes', and also more variable as a function of time. Figure 4 shows the observed dark rate as a function of time together with a spline fit function to the data. The dark rate has risen from approximately 50 to about 100 Hz over the first 2 years of the mission. There are two identical SOLSTICE instruments on SORCE, hereafter referred to as SOLSTICE A and SOLSTICE B (McClintock, Rottman, and Woods, 2005). The behavior of SOLSTICE B's MUV channel is very similar to SOLSTICE A's. Why these MUV tubes have such a large mean and variance in their dark rates is an ongoing topic of research.

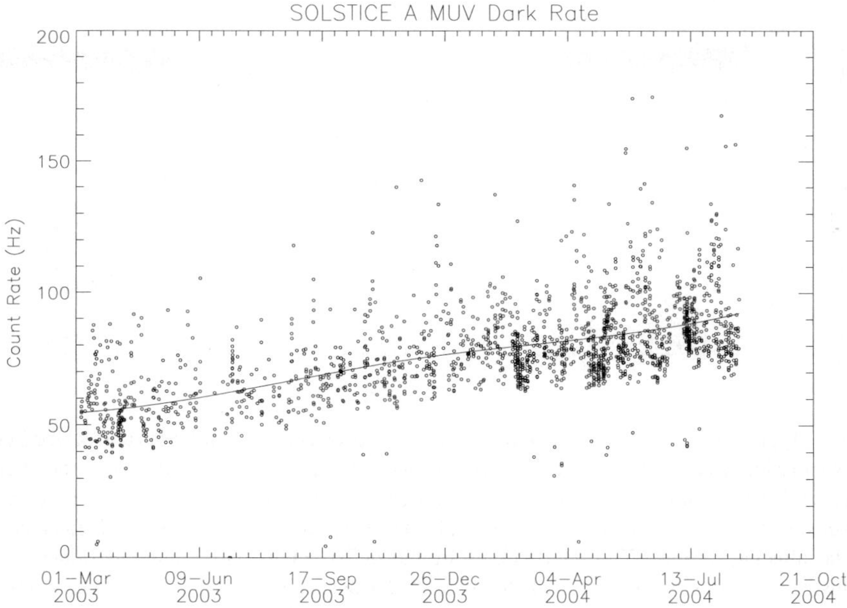

Figure 4. SOLSTICE A MUV channel dark rate and spline fit. The data shown in this plot are observations of dark regions used to determine the in-flight dark rate. The *circles* are the average rate over a 200 s observation, and the *solid line* is a best-fit spline model. The dark rate is steadily rising, but it has a large variance, leading to increased variance in the stellar irradiance time series.

Beginning in August 2004, the zero-order scans which are performed before and after every stellar observation include a brief measurement of the dark rate. The grating is rotated to a wavelength well outside the sensitivity range of the instrument (50 nm in the FUV channel and 135 nm in the MUV), and the count rate is measured for 15 s. For a typical MUV dark rate during this measurement of about 75 Hz, then the 15 s measurement will determine the background to within 3% (i.e., <3 Hz) based on counting statistics. The FUV dark rate determination will always be better than 1 Hz.

3. Stellar Irradiance Time Series

3.1. Irradiance Correction to Standard Wavelengths

In order to combine stellar observations at different epochs, the individual irradiance measurements must be corrected to a common grid of wavelengths. These standard wavelengths were chosen to adequately sample the sensitivity curve derived from preflight calibration as described by McClintock, Snow, and Woods (2005) and the wavelengths are shown in Figures 5 and 6.

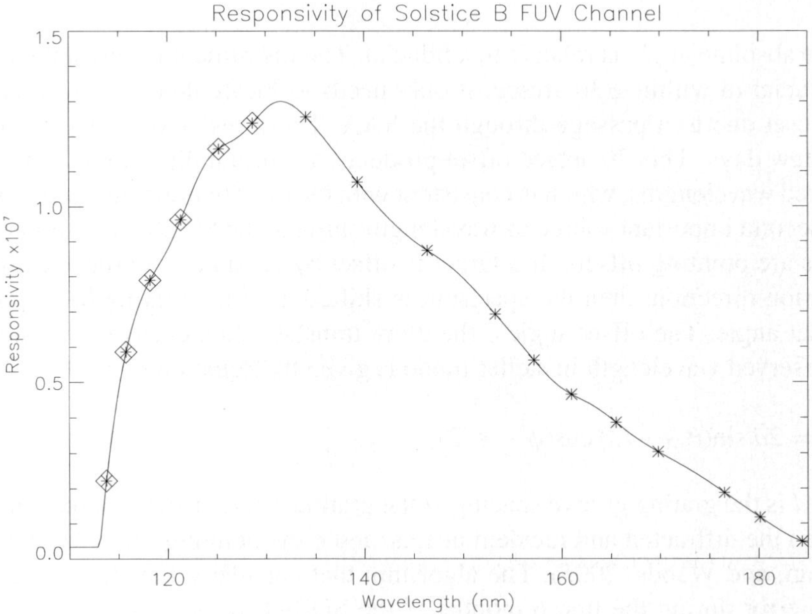

Figure 5. FUV responsivity curve showing standard stellar wavelengths. The sensitivity of the FUV channel is sampled at these 18 wavelengths by the standard stellar observations. Wavelengths below 130 nm must use the companion star technique described in the text to correct for geocoronal Lyman α airglow emission.

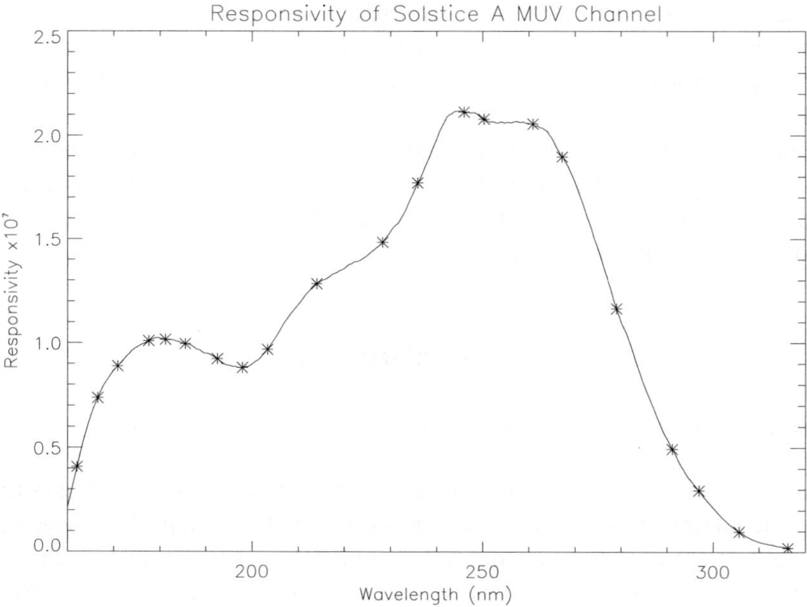

Figure 6. MUV responsivity curve showing standard stellar wavelengths. The MUV responsivity is monitored at 22 wavelengths as shown by the asterisks.

The absolute angle is relative to a fiducial. The instrument is only able to locate the fiducial to within ±30 arcsec. It only needs to locate this fiducial after it has been reset due to a passage through the SAA. These resets typically occur once every few days. This 30 arcsec offset produces a repeatability of 0.075 nm in the observed wavelengths, which is consistent with the distributions shown in Figure 7.

A second important source of wavelength shifts in the SOLSTICE stellar observations are pointing offsets. If a target is offset by an angle ϵ in the spectrometer dispersion direction, then the spectrum is shifted as if the grating had moved by half that angle. The offset angle ϵ therefore translates to a change in wavelength. The observed wavelength in stellar mode is given by Equation (1):

$$\lambda = 2d \sin(\theta + \epsilon/2) \cos(\phi - \epsilon/2), \qquad (1)$$

where d is the grating groove spacing, θ the grating rotation angle, ϕ the half-angle between the diffracted and incident beams, and ϵ the pointing offset (McClintock, Rottman, and Woods, 2005). The algorithm that calculates the stellar precession was in error during the first 6 months of the SORCE mission, and led to routine spacecraft pointing offsets of up to a few arc minutes. After that error was corrected, offsets in pointing are typically no larger than a few arc seconds, which is less than the uncertainty in the measurement of the position of zero order (described later). Each set of stellar observations begins and ends with a measurement of the position

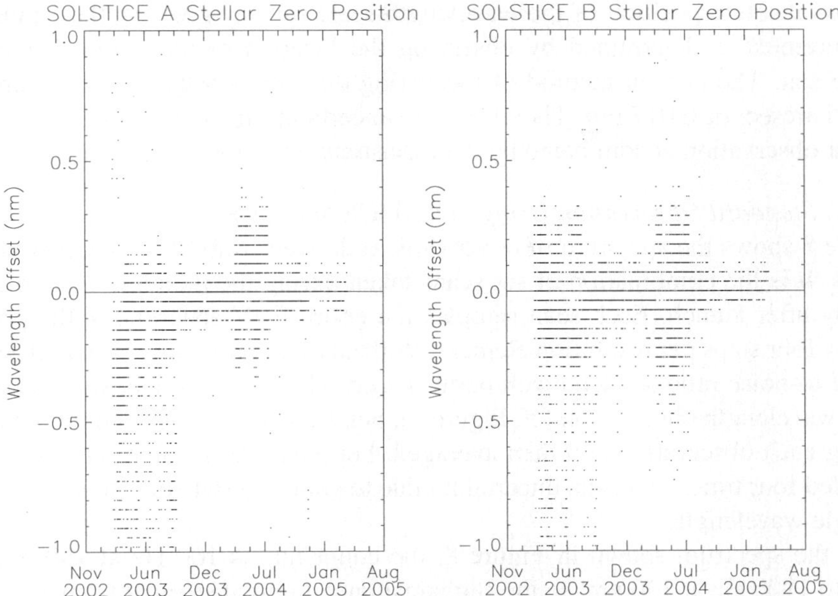

Figure 7. Location of zero order relative to nominal zero position. Relatively large variations during first 6 months of the mission were due to large pointing offsets caused by an error in the stellar precession software. This error reoccurred during the summer of 2004. Apart from these times, the spread in wavelength offsets (±0.08 nm) is primarily due to errors in locating the grating drive fiducial after a reset.

of zero order. The image of the slit is scanned with 13.5 arcsec grating steps, so the position of the slit is known to within 6.25 arcsec, i.e., 0.017 nm.

Figure 7 shows the result of these zero-order scans for both SOLSTICE A and SOLSTICE B. The large scatter in zero position during the first part of the mission is due to a software error in calculating each star's precession and pointing the spacecraft. The onboard computer was expecting J2000 stellar coordinates and calculating the precession to the current date. The ground system assumed that the spacecraft was expecting stellar coordinates that had already been precessed. Therefore, the precession correction was being applied twice, and the spacecraft pointed up to 2 arcmin away from the true position of the star.

Once the precession error was corrected, the zero-order position became much more stable. The only exception is the period of early 2004 where the precession error was inadvertently reintroduced into the planning software. It was corrected in July 2004, and the stellar wavelength scale has been very well behaved since then.

The relationship between grating angle and wavelength is that a change in angle of 1 arcsec corresponds to a wavelength shift of 0.0025 nm. As described by McClintock, Snow, and Woods (2005), the grating drive control system positions the grating to within ∼1 arcsec over its 40° range. So the accuracy of each stellar observation is 0.0025 nm, since the grating must slew from the position of zero order to the desired wavelength.

The absolute position of the wavelength scale for any given stellar irradiance measurement is determined by observing the location of the zero-order image of the star. The current method of measuring this position has an uncertainty of ±6.25 arcsec, or 0.017 nm. Therefore, the uncertainty in wavelength for any particular observation is dominated by the measurement of the position of zero order.

3.1.1. *Standard Observation Using SOLSTICE Spectrum*

Figure 8 shows the spectrum of α Vir (Spica) derived from SOLSTICE measurements. It is the combination of six scans taken during the month of March 2003, shortly after launch. Each scan samples the entire FUV spectrum with approximately four steps per resolution element. With an integration time of only 0.5 s, the signal-to-noise ratio at each dwell point is somewhat reduced relative to a normal fixed wavelength observation (cf, Figure 1), but the scan is repeated multiple times during each observation and then averaged. For example, a count rate of 100 Hz sampled four times yields an uncertainty due to counting statistics of about 7% for a single wavelength.

In the spectrum shown in Figure 8, the count rate is 100 Hz at 180 nm, but rises to 1900 Hz at 130 nm, so the statistical uncertainty decreases from 7% at the long wavelength end to less than 2% at the short wavelength end. Merging six such

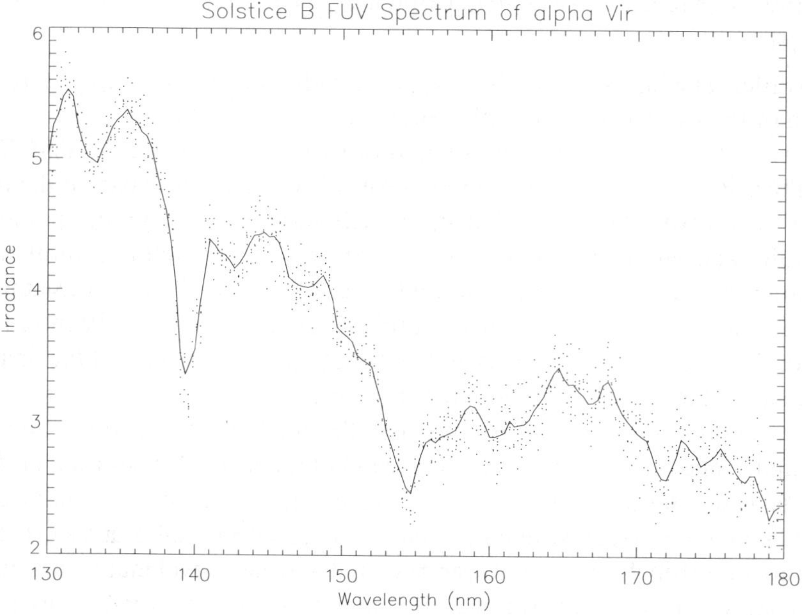

Figure 8. SORCE SOLSTICE stellar spectrum of α Vir in the FUV wavelength range. This spectrum is the combination of six individual scans taken during March 2003. The dots represent the irradiances from the six scans, and the solid line is the spectrum smoothed to the 1 nm bandpass of the instrument in stellar mode. Irradiances are given in units of 10^4 photons s^{-1} cm^{-2} nm^{-1}.

scans as shown in Figure 8 produces a spectrum with less than 3% uncertainty at all wavelengths.

The spread in measured wavelengths from one stellar observation to the next is typically less than 1 nm. This 1 nm scatter is due to a combination of grating drive accuracy and spacecraft pointing offsets. Pointing offset is a systematic error source that has been discovered and fixed, while the grating drive accuracy introduces a random offset in the observed wavelength. The total of these two effects is shown in Figure 7.

Each fixed-wavelength stellar irradiance observation must be corrected to one of the standard wavelengths shown in Figures 5 and 6 before further analysis. The irradiance at the measured wavelength is simply multiplied by the ratio of the irradiance at the measured and standard wavelengths on the reference spectrum (I_{ref}).

$$I_{\text{corrected}}(\lambda_0) = I_{\text{observed}}(\lambda) \times \frac{I_{\text{ref}}(\lambda_0)}{I_{\text{ref}}(\lambda)}, \tag{2}$$

where λ is the observed wavelength and λ_0 is one of the standard wavelenths. Depending on the details of the shape of the stellar spectrum near a given wavelength, this correction for spectral shape is generally on the order of 5% or less. If the reference spectrum is determined to the 3% level, then this correction factor due to spectral shape would typically introduce an additional 0.2% uncertainty in the corrected stellar irradiance. At some wavelengths for particular stars, i.e., near strong spectral features, this correction can be larger. In general, the irradiances for those target/wavelength combinations are excluded from the degradation analysis.

3.1.2. *Airglow Observation Using IUE Spectrum*

As described in Section 2.2, background signal from geocoronal Lyman α emission airglow complicates SOLSTICE stellar observations below 130 nm. The airglow emission varies as a function of time and wavelength. Without the offset pointing used in the companion observations described earlier, the stellar component cannot be reliably extracted from a spectral scan alone. Therefore, in this region of the spectrum, instead of using a spectral scan from SOLSTICE, a spectrum from another instrument must be used. The processing system uses archival IUE spectra, degraded to the SOLSTICE FUV stellar resolution of 1 nm as shown in Figure 9.

The diamonds in Figure 9 show the measured stellar irradiances for η UMa from SOLSTICE as a function of wavelength in the region of the spectrum affected by airglow. There is so much structure in the stellar spectrum (even at the moderately low resolution of SOLSTICE's stellar mode), that small wavelength shifts lead to a relatively large change in the irradiance. The SOLSTICE observations near 122 nm are typically within 0.5 nm of each other (Figure 9), but the stellar irradiance varies by a factor of 3 over this small wavelength range. The IUE reference spectrum allows these SOLSTICE irradiances to be corrected back to standard wavelengths using Equation (2).

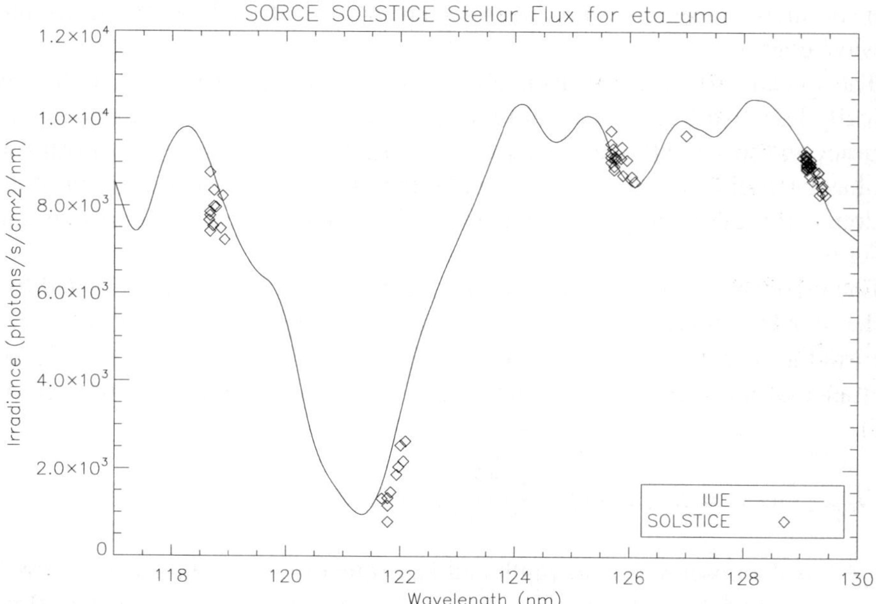

Figure 9. Spectrum of η UMa from IUE reduced to SOLSTICE spectral resolution along with companion observation irradiances. Wavelength shifts in the SOLSTICE stellar measurements can introduce a significant spread in the observed irradiances. The shape of the spectrum can be used to correct the individually measured irradiances to one of the standard SOLSTICE wavelengths using Equation (2).

The absolute accuracy of the IUE spectrum is not relevant in this analysis, since it is only used to make a relative correction to the irradiance based on wavelength. The absolute irradiance of the IUE spectrum does not affect the wavelength correspondence between SOLSTICE and IUE spectral features. The relative accuracy of an IUE spectrum is better than about 5% in this wavelength region (Gonzales-Riestra, Cassatella, and Wamsteker, 2001).

3.2. Normalization of Ensemble of Stellar Observations

After each observation has been corrected to one of the standard wavelengths, the measured irradiance from multiple epochs and multiple targets are combined to produce the observed change in responsivity of the instrument as a function of wavelength and time. The time series irradiances for two typical stars are shown in Figure 10. These two plots show the irradiances before correction for wavelength shifts. Figure 10 illustrates the typical observing frequency for stellar calibration targets. Table I of McClintock, Rottman, and Woods (2005) lists the full catalog of stars used by SOLSTICE. For each star, there are 40 wavelengths to observe (18 in the FUV channel, 22 in the MUV). An expert planning system determines which targets are available each orbit, then ranks them based on a priority assigned to

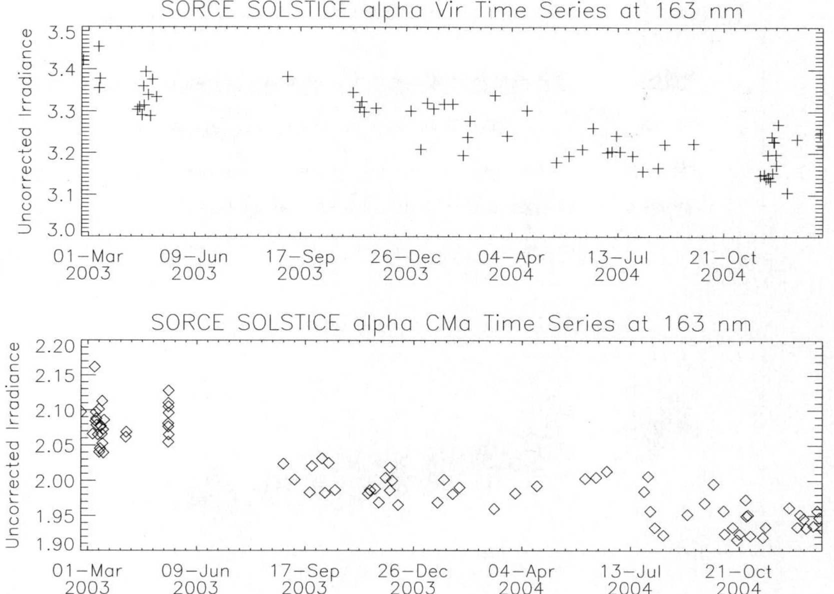

Figure 10. Stellar irradiance time series for two stars. The observing schedule averages about one observation per star per wavelength per week. The irradiance units are 10^4 photons cm^{-2} s^{-1} nm^{-1}.

each star/wavelength combination. The effective priority is a combination of a basic priority and a factor determined by the length of time since the last observation. The scheduler tries to prevent gaps in the data record for any particular target/wavelength pair. In general, each target is observed at each wavelength approximately once per week. The gap in observations for α Vir and α CMa during mid-2003 shown in Figure 10 was due to a database error in the planning system.

Observations from the ensemble of stars are combined to derive the instrumental degradation. The stars have different apparent magnitudes, so they must be normalized to a common intensity scale as well as fit to a function that varies in time. In the case of SOLSTICE, the simple exponential function given by Equation (3) is used to model the loss of sensitivity:

$$d = A_0(1 - \beta + \beta\, \mathrm{e}^{-t/\tau}). \tag{3}$$

The parameter A_0 is the normalization factor for each star. $1 - \beta$ is the asymptotic value of the degradation as t goes to infinity, and τ is the time constant for the exponential. The parameters (A_0 for each star, β, and τ) are determined at each standard wavelength. Figures 11 and 12 show the final reduction of the stellar measurements to produce SOLSTICE degradation curves at these two wavelengths. The top panel shows the irradiance time series for seven stars. The bottom panel shows those same measurements after correction to a common wavelength, normalization for apparent magnitude, and fit with an exponential function. For simplicity, the

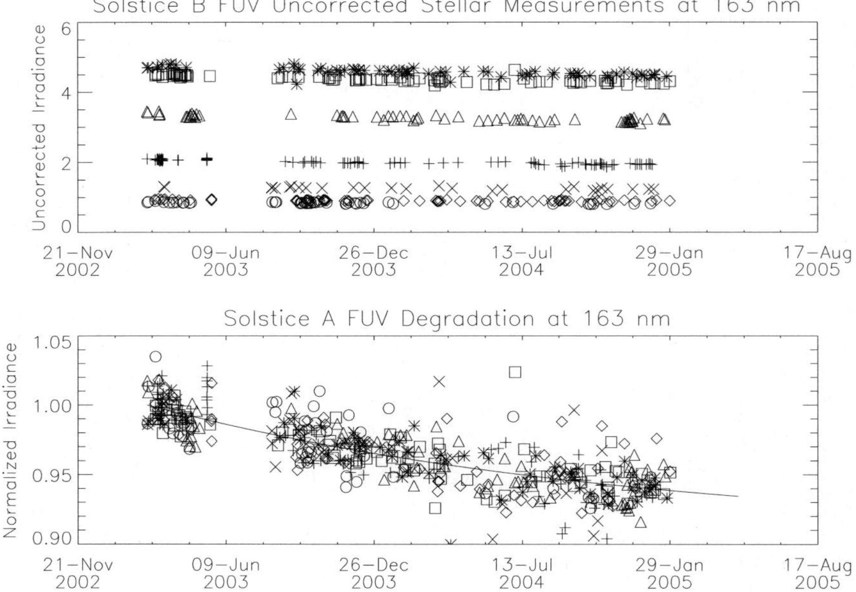

Figure 11. Final degradation curve from FUV stellar observations at 163 nm. The *top panel* shows the set of stellar irradiance measurements that meet the criteria described in the text. The different symbols indicate different stars as follows: (+) α CMa, ($*$) α Cru, ($\Diamond$) α Pav, ($\triangle$) α Vir, ($\Box$) β Cen, ($\times$) β CMa, ($\bigcirc$) η UMa. The *top panel* shows the uncorrected stellar irradiances over the course of the SORCE mission. The *bottom panel* shows the irradiances after having been normalized for stellar brightness and spectral shape. The *solid line* is the fit to the normalized irradiances. The irradiance units are 10^4 photons cm^{-2} s^{-1} nm^{-1}.

same set of stars are shown for both channels. The resolution of SOLSTICE in stellar mode in the MUV is 2 nm, so the stellar spectrum is nearly featureless. No correction for wavelength shape is therefore required or applied to the MUV stellar observations.

Notice that the stars are dimmer at 250 nm than at 163 nm. The spectral shape of these early-type stars has a gradually decreasing continuum throughout the UV, making SOLSTICE stellar measurements less precise at longer wavelengths. Figure 13 shows the measured stellar degradation as a function of wavelength for the full SOLSTICE range. The two curves for each channel show the degradation after 1 and 2 years. Above 250 nm, there has been no measurable change in stellar count rates. Nearly all of the observed responsivity change in the MUV channel happened within the first year of operation, so the two curves are nearly on top of each other. The diamond and asterisk symbols are at the standard wavelengths highlighted in Figures 5 and 6.

The relative brightness of each star and the parameters of the exponential fit are determined simultaneously using a nonlinear least-squares approach. Stars that turn out to be variable have a relatively large systematic deviation from the rest of

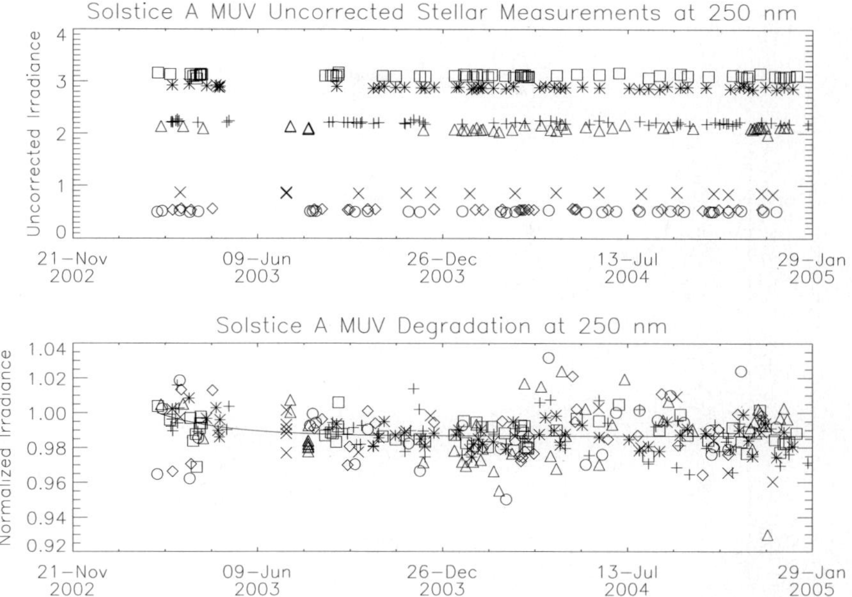

Figure 12. Final degradation curve from MUV stellar observations at 250 nm. The *top panel* shows the set of stellar irradiance measurements that meet the criteria described in the text. The different symbols indicate different stars as follows: (+) α CMa, (∗) α Cru, (◇) α Pav, (△) α Vir, (□) β Cen, (×) β CMa, (○) η UMa. The *top panel* shows the uncorrected stellar irradiances over the course of the SORCE mission. The *bottom panel* shows the irradiances after having been normalized for stellar brightness. The *solid line* is the fit to the normalized irradiances. The irradiance units are 10^4 photons cm^{-2} s^{-1} nm^{-1}.

the ensemble and are removed from the fit. Single observations that are statistical outliers are also removed. The final decaying exponential is therefore due entirely to sensitivity changes in the instrument. By allowing the stellar brightness to be a free parameter in the fit, the measurement of the SOLSTICE degradation is not dependent on the absolute value of the stellar measurements. The fit finds the brightness ratios among all the stars that minimizes the spread. The ratio of the sensitivity in solar mode to the sensitivity in stellar mode depends only on the ratio of the apertures and second-order differences in the illumination of the optics (McClintock, Rottman, and Woods, 2005; McClintock, Snow, and Woods, 2005). Correction for the difference in the FOV between solar and stellar modes is discussed in Section 5.

The stellar observations shown in Figure 13 indicate that the FUV channel is degrading faster than the MUV channel. This is particularly apparent at 180 nm where the two channels overlap. The results are consistent with the count-rate versus modal-gain measurements of Drake *et al.* (2003), which show a more rapid decline in counting efficiency for the FUV detectors relative to the MUV detectors with total photon dose.

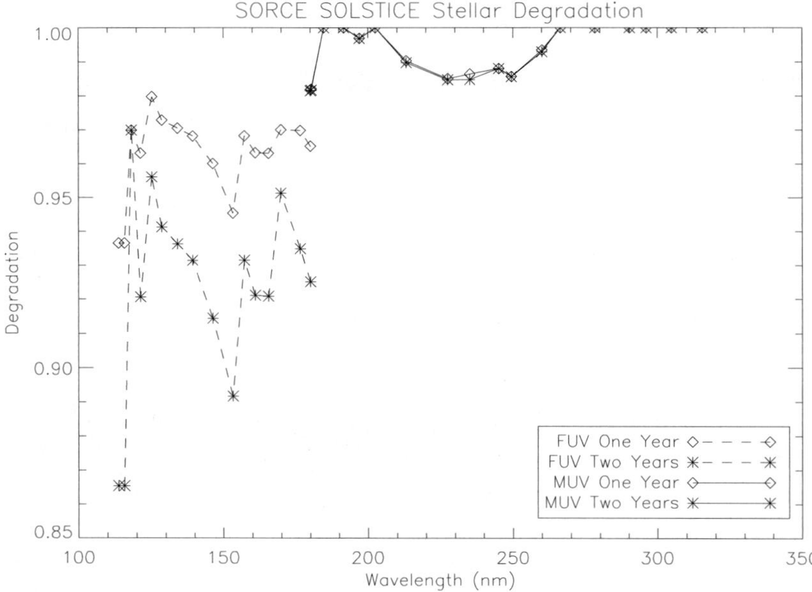

Figure 13. Summary SOLSTICE stellar degradation curves. The degradation over all wavelengths is shown here as a function of time. The *diamonds* show the degradation after one year (as of March 2004) and the *asterisks* show the result after 2 years (March 2005). The stellar degradation has been negligible above 250 nm. The stellar count rates are very low below 120 nm, so the uncertainty of the degradation correction is corresponding large.

4. Absolute Stellar Irradiances

Figure 14 shows the typical count rates and their uncertainties as a function of wavelength for two selected stars, β Cen and η UMa. These targets were chosen to show the wide range in stellar brightness and its effect on the uncertainty in the irradiance. β Cen is one of the brightest stars that SOLSTICE observes, while η UMa is relatively dim, particularly in the MUV.

The top panel shows the measured count-rate spectrum for the SOLSTICE standard wavelengths for both FUV and MUV channels. Each individual observation for the full mission has been corrected for degradation and then averaged. The lower panel shows the standard deviation divided by the mean irradiance at each wavelength for each star.

The uncertainty rises at the long wavelength end due to the low count rates. The stellar spectrum is decreasing and so is the detector responsivity, so the uncertainty for even the brightest stars rises sharply above 280 nm. Below that wavelength, the uncertainty in stellar irradiance is less than 0.5%.

The spike in the uncertainty for η UMa at 122 nm is easily understood after a quick comparison to Figure 9. The stellar irradiance is very low due to the deep Lyman α feature, and the observations sit on the side of a very steep absorption.

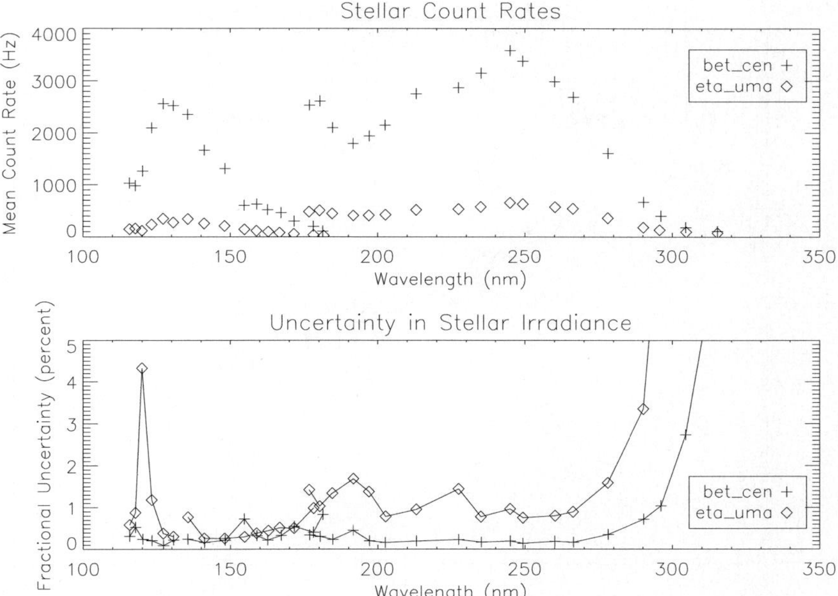

Figure 14. Uncertainty in stellar irradiances. The *top panel* shows the mean count rate over the full mission after correction for degradation. The *lower panel* shows the standard deviation of the irradiances divided by the irradiance for each standard wavelength. Below 270 nm, uncertainty in the stellar irradiance for the bright stars is less than 0.5%.

Small errors in the wavelength scale will have significant effects on the irradiance. The effect is much less pronounced at all other wavelengths.

Count rates are converted to absolute stellar irradiances using the instrument responsivity function described by McClintock, Snow, and Woods (2005). Absolute irradiance for the brightest stars is determined to about 5% for wavelengths less than 280 nm. The largest contributor to the irradiance uncertainty is the uncertainty in the responsivity. Figure 9 of McClintock, Snow, and Woods (2005) shows that the uncertainty in solar mode is about 4.5% below 280 nm, and they also show that the responsivity uncertainty in stellar mode is the same as that for solar mode. That paper provides a full discussion of the solar/stellar ratio calculation.

The other significant contribution to the stellar mode responsivity uncertainty is the ratio of the solar and stellar apertures. The SOLSTICE apertures were measured very carefully and the uncertainty in the ratio is 0.5% (McClintock, Snow, and Woods, 2005).

Figure 15 compares the absolute irradiance measured at the standard wavelengths for both SORCE and UARS SOLSTICE. A 10% correction has been applied to reduce the UARS stellar irradiances. The aperture ratio measurement technique used on the UARS instrument was not as accurate as that used for SORCE. Initial measurements of the SORCE ratios using the old technique produced values that were 6% low in the FUV and 15% high in the MUV, so a 10% error in the UARS

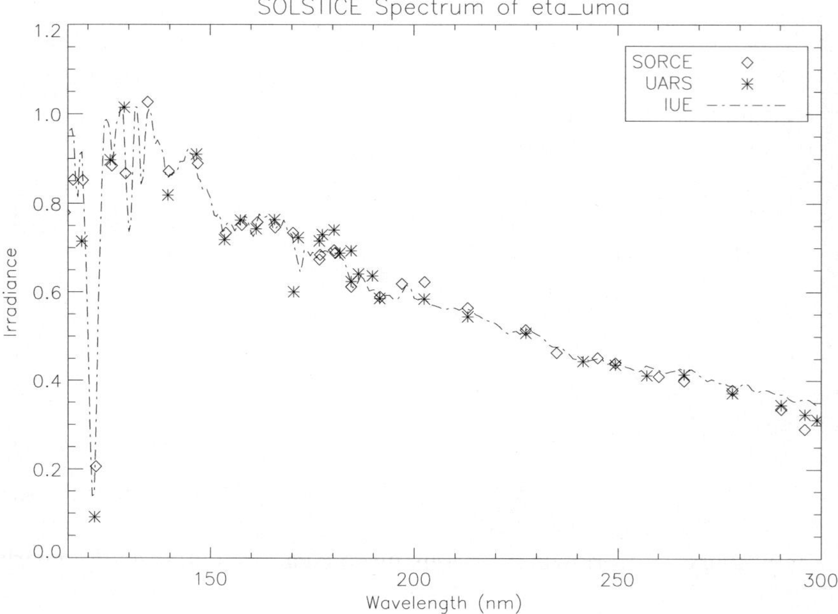

Figure 15. Spectrum of η UMa from both SORCE and UARS SOLSTICE. *Diamond symbols* indicate SORCE observations. The UARS irradiances have been reduced by 10% as discussed in the text. An IUE spectrum from the CDBS convolved with the SOLSTICE instrument profile is shown for comparison. Irradiances are given in units of 10^4 photons s^{-1} cm^{-2} nm^{-1}.

aperture ratio is entirely possible. Further analysis will be required to determine the exact correction factor for each UARS channel.

The IUE reference spectrum shown in Figure 15 was obtained from the Calibration Database System (CDBS) maintained by the Space Telescope Science Institute (Bohlin and Lindler, 1992; Bohlin, Dickinson, and Calzetti, 2001). It has been convolved with the SOLSTICE stellar-mode instrument profile, which has a resolution of 1.1 nm in the FUV and 2.2 nm in the MUV. The IUE spectrum is not meant to be definitive, and is shown only to indicate the shape of the spectral features. In general, the agreement between SOLSTICE and IUE is very good below 280 nm. A more detailed comparison of the absolute stellar irradiances from SOLSTICE and other instruments will be discussed in a future publication. It will include the irradiances from all the SOLSTICE program stars listed in Table I of McClintock, Rottman, and Woods (2005).

5. Solar Stellar Ratio

As described by McClintock, Rottman, and Woods (2005), the optical path for solar and stellar modes is very similar, but not identical. Differences in illumination on the optical elements must be accounted for in the full degradation analysis. In particular, the size of the illumination on the grating in solar mode is a spot 9 mm in diameter, while the stellar-mode spot is 16 mm in diameter. The ratio of these two areas is about 3. The solar illumination on the first folding mirror, M1, is a spot 4.75 mm across. In stellar mode, the spot is 16 mm in diameter. The ratio of these two areas is about 10.

The solar and stellar illumination on the optics following the grating are much more closely matched, but more importantly, the irradiance hitting each of these elements is dispersed, so they see orders of magnitude less energy than the first two elements. Therefore, it is reasonable to assume that all exposure-dependent degradation of the optics is confined to some combination of M1 and the grating. The following section derives a FOV correction assuming that the grating is the only relevant optical element. This assumption will be justified in Section 5.3.

McClintock, Rottman, and Woods (2005) give a more detailed explanation of the optical path for solar and stellar modes, but a brief summary is reproduced here to help make the following sections more clear. In stellar mode, nearly parallel rays from the target uniformly fill a spot equal in diameter to the 16 mm stellar aperture on both the first folding mirror and the grating. The beam is brought to a focus by the camera mirror and imaged on the exit slit. In solar mode, the sunlight diverges from the tiny entrance aperture (0.1 × 0.1 mm). The diameter of this beam expands from 4.75 mm at the first folding mirror to 9 mm at the grating and is then brought to a focus on the exit slit by the same camera mirror as in stellar mode. Figure 16 shows these two configurations graphically.

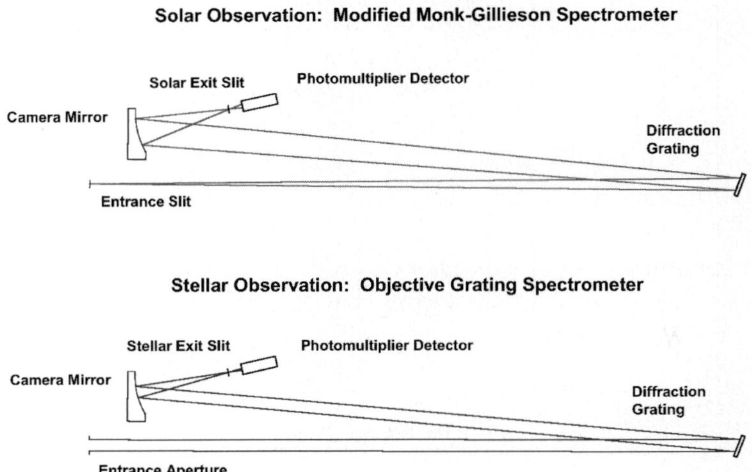

Figure 16. Schematic diagrams of the SOLSTICE optical system for solar and stellar observing modes.

5.1. Derivation of FOV Correction Factor

Measurements at SURF and in-flight have confirmed that there is no measurable change in responsivity across the FOV in the FUV channel after 2 years of operations. Therefore, the difference in illumination between the solar and stellar does not produce a significant difference between the degradation of the whole optical system in the two modes (solar and stellar) for the FUV. The measured stellar degradation is the same as the solar degradation.

However, in the MUV channel, there is a detectable change in the responsivity function with angle. Figure 17 shows a measurement of the responsivity with angle as measured at SURF in 2002. The solar and stellar modes sample different portions of this sensitivity map and appropriate corrections must be made.

The difference is quantified as follows. Divide the FOV into two pieces, one part exposed in solar mode ("sun") and the other exposed during stellar mode but excluding the solar mode region ("non-sun"). Let $\bar{r}$ be the mean responsivity of the region exposed in stellar mode but excluding the region exposed in solar mode. Let x be the difference between $\bar{r}$ and the responsivity in the "sun" portion of the FOV. These quantities are indicated in Figure 17. The center point responsivity, i.e., the responsivity averaged over the angular size of

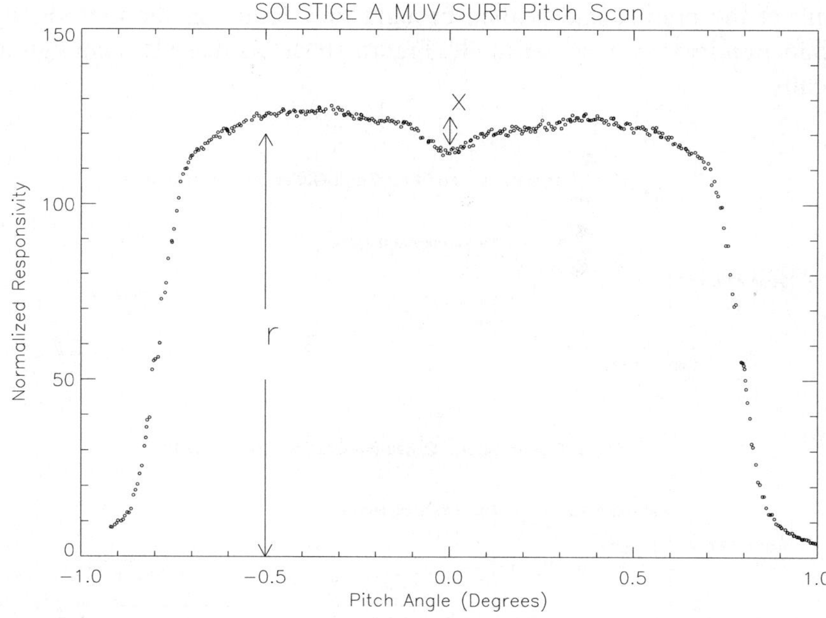

Figure 17. Scan of SURF beam in cross-dispersion direction. The quantity r is the mean responsivity for angles larger than 0.25°. x is the mean reduction in responsivity in the center of the FOV.

the target (from McClintock, Snow, and Woods, 2005, Equation (3)), is defined as

$$R_{\mathrm{C}}(\lambda, T, \Omega) = \frac{\int_{\Omega_{\mathrm{Target}}(0,0)} r(0, \lambda, T, \theta', \phi')\,\mathrm{d}\Omega'}{\int_{\Omega_{\mathrm{Target}}(0,0)} \mathrm{d}\Omega}, \tag{4}$$

where Ω is the angular size of the target. The responsivity for the stellar FOV can be expressed as

$$R_{\mathrm{star}} \int_{\mathrm{star}} \mathrm{d}\Omega = \int_{\mathrm{star}} r\,\mathrm{d}\Omega \tag{5}$$

$$= \int_{\mathrm{non-sun}} \bar{r}\,\mathrm{d}\Omega + \int_{\mathrm{sun}} r\,\mathrm{d}\Omega \tag{6}$$

$$= \int_{\mathrm{non-sun}} \bar{r}\,\mathrm{d}\Omega + \int_{\mathrm{sun}} (\bar{r} - x)\,\mathrm{d}\Omega. \tag{7}$$

x and $\bar{r}$ can be integrated over the solar FOV to get the following useful quantities:

$$\hat{r} = \int_{\mathrm{sun}} \bar{r}\,\mathrm{d}\Omega \quad \text{and} \quad \hat{x} = \int_{\mathrm{sun}} x\,\mathrm{d}\Omega. \tag{8}$$

The geometry of the instrument is such that the area of the solar-mode spot on the grating is one third of the area of the stellar-mode spot. Since $\hat{r}$ is defined in terms of the mean non-sun responsivity, the non-sun integrated responsivity is simply $\hat{r}$ times the area ratio, or $2\hat{r}$.

The stellar-mode responsivity can be written in terms of $\hat{r}$ and $\hat{x}$ as

$$R_{\mathrm{star}} \int_{\mathrm{star}} \mathrm{d}\Omega = \int_{\mathrm{non-sun}} \bar{r}\,\mathrm{d}\Omega + \int_{\mathrm{sun}} \bar{r}\,\mathrm{d}\Omega - \int_{\mathrm{sun}} x\,\mathrm{d}\Omega \tag{9}$$

$$= 2\hat{r} + \hat{r} - \hat{x} \tag{10}$$

$$= 3\hat{r} - \hat{x}. \tag{11}$$

The degradation of the instrument responsivity in stellar mode is therefore simply

$$d_{\mathrm{star}} = \frac{R_{\mathrm{star}}}{R_{\mathrm{star0}}} = \frac{3\hat{r} - \hat{x}}{3\hat{r}_0 - \hat{x}_0}, \tag{12}$$

where quantities with a 0 subscript represent pre-launch values. The solar-mode responsivity samples only the central portion of the FOV. Similarly,

$$R_{\mathrm{sun}} \int_{\mathrm{sun}} \mathrm{d}\Omega = \hat{r} - \hat{x} \tag{13}$$

and

$$d_{\mathrm{sun}} = \frac{R_{\mathrm{sun}}}{R_{\mathrm{sun0}}} = \frac{\hat{r} - \hat{x}}{\hat{r}_0 - \hat{x}_0}. \tag{14}$$

The general solar stellar FOV filling factor correction, $\Gamma_{\text{sun,star}}$ can be defined as follows (Equation (9) from McClintock, Snow, and Woods (2005)):

$$\Gamma_{\text{sun,star}} = \frac{\int_{\text{sun}} r \,\mathrm{d}\Omega}{\int_{\text{star}} r \,\mathrm{d}\Omega} \times \frac{\int_{\text{star}} \mathrm{d}\Omega}{\int_{\text{sun}} \mathrm{d}\Omega}. \tag{15}$$

Using Equations (11) and (13), $\Gamma_{\text{sun,star}}$ simplifies to

$$\Gamma_{\text{sun,star}} = \frac{(\hat{r} - \hat{x})}{3\hat{r} - \hat{x}} \times 3 \tag{16}$$

$$= \frac{1 - \hat{x}/\hat{r}}{1 - \frac{1}{3}\hat{x}/\hat{r}}. \tag{17}$$

The ratio $x' = \hat{x}/\hat{r}$ can be measured on orbit by comparing the irradiance at the center of the FOV to the edge of the FOV during an alignment maneuver. How this ratio changes as a function of time will be discussed in Section 5.2.

The significance of Equations (12) and (14) is that they are not equal if $\hat{x}$ is changing relative to $\hat{r}$ as a function of time. The ratio of the two degradation functions is the correction factor that must be applied to the degradation derived from the stellar measurements to get the proper solar degradation. Taking the ratio of Equations (12) and (14), we get:

$$\frac{d_{\text{sun}}}{d_{\text{star}}} = \frac{R_{\text{sun}}/R_{\text{sun}_0}}{R_{\text{star}}/R_{\text{star}_0}} \tag{18}$$

$$= \frac{\Gamma_{\text{sun,star}}(t)}{\Gamma_{\text{sun,star}}(0)}. \tag{19}$$

A more detailed derivation of this result is given in the Appendix.

This relationship can be approximated as follows. In the FUV channel, x' is not changing, i.e., there is no change in the FOV correction, $\Gamma_{\text{sun,star}}(t) = \Gamma_{\text{sun,star}}(0)$ and $d_{\text{star}} = d_{\text{sun}}$. *The loss of responsivity in the FUV measured from the stellar irradiances is the same as the responsivity loss for the solar irradiance and the stellar degradation correction can be applied directly to the solar data.* The degradation in this case appears to be entirely due to changes in the detector and not in the optics. This is consistent with the Drake *et al.* (2003) prediction discussed at the end of Section 3.2. Figure 18 shows a 1 nm binned solar irradiance time series before and after applying the degradation correction shown in Figure 11.

The situation for longer wavelengths is slightly more complicated, since $\hat{x} \neq \hat{x_0}$, or alternatively, $\Gamma_{\text{sun,star}}(t) \neq \Gamma_{\text{sun,star}}(0)$. The stellar degradation in the MUV channel has been very small (less than a few percent at most wavelengths – cf. Figure 13 – indicating very little change in the detector). Fortunately, the depth of the central depression is a small fraction of the total responsivity, so the following approximations provide an estimate of the magnitude of the FOV correction required. If we

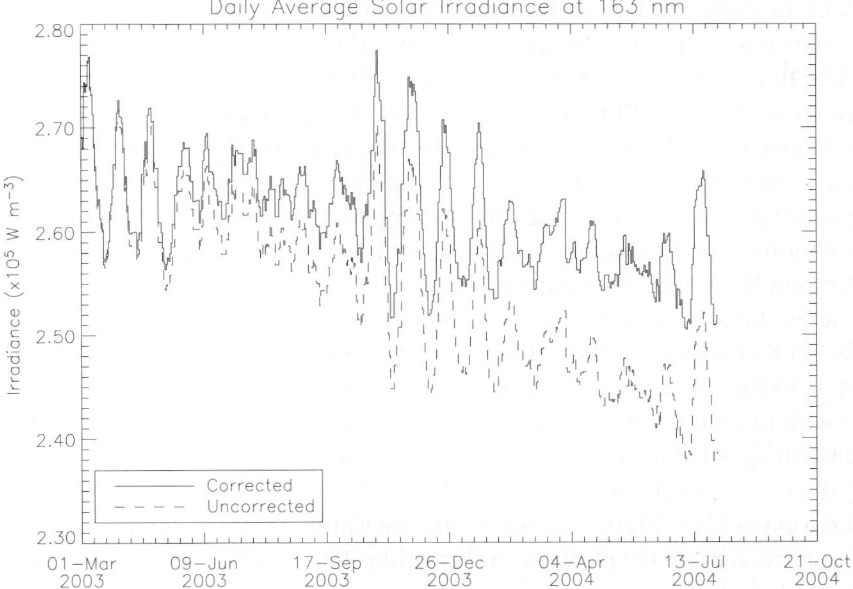

Figure 18. Time series of SOLSTICE solar observations after correction for degradation. The long-term decrease in uncorrected irradiance is due to instrument degradation. The ratio of solar to ensemble stellar irradiance has been removed in the "corrected" curve, and the remaining long-term trend is due to solar variability.

assume $r = r_0$ and $x_0 \ll r_0$, the stellar and solar degradation (Equations (12) and (14)) simplify to

$$d_{\text{starMUV}} \sim \frac{3\hat{r}_0 - \hat{x}}{3\hat{r}_0} \sim 1 - \frac{1}{3}\frac{\hat{x}}{\hat{r}_0} \tag{20}$$

and

$$d_{\text{sunMUV}} \sim \frac{\hat{r}_0 - \hat{x}}{\hat{r}_0} \sim 1 - \frac{\hat{x}}{\hat{r}_0}. \tag{21}$$

Therefore, the proper solar degradation correction factor (the difference from unity, or the second term in Equations (20) and (21)) is approximately a factor of 3 larger than the stellar degradation after correction for the FOV effect.

5.2. In-flight FOV measurements

Measurement of x' in-flight is more difficult than measuring it at SURF. The problem is that the solar irradiance does not come from a collimated beam with a flat spectrum. This presents a number of challenges. The first challenge is wavelength shift. As described by McClintock, Rottman, and Woods (2005), the observed wavelength depends on viewing angle. The magnitude of the shift in solar mode is less than in stellar mode (cf. Equation (1)) and it is further reduced by using the cross-dispersion direction for each detector. The problem is that if the observed wavelength is at the bottom of an absorption feature at the central angle, then the measured count rate will go up as a function of angle regardless of the true depth of the responsivity depression. In this case, the value of x' is overestimated.

The second challenge has to do with the optical layout of the instrument. Although the spot on M1 is relatively small compared to the diameter of the mirror, the illumination of the Sun nearly fills other optical elements. In particular, for angles greater than 40 arcmin, the image of the Sun falls off the camera mirror, preventing an accurate measurement of r at large angles.

The third challenge is integrating over 30 arcmin solar disk. The alignment scans shown in Figure 19 are measured as follows. The spacecraft slowly scans over a 4° range with the Sun at the center. Comparing Figures 17 and 19, one notes that the decrease in signal at the edge of the in-flight scan is much more gentle than the one using the nearly point source of the SURF FOV.

The top panel of Figure 20 shows the measured stellar degradation at 250 nm and the correction for the FOV effect (Equation (19)). The bottom panel contains the corrected and uncorrected 1-nm daily average solar irradiances for that wavelength. It is likely that these data are overcorrected for degradation by about 1% in the first

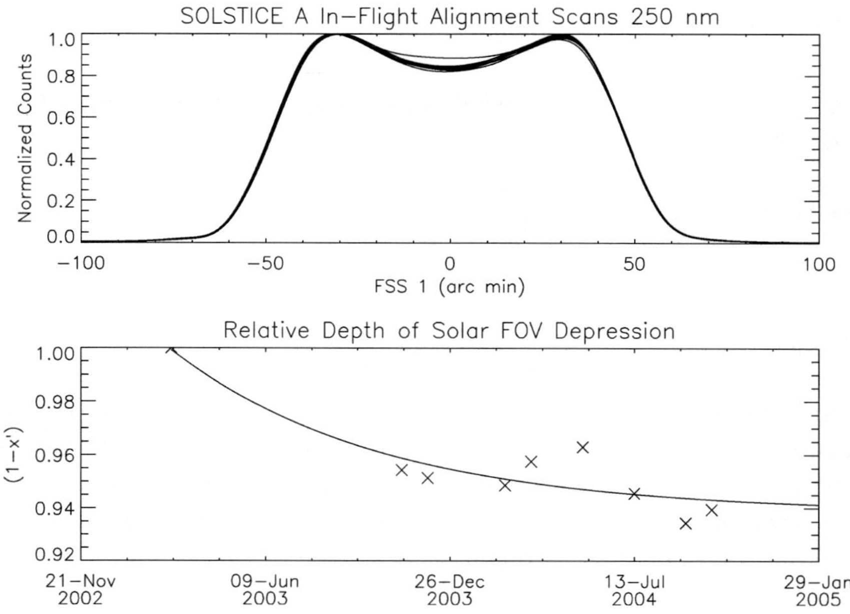

Figure 19. In-flight alignment scans and FOV correction. The *top panel* shows the solar alignment scans taken since launch at 250 nm. The spacecraft scans across the Sun to measure the centering and the evolution of the solar FOV. The *bottom panel* shows the measured depth of the central depression relative to the region outside the solar FOV as a function of time (i.e., $1 - x'$). The *curve* is a simple exponential fit to the data.

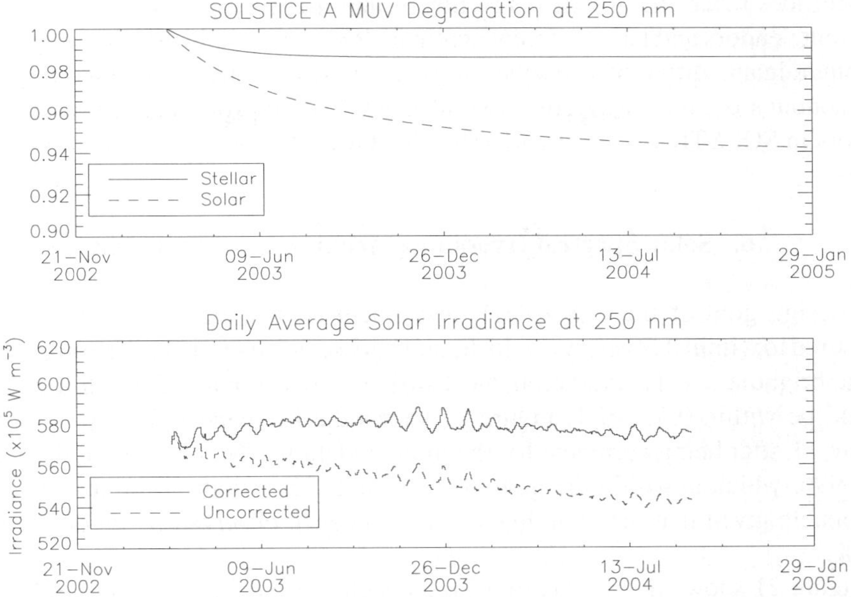

Figure 20. The FOV-corrected MUV degradation curve and corresponding solar time series. The *top panel* shows the stellar degradation measurement (*solid line*) and the correction for FOV (*dashed line*) at 250 nm. The *bottom panel* shows the measured solar irradiance before and after correction for degradation.

half of 2003 since the irradiance gently rises. As the SORCE mission continues, the measurement based on the stellar observations will improve. As of the end of 2004, the magnitude of the change in stellar irradiance is not much larger than the spread in the observations at many wavelengths in the MUV.

5.3. Grating or Mirror?

Is the assumption that the FOV effect is due entirely to illumination of the grating alone correct? If we were to assume that the majority of this effect were at M1, Equations (20) and (21) would indicate that the solar degradation should be a factor of 10 greater than the stellar degradation rather than a factor of 3. The upper limit on the solar degradation at 220 nm is 6.5%, assuming that there has been no change in the solar irradiance over the two years of the SORCE mission. The stellar count rates at that wavelength have declined by only 2% (Figure 13). Therefore, the FOV correction cannot multiply the stellar degradation function by more than about a factor of 4 at most, which is inconsistent with M1 alone producing this effect.

What might cause this hole to be burned in the grating but not the mirror? Both elements see the full, non-dispersed solar irradiance. M1 does see the hard radiation component of the solar output which the grating does not, but the total energy from the Sun at wavelengths below 100 nm is about equal to the energy in Lyman α

(which does reach the grating). The mirror and the grating were purchased from different vendors and have different coatings. It is conceivable that they might react to contaminants differently. Furthermore, the grating is enclosed within a housing that contains organic materials associated with its control electronics, while the mirrors in SOLSTICE are supported by clean metallic mounting structures.

6. Solar Spectral Irradiance Accuracy and Precision

The design goal of the SOLSTICE instrument is that the degradation is to be measured to within 0.5% per year. In the stellar case, it means that after correction for stellar brightness and degradation, the distribution of normalized stellar irradiances should be within 0.5% of 1. Figure 21 shows the irradiances for the six stars in Figure 11 after being corrected for the fit degradation. The distribution is centered on 0.996, which is within the goal of 0.5% of 1. The width of the distribution is reasonable given that the counting statistics on each observation are on the order of 1%.

Figure 21 shows the distribution for the stellar irradiances for the full SORCE mission from Figure 11, but the SOLSTICE technique should be able to correct irradiances over smaller time intervals as well. Figure 22 shows the centroid of

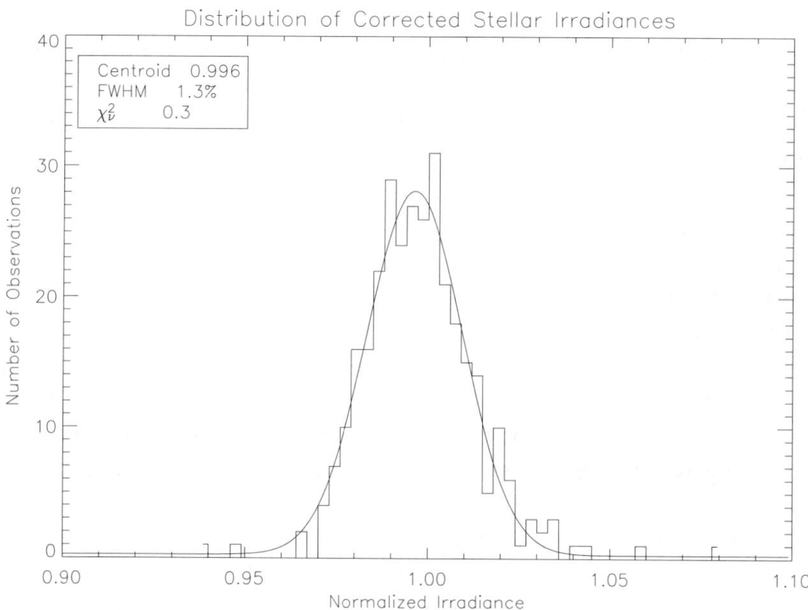

Figure 21. Histogram of normalized stellar irradiances at 163 nm for the stars shown in Figure 11 after correction for degradation. The irradiances are well modeled by a normal distribution centered on 99.6%.

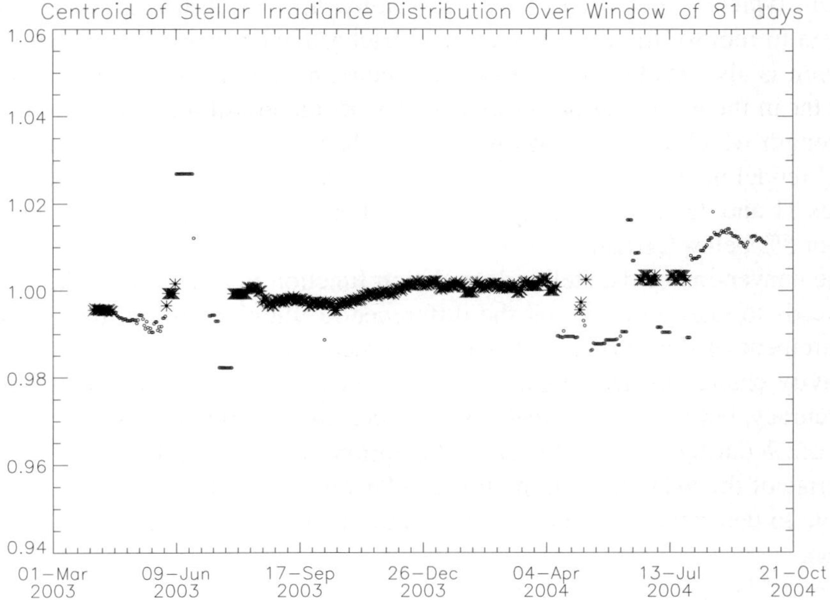

Figure 22. Centroid of distribution of corrected stellar irradiances from Figure 11 in 81-day windows. *Asterisks* mark days with a centroid within 0.5% of 1. The scheduling error discussed in the text explains the gap in the data in mid-2003.

the distribution of corrected stellar irradiances smoothed over an 81-day window. Asterisks indicate time periods where the centroid is within the target 0.5% of 1.

The fit to the normalized irradiances does indeed produce a centroid within the desired range nearly all the time. The only time period where the centroid consistently falls outside the 0.5% band is the period of low data volume in mid-2003 as described earlier. Since the total degradation in the FUV channel is the same as the stellar degradation, the design goal of 0.5% per year knowledge of the instrument response is achieved.

In the MUV channel, the change in the FOV must be included. Over the first 2 years of the mission, the total degradation is less than about 5%. The upper limit of 5% comes from assuming that there has been no intrinsic change in the solar irradiance, and that all the decrease in the observed irradiance is due to degradation alone. At many wavelengths, the upper limit on degradation is significantly less than that.

7. Conclusion

The SOLSTICE technique uses an ensemble of bright, early-type stars as an irradiance reference to track changes in instrument responsivity. Individual stellar count rates are typically measured to 1% uncertainty. Special observing techniques are required to remove the Lyman α airglow background below 135 nm, but the 1% precision is also easily achieved in this spectral region. Multiple observations of each star in the ensemble produces a stellar degradation function at each chosen wavelength which has a precision of better than 0.5% per year. A simple exponential model fits these observations to within the desired uncertainty as shown in Figures 21 and 22. The uncertainty in the stellar absolute irradiance measurement is about 5% below 280 nm.

The conversion of the stellar degradation function to a solar degradation function needs to take into account the difference in illumination on the optics. The measurement of the FOV effect is not as precise as the measurement of the responsivity change using the stars. New observing strategies will likely improve the accuracy, but the current analysis produces an x', which has an uncertainty of about $\pm 2\%$ during the first 2 years of operations as shown in Figure 19. The rate of change of the FOV correction during 2004 was relatively small, so it should be possible to determine this correction to better than 1% throughout the rest of the mission.

Acknowledgements

This research is supported by NASA contract NAS5-97045 at the University of Colorado. The authors would like to thank Linton Floyd for his valuable review comments which greatly improved the paper. We would also like to thank Craig Markwardt (U. Wisconsin) for his MPFIT software that was used throughout this analysis and Ken Griest and Cindy Russell (CU/LASP) for their assistance in planning the stellar observations.

Appendix: Full Derivation of the Solar Stellar Correction Factor

The relationship between the solar and stellar degradation functions (Equation (19)) can be derived from the irradiance ratio equations. The observed solar irradiance at time t must be corrected for instrumental degradation. This factor depends on both the degradation in stellar mode and a correction for the difference in FOV. Since this is such a fundamental aspect of the SOLSTICE technique, the full derivation is presented here. As discussed in Section 3.2.2 of McClintock, Snow, and Woods (2005), tracking the change in solar the solar/stellar irradiance ratio requires knowledge of changes in all the response functions (Equation (24) of that paper).

$$\Delta \frac{E_{\text{sun}}}{E_{\text{star}}} = \Delta \frac{C_{\text{sun}}}{C_{\text{star}}} \Delta \frac{G_{\text{star}}}{G_{\text{sun}} f_{\text{AU}}} \Delta \Gamma_{\text{DEG}}(t, \lambda, \Omega_{\text{sun,star}}), \tag{22}$$

where E is the irradiance, C the count rate, G the detector temperature gain factor, f_{AU} the correction to 1 AU, and finally, $\Delta\Gamma$ the change in the FOV:

$$\Delta\Gamma_{DEG}(t, \lambda, \Omega_{sun,star}) = \frac{\int_{star} r(t, \lambda, T, \theta, \phi)\, d\Omega / \int_{star} r(0, \lambda, T, \theta, \phi)\, d\Omega}{\int_{sun} r(t, \lambda, T, \theta, \phi)\, d\Omega / \int_{sun} r(0, \lambda, T, \theta, \phi)\, d\Omega}. \quad (23)$$

The Δ operator in Equation (22) means the ratio of a measured quantity at time t divided by that quantity at $t = 0$. The irradiance measurement equations for the Sun and stars become (dropping explicit dependence on wavelength and angle):

$$\Delta E_{sun} = \frac{E_{sun}(t)}{E_{sun}(0)} = \frac{C_{sun}(t)}{C_{sun}(0)} \frac{G_{sun}(0)}{G_{sun}(t)} \frac{f_{AU}(t)}{f_{AU}(0)} \frac{\Gamma_{std,sun}(0)}{\Gamma_{std,sun}(t)} \quad (24)$$

and

$$\Delta E_{star} = \frac{E_{star}(t)}{E_{star}(0)} = \frac{C_{star}(t)}{C_{star}(0)} \frac{G_{star}(0)}{G_{star}(t)} \frac{\Gamma_{std,sun}(0)}{\Gamma_{std,sun}(t)}. \quad (25)$$

$\Gamma_{std,sun}$ is the geometric correction factor to the responsivity between the FOV of the SURF beam and the Sun (Equation (9) of McClintock, Snow, and Woods (2005)). $\Gamma_{std,star}$ is similarly defined as the correction to the stellar FOV.

We can then take the ratio of these two equations to derive the relationship between the solar and stellar irradiances as a function of time. We will also simplify the algebra by correcting the count rates for gain and solar distance, i.e., $C' = C f_{AU}/G$.

$$\Delta\frac{E_{sun}}{E_{star}} = \frac{\Delta E_{sun}}{\Delta E_{star}} \quad (26)$$

$$= \frac{C'_{sun}(t)}{C'_{sun}(0)} \frac{C'_{star}(0)}{C'_{star}(t)} \frac{\Gamma_{std,sun}(0)}{\Gamma_{std,sun}(t)} \frac{\Gamma_{std,star}(t)}{\Gamma_{std,star}(0)}. \quad (27)$$

The two terms involving the ratio of gammas reduces to $\Delta\Gamma_{DEG}$ defined in Equation (23). Note also that $\Delta\Gamma_{DEG}$ can be simplified using Equation (15) to

$$\Delta\Gamma_{DEG} = \frac{\Gamma_{sun,star}(t)}{\Gamma_{sun,star}(0)}. \quad (28)$$

The stellar count rate ratio is the measured stellar degradation function, d_{star}, shown in Figures 11 and 12. It is assumed that the stars used in the SOLSTICE analysis are non-varying, i.e., $\Delta E_{star} = 1$. Therefore,

$$\Delta E_{sun} = \frac{C'_{sun}(t)}{C'_{sun}(0)} \frac{1}{d_{star}\Delta\Gamma_{DEG}}. \quad (29)$$

The observed change in the solar count rate must be corrected by the change in the stellar count rate times the change in the FOV filling factor to get the true change in the solar irradiance.

References

Arp, U., Friedman, R., Furst, M. L., Makar, S., and Shaw, P. S.: 2000, *Metrologia* **37**, 357.
Bohlin, R. C., Dickinson, M. E., and Calzetti, D.: 2001, *Astron. J.* **122**, 2118.
Bohlin, R. C. and Lindler, D.: 1992, *STSci Newslett.* **9**, 19.
Drake, V. A., McClintock, W. E., Woods, T. N., and Rottman, G. J.: 2003, *SPIE Proc.* **4796**, 107.
Gonzales-Riestra, R., Cassatella, A., and Wamsteker, W.: 2001, *Astron. Astrophys.* **373**, 730.
Mihalas, D. and Binney, J.: 1981, *Galactic Astronomy Structure and Kinematics*, Freeman, New York, p. 135.
McClintock, W., Rottman, G., and Woods, T.: 2005, *Solar Phys.*, this volume.
McClintock, W., Snow, M., and Woods, T.: 2005, *Solar Phys.*, this volume.
Rottman, G. J., Woods, T. N., and Sparn, T. P.: 1993, *J. Geophys. Res.* **98**, 10667.
Woods, T., Ucker, G. J., and Rottman, G.: 1993, *J. Geophys. Res.* **98**, 10679.
Woods, T., Rottman, G., Russell, C., and Knapp, B.: 1998, *Metrologia* **35**, 619.

Solar Physics (2005) 230: 325–344

THE Mg II INDEX FROM SORCE

MARTIN SNOW, WILLIAM E. McCLINTOCK, THOMAS N. WOODS,
ORAN R. WHITE, JERALD W. HARDER and GARY ROTTMAN
Laboratory for Atmospheric and Space Physics, University of Colorado, Boulder, CO 80309, U.S.A.
(e-mail: snow@lasp.colorado.edu)

(Received 22 December 2004; accepted 20 April 2005)

Abstract. The Solar–Stellar Irradiance Comparison Experiment (SOLSTICE) and the Spectral Irradiance Monitor (SIM) on the Solar Radiation and Climate Experiment (SORCE) both measure the solar ultraviolet irradiance surrounding the Mg II doublet at 280 nm on a daily basis. The SIM instrument's resolution (1.1 nm) is similar to the Solar Backscatter Ultraviolet instruments used to compute the standard NOAA Mg II index, while SOLSTICE's resolution is an order of magnitude higher (0.1 nm). This paper describes the technique used to calculate the index for both instruments and compares the resulting time series for the first 18 months of the SORCE mission. The spectral resolution and low noise of the SOLSTICE spectrum produces a Mg II index with a precision of 0.6%, roughly a factor of 2 better than the low-resolution index measurement. The full-resolution SOLSTICE index is able to measure short-timescale changes in the solar radiative output that are lost in the noise of the low-resolution index.

1. Introduction

The Mg II core-to-wing ratio developed by Heath and Schlesinger (1986) is an important measure of variability of radiation from the solar chromosphere. Radiation from this region of the solar atmosphere drives many processes in the middle atmosphere of Earth and influences the Earth's climate. Accurate monitoring of the solar variability through the Mg II index is one of the goals of the Solar Radiation and Climate Experiment (SORCE) mission. Two of the instruments on SORCE, the Solar-Stellar Irradiance Comparison Experiment (SOLSTICE) and the Spectral Irradiance Monitor (SIM) make frequent measurements of the solar spectral irradiance near the Mg II h and k resonance lines at 280 nm. This paper describes the technique of producing a Mg II index from both instruments and compares them with other measures of solar radiative variability.

The value of the Mg II index is that it is a ratio of irradiances at nearby wavelengths and therefore less susceptible to instrumental effects and long-term degradation. Variation in the index is an excellent proxy for the variability of the solar irradiance between 200 and 300 nm (e.g., Heath and Schlesinger, 1986) as well as for the He II 304 Å emission (Viereck *et al.*, 2001) and solar faculae (Lean *et al.*, 1997). Tobiska (1991) shows that 30% of the atmospheric heating in the 200–600 km range is due to solar He II 304 Å emission.

White *et al.* (1998) examined characteristics of the Mg II index produced at 1.1 nm resolution (SBUV) and one derived from a spectrum with 0.2 nm resolution (UARS SOLSTICE). This paper discusses the effect of increasing the resolution yet again to 0.1 nm using spectra from SORCE SOLSTICE. The comparisons to the NOAA Mg II index are based on the publically available version 9.1 data from the SEC web page. During the timeframe of the SORCE mission, the NOAA index is produced from SBUV data aboard the NOAA 16 satellite, scaled to the original NOAA 9 dataset as described by Viereck and Puga (1999) and Viereck *et al.* (2004).

2. Solar Spectrum Near 280 nm

Figure 1 shows the portion of the solar irradiance spectrum containing the Mg II cores and the much less-variable line wings well outside the line cores. The h and k emission features sit at the bottom of a broad absorption feature produced by the two transitions.

A partial term diagram for the Mg II ion is shown in Figure 2. The h and k transitions are from the ^{2}P configuration to ^{2}S. The emission core is formed in the upper chromosphere, at a temperature of around 7000 K. The wings of the

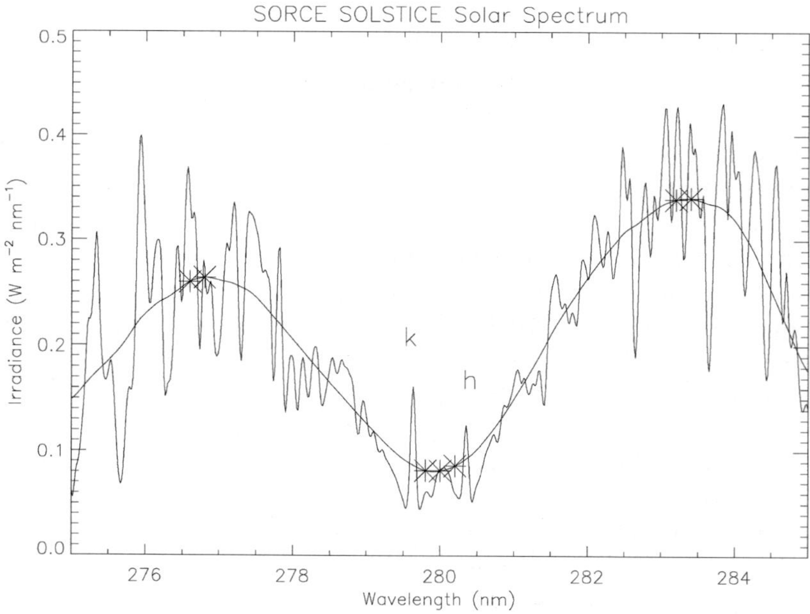

Figure 1. Solar spectrum showing Mg II h and k lines. The *smooth curve* is the SOLSTICE spectrum convolved with the SBUV instrument profile. In this wavelength region, the resolution of the SIM instrument is virtually identical to SBUV. The *asterisks* indicate the wavelengths of the core and wing points used to compute the standard NOAA Mg II index from the SBUV data.

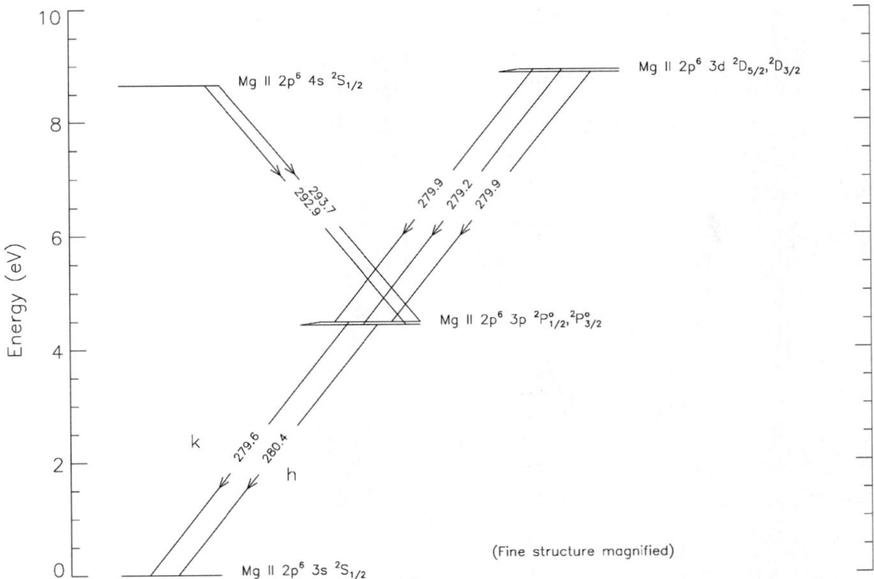

Figure 2. Partial term diagram for Mg II model atom. The h and k lines are resonance transitions from ^{2}P to ^{2}S configurations. Model atom *courtesy* of P. Judge (HAO/NCAR).

broad absorption feature are from the upper photosphere. The variability of the photosphere is quite small, while the chromospheric emission varies by about 30% as active regions evolve and rotate in and out of view on the solar disk. An active region may persist for a few months before merging back into the quiet network and therefore it may be observed on more than one solar rotation.

2.1. THE Mg II INDEX FROM SIM

The SIM instrument is a prism spectrometer that measures the solar spectral irradiance from 200 nm to 3 μm. For a full description of the instrument design, see Harder *et al.* (2005). The irradiances used in this analysis are from its UV photodiode detector, which is calibrated by the Electrical Substitution Radiometer (ESR). This combination produces a spectrum with extremely high signal-to-noise characteristics.

The spectral resolution of the SIM instrument changes as a function of wavelength (Harder *et al.*, 2005), but near 280 nm it is about 1.1 nm. The SIM instrument profile is very similar to the instrument response of NOAA SBUV. The SIM spectrum looks very similar to the smoothed SOLSTICE spectrum with 1.1 nm resolution shown in Figure 1.

The Mg II index derived from SIM data is calculated as follows. The full-resolution irradiance spectrum is fit with a b-spline model and then

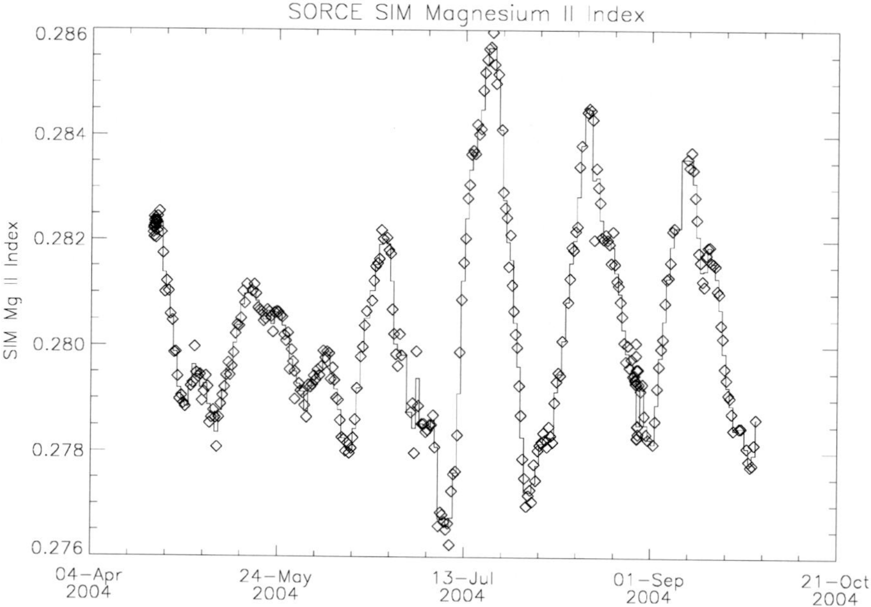

Figure 3. The daily average Mg II index from SIM measurements. The individual measurements are shown by the *diamonds*, and the *solid line* is the median of the measurements on each day.

evaluated at the SBUV discrete-mode wavelengths that define the SBUV Mg II index,

$$\text{Mg II}_{\text{SIM}} = \frac{4[I_{279.8} + I_{280.0} + I_{280.2}]}{3[I_{276.6} + I_{276.8} + I_{283.2} + I_{283.4}]}, \tag{1}$$

where $I_{279.8}$ is the solar irradiance at 279.8 nm, etc. This formula is exactly the original definition of the Mg II index from Heath and Schlesinger (1986). Figure 3 shows the derived Mg II index for the period of April 2004 through the end of September 2004. Irradiance data before April 2004 is being recalibrated and is currently unavailable. The available dataset does span more than five solar rotations, and is therefore enough to make preliminary comparisons to the NOAA index (see Section 3).

2.2. The Mg II index from SOLSTICE

SOLSTICE is a grating spectrometer that measures the solar spectral irradiance from 115 to 320 nm with a resolution of about 0.1 nm. The design and calibration of SOLSTICE are described more fully in McClintock, Rottman, and Woods (2005) and McClintock, Snow, and Woods (2005). The data used in constructing the Mg II index come from the MUV channel, which is sensitive from 160 to 320 nm. This channel is roughly equivalent in wavelength coverage to the “F” channel of the

UARS SOLSTICE instrument that was launched in 1991 (Rottman, Woods, and Sparn, 1993), but has a factor of 2 improvement in spectral resolution at 280 nm.

The 0.1 nm resolution of SOLSTICE allows the emission cores to be fully resolved, so they can be measured separately from the nearby absorption feature. Each emission feature is fit with a Gaussian. The *numerator* of the SOLSTICE index is the mean of the integrated Gaussian cores of both h and k lines.

The *denominator* of the SOLSTICE index is derived by convolving the full-resolution spectrum with a triangular response function with a 1.1 nm FWHM, matching SBUV's resolution and shown in Figure 1. The average irradiance at the four points in the wing used in the NOAA index (276.6, 276.8, 283.2, and 283.4 nm) is integrated over the smoothed 0.2 nm interval, and defines the SOLSTICE wing value. Variability in these wings will be discussed in Section 5.

$$\text{Mg II}_{\text{SOLSTICE}_{\text{FULL}}} = \frac{2[I_{\text{h}} + I_{\text{k}}]}{[I_{276.6} + I_{276.8} + I_{283.2} + I_{283.4}]}. \tag{2}$$

Figure 4 shows that during the first 18 months of the SORCE mission, the Mg II index has varied by ~30% as active regions rotate in and out of view on the solar disk. The largest oscillations were during the October–November 2003 time period. Those active regions produced many flares, and the general activity level of the Sun was more typical of solar maximum than of the declining phase of the solar cycle.

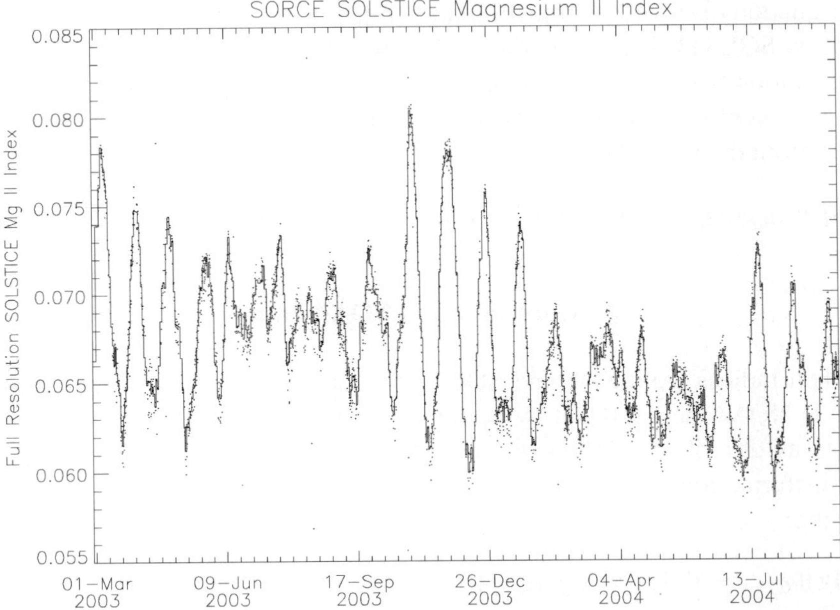

Figure 4. The daily average Mg II index from SOLSTICE. The individual measurements are shown by the *dots*, and the *solid line* is the median of all the measurements on each day.

Woods *et al.* (2004) discuss the Mg II observations from SOLSTICE during that time frame, including the Mg II variability during the flare on 28 October 2003. But that high activity level lasted only a few rotations. Since then, the Sun has behaved in a more typical fashion for this part of the solar cycle. The 81-day running mean of the Mg II index peaked at 0.0697 in November 2003, and declined to a minimum of 0.0635 in May 2004. These trends are consistent with the current declining phase of the solar cycle. The general slow decrease in the Mg II index should continue for the next few years.

In order to better understand the role of spectral resolution in the SOLSTICE Mg II index, the resolution of the magnesium cores can be reduced to the nominal 1.1 nm bandpass of SBUV. The series of SBUV instruments each have slightly different bandpasses, and this shows up in the slope of the transformation equations described in Section 3. The Mg II index from the reduced-resolution spectrum replicates what a SBUV-type instrument would observe if it measured the solar spectrum many times per day. The noise properties of this index will be discussed in Section 5. In this context, "SBUV-type" means a spectrograph with a 1.1 nm triangular bandpass that samples the spectrum at 0.2 nm intervals. This quantity will be referred to as $\mathrm{SOLSTICE_{LOW}}$ hereafter. The advantage of this comparison technique is that it eliminates the effect of any short-term solar variation, as well as any instrument-to-instrument differences. The two index measurements are at identical times since they are the same data just smoothed in different ways. The wing measurement is the same in both cases, so any differences in the two indices are due entirely to differences in the smoothed and unsmoothed core emission. Equation (3) gives the transformation relationship between the full- and low-resolution SOLSTICE indices. Figure 5 shows the linear correlation between the full-resolution $\mathrm{SOLSTICE_{FULL}}$ index and the $\mathrm{SOLSTICE_{LOW}}$ 1.1 nm version. The correlation coefficient is over 99% as can be expected since the two quantities are derived from the same data.

$$\mathrm{Mg\,II_{SOLSTICE_{LOW}}} = 0.665 \times \mathrm{Mg\,II_{SOLSTICE_{FULL}}} + 0.235. \tag{3}$$

3. Correlation to NOAA Index

The Mg II indices from SIM and SOLSTICE can be compared to the publically available NOAA data product by performing a linear regression between the two datasets as described in Viereck and Puga (1999) and De Toma *et al.* (1997). The transformation from the SIM Mg II index to the NOAA index is given by the following:

$$\mathrm{Mg\,II_{NOAA}} = 0.921 \times \mathrm{Mg\,II_{SIM}} + 0.013. \tag{4}$$

This relationship is determined by a linear regression between the two indices as shown in Figure 6 and has a correlation coefficient of 0.99. The extremely high

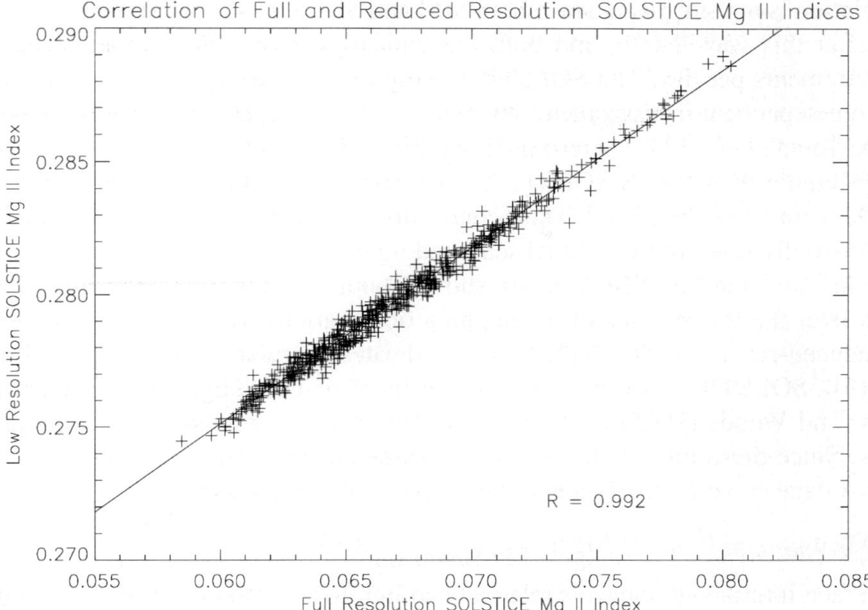

Figure 5. Relationship between the SOLSTICE Mg II full-resolution index daily average and the daily index derived from SOLSTICE spectra convolved with the SBUV instrument function (1.1 nm).

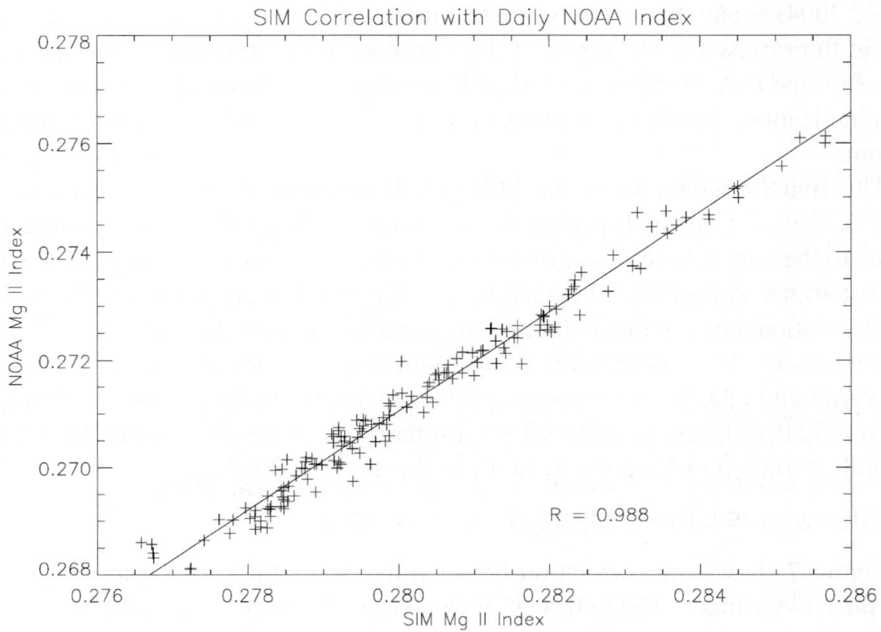

Figure 6. Relationship between the SIM Mg II index daily median and the NOAA daily index.

correlation is not surprising since both SIM and SBUV have very similar resolving power at this wavelength, and both are typically derived from just one or two measurements per day. The SOLSTICE daily index is determined from at least a dozen independent measurements each day.

De Toma *et al.* (1997) determined that the SBUV Mg II core values were about 6–7% higher than UARS SOLSTICE data convolved with the SBUV instrument profile would have predicted. Their conclusion was that the SBUV core may have been partially filled in by residual scattered light. Comparison with the SIM index also indicates that the SBUV index shows about 7% less variability than SIM's index over the few months of overlapping observations. A regression analysis of the reduced-resolution SOLSTICE index indicates a similar 7% discrepancy (Equation (5)). SOLSTICE also has very low levels of scattered light, and McClintock, Snow, and Woods (2005) contains a detailed analysis of its scattered light properties. Since these three independent measurements all give a consistent 7%, the NOAA dataset would then seem to be source of the discrepancy.

$$\text{Mg II}_{\text{NOAA}} = 0.921 \times \text{Mg II}_{\text{SOLSTICE}_{\text{LOW}}} + 0.015. \quad (5)$$

In fact, it turns out to have a relatively straightforward explanation. These comparisons all assume that the SBUV instruments have exactly a 1.1 nm bandpass. Numerical experiments with the SOLSTICE data show that a small change in the assumed bandpass of SBUV reproduces this reduction in core emission relative to the wings. The method of producing a long-term composite index (e.g., Viereck *et al.*, 2004) scales the data from later instruments to the resolution of the first one in the series. The bandpass of the first instrument in the NOAA time series (Nimbus7/NOAA 9) (Viereck *et al.*, 2004) is not well known (R. Viereck, private communication), but these comparisons indicate that it was larger than the nominal 1.1 nm.

The transformation from the SOLSTICE full-resolution Mg II index to the NOAA index is given by Equation (6). The relatively large slope in the relationship is due to the very different way of calculating the core irradiance and, as expected, is close to the slope value of Equation (3). The dynamic range of the SOLSTICE full-resolution index is about 20 times greater than the intra-day spread (~1% daily spread out of ~25% monthly variation). With only a single value per day, the intra-day variation in the NOAA measurement is unknown, but the similar quantities for the SOLSTICE data reduced to SBUV resolution are: intra-day spread of 0.2% and monthly variation of 5% for a dynamic range of about 20 also.

$$\text{Mg II}_{\text{NOAA}} = 0.615 \times \text{Mg II}_{\text{SOLSTICE}_{\text{FULL}}} + 0.232. \quad (6)$$

Figure 7 shows the correlation between the SOLSTICE full-resolution index and the NOAA index. The SOLSTICE value is the median value of the Mg II index observations on each day. White *et al.* (1998) recommend using the median value rather than the mean, and their suggestion is used in the analysis of this paper. The NOAA index is produced from a single measurement each day, so one would not

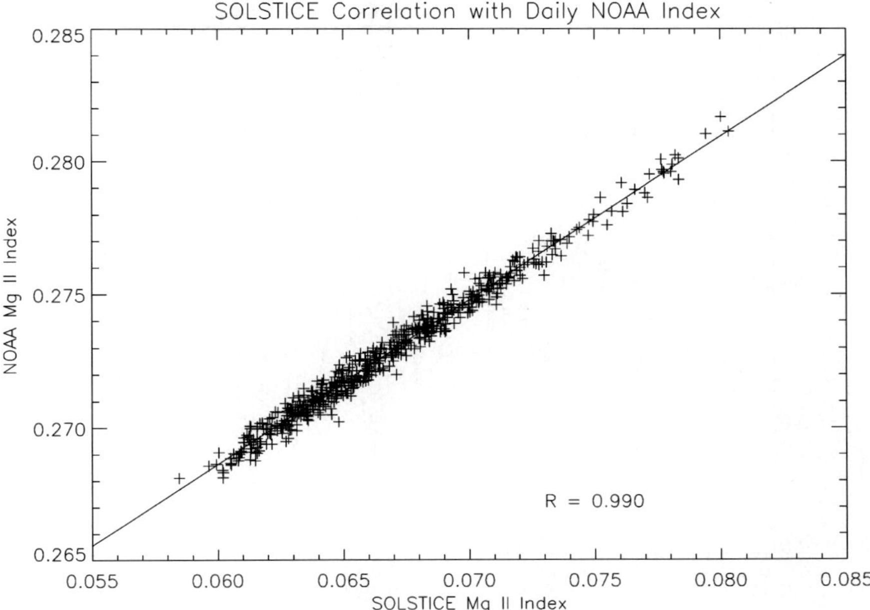

Figure 7. Relationship between the SOLSTICE Mg II index daily average and the NOAA daily index from 1 March 2003 through 31 August 2004.

expect the scatter between the SOLSTICE and NOAA indices to be less than the intra-day variation in the index.

4. Correlation to He II

Viereck *et al.* (2001) showed that the Mg II index as calculated by NOAA is a better proxy for He II 304 Å emission (based on the SOHO SEM 28-32 nm band) than the F10.7 index. Figure 8 shows the daily averaged 1-nm binned solar irradiance from 30 to 31 nm from the TIMED SEE instrument along with the scaled SORCE SOLSTICE Mg II index for the first year of SORCE observations. The SEE 30–31 nm data correlates well with the broadband SEM measurement (Woods *et al.*, 2005; Woodraska, Woods, and Eparvier, 2004), so the correlation with the Mg II index presented here would be similar for either EUV dataset.

The scaling factor is based on the linear correlation shown in Figure 9. Version 7 of the SEE data is used in this analysis, which has been validated only through early 2004. A longer timespan of good SEE data will certainly improve the degree of correlation. This 1 nm interval in the solar spectrum contains strong contributions from He II and Si XI. As a reminder that the EUV irradiance is a 1-nm binned data product and not a measurement of the emission from a single species, it will be referred to as just “30.5 nm emission.”

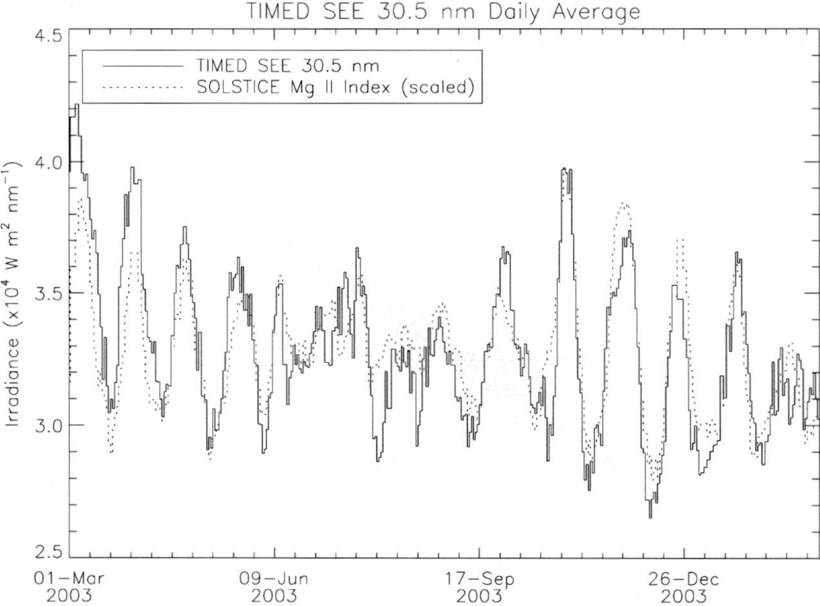

Figure 8. He II 30.4 nm irradiance from the TIMED SEE instrument. The SORCE SOLSTICE Mg II index has been scaled by the relationship shown in Figure 9 and overplotted.

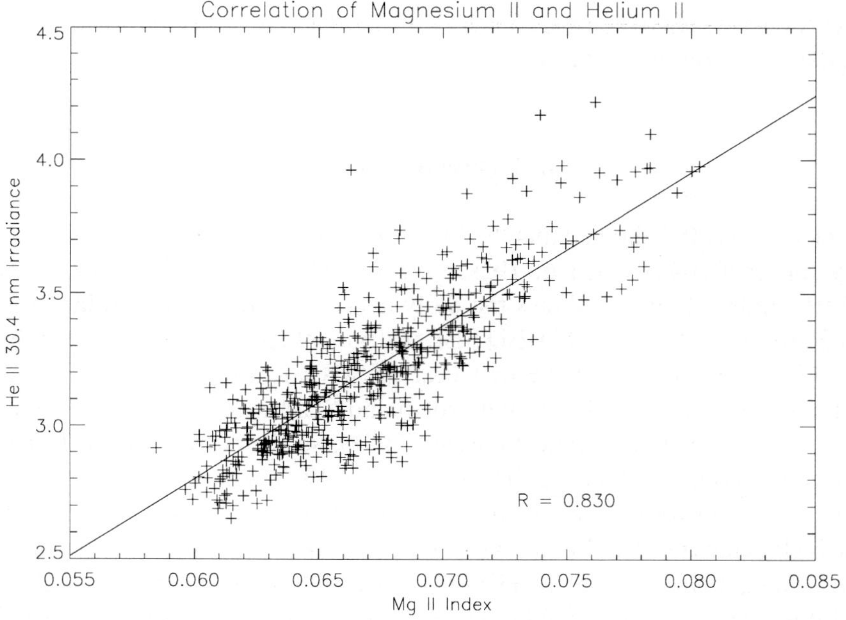

Figure 9. Relationship between the SOLSTICE Mg II index daily average He II 30.4 nm daily average for the time period shown in Figure 8.

In general, the Mg II index correlates very well with 30.5 nm emission, although there are certain solar rotations where the correlation is poor, such as the one in mid 2003. During this rotation, the Mg II index decreases to a minimum while the 30.5 nm irradiance remains relatively flat.

The correlation coefficient over a series of 27-day windows is shown in Figure 10. The solid line is the correlation with the SORCE SOLSTICE full-resolution daily index. The dashed curve is the correlation of the NOAA daily index to the same 30.5 nm data. Since the SOLSTICE and NOAA indices are highly correlated, it is not surprising that in general the correlation to 30.5 nm is basically the same. With the exception of the June 2003 event, the SOLSTICE index has about a 10% better correlation than the NOAA index during episodes of low correlation. Further investigation will be necessary to understand differences in how each index tracks the EUV emission, and also whether the differences observed in 2003 are significant or not. It is quite possible that the higher time cadence of observations makes the SOLSTICE index a slightly better predictor of 30.5 nm irradiance. It is also true that the full-resolution SOLSTICE index samples a more narrow range of heights in the solar atmosphere. Future studies will be required to develop a greater understanding of the differences in correlation between the NOAA and SOLSTICE Mg II indices and the 30.5 nm irradiance.

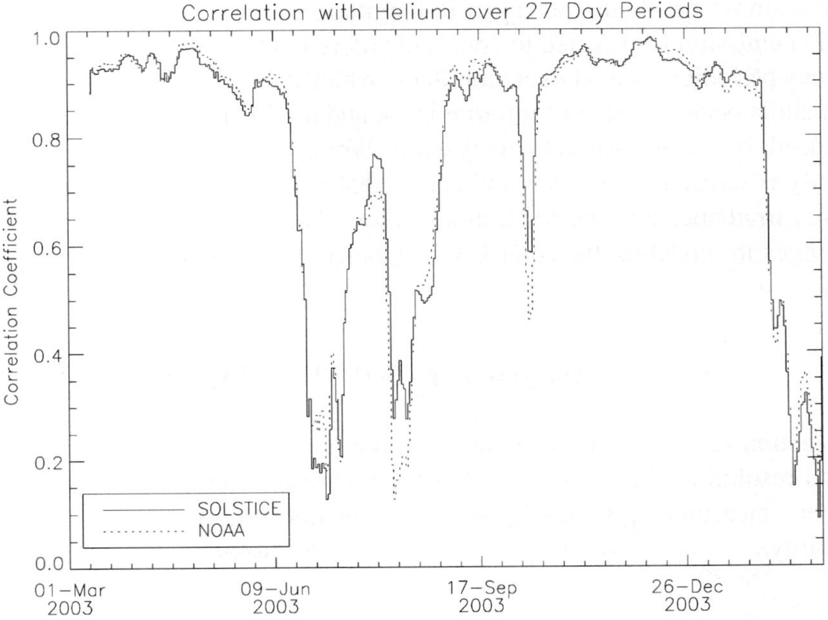

Figure 10. Relationship between the SOLSTICE Mg II index daily average 30.5 nm daily average for 27-day intervals. The *solid line* is the correlation with the SOLSTICE full-resolution index and the *dashed line* is the equivalent correlation for the NOAA index. The correlation is poor when the 13.5-day periodicity dominates the time series, which causes a phase shift between the 30.5 nm irradiance and the Mg II index.

Why is the correlation between the 30.5 nm irradiance and the Mg II index (whether NOAA or SOLSTICE) usually very high and yet very low or even negative at other times? Woods *et al.* (2005) showed that phase shifts between EUV irradiance and the Mg II index can be several days, both lagging and leading in phase, when 13.5-day periodicity is dominant in the time series. Phase shifts can occur between the coronal emissions and the non-coronal emissions because the coronal emissions have strong limb brightening and thus exhibit peaks when the active regions are near the limb. Other UV emissions have weak limb brightening and exhibit stronger peaks when active regions are near disk center. The 30.5 nm irradiance is a blend of He II 303.4 Å emission from the transition region and coronal Si XI 303.3 Å emission. While the He II emission is expected to be in phase most of the time with the Mg II index since it has similar center-to-limb variation, the Si XI emission is expected to be 2–7 days out of phase with the Mg II index since it is strongly limb-brightened. In other words, an active region on the limb will be bright in Si XI, but will then become dimmer when it reaches disk center. That same region will be dim in Mg II when it is on the limb, and then peak when it is at disk center a few days later.

Furthermore, Woods *et al.* (2005) showed that the coronal emissions have enhanced intensity during intervals when the 13.5-day periodicity dominates the time series. The 13.5-day periodicity arises when two active regions (or two groups of active regions) are about 180° apart in longitude on the Sun. Therefore, the coronal Si XI emission is expected to contribute more to the blend at 30.5 nm when the 13.5-day periodicity has a larger amplitude, which in turn will cause possibly larger phase shifts between the 30.5 nm irradiance and the Mg II index.

Indeed, based on SEE data analyzed in Woods *et al.* (2005), the 13.5-day periodicity is larger in mid-2003 and early 2004 when the correlation between the 30.5 nm irradiance and the Mg II index is low. This result illustrates some of the challenges in modeling the solar UV irradiance using another UV emission as a proxy.

5. Error Analysis of the SOLSTICE Mg II Index

This section discusses the uncertainty in the Mg II index measurement for both the full-resolution algorithm as well as the low-resolution method. Both the wing and core measurements include measurement uncertainty as well as real solar variability.

5.1. Uncertainty in the SOLSTICE wing measurement

A typical number of counts per grating dwell point in the wing is about 20 000, so the fractional uncertainty due to counting statistics alone for each point is about 0.7%. The SOLSTICE full-resolution spectrum is then smoothed with a 34-point triangular filter that approximates the SBUV instrument response. The general formula for the uncertainty reduction by a triangular smooth over N points is

$$\sigma_{\Delta}^2 = \frac{\frac{2}{6}N(N+1)(2N+1) - N^2}{(N(N+1)-N)^2}\sigma_{\text{point}}^2, \tag{7}$$

where σ_{point} is the uncertainty of each point in the smoothing operation. It is assumed that the uncertainty of each point is about the same and that errors are random and uncorrelated. The derivation of this formula is presented in the appendix. Evaluating this formula for $N = 34$ produces a reduction in the uncertainty of each wing value by a factor of $S_{34} = 0.14$ to 0.1%.

The denominator of the Mg II index is the mean of four such quantities, but they are not all independent. The two values on the blue wing are from overlapping observations, as are the two red wing values. In the case of a 0.2 nm shift at .03 nm sampling, only 226 of the 1156 weighted samples do not overlap with the adjacent measurement. Therefore, taking two measurements in the wing only increases the sample size by a factor of 0.1955 on each side. The mean of the two measurements decreases the uncertainty by only $1/\sqrt{1.391}$. The benefit of taking the mean of another pair of measurements in the other wing is a decrease in the uncertainty by $1/\sqrt{2.782}$. This quantity will be referred to as M_4 in the analysis below. The final uncertainty in the denominator of the SOLSTICE Mg II index is therefore 0.059%.

Figure 11 shows the mean of the four-wing measurements as a function of time (corrected to 1 AU). Any long-term trends or short-term variations in instrument responsivity, which vary linearly with wavelength in this region, will appear in both the numerator and denominator of the Mg II index and will therefore cancel out. In order to highlight the reproducibility of the SOLSTICE wing irradiance measurement, the observations shown in Figure 11 have been normalized to the daily mean.

Figure 12 shows the histogram distribution of the ratios of the individual measurements of the mean wing irradiance to the average wing irradiance for the day. If these deviations are due to random error, they can be modeled with a Gaussian distribution with a FWHM of 0.055%. This compares very favorably with the theoretical estimate of the uncertainty in the wing measurement of 0.059% due to counting statistics and smoothing algorithm. The theoretical estimate assumes that all points in the measurement had the same count rate, which is clearly not the case (cf. Figure 1). Therefore, there appear to be no significant sources of systematic error in the SOLSTICE measurement.

5.2. Uncertainty in the Low-Resolution SOLSTICE Mg II Index Measurement

The count rate in the emission core for the SOLSTICE spectrum convolved with the 1.1 nm bandpass is about a factor of 4 lower than in the wings, so the uncertainty in

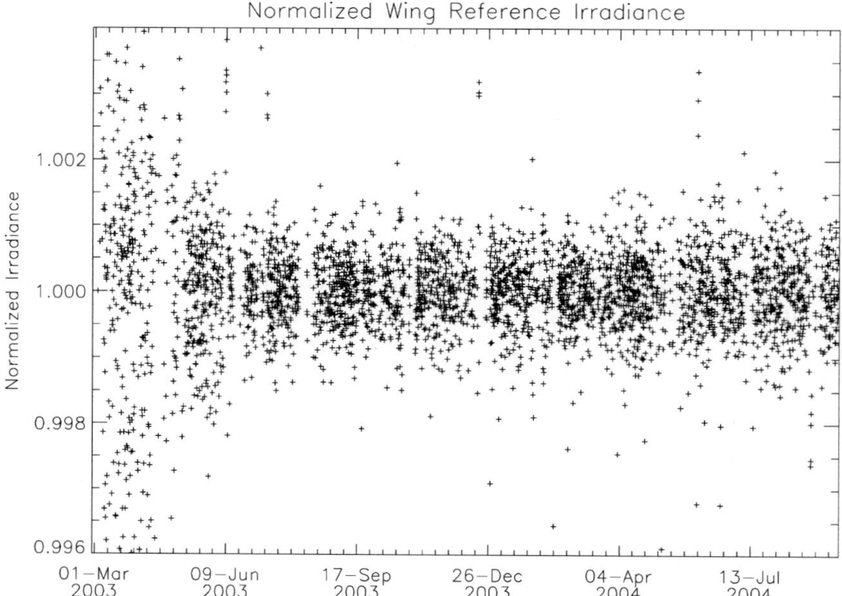

Figure 11. Normalized mean wing irradiance for Mg II index values shown in Figure 4. Each SOLSTICE scan is convolved with the SBUV instrument profile (1.1 nm) and then the mean of the four-wing irradiances is calculated. The irradiances in this plot have been normalized to the daily mean value. A plot of the distribution is shown in Figure 12.

each point is from counting statistics alone is 1.4%. The same triangular-smoothing factor, S_{34}, applies to the core measurement, so the uncertainty in each of the three points is 0.198%.

As for the case of the wings, the three core measurements are not all independent. The quantity M_3 can be defined similarly to M_4 as the decrease in uncertainty from making three partially independent measurements rather than just one. The red and blue sides of the triple-measurement contribute 0.1955 each as in the wing above. The part of the central observation that does not overlap with the other two adds only $25/1156 = 0.0216$ to the total for a final reduction in the core uncertainty by $M_3 = 1/\sqrt{1.412} = 0.841$. The total uncertainty in the numerator of the low-resolution SOLSTICE Mg II index is therefore 0.167%.

Equation (8) gives the total uncertainty in the SORCE SOLSTICE low-resolution Mg II index measurement. C_c and C_w are the counts recorded by the detector in the core and wing, respectively, S_{34} is the reduction in uncertainty by a triangular smoothing over 34 points given by Equation (7):

$$\sigma_{\text{LOW}}^2 = \left(\frac{\sqrt{C_c}}{C_c S_{34} M_3}\right)^2 + \left(\frac{\sqrt{C_w}}{C_w S_{34} M_4}\right)^2. \tag{8}$$

Therefore, the intrinsic uncertainty of the calculated low-resolution Mg II index measurement is 0.175% assuming only photon counting statistics. In order to

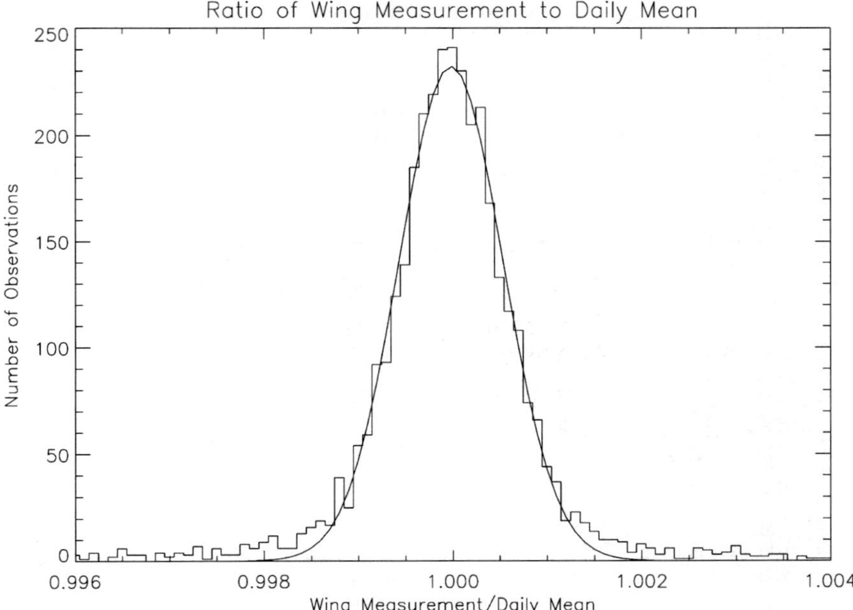

Figure 12. Distribution of wing values relative to their daily average. The FWHM of this distribution is 0.055%, which shows that there is no significant source of systematic errors in the SOLSTICE measurement.

compare this value to the full-resolution value, it must be scaled by the relation shown in Figure 5 (Equation (3)). The amplitude of the low-resolution index is smaller by about a factor of 7, i.e., the full-resolution index varies from 0.08 to 0.06 (33%), while the transformed low-resolution index varies from 0.288 to 0.275 (4.7%). So the 0.175% uncertainty in the low-resolution index corresponds to a 1.23% uncertainty in the index scaled to the full-resolution index. The FWHM variation in the scaled low-resolution Mg II index from scan to scan on any given day is indeed about 1.2% as shown in the bottom panel of Figure 13. Therefore, the low-resolution SOLSTICE Mg II index intra-day variation is primarily due to photon counting statistics and not from sources of systematic errors.

5.3. Uncertainty in the full-resolution SOLSTICE Mg II index

The uncertainty of the full-resolution SOLSTICE Mg II index is

$$\sigma_{\mathrm{FULL}}^2 = \frac{1}{M_2 \mathrm{df}}\left[\left(\frac{\sigma_{A_2}}{A_2}\right)^2 + \left(\frac{\sigma_{A_0}}{A_0 + A_3}\right)^2 + \left(\frac{\sigma_{A_3}}{A_0 + A_3}\right)^2\right] + \\ + \left(\frac{\sqrt{C_{\mathrm{w}}}}{C_{\mathrm{w}} S_{34} M_4}\right)^2, \tag{9}$$

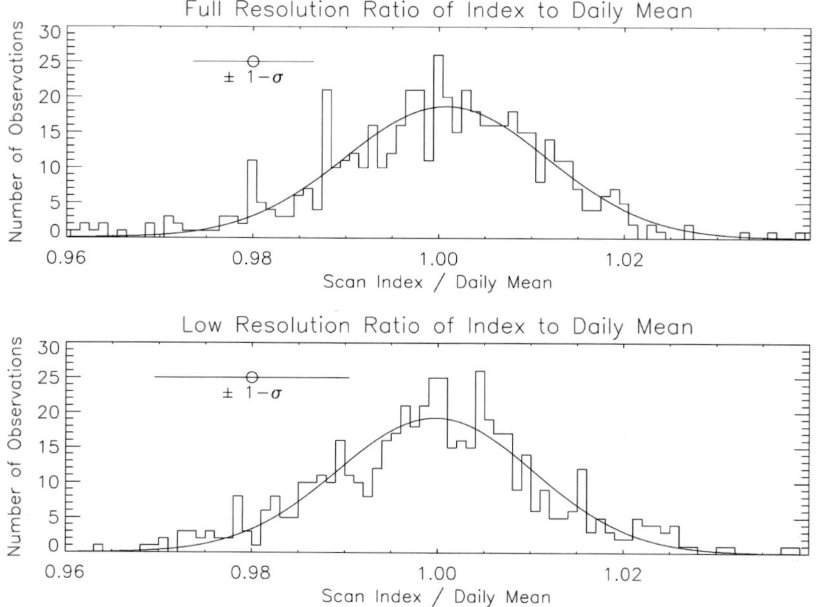

Figure 13. Ratio of Mg II index measurement to daily average for the full-resolution method (*top*) and the spectrum convolved to 1.1 nm resolution method (*bottom*). The error bar in the upper left corner of each panel shows the calculated measurement uncertainty for each method.

where M_2 is the reduction in uncertainty from taking the mean of the two emission lines, A_2 the width of the emission line, A_0 the height of the Gaussian, and A_3 the baseline. df is the number of degrees of freedom, i.e., the number of measurements used to fit the Gaussian minus the number of parameters in the fit. This number is typically 12–20 depending on the number of co-added spectra during a scan. Unlike the case of the low-resolution core, these two core measurements are indeed independent, so $M_2 = 1/\sqrt{2}$.

Figure 13 shows the ratio of the index from each individual measurement to the average for the day. The top panel is the result for the full-resolution index, while the bottom panel shows the result after convolving the SOLSTICE spectrum with the SBUV instrument profile. The low-resolution ratio has already been scaled to the full-resolution index using the transformation relationship shown in Figure 5 (Equation (3)) for ease of comparison.

The total uncertainty in the full-resolution measurement is 0.65% which, unlike the case for the low-resolution measurement, is significantly *less* than the typical fluctuation of the index during the day. Therefore the distribution of index measurements shown in the top panel of Figure 13 is primarily due to real solar variability. The low-resolution index has a measurement uncertainty that is the same size as this variation, and consequently does not measure it uniquely. In other words, the higher resolution measurement of the Mg II index provides a more precise measurement of solar variability.

Monte Carlo simulations of small changes in the emission cores confirm these formal error estimates. Taking a single SOLSTICE spectral scan as the baseline and adding noise appropriate for counting statistics, the measurement errors for the wing, low-resolution index, and full-resolution index are $\sigma_{\mathrm{WING}} = 0.068 \pm 0.0057\%$, $\sigma_{\mathrm{LOW}} = 0.182 \pm 0.014\%$, and $\sigma_{\mathrm{FULL}} = 0.773 \pm 0.060\%$. These values and their uncertainty estimates come from 100 trials of 500 simulated spectra. To simulate solar variability, the emission cores are increased by 1% for half of the spectra in each trial. The resulting distribution widths become $\sigma_{\mathrm{LOW}} = 0.217 \pm 0.015\%$ and $\sigma_{\mathrm{FULL}} = 1.07 \pm 0.068\%$. The 1% change in the underlying emission shows up in the full-resolution index as a change in the distribution width by 5 standard deviations. In the low-resolution index, the change is only 2.5 standard deviations.

For a 0.5% change in the emission cores, the width distribution of the low-resolution index became $\sigma_{\mathrm{LOW}} = 0.189 \pm 0.017$, and the full-resolution distribution width became $\sigma_{\mathrm{FULL}} = 0.834 \pm 0.068$. In other words, the change in the underlying spectrum resulted in a change in the low-resolution distribution of only 0.3 standard deviations, but became a nearly 1 standard deviation change in the full-resolution distribution.

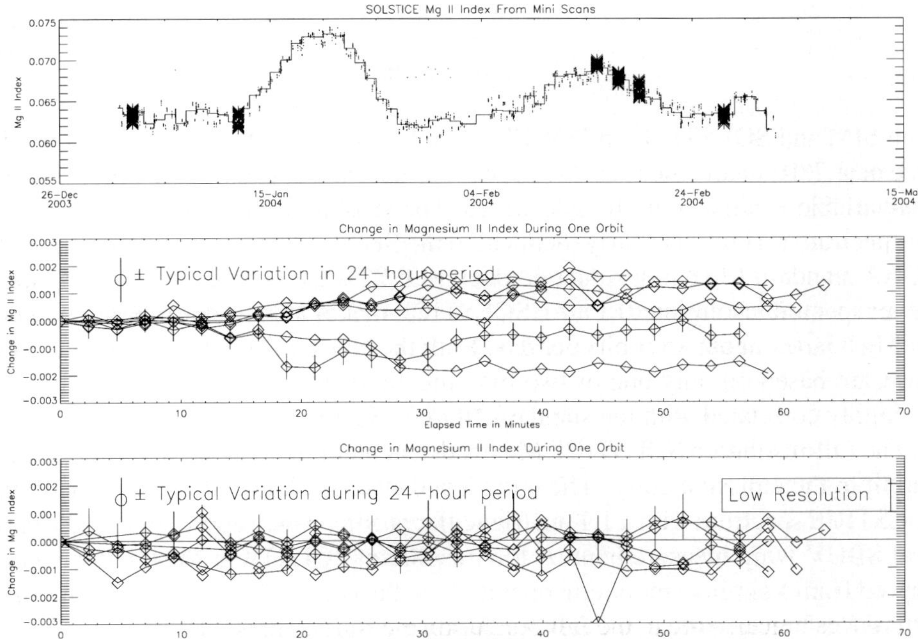

Figure 14. Changes in the Mg II index on timescales of minutes. The *top panel* indicates the dates of the miniscan orbits plotted in the lower panels. The *middle panel* and *bottom panel* show the change in index relative to the first observation of the orbit. The typical FWHM of daily variation from Figure 13 is shown at *upper left*. Measurement uncertainty error bars are given for just one orbit for clarity (same orbit in both panels). The short-term variation for the low-resolution method (*bottom panel*) is within the measurement uncertainty. The full-resolution method (*middle panel*) resolves the short timescale variations.

6. Short-Term Solar Variability

How quickly does the Mg II index change in a measurable way? SORCE SOLSTICE frequently operates in a mode where it makes repeated measurements of just the region near 280 nm. One entire orbit on alternate days is devoted to these "mini-scans." Stepping through the mini-scan command sequence takes about 2.5 min. Figure 14 shows the change in the Mg II index during a few such orbits. The measurement error bars are shown on just one of the curves for clarity. A $\pm 1\%$ error bar is also shown in the upper left as a reminder of the typical 1-σ variation in a 24-h period determined from the full-mission of normal scans shown in Figure 13.

On timescales of an hour, the Mg II index varies by about 1%, which is similar to the amount of variation seen over a day. The low-resolution method of calculating the index is inherently less precise, and the short-term changes are hard to distinguish from measurement error. The uncertainty of the full-resolution index measurement is lower, so the minute-to-minute differences in the index can be reliably detected. The Mg II index can also change by $\sim$10% during large flares as observed for the X17 flare on 28 October 2003 (Woods *et al.*, 2004).

7. Conclusions

Both SIM and SOLSTICE on SORCE are able to observe the solar spectral irradiance near 280 nm and measure the strength of the Mg II h and k cores relative to the less-variable irradiance in the line wings. The resolution of SIM in this region of the spectrum (1.1 nm) is nearly identical to the SBUV instrument used to create the NOAA standard Mg II index data product. SOLSTICE observes with significantly higher spectral resolution (0.1 nm). SOLSTICE makes at least one measurement of the Mg II index about 15 orbits per day. Both the NOAA data product and the SIM index are based on only one or two measurements on a typical day. Both indices are highly correlated with the standard NOAA Mg II index.

The full-resolution SOLSTICE Mg II index numerator is calculated from a Gaussian fit to the emission cores. The wing irradiance is calculated by convolving the SOLSTICE spectrum with a 1.1 nm triangular profile, which approximates the classical SBUV wing measurement. A low-resolution SOLSTICE Mg II index can be derived from a similar smoothing operation on the cores. There are some important differences between the two index formulations, most notably the precision.

The precision of the full-resolution index measurement is 0.6%. The low-resolution measurement has a precision of 0.175% on its native scale, but that is relative to an amplitude that is a factor of 7 smaller than the full-resolution scale, so the scaled uncertainty of the low-resolution index is equivalent to an uncertainty of 1.23%, roughly twice as large as the uncertainty of the full-resolution index.

The variation of the Mg II index relative to the daily mean is well fit by a Gaussian with a FWHM of 1%. So the 1% precision of the low-resolution index is not quite adequate to measure the Mg II index on timescales of a day or less. The full-resolution index, on the other hand, has a precision of less than a percent, so the observed 1% daily spread is likely due to real solar variability.

Higher time cadence observations of the Mg II index (every 2.5 min) show a statistically significant change in the index over the course of an hour using the full-resolution spectrum. The low-resolution index measurement has a lower precision and cannot resolve these minute-to-minute changes. Understanding the source of this short-timescale solar irradiance variability will require detailed model analysis of the solar chromosphere.

Correlation of the Mg II index with solar EUV irradiance in the 30–31 nm band is very high when the EUV timeseries is dominated by 27-day periodicity. The correlation decreases when 13.5-day EUV periodicity is significant. The 13.5-day periodicity is significant when there are active regions separated by $\sim 180°$ solar longitude. The correlation decreases because of the very different limb brightening characteristics of chromospheric and coronal emissions.

Acknowledgements

This research is supported by NASA contract NAS5-97045 at the University of Colorado. The authors would like to thank the referee, Dr Rodney Viereck, for his insightful comments that greatly improved this manuscript. The authors would also like to thank Craig Markwardt (U. Wisconsin) for his MPFIT software that was used throughout this analysis.

Appendix: Derivation of Uncertainty for Triangular Smoothing

The derivation of Equation (7) uses the following results.

The sum of the first N integers, S_N is given by

$$S_N = 1 + 2 + 3 + \cdots + N = \frac{1}{2} N(N + 1), \tag{10}$$

and the sum of the squares of the first N integers, S_{N^2},

$$S_{N^2} = 1 + 4 + 9 + \cdots + N^2 = \frac{1}{6} N(N + 1)(2N + 1). \tag{11}$$

The triangular smoothing method used in this analysis is that a second boxcar filter is applied to the result of an initial boxcar smooth of the same size. For a filter width of N, the result is

$$\begin{aligned} \Delta &= \frac{p_1 + 2p_2 + \cdots + Np_N + \cdots + 2p_{2N-2} + p_{2N-1}}{1 + 2 + \cdots + N + \cdots + 2 + 1} \\ &= \frac{p_1 + 2p_2 + \cdots + Np_N + \cdots + 2p_{2N-2} + p_{2N-1}}{2S_N - N}, \end{aligned}$$

where the p_i's are the independent measurements to be smoothed.

The uncertainty in this quantity is given by

$$\sigma_\Delta^2 = \frac{\sigma_1^2 + 4\sigma_2^2 + \cdots + N^2\sigma_N^2 + \cdots + 4\sigma_{2N-2}^2 + \sigma_{2N-1}^2}{(2S_N - N)^2}. \tag{12}$$

If we assume that all the σ_i's are the same, then this reduces to

$$\sigma_\Delta^2 = \frac{2S_{N^2} - N^2}{(2S_N - N)^2}\sigma_i^2 \tag{13}$$

$$= \frac{\frac{2}{6}N(N+1)(2N+1) - N^2}{(N(N+1) - N)^2}\sigma_i^2. \tag{14}$$

For $N = 34$, the uncertainty of the smoothed point is as follows:

$$\sigma_\Delta^2 = 0.0196\sigma_i^2 \tag{15}$$

and

$$\sigma_\Delta = 0.14\sigma_i. \tag{16}$$

References

De Toma, G., White, O. R., Knapp, B. G., Rottman, G. J., and Woods, T. N.: 1997, *J. Geophys. Res.* **102**, 2597.

Harder, J., Lawrence, G., Fontenla, J., Rottman, G., and Woods, T.: 2005, *Solar Phys.*, this volume.

Heath, D. F. and Schlesinger, B. M.: 1986, *J. Geophys. Res.* **91**, 8672.

Lean, J. L., Rottman, G., Kyle, H. L., Woods, T. N., Hickey, J. R., and Puga, L. C.: 1997, *J. Geophys. Res.* **102**, 29,939.

McClintock, W., Rottman, G., and Woods, T.: 2005, *Solar Phys.*, this volume.

McClintock, W., Snow, M., and Woods, T.: 2005, *Solar Phys.*, this volume.

Rottman, G. J., Woods, T. N., and Sparn, T. P.: 1993, *J. Geophys. Res.* **98**, 10,667.

Tobiska, W. K.: 1991, *J. Atmos. Terr. Phys.* **53**, 1005.

Viereck, R. and Puga, L.: 1999, *J. Geophys. Res.* **104**, 9995.

Viereck, R., Puga, L., McMullin, D., Judge, D., Weber, M., and Tobiska, W. K.: 2001, *Geophys. Res. Lett.* **28**, 1343.

Viereck, R. A., Floyd, L. E., Crane, P. C., Woods, T. N., Knapp, B. G., Rottman, G., Weber, M., Puga, L. C., and DeLand, M. T.: 2004, *Space Weather*, **2**, doi: 10.1029/2004SW000084, S10005.

White, O. R., de Toma, G., Rottman, G. J., Woods, T. N., and Knapp, B. G.: 1998, *Solar Phys.* **177**, 89.

Woodraska, D. L., Woods, T. N., and Eparvier, F. G.: 2004, *Proc. SPIE Int. Soc. Opt. Eng.* **5660**, 36.

Woods, T. N., Eparvier, F. G., Bailey, S. M., Chamberlin, P. C., Lean, J., Rottman, G. J., Solomon, S. C., Tobiska, W. K., and Woodraska, D. L.: 2004a, *J. Geophys. Res.* **110**, A01312.

Woods, T. N., Eparvier, F. G., Fontenla, J., Harder, J., Kopp, G., McClintock, W. E., Rottman, G., Smiley, B., and Snow, M.: 2004b, *Geophys. Res. Lett.* **31**, 801.

Solar Physics (2005) 230: 345–374

XUV PHOTOMETER SYSTEM (XPS): OVERVIEW AND CALIBRATIONS

THOMAS N. WOODS[1], GARY ROTTMAN[1] and ROBERT VEST[2]
[1]*Laboratory for Atmospheric and Space Physics, University of Colorado, Boulder, CO, U.S.A.*
(e-mails: woods@lasp.colorado.edu, rottman@lasp.colorado.edu)
[2]*National Institute of Standards and Technology, Gaithersburg, MD, U.S.A.*
(e-mail: robert.vest@nist.gov)

(Received 9 December 2004; accepted 16 March 2005)

Abstract. The solar soft X-ray (XUV) radiation is highly variable on both short-term time scales of minutes to hours due to flares and long-term time scales of months to years due to solar cycle variations. Because of the smaller X-ray cross sections, the solar XUV radiation penetrates deeper than the extreme ultraviolet (EUV) wavelengths and thus influences the photochemistry and ionization in the mesosphere and lower thermosphere. The XUV Photometer System (XPS) aboard the Solar Radiation and Climate Experiment (SORCE) is a set of photometers to measure the solar XUV irradiance shortward of 34 nm and the bright hydrogen emission at 121.6 nm. Each photometer has a spectral bandpass of about 7 nm, and the XPS measurements have an accuracy of about 20%. The XPS pre-flight calibrations include electronics gain and linearity calibrations in the laboratory over its operating temperature range, field of view relative maps, and responsivity calibrations using the Synchrotron Ultraviolet Radiation Facility (SURF) at the National Institute of Standards and Technology (NIST). The XPS in-flight calibrations include redundant channels used weekly and underflight rocket measurements from the NASA Thermosphere-Ionosphere-Mesosphere-Energetics-Dynamics (TIMED) program. The SORCE XPS measurements have been validated with the TIMED XPS measurements. The comparisons to solar EUV models indicate differences by as much as a factor of 4 for some of the models, thus SORCE XPS measurements could be used to improve these models.

1. Introduction

Solar ultraviolet (UV) radiation at wavelengths less than 320 nm is an important source of energy for atmospheric processes. Solar UV photons are absorbed in Earth's atmosphere via photodissociation of molecules, photoionization of molecules and atoms, and photoexcitation including resonance scattering (e.g., see Chamberlain, 1978). Nominal subdivisions of the UV spectral range are: near ultraviolet (NUV) from 300 to 400 nm, middle ultraviolet (MUV) from 200 to 300 nm, vacuum ultraviolet (VUV) for wavelengths shortward of 200 nm, far ultraviolet (FUV) from 120 to 200 nm, extreme ultraviolet (EUV) from 30 to 120 nm, X-ray ultraviolet (XUV) from 1 and 30 nm, and X-rays at wavelengths less than 1 nm. Solar EUV and XUV radiation photoionizes the neutral constituents of the atmospheres and participates in the formation of the ionosphere. The photoelectrons created in this process interact further with the neutrals, leading to excitation, dissociation, and additional ionization. The excess energy from the absorption processes heats

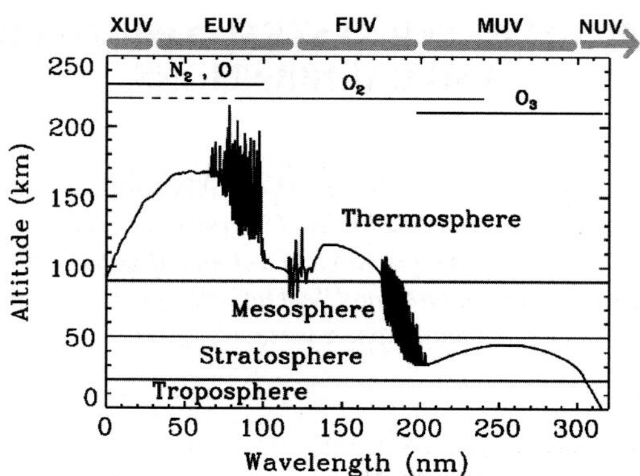

Figure 1. Solar absorption in Earth's atmosphere. The unit optical depth for the solar radiation with a solar zenith angle of 90° is given for solar minimum conditions. The principal absorbers are O, O_2, O_3, and N_2. The definitions of the UV ranges are indicated.

the atmosphere. The unit optical depth for the solar radiation with a solar zenith angle of 90° is shown in Figure 1 for solar minimum conditions. The absorption height during solar maximum is higher because the additional solar heating from the increased solar UV irradiance levels causes the atmosphere to expand outward. Solar MUV radiation heats the stratosphere, and solar UV radiation shortward of 170 nm heats the thermosphere. Atmospheric absorption of the solar UV radiation also initiates many chemical cycles, such as those involving water vapor, ozone, nitric oxide, and chlorofluorocarbons in Earth's atmosphere (e.g., see Brasseur and Solomon, 1986). The shortest wavelengths in the XUV range penetrate deeper into the atmosphere than most of the EUV wavelengths and thus can directly influence the mesosphere and lower thermosphere. All of these solar driven atmospheric processes are wavelength dependent and are expected to be as variable as the intrinsic solar variability at the appropriate wavelengths.

Accurate measurements of the solar UV spectral irradiance, along with an understanding of its variability, are important for detailed studies of the atmospheric processes. The Sun varies on all time scales and the amount of variability is a strong function of wavelength. The amount of 11-year solar cycle variability in the MUV range increases from perhaps 0.1% near 300 nm to a few percent near 200 nm. The solar cycle variability continues to increase into the FUV and EUV ranges with the magnitude of the variation approaching a factor of 2, for example at the H Lyman-α emission at 121.6 nm. The XUV region is dominated by emission lines of primarily coronal origin that may vary by an order of magnitude during a solar cycle. Short term variations, lasting from minutes to hours, are related to eruptive phenomena on the Sun; intermediate term variations, modulated by the 27-day rotation period of the Sun, are related to the appearance and disappearance of active regions on the solar disk, and the more elusive long term variability is related to the 22-year magnetic field cycle of the Sun. The long-term variations in the XUV and EUV ranges are poorly determined due to the lack of measurements and to the inadequate long-term relative accuracy of previous satellite solar instruments. Recent reviews about the solar EUV and UV variability with more details include those by White (1977), Rottman (1987), Lean (1987, 1991), Tobiska (1993), Pap *et al.* (1994), and Woods *et al.* (2004).

Studies of the solar XUV radiation began in the 1950s with space-based rocket experiments, but the knowledge of the solar XUV irradiance, both in absolute magnitude and variability, has been questionable due largely to the very limited number of observations. With the launch of Solar and Heliospheric Observatory (SOHO) in 1995, Student Nitric Oxide Explorer (SNOE) in 1998, and Thermosphere-Ionosphere-Mesosphere-Energetics-Dynamics (TIMED) spacecraft in 2001, there is now a continuous data set of the solar XUV irradiance, and advances in the understanding of the solar XUV irradiance have begun. The XUV Photometer System (XPS) aboard the Solar Radiation and Climate Experiment (SORCE) spacecraft has evolved from earlier XUV photometer versions flown on SNOE and TIMED and will continue these earlier solar XUV irradiance measurements with improvements to accuracy, spectral coverage, and time cadence.

The SORCE XPS measures the solar spectral irradiance in the XUV range from 0.1 to 34 nm and also the bright H Lyman-α emission at 121.5 nm. The SORCE satellite was launched on 25 January 2003, and daily measurements of the solar irradiance by the SORCE instruments began routinely in March 2003. The SORCE XPS measurements have partial spectral overlap with the concurrent solar XUV irradiance measurements from the Solar EUV Experiment (SEE) aboard the TIMED satellite (Woods *et al.*, 1998) and the Solar EUV Monitor (SEM) aboard the SOHO spacecraft (Judge *et al.*, 1998). A summary of the SORCE XPS instrument, including calibrations and irradiance algorithms, are the focus of this paper. A companion paper (Woods and Rottman, 2005) discusses the solar XUV irradiance results from the SORCE XPS measurements during the first 18 months of the SORCE mission.

2. XPS Instrument Overview

The XPS is a set of filter photometers to measure the solar irradiance from 0.1 to 34 nm and at 121–122 nm with each filter having a spectral bandpass of about 7 nm. The SORCE XPS is almost identical to the XPS that is part of the TIMED SEE instrument. A more detailed description of the SEE instrument is given by Woods *et al.* (1998), and some pre-flight calibration results for the TIMED SEE XPS are given by Woods *et al.* (1999a). This section describes the SORCE XPS instrument.

The XPS includes twelve Si photodiodes of which eight photodiodes have thin-film XUV filters deposited directly on them, one photodiode has a 121 nm interference filter in front of it, and three photodiodes are bare Si photodiodes for tracking visible light transmission of the XPS fused silica filters. The thin-film XUV filters

TABLE I
SORCE XPS photometer parameters.

XP no.	Filter coating	Bandpass (nm)	Design thickness (nm)	Modeled thickness (nm)	Gain[a] factor (pA/Hz)
1	Ti/C	0.1–7	500/50	387.5/50	0.0633
2	Ti/C	0.1–7	500/50	387.5/50	0.0635
3	Al/Sc/C	17–23	270/50/50	179.1/50/25	0.0630
4	None	160–1000	–	–	0.470
5	Al/Nb/C	17–21	250/50/50	208.9/39.2/47.3	0.0635
6	Ti/Mo/Au	0.1–11	40/200/100	45.2/111.3/74.1	0.0629
7	Ti/Mo/Si/C	0.1–7, 7–18	40/200/100/50	34.1/131.3/103.5/46.1	0.0637
8	None	160–1000	–	–	0.477
9	Al/Cr	0–7, 27–37	270/100	175/114.4	0.0628
10	Al/Mn	0–7, 25–34	270/100	175/144.7	0.0642
11	Acton Ly-α (x2)	121–122	–	–	0.0631
12	None	160–1000	–	–	0.476

[a]Gain factor is for 15 °C. There is ~200 ppm relative change per °C thermal effect on the gain.

are deposited directly on the photodiode to avoid using metal foil filters which are more difficult to handle, prone to develop pin holes, and degrade with time. R. Korde of International Radiation Detectors, Inc. (IRD)[1] developed the XUV photodiodes with thin-film filters to have low-noise and good long-term stability (Korde and Geist, 1987; Korde, Canfield, and Wallis, 1988; Canfield, Kerner and Korde, 1989; Korde and Canfield, 1989; Canfield *et al.*, 1994). The SORCE XPS photodiodes are the IRD AXUV-100 devices that have an active area of 100 mm^2, but the XPS apertures limit the illumination on the photodiodes to about 10 mm^2. The National Institute for Standards and Technology (NIST) employs these Si photodiodes as XUV standard detectors.

Several materials are suitable for use as XUV filters, and multiple coatings on the same diode provide a way to narrow the bandpass of each diode. Powell *et al.* (1990) discuss thin-film filters suitable for this wavelength range. In addition to selecting the spectral bandpass, the filter must also block solar visible radiation by a factor of 10^{10} or better; otherwise, solar visible radiation (instead of the XUV radiation) dominates the XUV photometer signal. The thin-film filters used on XPS photodiodes are Ti/C, Ti/Mo/Si/C, Ti/Mo/Au, Al/Sc/C, Al/Nb/C, Al/Cr, and Al/Nb. The nominal thicknesses of these filters are listed in Table I, and the responsivities

[1]Certain commercial equipment, materials, or software are identified in this paper to foster understanding. Such identification does not imply recommendation or endorsement by the Laboratory for Atmospheric and Space Physics (LASP) and National Institute of Standards and Technology (NIST), nor does it imply that the equipment, materials, or software identified are necessarily the best available for the purpose.

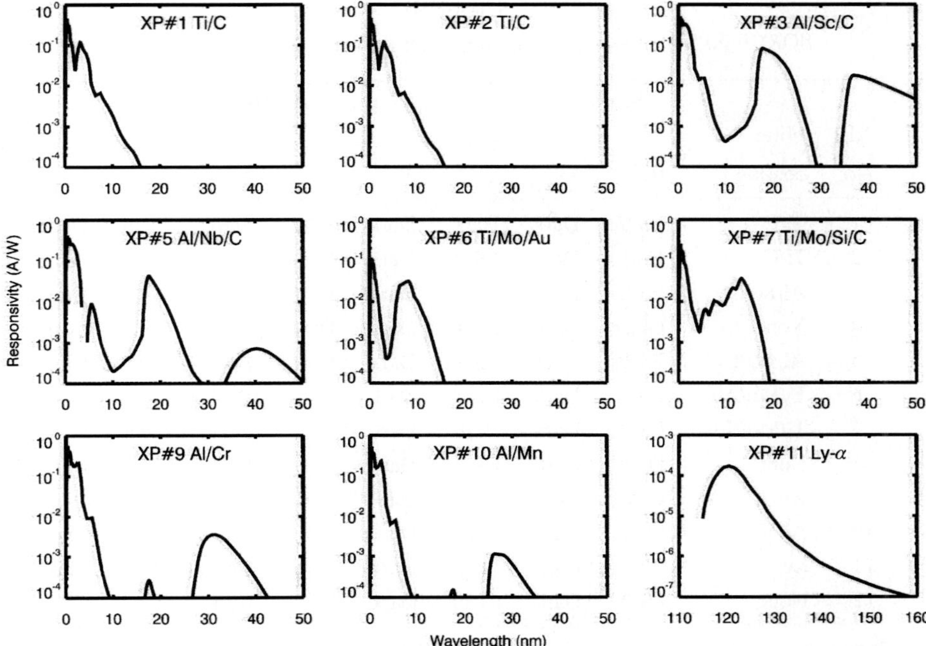

Figure 2. XPS photometer responsivities. The modeled responsivities are shown for each filtered photometer in XPS. The XP Nos. 4, 8, and 12 only have bare Si photodiodes and thus do not have filters.

for each filtered photodiode are shown in Figure 2. In addition, one bare Si photodiode is used with an Acton Lyman-α filter to provide a measurement of the important solar Lyman-α emission.

The XPS 12 photometers are packaged together with a common filter wheel mechanism as shown in Figure 3. The filter wheel is an eight-position filter mechanism using a Geneva gear system, optical encoders for position sensing, and a brushless DC motor. The filter wheel mechanism permits solar XUV measurements with a clear aperture, dark measurements with blocked apertures, and visible background measurements with fused silica windows. The 12 photometers consist of three groups of four photodiodes each on a common radius for the filter mechanism and sharing two fused silica windows within the group. Nine photodiodes have filters for making solar UV irradiance measurements, and three photodiodes are bare Si photodiodes to track the transmission of the fused silica windows. The XUV Photometers (XP) Nos. 1–4 are grouped in the inner ring. XP Nos. 5–8 are grouped in the middle ring. XP Nos. 9–12 are grouped in the outer ring. XP Nos. 1, 7, and 10 are redundant photometers as part of the in-flight calibration plan. XP Nos. 4, 8, and 12 are the bare photodiodes for measuring the fused silica window transmission. The configurations of the 12 photometers at the 8 filter wheel positions are listed in Table II.

TABLE II

SORCE XPS photometer configuration for the eight filter wheel positions.

XP No.	Filter coating	Filter wheel position 0	1	2	3	4	5	6	7
1	Ti/C	FS-1	Dark	Dark	Dark	FS-2	Dark	Clear	Dark
2	Ti/C	Dark	Dark	FS-2	Dark	Clear	Dark	FS-1	Dark
3	Al/Sc/C	FS-2	Dark	Clear	Dark	FS-1	Dark	Dark	Dark
4	None	Clear	Dark	FS-1	Dark	Dark	Dark	FS-2	Dark
5	Al/Nb/C	FS-2	Dark	Clear	Dark	FS-1	Dark	Dark	Dark
6	Ti/Mo/Au	Clear	Dark	FS-1	Dark	Dark	Dark	FS-2	Dark
7	Ti/Mo/Si/C	FS-1	Dark	Dark	Dark	FS-2	Dark	Clear	Dark
8	None	Dark	Dark	FS-2	Dark	Clear	Dark	FS-1	Dark
9	Al/Cr	Clear	Dark	FS-1	Dark	Dark	Dark	FS-2	Dark
10	Al/Mn	FS-1	Dark	Dark	Dark	FS-2	Dark	Clear	Dark
11	Ly-α	Dark	Dark	FS-2	Dark	Clear	Dark	FS-1	Dark
12	None	FS-2	Dark	Clear	Dark	FS-1	Dark	Dark	Dark

Photometers 1–4, 5–8, and 9–12 are grouped into common circle and share the same apertures and fused silica (FS) windows. Filter position 6 is for calibrations (redundant channels).

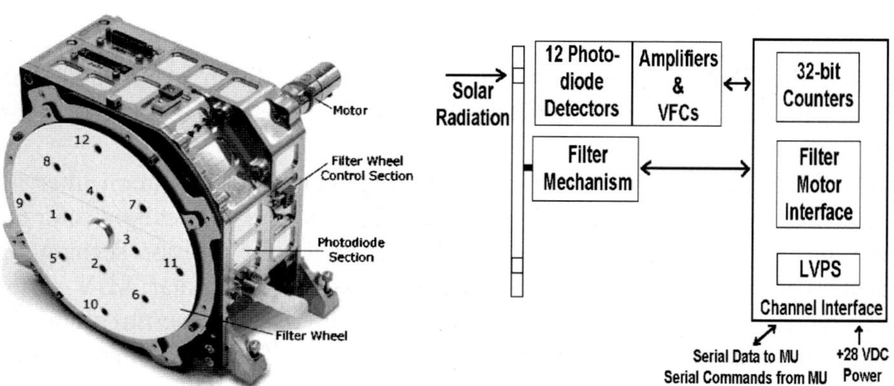

Figure 3. SORCE XPS instrument and block diagram. The XPS is a set of 12 photometers coupled with a common filter mechanism. The XPS Channel Interface electronics are not shown.

A typical measurement cycle for each XUV photometer measures the dark signal with a blocked aperture, measures the background signal with the window, and then measures the solar XUV radiation with a clear aperture. This measurement cycle takes about 5 min and is repeated throughout the day-lit side of the SORCE orbit, yielding about 70% duty cycle for the solar observations.

The electronics for each XPS photodiode are simple and include only a current amplifier and a voltage-to-frequency converter (VFC). The pulses from the VFC are fed into a 32-bit counter in the XPS channel interface electronics and the counts over its integration period, typically being 10 s, are buffered and sent to the SORCE instrument Microprocessor Unit (MU). The XPS channel interface electronics also contains filter wheel control electronics and low voltage power supplies (LVPS) that convert the unregulated 28 VDC input to regulated 5 VDC and ±15 VDC.

3. Pre-Flight Calibrations

The pre-flight photometric calibrations of XPS include transferring to the instrument the calibrations of the National Institute of Standards and Technology (NIST) radiometric standards, such as reference photodiodes, radioactive X-ray sources, and synchrotron radiation (Walker *et al.*, 1988; Canfield and Swanson, 1987; Parr and Ebner, 1987). The current XUV calibration techniques are able to achieve an accuracy of 5–20% (1-σ value). The primary photometric standard for the XPS calibrations is the Synchrotron Ultraviolet Radiation Facility (SURF-III) at NIST in Gaithersburg, Maryland. The unit level calibrations of the Si photodiodes are performed as a function of wavelength using a monochromator and reference photodiode on SURF Beam Line 9 (BL-9), and the system level calibrations for the XPS instrument are performed directly viewing the synchrotron source on SURF BL-2. Some of the Si photodiodes, mostly those used for TIMED SEE, were also calibrated at the Physikalisch-Technische Bundesanstalt (PTB) electron storage ring called BESSY using a monochromator and reference photodiode (Scholze, Thornagel, and Ulm, 2001).

Besides the fundamental responsivity calibrations, pre-flight characterizations are carried out for several other important instrument parameters. The aperture areas are determined using a precision microscope to measure the dimensions of the round apertures for XPS. The response of the fully assembled instrument is mapped over its field of view (FOV) to precisely determine its uniformity. A gimbal table at LASP with $12° \times 12°$ range of motion is used to obtain the visible FOV maps for XPS. The gain factor (linearity) for the XPS detector electronics uses a calibrated, adjustable current source in place of the photodiode and calibrated over 6 orders of magnitude. The gain factors of the detector electronics are also determined over the operating temperature range (−40 to 45 °C). As needed for the calibration analyses, the detector dark signals are measured for all of these characterizations.

3.1. Selection Process for Flight Photodiodes

The first characterization for XPS is the selection of the flight photodiodes. Following the design of the filter by LASP and the manufacturing of the filtered photodiode

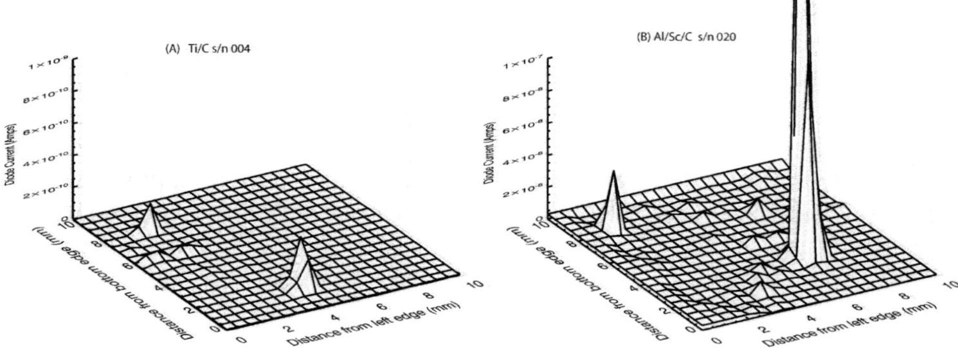

Figure 4. Example photodiode scan results. The *left scan map* is of a Ti/C coated photodiode without any pinholes, and the *right scan map* is of a Al/Sc/C coated photodiode with a pinhole in the filter. Note that there is two orders of magnitude difference in these two scales.

by IRD, each photodiode is scanned for the detection of filter pinholes with the photodiode mounted to a two-axis translation stage and using a red He–Ne laser as the light source. A couple of examples of the photodiode scans are shown in Figure 4. Then the photodiodes without any pinholes and with high shunt resistance (typically above 10 MΩ) are selected for masking. An Al mask is attached to the top of the photodiode using Ag epoxy to assure that light cannot directly scatter to the photodiode sides, which are sensitive to visible light. These photodiodes are rescanned after masking and those with the lowest visible light responsivity are selected as candidates for flight. Finally, the responsivity of these selected, masked photodiodes are calibrated at NIST before installing the best photodiodes in XPS.

3.2. Electronics Gain Calibrations

The gain of the photometer electronics is calibrated to obtain the conversion factor for the current input to the frequency output of the VFC. The gain calibrations are over the full dynamic range, 1 Hz–2 MHz, and over the operational temperature range of the photometers, −40 to 45 °C. These gain characterizations use a calibrated, adjustable current source in place of the photodiode, and the outputs from the electronics are counted by a calibrated frequency meter. The current source is adjusted over a range of more than 6 orders of magnitude to determine the current conversion factor and to determine the maximum current input before the VFC output exhibits non-linear behavior. An example of this gain calibration is shown in Figure 5, and the gain factors at 15 °C are listed in Table I for each photometer. The gain conversion factor has a 1-σ uncertainty of about 0.1%.

These gain calibrations are performed in a thermal oven at a constant temperature and are repeated at temperatures near −35, −15, 5, 25, and 40 °C. These results combined determine how much the gain factors change with temperature. The

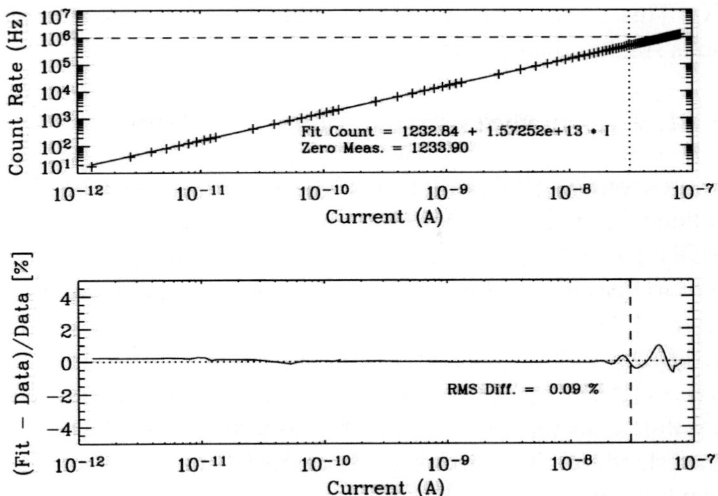

Figure 5. Example gain calibration. The gain calibration for XP No. 5 detector electronics is shown at 21 °C. The *top plot* shows the measurements, and the *bottom plot* shows the difference to the linear fit up to 30 nA. The VFC has an offset bias of about 1000 Hz so that measurements can be made even if the dark current drifts with temperature. The VFC maximum rate is designed for 1 MHz output (*dashed line* in top plot).

photometer electronics have a slight sensitivity to temperature (~200 ppm/°C) that arises primarily because of the thermal sensitivity of the feedback resistor and capacitor used with the current operational amplifier. The temperature is measured for both these laboratory calibrations and in-flight with a precision of about 0.1 °C. An example of the thermal sensitivity for the gain is shown in Figure 6, whereby the gain factor is fit as a quadratic function of temperature. The gain thermal parameters have a 1-σ uncertainty of about 0.2%.

Another small contribution to the electronics gain changes with temperature is the digital counters in the XPS Channel Interface electronics. The timing reference for all of the digital counters is a single crystal oscillator, which drifts by about

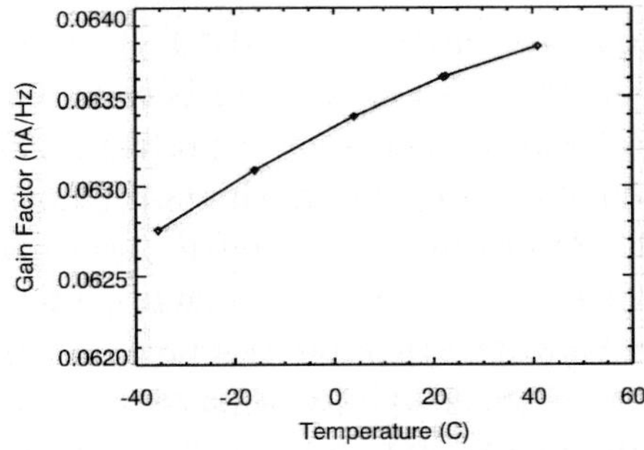

Figure 6. Example thermal sensitivity for the gain factor of XP No. 5 Detector Electronics.

5 ppm per °C. This effect over the XPS operational temperature range is about 500 ppm and is smaller than the uncertainty of the gain factors.

3.3. SURF BL-9 RESPONSIVITY CALIBRATIONS

The pre-flight responsivity calibrations for XPS in 2001 are unit level calibrations of the photodiodes using a monochromator and calibrated transfer standard photodiodes on SURF Beam Line 9 (BL-9). The EUV Detector Radiometry Beamline (BL-9) is used to disseminate the NIST scale of detector quantum efficiency. The photocurrent from the test detector is compared to the photocurrent from a working standard detector of known quantum efficiency. The calibration of the working standard is traceable to a cryogenic radiometer and an ionization chamber, both of which are absolute detectors that serve as the basis for the NIST EUV radiometric scale. Transfer standard photodiodes are calibrated with a relative uncertainty of 4–10%, depending on wavelength.

The beamline consists of a grazing-incidence, toroidal-grating monochromator that images the storage ring electron beam onto an exit slit. There are two gratings that can be interchanged: a 300 grooves per mm grating covering the spectral region from 17 to 50 nm and a 1200 grooves per mm grating covering the spectral region from 5 to 17 nm. The monochromator exit beam is fairly well collimated to eliminate angle-of-incidence effects. High diffraction orders are suppressed by the use of thin-film filters and the selection of appropriate SURF electron beam energies.

The working standard and test detectors are mounted on a rotatable platform and can be alternately placed in the monochromator exit beam. The photocurrent from each detector is measured in sequence. The comparison of photocurrents is made several times using a bracketing technique to correct for linear drifts in the incident beam intensity and to assure acceptable measurement statistics. The quantum efficiency of the test diode is calculated from the measured ratio of photocurrents and the known quantum efficiency of the working standard. A correction for out-of-band radiation is applied at some wavelengths.

3.4. SURF BL-2 RESPONSIVITY CALIBRATIONS

Some of the same photodiodes calibrated at SURF BL-9 in 2001 were also calibrated at PTB BESSY in 1998. The differences between the SURF and BESSY results are larger than a factor of 2 at some wavelengths for some photodiodes and this discrepancy required further study. This was accomplished by the direct calibration of the rocket XPS instrument viewing the synchrotron source without a monochromator. The SURF BL-2 facility has a large vacuum tank for instrument calibrations that directly view the synchrotron source. The test instrument is on a pitch-yaw gimbal to align its optical axis to the beam, and the entire tank translates horizontally and vertically to scan the instrument responsivity. By adjusting the synchrotron source beam energy, the synchrotron radiation has different spectral shapes and thus different spectral response of the XPS photometers based on its filter spectral shape. This technique is similar to the multiple beam energy technique used to sort higher order contributions in a grating spectrograph as described by Woods and Rottman (1990) and Chamberlin, Woods, and Eparvier (2004).

The beam energies used for the rocket XPS calibrations are 229, 285, 331, and 380 MeV, which have a peak radiance near 21, 11, 7, and 4.5 nm, respectively. The lower beam energies suppress the contribution to the photocurrent from short wavelengths and thus help to reduce the calibration uncertainty for long-wavelength channels. With the photometer current measured at each beam energy, a model of the filter coatings (Henke, Gullikson, and Davis, 1993) iterates the filter coating thicknesses until a minimum difference is found for all results from the four beam energies; that is, a least squares fit for the filter coatings. The SURF BL-2 calibrations of the rocket XPS were performed in 2003 and 2004, and the two sets of BL-2 results indicate no degradation or change between the two calibrations. These BL-2 model results are compared to the SURF BL-9 and BESSY calibrations in the next section.

3.5. Comparisons of Responsivity Results

The photodiodes for SORCE XPS were selected from two different batches of photodiodes made by IRD. The first batch was made in 1998 for the TIMED SEE XPS, and the second batch was made in 2000 for the SORCE XPS. In the second batch, the filters with Zr were changed to Mo due to a concern for the stability of Zr, and several filters had thicker coatings to improve the visible light rejection. The Ti/C and Al/Nb/C photodiodes from the first batch were suitable and were included in SORCE XPS, and the other photodiodes in SORCE XPS were from the second batch.

The calibration results from the first batch had calibrations at BESSY, SURF BL-9, and SURF BL-2. The photodiodes from the second batch do not have calibrations from BESSY. The responsivities for the Ti/C and Al/Nb/C photodiodes from the first batch are shown in Figure 7 from all three calibration facilities. The main differences are that the BESSY results are lower than the NIST results by about a factor of 2 between 4 and 10 nm for both types of photodiodes and that the SURF BL-2 results are higher than the results in monochromatic radiation by about 20% for the second peak of responsivity near 17 nm for the Al/Nb/C photodiode. These differences might be explained by (1) temporal changes of the responsivities as photodiodes are calibrated over the period of 1998–2004, (2) errors in modeling the filter transmission for the SURF BL-2 method, and/or (3) errors in the BESSY and SURF BL-9 corrections for their monochromators' scattered light and grating higher orders. These options are discussed next, but there is no conclusive evidence that any of these options is the main cause of the differences.

When modeling each result separately, the BESSY results indicate thicker coatings than the two SURF results. If calibrations are made off-axis or in a beam with significant divergence, then the modeled filter coating thicknesses would increase

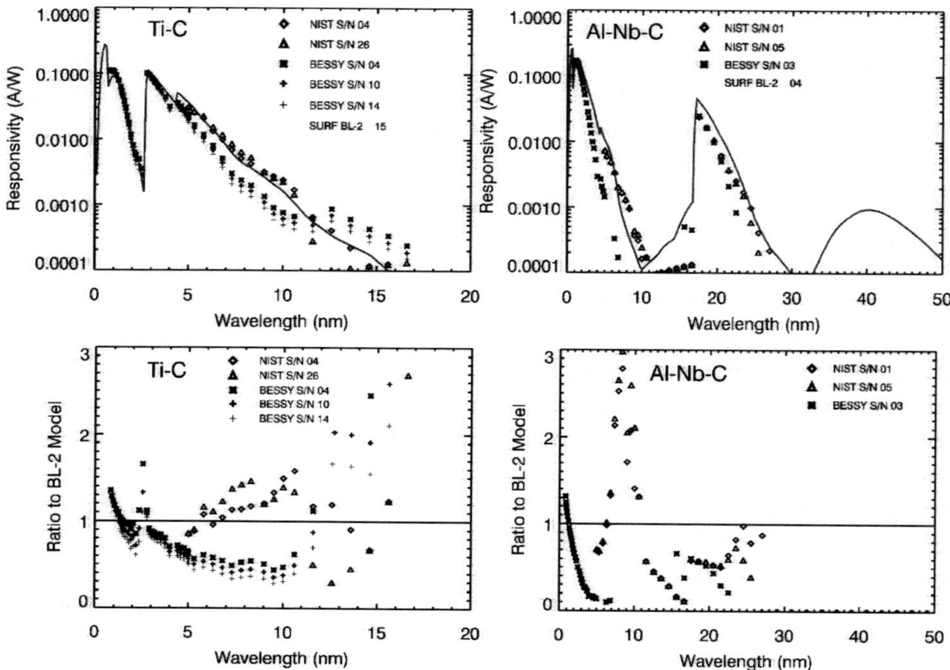

Figure 7. Responsivity for Ti/C and Al/Nb/C photodiodes. These photodiodes are from the first batch manufactured in 1998 and were calibrated at BESSY in 1998, SURF BL-9 in 2000, and SURF BL-2 in 2003 and 2004. The discrete points are for different photodiodes, identified on the legend by serial number, calibrated at either BESSY, SURF BL-9, or SURF BL-2.

proportional to the cosine of the angle of incidence. But the BESSY calibrations are made with its angle of incidence on the photodiodes less than 2°. Furthermore, the BESSY beam divergence is much smaller than a ~30° beam divergence that would be required to account for the observed difference. Another possibility is that the coatings became thinner between 1998 and 2000 when the SURF BL-9 calibrations were made. It is more common that filters become thicker with time due to oxidation and any deposition of contaminates. Furthermore, the photodiodes have shown no degradation in-flight and no degradation between the 2003 and 2004 calibrations for SURF BL-2, so the differences cannot be adequately explained by the photodiodes changing with time. There is no conclusive evidence that can explain the calibration differences between BESSY and SURF BL-9 at some wavelengths.

Errors in modeling the filter transmission for the SURF BL-2 method likely contribute to the differences for the BL-2 results. In general, the SURF BL-2 results agree reasonably well with BESSY results shortward of 4 nm and with SURF BL-9 results at most wavelengths. However, the spectral shape of the BL-2 results is dependent on the Henke, Gullikson, and Davis (1993) atomic scattering factors that are used to calculate the filter transmission. Additionally, the BL-2 method is very dependent on where the majority of the signal is in the spectrum, which occurs at shorter wavelengths for all of the photodiodes for the beam energies of 229–380 MeV. New calibrations of the rocket XPS at SURF BL-2 using lower beam energies could improve the results in the 17–30 nm range. The mean differences between the BL-2 results and the BESSY and SURF BL-9 results is about 20–30%, except in the 4–10 nm range for the BESSY results. So the uncertainty for the BL-2 modeling results is considered to be about 20–30% and is consistent with the expected uncertainty for the Henke, Gullikson, and Davis (1993) atomic constants.

The BESSY and SURF BL-9 results are dependent on the corrections for their monochromators' scattered light and grating higher orders. These corrections are small (a few percent) at most wavelengths because they use foil filters to isolate a single order from the grating. Their results for wavelengths where these corrections are larger than 10% are excluded in the plots of the responsivity, so this option is not considered viable for attempting to explain the differences between BESSY and SURF results in the 4–10 nm range.

The responsivities of the Ti/Zr/Au and Ti/Mo/Au photodiodes, which have a similar bandpass of 5–12 nm, are displayed in Figure 8. A Ti/Zr/Au photodiode from the first batch is on TIMED SEE XPS, and a Ti/Mo/Au photodiode from the second batch is on SORCE XPS. The Ti/Zr/Au photodiodes have only been

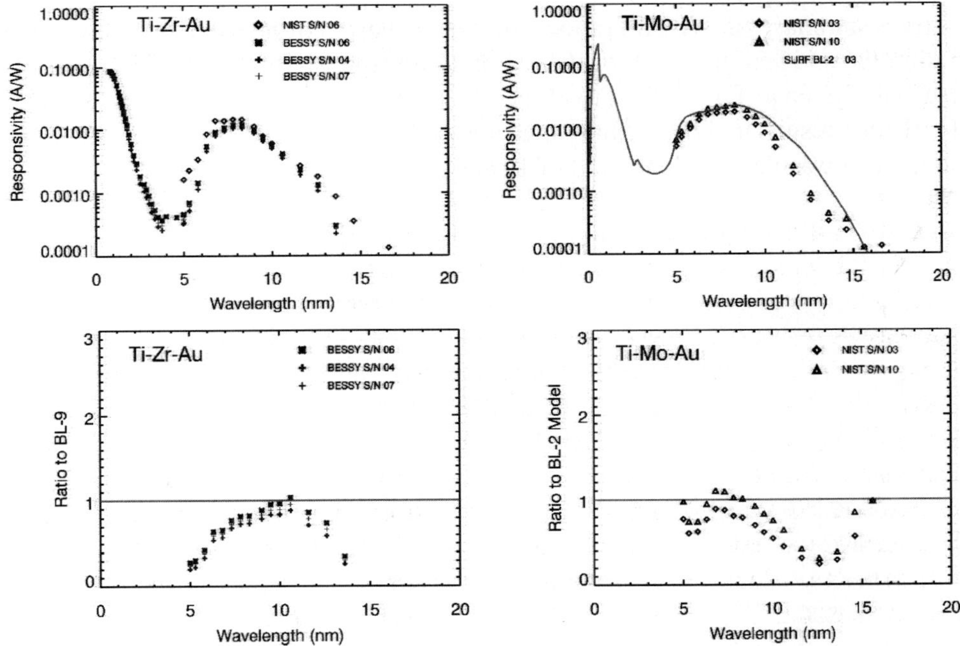

Figure 8. Responsivity for Ti/Zr/C and Ti/Mo/Au photodiodes. The Ti/Zr/C photodiodes are from the first batch and were calibrated at BESSY in 1998 and SURF BL-9 in 2000. The Ti/Mo/Au photodiodes are from the second batch and were calibrated at SURF BL-9 in 2000 and SURF BL-2 in 2003 and 2004. The discrete points are for different photodiodes, identified on the legend by serial number, calibrated at either BESSY, SURF BL-9, or SURF BL-2.

calibrated at BESSY and SURF BL-9. Those results agree to within 20% between 7 and 11 nm, but SURF BL-9 results are higher than BESSY at other wavelengths. There are also Ti/Pd photodiodes (0–10 nm bandpass) from the first batch that were calibrated at both BESSY and SURF BL-9, but not used on SORCE. The Ti/Pd responsivities agree between 5 and 11 nm to within 10%. So the BESSY and SURF BL-9 results compare best at wavelengths near the peak of the photodiode bandpasses.

The similar bandpass Ti/Mo/Au photodiodes have only been calibrated at SURF BL-9 and BL-2. Those results agree reasonably well, typically better than 30% at most wavelengths. The largest difference near 12 nm is probably due to errors in the atomic constants in modeling the filter transmission for the BL-2 results. There are also Ti/Mo/Si/C photodiodes (5–17 nm bandpass) from the second batch that were calibrated at both SURF BL-9 and BL-2. The results for these photodiodes agree best near the peak of the bandpass near 5 and 8 nm. So again, the photodiode calibrations compare best at wavelengths near the peak of the photodiode bandpasses.

The other photodiodes, namely the Al/Sc/C, Al/Cr, and Al/Mn photodiodes, show differences in the responsivities similar to the Al/Nb/C photodiode. That is, the BESSY results are lower than the SURF results in the 4–10 nm range, and the SURF BL-2 results are higher than the results in monochromatic radiation at wavelengths longward of 17 nm.

As a summary of the calibration differences between the three different calibration techniques, the ratios of the calibration responsivities to the SURF BL-2 results are given in Table III in broad spectral ranges. The primary concerns are that (1) BESSY results in the 4–10 nm range are too low and (2) SURF BL-2 results are too large at wavelengths longward of 10 nm.

TABLE III

Comparison of XPS calibrations.

Filter	BESSY/SURF BL-2 ratio			SURF BL-9/BL-2 ratio		
coating	1–5 nm	5–15 nm	15–25 nm	5–15 nm	15–25 nm	25–35 nm
Ti/C	1.06	0.50	–	0.95	1.32	–
Al/Sc/C	–	–	–	0.48	0.65	–
Al/Nb/C	1.01	–	0.59	0.77	0.63	–
Ti/Mo/Au	–	–	–	0.86	–	–
Ti/Mo/Si/C	–	–	–	0.71	–	–
Al/Cr	–	–	–	3.20	–	0.56
Al/Mn	–	–	–	6.24	–	1.02

The ratios are the calibration responsivities to the SURF BL-2 results, and the average ratio in each spectral range is obtained using a weighting of the responsivity. A "–" symbol is displayed for spectral ranges where there are no or limited calibration measurements.

Currently the responsivities from the SURF BL-2 results are used in Version 6 of the SORCE XPS data processing software. The future version of the XPS data products will be improved by merging the results from all three calibration facilities. For this future version, the responsivity shortward of 4 nm will be from the BESSY results, the SURF BL-9 results will be used longward of 5 nm, and the SURF BL-2 model results will be used to fill in spectral gaps and to extend to long wavelengths.

4. In-Flight Calibrations

In addition to measuring the absolute value of the solar irradiance, determining the long-term variation of the irradiance is a fundamental scientific goal; therefore, in-flight tracking of XPS's instrument responsivity is required. The in-flight tracking procedures include on-board redundant channels that are augmented by direct calibrations transferred from instruments underflown on rockets (discussed below). The goal of this variety of techniques is to achieve long-term relative accuracy of XPS's solar irradiance of 10% uncertainty (1-σ value).

The basic assumption for the redundant channel technique is that exposure to the space environment and to solar radiation is the major factor causing instrument degradation. The use of different duty cycles permit an evaluation of the instrument responsivity changes (typically degradation) as related to solar exposure rate. Maintaining a high level of cleanliness for the instruments greatly reduces the degradation related to contaminants on the optics (e.g., Woods *et al.*, 1999b).

The XPS instrument has three redundant channels. Six channels are utilized for daily measurements, and the three channels are used once a week to regularly provide a relative calibration for channels with similar bandpasses. As listed in Table I, the calibration XP No. 1 (Ti/C) serves as calibration for XP No. 2 (Ti/C) in the 0–7 nm range, the XP No. 7 (Ti/Mo/Si/C) as calibration for XP No. 6 (Ti/Mo/Au) in the 5–18 nm range, and XP No. 10 (Al/Mn) as calibration for XP No. 9 (Al/Cr) in the 27–34 nm range. In addition, the XP No. 3 (Al/Sc/C) and XP No. 5 (Al/Nb/C) have similar bandpasses and serve as redundant measurements in the 17–27 nm range. The XP No. 11 (Lyman-α) channel does not have a corresponding redundant channel in XPS, but there are two SOLSTICE channels aboard SORCE that also measure the solar Lyman-α irradiance that serve as in-flight calibration for XP No. 11. The comparison of daily channel XP No. 2 and the weekly redundant channel XP No. 1 in Figure 9 indicates very little, if any, degradation for XP No. 2, and this result is typical for all of the photometers except XP No. 5 and XP No. 11. The derived degradation rate for XP No. 11 is shown in Figure 10 and indicates initial, more rapid degradation and then a slowing down of the degradation rate.

Other in-flight characterizations include cruciform scans once a week to check optical alignment to the spacecraft reference center and field of view maps every 3 months to check responsivity changes as a function of angle from optical center.

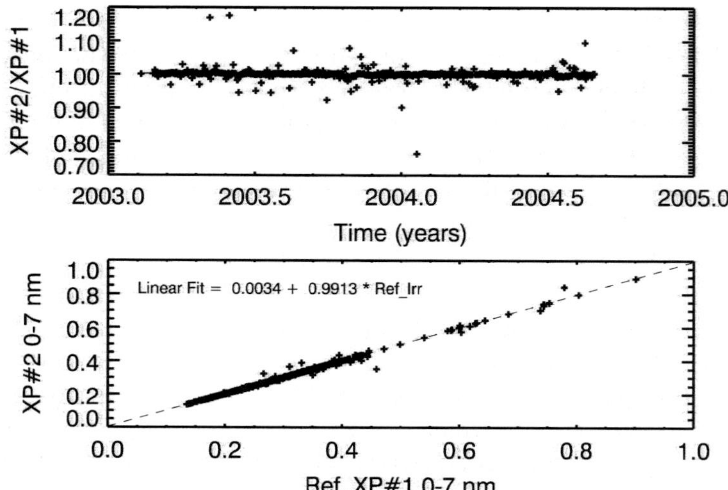

Figure 9. Example redundant channel calibration. The daily XP No. 2 results are compared to the redundant (weekly) XP No. 1 results to determine the degradation rate for XP No. 2. The large deviations from unity in the top plot are due primarily to flare events as the XP No. 1 and XP No. 2 measurements are separated by about 1 min. As with most other photometers, the XP No. 2 channel does not indicate any degradation with time.

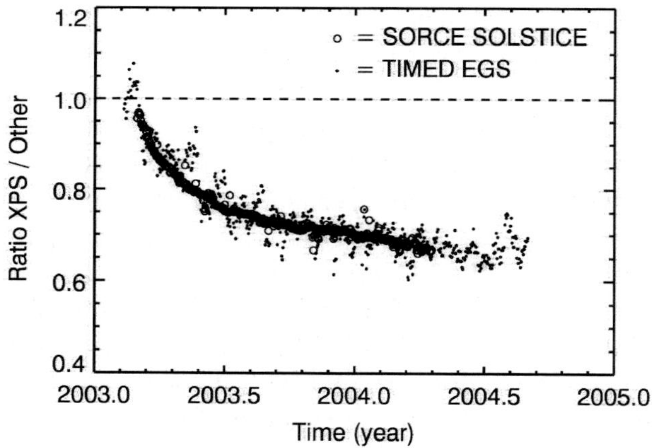

Figure 10. Degradation rate for XP No. 11 (Lyman-α). The degradation rate for XP No. 11 is determined from the SORCE SOLSTICE and TIMED SEE EGS spectral measurements of the H Lyman-α emission at 121.6 nm. The spread in the ratio with the EGS result is due to the precision of the EGS measurement being about ~4%.

These in-flight tests confirm similar pre-flight calibration measurements and ensure that the XPS data processing utilizes the most accurate instrumental parameters.

Independent measurements of the solar VUV spectral irradiance by TIMED SEE prototype instruments flown on sounding rockets provide a calibration for the in-flight SEE and SORCE XPS observations. These underflight measurements on about an annual basis are crucial because the redundant channels can also degrade. The rockets are launched from the NASA facility at the White Sands Missile Range in New Mexico and are supported by the NASA Sounding Rocket Operations Contract (NSROC). Pre-flight and post-flight calibrations of the prototype instrument are performed at NIST SURF-III. The second calibration rocket for SEE was launched on 12 August 2003 and is the first underflight calibration for SORCE XPS, and those results are included in the SORCE XPS Version 6 data products. The third calibration rocket for SEE was launched on 15 October 2004, and those results will be included in the SORCE XPS Version 7 data products.

Two of the photometers have in-flight anomalies. The XP No. 10 developed a large visible light leak early in the SORCE mission, and it is thought that the Ag epoxy holding the photodiode mask developed cracks while XPS was about −40 °C during the first 2 weeks of operations. A similar anomaly had occurred for XP No. 9 and XP No. 10 during a ground based thermal vacuum test, but those two photodiodes were replaced before launch. The second anomaly is with XP No. 5, which has developed excessive noise and significant degradation. It is thought that this photometer might have charging effects on the photodiode surface or failing detector electronics. While the XPS data processing is producing irradiances from these photometers, the uncertainties for these irradiances are large (over 50%); therefore, the XP No. 5 and XP No. 10 irradiance results are excluded from the released XPS data products.

5. Data Processing Algorithms

The SORCE XPS data are processed from raw units of counts per second into irradiance units using straightforward conversion equations. There are four primary XPS data products: Level 1, Level 2, Level 3, and Level 4. These data products are produced on a daily basis. The Level 1 data products are the result of merging the raw science data (in units of counts) with the XPS and spacecraft housekeeping data (voltage, temperature, solar position) needed for science analysis. The Level 2 data products are the irradiances of each measurement usually taken with a 10 s integration period and at instrument spectral resolution. The Level 3 data products are the irradiances averaged over each 6 h period and averaged over the day, corrected to a mean distance of 1 AU, and at instrument spectral resolution. These averages are taken as the median values, which are less sensitive than the mean to large, short-term deviations such as from flares. Only the highest quality data are used for the Level 3 data products. Known anomalous data are discarded, e.g. the data taken in the South Atlantic Anomaly (SAA), data taken near the FOV edges, solar occultation data, and any data identified outside calibrated ranges such as temperature. The Level 4 product includes the irradiances from the individual XPS photometers on a time cadence of 5 min and the solar spectrum in 1 nm intervals on 0.5 nm centers by scaling a model spectrum to match the XPS results.

The SORCE XPS data products are stored as NetCDF files on the LASP computers, and the XPS results are integrated into the SORCE Solar Spectral Irradiance (SSI) data product that is available from the Goddard Earth Sciences (GES) Distributed Active Archive Center (DAAC) (*http://daac.gsfc.nasa.gov/upperatm/sorce/*). In addition, the XPS data are available as an Interactive Data Language (IDL) save set and as a text data file from the SORCE web site (*http://lasp.colorado.edu/sorce/ssi_data.html*). So far, there are no full-day gaps in the XPS data products. The SORCE XPS data are also incorporated into the TIMED SEE Level 3 and Level 3A data products after day 2003/070, and these SEE data are available from the SEE web/ftp site (*http://lasp.colorado.edu/see/*). Due to the on-going analysis of instrument degradation, one should exercise caution in using the most recent XPS data until they are validated using the in-flight calibration experiments, some taken on a weekly basis, some taken on a 3-month basis, and the underflight rocket calibration experiment flown annually. The following algorithm descriptions are for Version 6 of the SORCE XPS data processing code.

5.1. XPS Level 2 Algorithms

The following algorithm describes the irradiance derivation used in SORCE XPS Level 2 data processing code. The algorithm to calculate the solar irradiance, E, is a straightforward calculation using Equation (1) for the XPS photometer "*i*" when it is at a filter position with a clear aperture. Not shown in this equation is the conversion of the XPS raw data of counts per second to current, *I*, in units of nA. This current conversion includes pre-flight gain factors for the electrometer, and these parameters are a function of temperature.

$$E_i = \frac{(I_{i,\text{total}} - I_{i,\text{dark}} - I_{i,\text{visible}})}{f_{i,E\_\text{total}} \cdot \langle T_{i,\text{xuv}} \rangle \cdot A_i \cdot f_{i,\text{xuv\_fov}}} \cdot k_E \cdot f_{\text{Degrade}}, \tag{1a}$$

$$I_{i,\text{visible}} = \frac{(I_{i,\text{window}} - I_{i,\text{dark}})}{T_{\text{window}}} \cdot \frac{f_{i,\text{clr\_fov}}(\alpha_{\text{xuv}}, \beta_{\text{xuv}})}{f_{i,\text{vis\_fov}}(\alpha_{\text{window}}, \beta_{\text{window}})}, \tag{1b}$$

$$T_{\text{window}} = \frac{(I_{b,\text{window}} - I_{b,\text{dark}})}{(I_{b,\text{clear}} - I_{b,\text{dark}})} \cdot \frac{f_{b,\text{clr\_fov}}(\alpha_{\text{clear}}, \beta_{\text{clear}})}{f_{b,\text{vis\_fov}}(\alpha_{\text{window}}, \beta_{\text{window}})}, \tag{1c}$$

$$f_{i,E\_\text{total}} = \frac{\int_0^{\infty} T \cdot E \cdot \mathrm{d}\lambda}{\int_{\lambda_1}^{\lambda_2} T \cdot E \cdot \mathrm{d}\lambda}, \tag{1d}$$

$$\langle T_{i,\text{xuv}} \rangle = \frac{\int_{\lambda_1}^{\lambda_2} T \cdot E \cdot \mathrm{d}\lambda}{\int_{\lambda_1}^{\lambda_2} E \cdot \mathrm{d}\lambda}. \tag{1e}$$

Basically, the irradiance is the photometer XUV current, being the total current corrected for its dark current and visible light current, divided by the responsivity factors, f and T, and aperture area, A. The dark current and visible light current are measured when the photometer is at a filter position with a blocked (dark) aperture and an aperture with a fused silica window (visible), respectively. The visible light current is also corrected for the window transmission, which is also measured as part of the solar observation with a bare Si photodiode. The fractional parameters indicated by "f" are ratios that are mostly near unity. For example, the field of view (FOV) parameters are the ratios of the responsivity at offset angles α and β to the responsivity at the center of the FOV. The constant k_E is included to convert the measured photocurrent and aperture area into energy units of W m^{-2}. The constant k_E includes the intrinsic Si responsivity R of 0.272 A/W which is assumed to be constant with respect to wavelength.

There are three responsivity parameters used in Equation (1). As given by Equation (1d), the f_{E_total} is the inverse fraction of the photometer signal in the photometer bandpass, $\lambda_1 - \lambda_2$, and is a number greater than unity. The calculation of this fraction uses the pre-flight calibration transmission, T, and a modeled solar irradiance spectrum, E, and it is assumed that the Si responsivity ($R = 0.272$ A/W) is constant with respect to wavelength. This fraction changes depending on which solar spectrum is applied. This fraction is averaged over different solar conditions by using the Woods and Rottman (2002) reference spectra for solar minimum and maximum conditions and the flare spectrum from the NRLEUV model (Meier *et al.*, 2002; Warren, Mariska, and Lean, 2001). The uncertainty of this fraction is assumed to be the range of values of the fraction using different modeled spectra in the calculation. As given by the Equation (1e), the $\langle T_{\text{XUV}} \rangle$ is the transmission of the XUV filter weighted with the modeled irradiance spectrum. The uncertainty for the weighted transmission is assumed to be the range of values derived using different modeled spectra. The $f_{\text{xuv\_fov}}$ is the XUV field of view (FOV) factor relative to the center point of the optics and is a value near unity. The FOV factor is derived from the in-flight FOV map experiments and has an uncertainty proportional to the measurement precision. The values for the $\langle T \rangle$ and f responsivity parameters for the 9 XPS photometers are given in Table IV. The XP Nos. 9 and 10 photometers also have contributions in the 26–34 nm range, but only the 0.1–7 nm irradiances are derived from these diodes in the current version of XPS data processing.

The measurement precision, being the uncertainty for a single measurement, is given by Equation (2). These uncertainties are stated as relative uncertainties (1σ) and are thus unitless. The uncertainty for the current conversion is less than 1%, and the typical measurement precision is about 1%.

$$\sigma_{\text{meas}} = \frac{\sqrt{\sigma_{\text{total}}^2 \cdot I_{\text{total}}^2 + \sigma_{\text{dark}}^2 \cdot I_{\text{dark}}^2 + \sigma_{\text{visible}}^2 \cdot I_{\text{visible}}^2}}{I_{\text{total}} - I_{\text{dark}} - I_{\text{visible}}}, \tag{2a}$$

$$\sigma_{\text{visible}} = \sqrt{\sigma_{w-d}^2 + \sigma_{T_w}^2 + \sigma_{\text{fov}_{\text{vis}}}^2(\text{xuv}) + \sigma_{\text{fov}_{\text{vis}}}^2(\text{window})}, \tag{2b}$$

TABLE IV

SORCE XPS responsivity parameters.

XP No.	Filter coating	Bandpass (nm)	f_{E_total}	$\langle T_{XUV} \rangle$	Aperture[a] area (mm^2)
1	Ti/C	0.1–7	1.03	0.135	6.187
2	Ti/C	0.1–7	1.03	0.140	6.176
3	Al/Sc/C	17–23	1.74	0.238	5.998
5	Al/Nb/C	17–21	2.28	0.0368	6.153
6	Ti/Mo/Au	0.1–11	1.01	0.0486	6.261
7	Ti/Mo/Si/C	0.1–7 (7–18)	1.17 (7.6)	0.0615 (0.013)	6.198
9	Al/Cr	0–7 (27–37)	1.13 (9.8)	0.448 (0.0090)	6.266
10	Al/Mn	0–7 (25–34)	1.06 (21.7)	0.144 (0.0031)	6.390
11	Acton Ly-α (x2)	121–122	1.02	0.00156	15.76

[a]Aperture for bare photodiodes (XP Nos. 4, 8, and 12) is a plate with 10 μm pinholes on a 1 mm grid.

$$\sigma_{w-d} = \frac{\sqrt{\sigma_{\text{window}}^2 \cdot I_{\text{window}}^2 + \sigma_{\text{dark}}^2 \cdot I_{\text{dark}}^2}}{I_{\text{window}} - I_{\text{dark}}}. \tag{2c}$$

The accuracy, or combined standard uncertainty from a single measurement, is given by Equation (3). The largest source of uncertainty for the irradiance is the uncertainty of the transmission parameter, $\langle T_{\text{XUV}} \rangle$, which ranges from 5 to 20%. The uncertainty of the fraction factor, f_{E_total}, is also large for photometers with dual bandpasses, such as for XP No. 3 and XP No. 5, and the photometers with significant visible light corrections, such as for XP No. 6, have large uncertainties for the measurement precision. The end result is that the uncertainty for the solar XUV irradiance ranges from 12 to 30%.

$$\sigma_E = \sqrt{\sigma_{\text{meas}}^2 + \sigma_A^2 + \sigma_{\text{fov}_{\text{xuv}}}^2 + \sigma_{f_{E\_\text{total}}}^2 + \sigma_{f_{\text{Degrade}}}^2 + \sigma_{\langle T_{\text{XUV}} \rangle}^2}\,. \tag{3}$$

The degradation of the XPS is tracked through two functions: weekly degradation (f_{Degrade}) and annual rocket calibrations (adjusted $\langle T_{\text{XUV}} \rangle$). The weekly degradation function is derived using trends in the ratios of the daily photometer irradiances to the calibration photometer irradiances. For transfer of the rocket XPS results to SORCE XPS, the transmission parameter, $\langle T_{\text{XUV}} \rangle$, is adjusted for the SORCE XPS so that its irradiance on day 2003/224 matches the rocket XPS irradiance.

The irradiance uncertainties for the four primary bandpasses measured by XPS are listed in Table V. The XUV channels XP Nos. 2, 3, and 6 are the ones used for the XPS Level 4 data products that are described in the next section. The measurement precision (σ_{meas}) is very dependent on the ratio of the visible light signal to the

TABLE V
SORCE XPS irradiance uncertainties on day 2003/224.

Parameter	XP No. 2 (0.1–7 nm) (%)	XP No. 6 (0.1–11 nm) (%)	XP No. 3 (17–23 nm) (%)	XP No. 11 (121–122 nm) (%)
σ_{total}	0.7	0.5	0.5	0.7
σ_{dark}	0.9	0.9	0.9	0.9
$\sigma_{visible}$	8.7	1.1	8.2	58
$I_{visible}/I_{total}$	15	76	5.4	3.0
σ_{meas}	6.1	23	1.4	11
σ_A	0.9	0.9	0.9	0.6
$\sigma_{FOV_{xuv}}$	0.2	0.4	0.5	0.1
$\sigma_{f_{E\_total}}$	0.6	0.1	14	4.6
$\sigma\langle T_{XUV}\rangle$	11	8.1	5.1	10
σ_E	12	24	16	16

total signal; that is, a photometer, such as XP No. 6, with a large visible light signal has a low measurement precision (large percentage). The accuracy of the irradiance (σ_E) is driven largely by the responsivity uncertainties and also by the measurement precision for the photometers with larger visible light signals. There are two responsivities uncertainties: f_{E_total} is the inverse fraction of the photometer signal in band and $\langle T_{XUV}\rangle$ is the weighted transmission. The uncertainty of the fraction in band is relatively small except for the photodiodes with a second long wavelength band, such as XP No. 3. This result implies for these larger uncertainties that the solar spectrum changes significantly between the two different photometer bands in the solar models. The uncertainty for the weighted transmission is largely due to the uncertainties of the pre-flight responsivity calibrations.

5.2. XPS Level 4 Algorithms

A different algorithm is used for the SORCE XPS Level 4 data products in deriving the 1 nm XUV spectra from the XPS measurements. Because the XPS measurements are broadband, a spectral model of the solar irradiance at 1 nm resolution is scaled to match the XPS measurement. This algorithm is similar to the data processing technique used for TIMED SEE Level 3 data products and by the SNOE SXP. The scale factors are determined using the following equations for the bands at 0–4, 4–14, and 14–27 nm:

$$I_{predict} = \int_0^\infty R(\lambda) \cdot E(\lambda) \cdot d\lambda, \tag{4a}$$

$$I_{\text{measure}} = \int_0^{\lambda_1} R(\lambda) \cdot E(\lambda) \cdot \mathrm{d}\lambda + \mathrm{SF} \cdot \int_{\lambda_1}^{\lambda_2} R(\lambda) \cdot E(\lambda) \cdot \mathrm{d}\lambda + \\ + \int_{\lambda_2}^{\infty} R(\lambda) \cdot E(\lambda) \cdot \mathrm{d}\lambda. \quad (4b)$$

The responsivity function, R, of a photodiode, that includes all of the instrument corrections such as field of view and aperture area, is convolved with the model solar spectrum to determine the predicted current in Equation (4a). The scale factor, SF, for a spectral band, $\lambda_1 - \lambda_2$, is determined using the measured XUV current as compared to the predicted current but with a scale factor applied as shown in Equation (4b). The measured XUV current is the total current minus the dark current and the visible current. The scale factor for the 0 – 4 nm band is first determined. For the scale factors at the other bands, the scale factors from the shorter wavelength bands are first applied to the solar irradiance model spectrum. That is, the 4–14 nm scale factor is calculated using the 0 – 4 nm scale factor, and the 14–27 nm scale factor is calculated using the other two scale factors. The SORCE XPS photometers that are used for the scale factors at 0 – 4, 4–14, and 14–27 nm are XP No. 2 (Ti/C), No. 6 (Ti/Mo/Au), and No. 3 (Al/Sc/C), respectively. These SORCE photodiodes have very similar bandpasses as the corresponding TIMED SEE XPS photodiodes.

This scale factor algorithm, which is only used in Level 4 processing, is especially valuable for determining the flare spectra because it permits the model spectrum to change in broad bands. During non-flare periods, the two different algorithms provide the same irradiances to within 10%. But during flare events the direct irradiance algorithm represented by Equation (1) over predicts the irradiance at the longer wavelengths, namely in the 14–27 nm range. The selection of the XUV bands is based partially on the prediction of significant hot corona emissions during flares in the 0 – 4 and 9–13 nm regions (Mewe and Gronenschild, 1981; Mewe, Gronenschild, and van den Oord, 1985; Mewe, Lemen, and van den Oord, 1986).

6. Validation and Comparisons

Validation with solar XUV measurements during the SORCE mission is limited to the 0.1–10 nm measurements by TIMED SEE XPS and to the bands 26 –34 and 0 –50 nm by the Solar and Heliospheric Observatory (SOHO) Solar EUV Monitor (SEM). The TIMED SEE XPS and SORCE XPS measurements can be directly compared as their bandpasses are very similar. The SOHO SEM measurements can not directly be compared to the SORCE XPS measurements because the SORCE XPS is limited to wavelengths shortward of 27 nm, but SORCE XPS combined with the TIMED SEE EGS measurements can be compared to the SOHO SEM measurements. The XPS Lyman-α channel (XP No. 11) does overlap with TIMED SEE and SORCE SOLSTICE measurements by grating spectrometers, and those comparisons have been used to determine the XP No. 11 degradation as shown in Figure 10. The comparisons for SORCE XPS shown in this section are those to the TIMED SEE XPS and SOHO SEM measurements and to predictions from models of the solar XUV irradiance.

6.1. Comparisons to TIMED SEE XPS measurements

Both TIMED SEE and SORCE XPS are calibrated to the underflight experiment flown on 12 August 2003 (day 2003/224), so the absolute scale for both irradiance results is the same by design. A more interesting comparison is that of the daily average from the two different instruments. As shown in Figure 11, the daily averages of the SORCE XPS Level 4 irradiances at 0–7 nm are compared to the daily average of the irradiance from the TIMED SEE XP No. 1 (Ti/C with bandpass of 0.1–7 nm). While the mean and median of the ratio of SORCE to SEE irradiances are within a couple percent of unity, there are significant differences of 40% or more. The SORCE results indicate more variability mainly because the TIMED SEE duty cycle for solar observations is only 3% while the SORCE XPS has a duty cycle of ∼70% and captures more flares. That is, the flares do have a significant impact on the daily average for the XUV irradiance.

Some of the differences are also related to the different algorithms for these data. The SORCE XPS irradiances in this comparison are from the Level 4 data product that is described in Section 5.2, and the TIMED SEE XPS irradiances

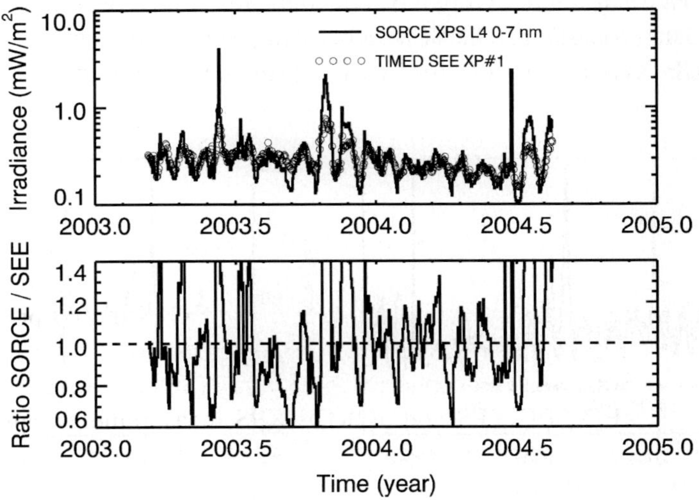

Figure 11. Comparison of SORCE XPS Level 4 Irradiance to TIMED SEE XP No. 1 Irradiance. The daily averages of the SORCE XPS Level 4 irradiances at 0–7 nm are compared to the daily averages of the irradiances from the TIMED SEE XP No. 1 (Ti/C with bandpass of 0.1–7 nm). The differences arise largely from TIMED SEE having a 3% duty cycle and SORCE XPS having a 70% duty cycle.

are derived using the same type algorithm described for the SORCE Level 2 data product that is described in Section 5.1. The Level 4 algorithm produces model scaling factors based on three photometers in bands of 0–4, 4–14, and 14–27 nm; therefore, the 0–7 nm irradiances in the Level 4 data products are the result of the measurements from multiple photometers. Consequently, this comparison is not a direct assessment between single photometers.

6.2. Comparisons to SOHO SEM measurements

While there are a few different solar EUV measurements made by SOHO, only the Solar EUV Monitor (SEM) provides irradiances on a daily basis and with high accuracy (Judge *et al.*, 1998). The SEM measures the solar XUV irradiance at 26–34 nm (1st grating order) and at 0–50 nm (0th order). The SEM irradiances are reported in photon units instead of energy units and are derived by scaling a reference spectrum (Woods, 1992) to match the SEM photodiode currents. The ratio of this reference spectrum over the desired bandpass in energy units to itself in photon units is used to convert the SEM irradiances to energy units. The combination of the SORCE XPS Level 4 results at 0–27 nm and the TIMED SEE EGS results at 27–50 nm are compared to the SEM 0–50 nm irradiance in Figure 12. This comparison indicates good agreement of ~10% in the absolute values of the irradiances. This comparison also indicates good agreement in their relative long-term variations, which in turn indicate good understanding of instrument degradation functions for all three instruments. The largest differences are in the magnitude of the solar rotation variation with the SORCE/TIMED variation being larger than the SEM variation. These differences might be due to the reference spectra used in the SEM and XPS data processing. The SEM processing uses a single reference spectrum, and SORCE XPS processing uses a daily reference spectrum based on a F10.7

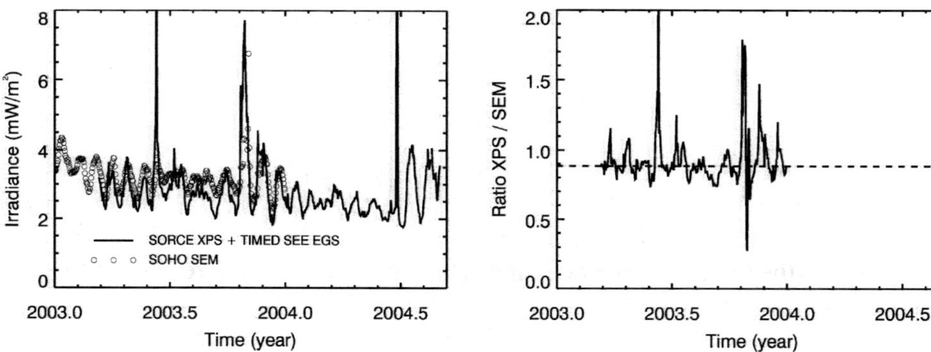

Figure 12. SOHO SEM comparison. The combination of the SORCE XPS Level 4 data from 0–27 nm and the TIMED SEE EGS Level 3 data from 27–50 nm are compared to the SOHO SEM measurements at 0–50 nm. The *left plot* shows the irradiance time series, and the *right plot* shows the ratio of XPS to SEM with the median value of 0.87 indicated as the *dashed line*.

model (Woods and Rottman, 2002). In both cases, the reference spectra are scaled to match the photodiode currents. Some of these differences might be resolved if SEM and SORCE used the same reference spectra. Nonetheless, relative changes in the 0–50 nm irradiances are in good agreement throughout the SORCE mission.

6.3. Comparisons to Solar Irradiance Models

Because of the limited amount of actual solar data, especially for the XUV and EUV regions, models of the solar variability are widely used in aeronomic studies. Commonly used solar irradiance models are empirical models, frequently called proxy models, that are derived using linear relations between one or two solar proxies and extant observations of the solar VUV irradiance. These models use commonly available solar measurements, such as the ground-based 10.7 cm radio solar flux (F10.7) and the NOAA Mg II core-to-wing index (Mg C/W), to represent solar irradiance variations in the VUV spectral range. Hinteregger, Fukui, and Gilson (1981) developed the first, and still, widely used proxy model based on the AE-E satellite observations and several sounding rocket measurements. The original proxies for this model were the chromospheric H Lyman-β (102.6 nm) and the coronal Fe XVI (33.5 nm) emissions. As measurements of these emissions are not generally available, they are constructed from correlations with the daily F10.7 and its 81-day average, which have been available on a daily basis since 1947. The Hinteregger, Fukui, and Gilson (1981) model is also referred to as EUV81 and SERF 1 by the Solar Electromagnetic Radiation Flux (SERF) subgroup of the World Ionosphere–Thermosphere Study. Richards, Fennelly, and Torr (1994) developed a different F10.7 proxy model called EUVAC in which the solar soft X-ray irradiances were increased by a factor of 2–3 compared with the SERF 1 model. W. K. Tobiska has developed several proxy models of the solar EUV irradiance: SERF 2 by Tobiska and Barth (1990), EUV91 by Tobiska (1991), EUV97 by Tobiska and Eparvier (1998), and the latest version, SOLAR2000, by Tobiska *et al.* (2000). Augmenting these simple proxy models are physical and semi-empirical models of the solar EUV irradiance: Fontenla *et al.* (1999), Warren, Mariska, and Lean (1998a,b), and Lean *et al.* (1982). Of these models, the NRLEUV model has been parameterized to use the F10.7 and Mg II core-to-wing index as solar proxy inputs (Warren, Mariska, and Lean, 2001).

The comparison of four different solar EUV irradiance models to the SORCE XPS measurement on 12 August 2003 (day 2003/224) is listed in Table VI in the bands used in the Level 4 data processing. The four models compared are the EUV81 (Hinteregger, Fukui, and Gilson, 1981), the NRLEUV (Warren, Mariska, and Lean, 2001), the EUVAC (Richards, Fennelly, and Torr, 1994), and the SOLAR2000 version 2.23 (Tobiska *et al.*, 2000). The NRLEUV and EUVAC models do not have irradiances shortward of 5 nm, so no ratios are listed in Table VI for those models for the 0–4 nm band. The EUVAC and SOLAR2000 models agree best with the XPS measurements. The latest SOLAR2000 version 2.23 model did include the

TABLE VI
Comparison of SORCE XPS to models on day 2003/224.

Model	XPS/model 0 – 4 nm	XPS/model 4–14 nm	XPS/model 14–27 nm
EUV81	**4.1**	**1.8**	**1.4**
NRLEUV	n/a	**2.9**	1.1
EUVAC	n/a	0.8	0.7
SOLAR2000	0.8	1.1	1.0

TIMED SEE measurements to determine the model coefficients, so it is expected to agree with the SORCE XPS measurements. The EUVAC model is based on some of the same rocket and AE-E data that was used to develop the EUV81 model but increased the 0–20 nm irradiance by about a factor of 2 to agree better with photoelectron measurements (Richards, Fennelly, and Torr, 1994). So as expected, the EUV81 model is about a factor of 2 lower than the XPS irradiance. The EUV81 difference in the 0 – 4 nm band is larger (factor of 4) mainly because the EUV81 model does not have any irradiances listed shortward of 1.8 nm. The NRLEUV model agrees well with XPS in the 14–27 nm band but is a factor of 3 lower than XPS in the 4–14 nm band.

Another important aspect for model comparisons is the relative variability from day to day and over the solar cycle. A couple of comparisons are shown here to illustrate some of the issues. The comparison of the XPS measurements and models of the 5–25 nm is shown in Figure 13. The obvious results are that the XPS 5–25

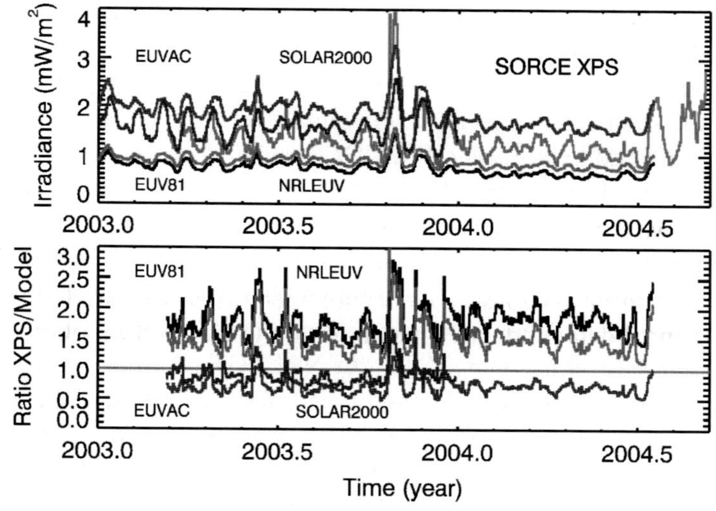

Figure 13. Comparison of models to SORCE XPS measurements at 5–25 nm.

nm irradiances are 80% higher than the EUV81 model, 50% higher than NRLEUV, 10% lower than SOLAR2000, and 30% lower than EUVAC.

A more subtle result is that the XPS 5–25 nm irradiances vary more than any of the model predictions. One contributing factor for the XPS 5–25 nm variations being larger than the models is that the flare effects in the XPS data are not well represented in the F10.7 or Mg C/W proxies used by the models. Nonetheless, even when flares are not significant, the XPS data have larger solar rotation variations than the models. This result is similar as the NRLEUV model having less variability than the SOHO SEM and SNOE measurements (Lean *et al.*, 2003). Part of this difference is in how the models are parameterized, but physical differences between the proxy variations and the XUV irradiance variations are probably contributing to the model relative day-to-day errors.

With more analysis, the XPS measurements will be used to improve the existing models and to develop new models of the solar XUV irradiance. The solar UV irradiance above 115 nm has been measured more extensively; consequently, models of the solar FUV and MUV irradiance are much more accurate than the solar EUV models, due primarily to the lack of solar EUV measurements during the past two solar cycles.

7. Summary

A challenge for SORCE XPS has been the establishment of the photometer's responsivity due to conflicting results at some wavelengths from BESSY and SURF pre-flight calibrations. The differences between the BESSY and SURF results in the 4–12 nm range for some photometers are as large as a factor of 2. Because the SURF BL-9 and BL-2 results are consistent in this wavelength range, the adopted responsivities for SORCE XPS data processing are currently the SURF BL-2 results. The current XPS data processing includes the uncertainty for the responsivities in the range of 5–15% based on the SURF calibrations, but the uncertainty for these responsivities is possibly as large as 20–30% due primarily to the uncertainties of the atomic scattering constants used in modeling the XPS filter transmissions.

There are two different algorithms used in the SORCE XPS data processing software to generate the solar XUV irradiances. The direct calculation of the irradiance for the primary bandpass of each photometer is used to produce the SORCE XPS Level 2 and 3 data products. This algorithm works well for photometers with a single bandpass and for applications that only need low spectral resolution. For dual bandpass photometers, such as XP No. 3 photometer with bands at 0.1–7 and 17–23 nm, the direct calculation of irradiance overpredicts the irradiance in the long wavelength band during flare events because the spectral shape of the solar spectrum changes dramatically with a major increase in the irradiance shortward of 5 nm and only moderate increases at other wavelengths. To address this issue and to provide a higher spectral resolution of 1 nm, the algorithm of scaling a model spectrum (at 1 nm resolution) in multiple broad bands is used to produce the SORCE XPS Level 4 data product. The XPS results from both algorithms are consistent for solar quiet periods (non-flaring times), but the XPS Level 4 data products should be used to study flare events.

The SORCE XPS and TIMED SEE solar XUV irradiances agree well, as expected because they both utilize the rocket underflight calibration results on 12 August 2003. There are temporal differences between SORCE XPS and TIMED SEE daily average results though because SORCE has a 70% duty cycle for solar observations while the TIMED SEE with its 3% duty cycle samples fewer of the flare events during a day. The SORCE XPS measurements, coupled with the TIMED SEE EGS results longward of 27 nm, are about 10% lower than the SOHO SEM solar irradiance in the 0–50 nm band. This degree of agreement is considered excellent for this wavelength range due to the use of different calibration techniques, different algorithms, and different reference solar models used in the analysis of these broadband measurements.

The comparison of the SORCE XPS results to the solar XUV irradiances predicted by models indicates good agreement longward of 14 nm. Shortward of 14 nm, the EUVAC and SOLAR2000 models agree better with XPS measurements than the EUV81 and NRLEUV models, which indicate differences of 2 to 4 lower irradiance values than XPS. The comparison of these models to XPS in the 5–25 nm band indicate a systematic offset but good long-term tracking of the XPS time series. The solar XUV irradiance from XPS measurements indicate more solar rotation variability than the model predictions, thus indicating that the models are predicting the long-term variations well but under-estimating the solar rotation effects. Because all of these models only provide daily predictions, all of these models are not able to predict the large flare events that are observed in the XUV range. While a couple of time series were presented here, the solar XUV variability during the SORCE mission is discussed in more detail in the XPS companion paper included in this special issue (Woods and Rottman, 2005).

Data Archive

The SORCE Solar Spectral Irradiance (SSI) data products contain the solar XUV irradiance results from XPS. The SORCE data products are available from the Goddard Earth Sciences (GES) Distributed Active Archive Center (DAAC) at *http://daac.gsfc.nasa.gov/upperatm/sorce/*. The SORCE XPS data are also available as an Interactive Data Language (IDL) save set and as a text data file from the SORCE web site at *http://lasp.colorado.edu/sorce/ssi_data.html*. In addition, the SORCE XPS data are also ingested into the TIMED SEE Level 2 and 3 data products that are available at *http://lasp.colorado.edu/see/*.

Acknowledgements

This research was supported by NASA contract NAS5-97045 to the University of Colorado. We are grateful to many engineers, technicians, and managers at Laboratory for Atmospheric and Space Physics (LASP) for supporting the development of the XPS instrument, especially noting Terry Leach for much of the original design of the XPS instrument. We are grateful to Raj Korde of the International Radiation Detectors Inc. (IRD) for the collaborative development of the XUV filters on the Si photodiodes that are the basis for the XPS instrument. We are grateful to many engineers, technicians, and managers at Orbital Sciences Corporation for the development of the excellent SORCE spacecraft bus that is exceeding all requirements. We are grateful to the SORCE operations, planning, and data processing teams at LASP for a very successful mission, especially noting Ann Windnagel for the routine processing of the SORCE XPS data. Finally last but not least, we are grateful to the managers at NASA HQ and GSFC for their excellent support for the SORCE development and mission.

References

Brasseur, G. and Solomon, S.: 1986, *Aeronomy of the Middle Atmosphere: Chemistry and Physics of the Stratosphere and Mesosphere*, D. Reidel, Dordrecht, Boston.

Canfield, L. R. and Swanson, N.: 1987, *J. Res. Natl. Bureau Standards* **92**, 97.

Canfield, L. R., Kerner, J., and Korde, R.: 1989, *Appl. Opt.* **28**, 3940.

Canfield, L. R., Vest, R., Woods, T. N., and Korde, R.: 1994, *SPIE Proc.* **2282**, 31.

Chamberlain, J. W.: 1978, *Theory of Planetary Atmospheres: An Introduction to Their Physics and Chemistry*, Academic Press, New York.

Chamberlin, P. C., Woods, T. N., and Eparvier, F. G.: 2004, *SPIE Proc.* **5538**, 31.

Fontenla, J. M., White, O. R., Fox, P. A., Avrett, E. H., and Kurucz, R. L.: 1999, *Astrophys. J.* **518**, 480.

Henke, B. L., Gullikson, E. M., and Davis, J. C.: 1993, *At. Data Nucl. Data Tables* **54**, 181.

Hinteregger, H. E., Fukui, K., and Gilson, G. R.: 1981, *Geophys. Res. Lett.* **8**, 1147.

Judge, D. L., McMullin, D. R., Ogawa, H. S., Hovestadt, D., Klecker, B., Hilchenbach, M., Mobius, E., Canfield, L. R., Vest, R. E., Watts, R., Tarrio, C., Kuehne, M., and Wurz, P.: 1998, *Solar Phys.* **177**, 161.

Korde, R. and Canfield, L. R.: 1989, *SPIE Proc.* **1140**, 126.

Korde, R. and Geist, J.: 1987, *Appl. Opt.* **26**, 5284.

Korde, R., Canfield, L. R., and Wallis, B.: 1988, *SPIE Proc.* **932**, 153.

Lean, J.: 1987, *J. Geophys. Res.* **92**, 839.

Lean, J.: 1991, *Rev. Geophys.* **29**, 505.

Lean, J. L., Livingston, W. C., Heath, D. F., Donnelly, R. F., Skumanich, A., and White, O. R.: 1982, *J. Geophys. Res.* **87**, 10307.

Lean, J. L., Warren, H. P., Mariska, J. T., and Bishop, J.: 2003, *J. Geophys. Res.* **108**, 1059, doi: 10.1029/ 2001JA009238.

Meier, R. R., Warren, H. P., Nicholas, A. C., Bishop, J., Huba, J. D., Drob, D. P., Lean, J., Picone, J. M., Mariska, J. T., Joyce, G., Judge, D. L., Thonnard, S. E., Dymond, K. F., and Budzien, S. A.: 2002, *Geophys. Res. Lett.* **29**, doi: 10.1029/2001GL013956.

Mewe, R. and Gronenschild, E. H. B. M.: 1981, *Astron. Astrophys. Suppl. Ser.* **45**, 11.
Mewe, R., Gronenschild, E. H. B. M., and van den Oord, G. H. J.: 1985, *Astron. Astrophys. Suppl. Ser.* **62**, 197.
Mewe, R., Lemen, J. R., and van den Oord, G. H. J.: 1986, *Astron. Astrophys.* **65**, 511.
Pap, J. M., Fröhlich, C., Hudson, H. S., and Solanki, S. K. (eds.): 1994, *The Sun as a Variable Star: Solar and Stellar Irradiance Variations*, Cambridge University Press, Cambridge.
Parr, A. C. and Ebner, S.: 1987, *SURF II User Handbook*, NBS Special Publication, Gaithersburg, MD.
Powell, F. R., Vedder, P. W., Lindblom, J. F., and Powell, S. F.: 1990, *Opt. Eng.* **26**, 614.
Richards, P. G., Fennelly, J. A., and Torr, D. G.: 1994, *J. Geophys. Res.* **99**, 8981.
Rottman, G. J.: 1987, in P. Foukal, (ed.), *Solar Radiative Output Variation*, Cambridge Research and Instrumentation Inc., Boulder, Colorado, p. 71.
Scholze, F., Thornagel, R., and Ulm, G.: 2001, *Metrologia* **38**, 391.
Tobiska, W. K.: 1991, *J. Atmos. Terr. Phys.* **53**, 1005.
Tobiska, W. K.: 1993, *J. Geophys. Res.* **98**, 18879.
Tobiska, W. K. and Barth, C. A.: 1990, *J. Geophys. Res.* **95**, 8243.
Tobiska, W. K. and Eparvier, F. G.: 1998, *Solar Phys.* **177**, 147.
Tobiska, W. K., Woods, T. N., Eparvier, F. G., Viereck, R., Floyd, L., Bouwer, D., Rottman, G. J., and White, O. R.: 2000, *J. Atmos. Sol.-Terr. Phys.* **62**, 1233.
Walker, J. H., Saunders, R. D., Jackson, J. K., and McSparron, D. A.: 1988, *J. Res. Natl. Bureau Standards* **93**, 7.
Warren, H. P., Mariska, J. T., and Lean, J.: 1998a, *J. Geophys. Res.* **103**, 12077.
Warren, H. P., Mariska, J. T., and Lean, J.: 1998b, *J. Geophys. Res.* **103**, 12091.
Warren, H. P., Mariska, J. T., and Lean, J.: 2001, *J. Geophys. Res.* **106**, 15745.
White, O. R. (ed.): 1977, *The Solar Output and Its Variation*, Colorado Associated University Press, Boulder.
Woods, T.: 1992, in D. Donnelly (ed.), *Working Group 4 and 5 Report for 1991 SOLERS 22 Workshop*, Proc. of SOLERS 22 Workshop, NOAA, Boulder, Colorado, pp. 460 – 467.
Woods, T. N. and Rottman, G. J.: 1990, *J. Geophys. Res.* **95**, 6227.
Woods, T. N. and Rottman, G. J.: 2002, in M. Mendillo, A. Nagy, and J. Hunter Waite, Jr. (eds.), *Comparative Aeronomy in the Solar System*, Geophysics Monograph Series, Washington, DC, pp. 221–234.
Woods, T. N. and Rottman, G. J.: 2005, *Solar Phys.*, this volume.
Woods, T., Eparvier, F., Bailey, S., Solomon, S. C., Rottman, G., Lawrence, G., Roble, R., White, O. R., Lean, J., and Tobiska, W. K.: 1998, *SPIE Proc.* **3442**, 180.
Woods, T., Rodgers, E., Bailey, S., Eparvier, F., and Ucker, G.: 1999a, *SPIE Proc.* **3756**, 255.
Woods, T., Rottman, G., Russell, C., and Knapp, B.: 1999b, *Metrologia* **35**, 619.
Woods, T. N., Acton, L. W., Bailey, S., Eparvier, F., Garcia, H., Judge, D., Lean, J., McMullin, D., Schmidtke, G., Solomon, S. C., Tobiska, W. K., and Warren, H. P.: 2004, in J. Pap, C. Fröhlich, H. Hudson, J. Kuhn, J. McCormack, G. North, W. Sprig, and S. T. Wu (eds.), *Solar Variability and Its Effect on Climate*, Geophysics Monograph Series, **141**, Washington, DC, pp. 127–140.

Solar Physics (2005) 230: 375–387

XUV PHOTOMETER SYSTEM (XPS): SOLAR VARIATIONS DURING THE SORCE MISSION

THOMAS N. WOODS and GARY ROTTMAN
Laboratory for Atmospheric and Space Physics, University of Colorado, Boulder
(e-mails: tom.woods@lasp.colorado.edu; gary.rottman@lasp.colorado.edu)

(Received 9 December 2004; accepted 11 February 2005)

Abstract. The solar soft X-ray (XUV) radiation is important for upper atmosphere studies as it is one of the primary energy inputs and is highly variable. The XUV Photometer System (XPS) aboard the Solar Radiation and Climate Experiment (SORCE) has been measuring the solar XUV irradiance since March 2003 with a time cadence of 10 s and with about 70% duty cycle. The XPS measurements are between 0.1 and 34 nm and additionally the bright hydrogen emission at 121.6 nm. The XUV radiation varies by a factor of ~2 with a period of ~27 days that is due to the modulation of the active regions on the rotating Sun. The SORCE mission has observed over 20 solar rotations during the declining phase of solar cycle 23. The solar XUV irradiance also varies by more than a factor of 10 during the large X-class flares observed during the May–June 2003, October–November 2003, and July 2004 solar storm periods. There were 7 large X-class flares during the May–June 2003 storm period, 11 X-class flares during the October–November 2003 storm period, and 6 X-class flares during the July 2004 storm period. The X28 flare on 4 November 2003 is the largest flare since GOES began its solar X-ray measurements in 1976. The XUV variations during the X-class flares are as large as the expected solar cycle variations.

1. Introduction

The recent solar activity has been an interesting period during the declining phase of solar cycle 23. The solar storms in May–June 2003, October–November 2003, and July 2004 have unleashed an extraordinary number of flares, energetic proton events, and coronal mass ejections (CMEs), as well as several record breaking large flares. The X28 flare on 4 November 2004 is the largest X-ray flare since GOES began its solar X-ray measurements in 1976. The X17 flare on 28 October 2003 is the fourth largest X-ray flare. The other two large X-ray flare records are X20 flares on 2 April 2001 and 16 August 1989. Garcia (2000) reports that there are, on average, one M-class (modest) solar flares every 2 days and one X-class (extreme) flare per month as determined from a survey of flares during solar cycle 21 and 22. Therefore, it is extraordinary to observe 44 M-class flares and 11 X-class flares during the ~2 week period of 19 October 2003 to 5 November 2003. The other interesting period in 2003 is 26 May 2003 to 17 June 2003 when 48 M-class flares and 7 X-class flares were observed. The more recent period between 12 July 2004 and 28 July 2004 was equally interesting with 33 M-class flares and 6 X-class flares.

The October–November 2003 period has obtained the most attention so far from both the scientific community and the public, as this period exhibited the most intense flares, proton events, and CMEs and thus had the most societal impacts. For example, the large X17 flare and proton storm on October 28 disrupted communications and caused airline traffic to be re-routed away from flying over the poles. The very large and very fast CMEs on 28 and 29 October 2003 created aurora as far south as Florida and Mexico, and these storms have been dubbed as the Halloween 2003 storm. An even larger CME impacted Earth on 20 November 2003 with a repeat performance of very southern aurora. In response to the NOAA space weather alerts to these storms, many satellites were commanded to a safe operational mode and did not return data for several days. The most significant damage from these storms was the loss of two Japanese satellites due possibly to electronics and cables on-board these satellites being over-charged by the solar particles impinging on the satellites.

The NASA Solar Radiation and Climate Experiment (SORCE) was launched on 25 January 2003, and so these interesting periods of solar activity have been observed by the SORCE solar irradiance instruments. The XUV Photometer System (XPS) is the most sensitive of the SORCE instruments to the solar flares because XPS measures the highly, variable solar XUV radiation. For example, the solar XUV radiation can vary by more than an order of magnitude during the large X-class flares, and XPS observed a factor of almost 100 increase of the 0.1–4 nm irradiance during the X28 flare on 4 November 2003. The XPS has been measuring the solar XUV irradiance since March 2003 with a time cadence of 10 s. Observing throughout most of the sunlit portion of each orbit it achieves a duty cycle of about 70%. The XPS measurements are between 0.1 and 34 nm and additionally include the bright hydrogen emission at 121.6 nm.

The companion (first) paper about XPS (Woods, Rottman, and Vest, 2005) describes the details about the XPS instrument design, calibration, and validation; and this paper focuses on the solar XUV variability observed so far during the SORCE mission. The variations discussed are the short-term, intermediate-term, and then the long-term variations. The short-term irradiance variations, lasting from minutes to hours, occur during eruptive events on the Sun. The intermediate-term variations, modulated by the 27-day rotation period of the Sun, are related to the appearance and disappearance of active regions on the solar disk. The more elusive long-term variability is related to the 22-year magnetic field cycle of the Sun and the corresponding 11-year sunspot cycle.

2. Short-Term Variations

The short-term variations in the XUV range are dominated by flare events. Most solar flares affect the solar irradiance primarily in the X-ray spectrum, but sometimes a large flare can affect the solar VUV irradiance over a broad wavelength range up to 180 nm (e.g., Brekke *et al.*, 1996; Meier *et al.*, 2002; Woods *et al.*, 2004a; Woods *et al.*, 2004b). Flares and the associated response of Earth's upper atmosphere to these abrupt events are an important aspect of space weather studies such as space-based communication and navigation systems. The solar EUV irradiance is needed on time scales of seconds to hours to improve the understanding of how flares cause abrupt space weather changes (e.g., sudden ionospheric disturbances or SIDs). The XPS flare measurements provide new information about the solar spectral irradiance changes during a flare and are contributing new understanding of the effects that flares could have on Earth's atmosphere.

Because of its high duty cycle (70%), SORCE XPS observes the majority of the solar flares but not all of the flares. SORCE XPS was fortunate in observing all of the large X-class flares in the three interesting solar storm periods: May–June 2003, October–November 2003, and July 2004. The October–November 2003 period was more intense with the X17 flare on 28 October and the X28 flare on 4 November. The dates and times of the 20 largest flares observed by SORCE XPS are listed in Table I. It is important to note that the XPS flare maximum time listed in this table is not the flare peak time but is instead the observation time for XPS, which has a time cadence of about 5 min.

The time series of the solar 0.1–4 nm irradiance measurements by SORCE XPS, which are shown in Figure 1, illustrate the high frequency of flares during the SORCE mission. There are small flares that occur on almost every day, and there are periods when there is a sequence of large flares. The measurements from the XPS channel 11 (121–122 nm) are also shown in this figure and indicate much less variability than the XUV irradiance. It is interesting to note that the 0.1–4 nm irradiance, which is usually more than an order of magnitude dimmer than the H Lyman-α (121.6 nm) emission, is sometimes more intense than the H Lyman-α emission, being the brightest solar line shortward of 200 nm.

The XPS flare index is calculated as the ratio of the 0.1–4 nm irradiance to the minimum irradiance on each day. The detection of a new flare event is easily determined using the XPS flare index by noting a sudden rise from the previous sample. The 358 flare events detected by XPS, when the flare index is more than a factor of 2, are shown in the bottom panel of Figure 1. The periods of intense solar storms are 26 May 2003 (day 2003/146) to 17 June 2003 (day 2003/168), 18 October 2003 (day 2003/291) to 20 November 2003 (day 2003/324), and 11 July 2004 (day 2004/193) to 19 August 2004 (day 2004/232). These three periods are indicated in Figure 1 as A, B, and C and also are the time periods expanded for the three plots in Figure 2. The total number of M class (moderate) and X class (large) flares during the A, B, and C periods are 55, 68, and 48, respectively.

The majority of the flares were emitted from the large active regions that persisted during these periods. The intense magnetic fields and complexity of the magnetic fields from these large active regions are ultimately the sources for the flare events. It is interesting to note that the large active regions elevated the background level of the X-ray irradiance by an order of magnitude more than the usual, quieter level. If the background level had not been elevated, then the flare index (ratio) would had

TABLE I
Summary of 20 large flares in 2003 and 2004.

	GOES		XPS	
Date (DOY)	Flare class	Peak time (UT)	Max. time (UT)	Flare ratio
17 Mar 2003 (076)	X1.5	19:05	19:08	14.5
18 Mar 2003 (077)	X1.5	12:08	12:14	13.0
9 Jun 2003 (160)	X1.7	21:39	21:38	12.6
15 Jun 2003 (166)	**X1.3**	**23:56**	**24:32**	**9.7**
17 Jun 2003 (168)	**M6.8**	**22:55**	**22:55**	**8.8**
23 Oct 2003 (296)	X5.4	08:35	08:55	12.7
26 Oct 2003 (299)	X1.2	06:54	06:56	8.9
26 Oct 2003 (299)	X1.2	18:19	18:37	9.6
28 Oct 2003 (301)	**X17**	**11:10**	**11:13**	**66.2**
29 Oct 2003 (302)	**X10**	**20:49**	**20:51**	**35.3**
2 Nov 2003 (306)	**X8.3**	**17:25**	**17:29**	**29.2**
3 Nov 2003 (307)	**X2.7**	**01:30**	**01:36**	**8.9**
3 Nov 2003 (307)	**X3.9**	**09:55**	**09:57**	**13.9**
4 Nov 2003 (308)	**X28**	**19:44**	**19:46**	**83.7**
20 Jan 2004 (020)	M6.1	07:43	07:43	8.6
26 Feb 2004 (057)	M5.7	22:30	22:30	10.4
15 Jul 2004 (197)	**X1.6**	**18:24**	**18:23**	**8.8**
16 Jul 2004 (198)	**X3.6**	**13:55**	**13:55**	**23.3**
18 Aug 2004 (231)	X1.8	17:40	17:45	12.2
12 Sep 2004 (256)	M4.8	00:56	01:01	10.5

The flare class from the GOES X-ray measurement (0.1–0.8 nm) is shown along with the SORCE XPS flare result in the 0–4 nm range. The flare ratio is the flare irradiance divided by the pre-flare irradiance. These flares were selected as the largest ones observed by SORCE XPS, and all have a flare ratio of more than 8.0. Because SORCE does not have continuous solar observations, the XPS flare maximum time can be 30 min after the GOES flare peak time. The flares shown in **bold** have their time series shown in Figure 3.

been a factor of 10 higher during this period. Because the flare index is sensitive to the background level, the flare index is useful primarily for identifying flare events; whereas, the X-ray classification or the irradiance value itself is a more accurate representation of the flare's total energy for studies involving flare physics and influences on Earth's atmosphere.

The time variation during a flare can usually be separated into the impulsive component and the gradual (slow) component that follows the impulsive phase

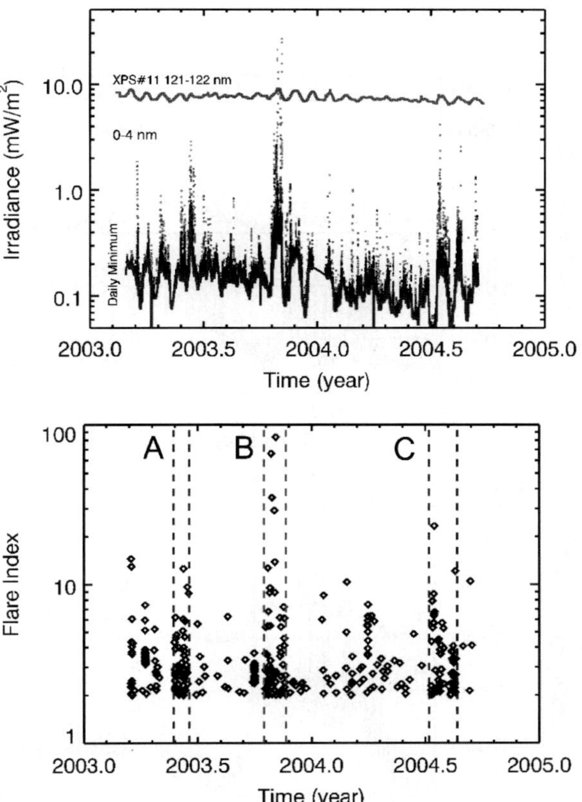

Figure 1. SORCE XPS 0–4 nm time series and flare index. The *top panel* shows the 0.1–4 nm irradiance time series along with the daily minimum value used to make the flare index. The *bottom panel* shows the flare index that is defined as the irradiance divided by the daily minimum value. Only the flare events when the flare indices are above a factor of 2.0 are shown in this plot. The time periods labeled A, B, and C are plotted in more detail in Figure 2 (panels A, B, and C).

(e.g., Donnelly, 1976). The gradual component normally peaks several minutes after the impulsive phase and can usually be described as the time integral of the impulsive component. This relation is well known as the Neupert flare effect (Neupert, 1968) that was derived from studying the impulsive component of the hard X-rays (<0.1 nm) and the gradual component of the soft X-rays (0.1–10 nm). The Neupert effect is typically explained by the thick-target flare model (e.g., Kopp and Pneuman, 1976) that has the same energetic electrons that produced the hard X-ray emissions (bremsstrahlung) near the magnetic reconnection footpoints in the chromosphere, also heating the plasma that rises and emits the soft X-rays at a delayed time.

The more intense and longer duration flares are the most effective in changing the Earth's upper atmosphere, primarily by increasing the thermosphere's temperature and the density of the neutral atmosphere and ionosphere above 100 km. The time

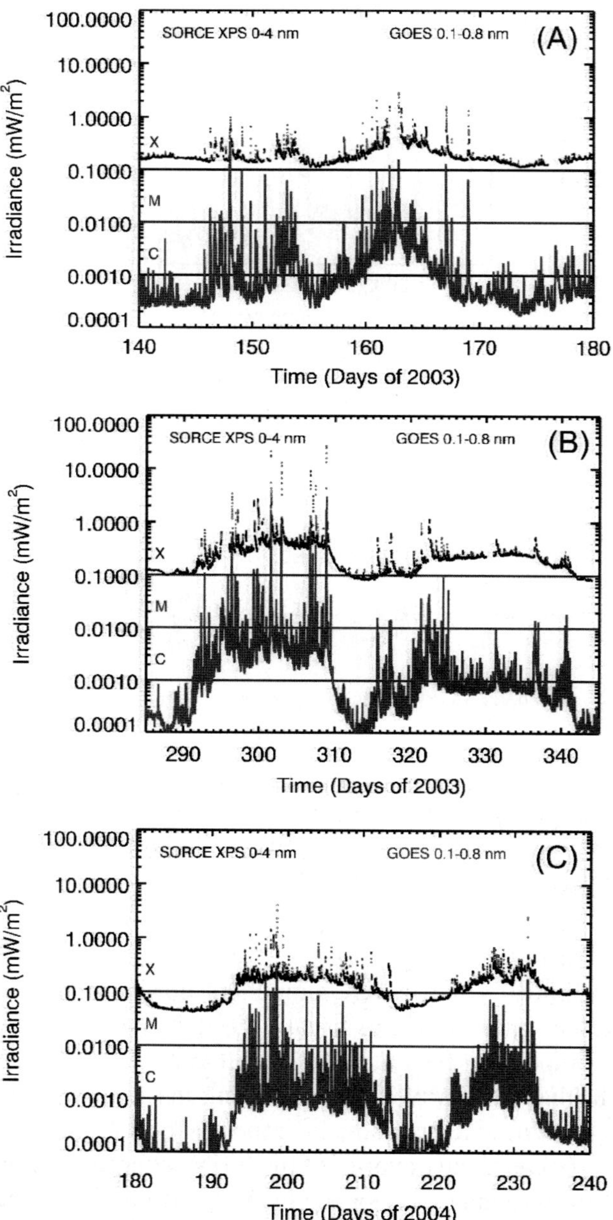

Figure 2. Three time periods of active solar storms. The interesting solar storm periods of May–June 2003, October–November 2003, and July–August 2004 are shown in panels A, B, and C, respectively. The GOES X-ray measurements in the 0.1–0.8 nm range are included, and the GOES X-ray flare classifications of the C, M, and X ranges are indicated in these plots.

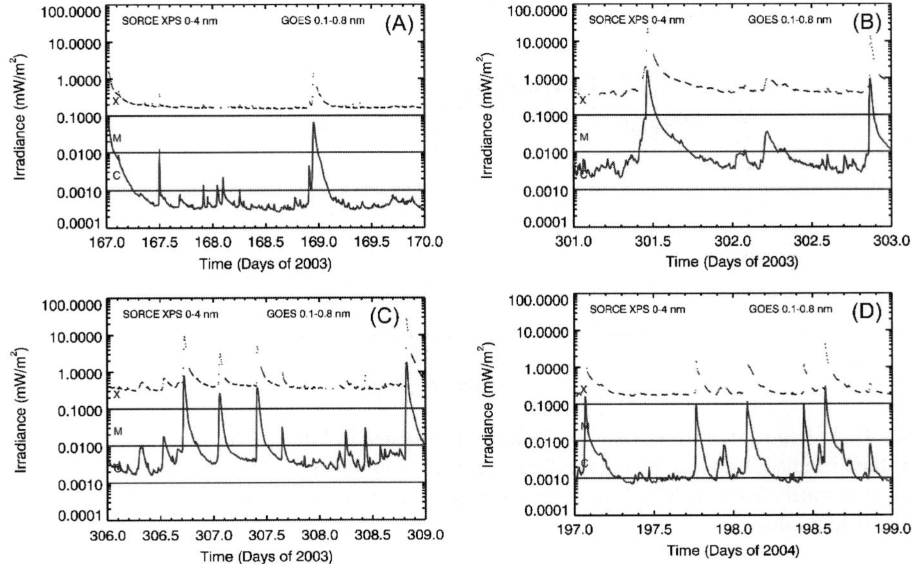

Figure 3. Four time periods of largest flares. The largest flares observed by SORCE XPS are included for 16–18 June 2003, 28–29 October 2003, 2–4 November 2003, and 15–16 July 2004 and are shown in panels A, B, C, and D, respectively. The GOES X-ray measurements in the 0.1–0.8 nm range are included, and the GOES X-ray flare classifications of the C, M, and X ranges are indicated in these plots.

profiles of some of the most geo-effective flares during these periods are displayed in Figure 3. These flares are generally characterized as having a fast rise during the impulsive phase and then a slow decline during the gradual phase. There are also several flares with a fast rise time and fast decline, such as near day 167.5 in Figure 3A, and there are a few flares with a slow rise and slow decline and with lower intensity, such as near day 302 in Figure 3B. These other types of flares are less geo-effective due to the lower total energy released.

The location of the flare on the solar disk is also important as to how effective the flare is on the Earth's atmosphere. Donnelly (1976) had shown that the EUV radiation during a flare is more intense when the flare is near the solar disk center than when the flare is near the limb. Donnelly (1976) also showed that the XUV radiation during a flare is much less sensitive to the flare location on the solar disk. This difference is because the majority of the XUV radiation is more optically thin in the solar atmosphere than most of the EUV radiation. The effectiveness of the flare location is well illustrated by the X17 and X28 flares shown in Figures 3B and 3C, respectively. The X17 flare was near disk center on 28 October 2003, while the X28 flare was near the limb on 4 November 2003. While the X-ray irradiance is about a factor of two more for the X28 flare, the EUV irradiance is a factor of two less for the X28 flare (Woods *et al.*, 2004b). Consequently, the total energy of the X17 flare from disk center is more than the energy of the X28 flare near the

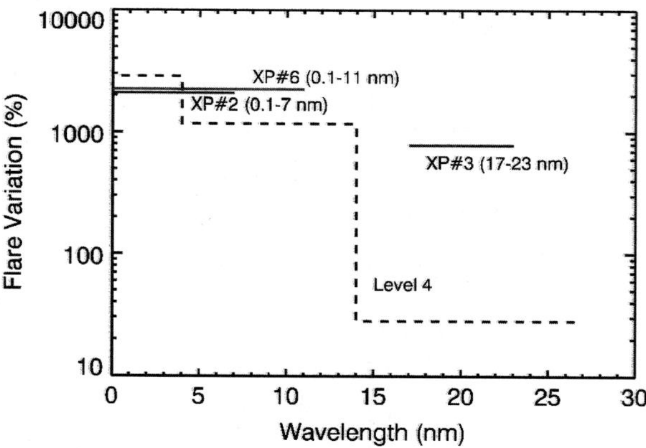

Figure 4. Average spectral variation for the large flares listed in Table I. The flare variation is defined as the flare irradiance divided by the pre-flare irradiance and minus 1.0. The *solid lines* are the results for three channels from the XPS Level 2 data product, and the *dashed line* is the result from the XPS Level 4 data product.

limb; therefore, the X17 flare is the most geo-effective flare that has been observed during the SORCE mission. For additional flare information, Woods *et al.* (2004b) give total and spectral solar irradiance results from SORCE and TIMED for the October–November solar storm period with special emphasis on the X17 flare on 28 October 2003.

The ratio of the flare irradiance to the pre-flare irradiance for different XPS channels clarifies how the XUV spectrum increases during the flare events. These ratios for the large X-class flares listed in Table I are averaged together, and the resulting flare variation (ratio minus 1.0) is shown in Figure 4 for the XPS channels #2 (0.1–7 nm), #6 (0–11 nm), and #3 (17–23 nm) as obtained from the XPS Level 2 data product. As shown in Figure 4, the average flare variations for these bands are about 2000% (factor of 20), 2000%, and 800%, respectively. The flare variations in the broad bands are fairly similar between all of these 20 flares. The largest flare observed by SORCE XPS was on 4 November 2003 and its variation is about 4 times larger than the results shown in Figure 4.

It is known that the photometers for the 17–34 nm range over estimate the flare variations in the Level 2 data product because they also have significant sensitivity at the shorter wavelengths in the XUV range where the flares have the largest increase. The XPS Level 2 processing algorithm, as described in more detail by Woods, Rottman, and Vest (2005), is a straightforward calculation of the irradiance from the photodiode current without regard to significant spectral changes in the solar spectrum. This algorithm is very accurate for photometers with a single bandpass, such as XP#2 and XP#6 shown in Figure 4, but has its limitations for photometers with dual bandpasses, such as XP#3 that has a bandpass at 17–23 nm and 0.1–7 nm. For example, the estimated uncertainty for the effects of the solar spectrum having relative changes over the 11-year solar cycle is 14% for XP#3 but is less than 1% for XP#2 and XP#6 (Woods, Rottman, and Vest, 2005).

It became clear during the first year of the TIMED mission that a different algorithm is needed to process flare events for a broadband instrument like XPS because the coronal emissions in the 33–37 nm range as measured by the EUV Grating Spectrograph (EGS) aboard TIMED show much less flare variations than the coronal emissions in the 17–23 nm range as measured by XPS using the Level 2 algorithm (Woods *et al.*, 2003). A new algorithm that is more robust for flare events was implemented as Level 4 processing software. The Level 4 algorithm derives the solar irradiance by scaling a reference spectrum in 1-nm intervals to agree with the broadband measurements by XPS (Woods, Rottman, and Vest, 2005). This approach for SORCE XPS produces the scaling factors of the reference spectrum in the 0.1–4 nm, 4–14 nm, and 14–27 nm bands using the signals from the XP#3, XP#6, and XP#3. This approach works significantly better for the XPS measurements, both flare and daily measurements, and is somewhat validated by the better agreement of the flare variations of the coronal emissions measured by EGS at longer wavelengths and the coronal emissions measured by XPS near 17 nm.

The flare variations from the XPS Level 4 data product are also shown in Figure 4 for comparison to the Level 2 results. Because the reference spectrum in the Level 4 processing is based on the daily 10.7 cm radio flux (F10.7), the flare variations are constant over the broad bands used in the Level 4 data processing algorithm, being 0.1–4 nm, 4–14 nm, and 14–27 nm. As shown in Figure 4, the average flare variations for these Level 4 bands are about 3000% (factor of 30), 1200%, and 30%, respectively. Because a flare model is not used, the flare results from the Level 4 data product could have significant errors at the 1-nm resolution. The channel results discussed above are from the XPS Level 2 data product that is independent of any solar model. While the results from the Level 2 and 4 data products agree at the shorter wavelengths in the XUV range, the difference in the 14–27 nm range is large as expected, and the Level 4 data product is considered more accurate for the flare variations at this wavelength. Even though the Level 4 algorithm improves the XPS flare results, additional work is planned to further advance the XPS processing algorithms with the goal to better understand the spectral variations of the solar XUV irradiance from the XPS measurements.

It is interesting to note that the flare variations shown in Figure 4 are similar in magnitude as the long-term variations, which are discussed later and shown in Figure 6. A detailed study of the several hundred flare observations by SORCE XPS promises new insight on the flare variations in the XUV range.

3. Intermediate-Term Variations

One of the dominant variations observed in the solar irradiance time series is due to the solar rotation. These variations are mainly observed with a period near 27 days, which is the mean rotation period of the Sun. Because of differential rotation (equator rotating faster than poles) and the inhomogeneous distribution of active regions on the Sun, solar rotation variations actually range in period from 25 to 30 days. In addition, there are epochs when the distribution of active regions on opposite sides of the solar disk produces a strong variation of about 13.5 days (half the solar rotation period) with its amplitude reduced below the 27- day variations. These 13- day variations depend on the center-to-limb variation of each emission and thus have a strong wavelength dependence that can differ from the 27-days variations (Donnelly and Puga, 1990).

The early part of the SORCE mission provides a good (typical) example of solar rotation variability that is not dominated by solar storms. The solar rotation variations from 22 March 2003 (day 2003/081) to 1 April 2003 (day 2003/091) are provided in Figure 5 for three XPS channels. This example is considered moderate variation because there are other periods when the solar rotation variation is larger, such as the October 2003 storm period, and when the variation is smaller, such as the early part of 2004. As expected, the hotter coronal emissions in the 0.1–7 nm range indicate the most variability of about 100%. Out of the XPS channels, the H Lyman-α measurement by XP#11 indicates the least amount of variability of about 20%. The H Lyman-α emission arises from the upper chromosphere and transition region and varies less than the coronal emissions that dominate in the bandpasses of the other XPS channels. Of course, the irradiances from the lower chromosphere and photosphere vary even less as expected and also measured by the other SORCE instruments.

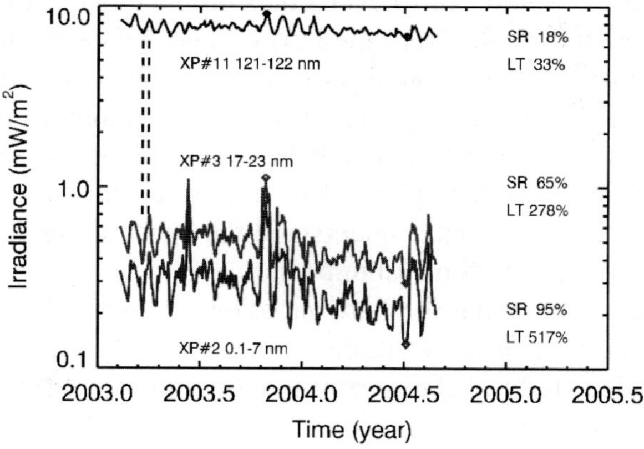

Figure 5. Solar irradiance time series from three SORCE XPS channels. The intermediate-term variations seen in these data are primarily caused by the rotation (~27 days) of the active regions on the Sun. The solar rotation (SR) variation listed is for the days indicated by the *dashed lines* on days 2003/081 and 2003/091. The long-term (LT) variation listed is for the days indicated by the *diamond symbols* on days 2004/187 and 2003/301. These variations are the ratio of maximum to minimum and minus 1.0 and expressed in percentage.

4. Long-Term Variations

The SORCE mission began with moderate solar conditions in early 2003, and there have been several large solar storms during the declining phase of solar cycle 23 that are comparable to solar maximum levels. Except for the recent outburst in July 2004, the solar activity has decreased to even lower moderate levels in 2004. The low solar activity levels associated with solar cycle minimum are not expected until sometime in 2006, so the long-term variations reported here during the SORCE mission do not yet represent the full range of variability expected over the solar cycle.

During the SORCE mission so far, the highest level of activity is during the October 2003 storm period with the maximum occurring on 28 October 2003 (day 2003/301). The lowest level of activity is on 5 July 2004 (day 2004/187), right before another solar storm period in July 2004. The long-term variations from these dates are provided in Figure 5. Similar to the intermediate-term variations, the 0.1–7 nm range has the most variability by about a factor of 5, and the H Lyman-α measurement has the least variability by about 30%. The full range of solar cycle variation is expected to be a factor of 2 to 3 times more than these values.

In addition to the results shown in Figure 5 from the XPS Level 3 data product, the 1-nm spectral irradiances from the XPS Level 4 data products are examined, and these variations are shown in Figure 6. It is important to note that the 1-nm resolution in this plot is from reference spectra that are scaled over the broadbands used in the XPS Level 4 data processing algorithm, being 0.1–4 nm, 4–14 nm, and 14–27 nm. That is, the XPS provides the absolute magnitude of the solar irradiance over its broad bands, and the reference spectrum provides the relative shape of the solar spectrum in 1-nm intervals. The reference spectrum is calculated using the daily time series of the solar 10.7 cm radio flux (F10.7) and is based on rocket and AE-E measurements (Woods and Rottman, 2002). Therefore, the relative shape

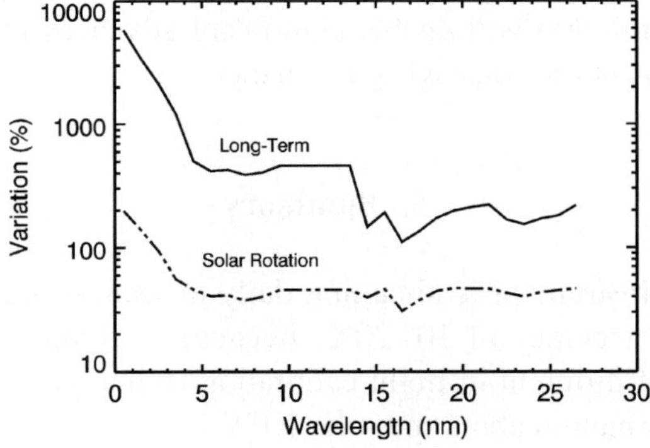

Figure 6. Variations over the SORCE mission. Comparison of the long-term variation over the SORCE mission (*solid line*) to solar rotation (*dot-dash line*) variations shows some similarity in wavelength dependence. These spectral results are from the XPS Level 4 data product that includes a spectral model of the solar irradiance in 1-nm intervals.

of the solar spectrum within the three XPS broad bands reflects the accuracy of the historical measurements at higher spectral resolution, and the absolute levels, both in irradiance and variation, of the broad bands reflect the XPS measurement accuracy and precision.

The long-term variations shown in Figure 6 represent primarily solar related changes during the first 18 months of the SORCE mission but could also include instrument degradation effects that might not be fully removed in the SORCE XPS Version 6 data set. The similarity of the spectral dependence of the long-term variation with the solar rotation variation (also shown in Figure 6) suggests that the instrument responsivity changes are reasonably corrected for in the XPS data.

The long-term (solar cycle) variations are expected to be a factor of 2–6 times more than the 27-day solar rotation variations as also learned from the analysis of the UARS solar FUV irradiances (Woods *et al.*, 2000). However, the long-term variations cannot simply be scaled from the short-term variations using a solar proxy as learned from the UARS measurements (Woods *et al.*, 2000). The relative differences between short-term and long-term variations are due to the differences in how the radiation at different wavelengths is manifested in the solar atmosphere. For example, the ratio of the solar cycle variation to the rotation variation for the H Lyman-α emission is about a factor of 2 larger than the same ratio for the Mg II core-to-wing ratio (Mg proxy). Using results from analyzing solar images, Woods *et al.* (2000) explained this difference as the result of the differences between the plages and the active network regions on the Sun for different emissions. The active network regions have higher contrast for transition region emissions, such as the H Lyman-α emission, than chromospheric emissions, such as the Mg proxy. Because the active network regions contribute more to the long-term variations than to the solar rotation variations (Worden, White, and Woods, 1998; Worden *et al.*, 1999), the transition region emissions have different long-term behavior than the chromospheric emissions. Just as the multi-year UARS mission has led to better understanding of the solar UV irradiance at the longer wavelengths, an extended SORCE mission will enable significant advances in understanding the long-term variations of the solar XUV irradiance.

5. Summary

The SORCE XPS instrument is obtaining daily measurements of the solar XUV irradiance with an accuracy of 10–20%. Because of rigorous pre-flight calibrations and the use of different in-flight calibration techniques, the SORCE XPS is providing new information about the solar XUV irradiance, both in the magnitude of the solar irradiance and the amount of the short-term variations caused by flare events and the intermediate-term variations caused by solar rotation of active regions. Continued SORCE observations will provide new determinations of solar cycle variability for the XUV range.

With just 18 months into the SORCE mission, the XPS has already observed over 300 solar flares. The most interesting solar storm periods during the SORCE mission are May–June 2003, October–November 2003, and July 2004. The October–November 2003 storms produced record solar flares with the X28 flare on 4 November 2003 and the X17 flare on 28 October 2003. The XUV variations during these large flares are larger than that expected for solar cycle variations for the XUV range.

Acknowledgement

This research was supported by NASA contract NAS5-97045 to the University of Colorado.

References

Brekke, P., Rottman, G. J., Fontenla, J., and Judge, P. G.: 1996, *Astrophys. J.* **468**, 418.

Donnelly, R. F.: 1976, *J. Geophys. Res.* **81**, 4745.

Donnelly, R. F. and Puga, L. C.: 1990, *Solar Phys.* **130**, 369.

Garcia, H.: 2000, *Astrophys. J. Suppl.* **127**, 189.

Kopp, R. A. and Pneuman, G. W.: 1976, *Solar Phys.* **50**, 85.

Meier, R. R., Warren, H. P., Nicholas, A. C., Bishop, J., Huba, J. D., Drob, D. P., Lean, J., Picone, J. M., Mariska, J. T., Joyce, G., Judge, D. L., Thonnard, S. E., Dymond, K. F., and Budzien, S. A.: 2002, *Geophys. Res. Lett.* **29** (10), doi: 10.1029/ 2001GL013956.

Neupert, W. M.: 1968, *Astrophys. J.* **153**, L59.

Woods, T. N. and Rottman, G. J.: 2002, in M. Mendillo, A. Nagy, and J. Hunter Waite (eds.), *Comparative Aeronomy in the Solar System, J. Geophys. Monograph Series,* Washington, DC, pp. 221–234.

Woods, T. N., Rottman, G., and Vest, R.: 2005, *Solar Phys.*, this volume.

Woods, T. N., Tobiska, W. K., Rottman, G. J., and Worden, J. R.: 2000, *J. Geophys. Res.* **105**, 27195.

Woods, T. N., Bailey, S. M., Peterson, W. K., Warren, H. P., Solomon, S. C., Eparvier, F. G., Garcia, H., Carlson, C. W., and McFadden, J. P.: 2003, *Space Weather* **1** (1), doi:10.1029/2003SW000010, 1001.

Woods, T. N., Acton, L. W., Bailey, S., Eparvier, F., Garcia, H., Judge, D., Lean, J., McMullin, D., Schmidtke, G., Solomon, S. C., Tobiska, W. K., and Warren, H. P.: 2004a, *Solar Variability and Its Effect on Climate*, Geophys. Monograph Series 141, Washington, DC, p. 127.

Woods, T. N., Eparvier, F. G., Fontenla, J., Harder, J., Kopp, G., McClintock, W. E., Rottman, G., Smiley, B., and Snow, M.: 2004b, *Geophys. Res. Lett.* **L10802**, doi: 10.1029/ 2004GL019571.

Worden, J. R., White, O. R., and Woods, T. N.: 1998, *Astrophys. J.* **496**, 998.

Worden, J., Woods, T. N., Neupert, W. M., and Delaboundiniere, J. P.: 1999, *Astrophys. J.* **511**, 965.

Solar Physics (2005) 230: 389–413

THE SORCE SCIENCE DATA SYSTEM

CHRISTOPHER K. PANKRATZ, BARRY G. KNAPP, RANDY A. REUKAUF, JUAN FONTENLA, MICHAEL A. DOREY, LILLIAN M. CONNELLY and ANN K. WINDNAGEL

Laboratory for Atmospheric and Space Physics, University of Colorado, Boulder, CO 80309, U.S.A.
(e-mail: pankratz@lasp.colorado.edu; knapp@lasp.colorado.edu)

(Received 23 March 2005; accepted 30 March 2005)

Abstract. The SORCE Science Data System produces total solar irradiance (TSI) and spectral solar irradiance (SSI) data products on a daily basis, which are formulated using measurements from the four primary instruments onboard the SORCE spacecraft. The Science Data System utilizes raw spacecraft and instrument telemetry, calibration data, and other ancillary information to produce and distribute a variety of data products that have been corrected for all known instrumental and operational effects. SORCE benefits from a highly optimized object-oriented data processing system in which all data are stored in a commercial relational database system, and the software itself determines the versions of data products at run-time. This unique capability facilitates optimized data storage and CPU utilization during reprocessing activities by requiring only new data versions to be generated and stored. This paper provides an overview of the SORCE data processing system, details its design, implementation, and operation, and provides details on how to access SORCE science data products.

1. Overview

The Solar Radiation and Climate Experiment (SORCE) consists of a small, free-flying satellite carrying four instruments to measure solar radiation incident at the top of the Earth's atmosphere. SORCE launched in January 2003, carrying the Total Irradiance Monitor (TIM), the SOlar Stellar Irradiance Comparison Experiment (SOLSTICE), the Spectral Irradiance Monitor (SIM), and the XUV Photometer System (XPS).

Solar irradiance is the dominant energy source to the Earth's atmosphere, establishing much of the atmosphere's chemistry and dynamics, and becomes the dominant term in the global energy balance and an essential determinant of atmospheric stability and convection. The SORCE measurements provide the requisite understanding of one of the primary climate system variables. SORCE provides daily measurements of total solar irradiance (TSI) and spectral solar irradiance (SSI) from 0.1 to 2700 nm and, in the case of the ultraviolet measurements, SORCE maintains calibration by comparison to bright, early-type stars. The SOLSTICE instrument measures spectral irradiance from 115 to 310 nm with a spectral resolution of 1 nm, the SIM measures spectral irradiance from 200 to 2700 nm with a spectral resolution varying from 1 to 34 nm, and the XPS measures six broadband samples from 0.1 to 34 nm, and at Lyman α (121.6 nm). Measurements from

34 to 115 nm are not made by the SORCE mission, but are available on a daily basis from the NASA Thermosphere–Ionosphere–Mesosphere Energetics and Dynamics (TIMED) program (Woods *et al.*, 2005).

The SORCE is a Principal Investigator-mode mission, for which the Laboratory for Atmospheric and Space Physics (LASP) has full responsibility and accountability for all aspects of the mission. LASP operates the spacecraft and instruments from facilities in Boulder, Colorado, USA, and maintains responsibility to capture, manage, process, analyze, validate, and distribute all science data products.

All mission aspects related to the SORCE ground system are facilitated by the Mission Operations and Information Systems (MO&IS) division of LASP at its facilities in Boulder, Colorado. SORCE MO&IS support activities can be broken down into three primary systems: Instrument Operations, Mission Operations, and the Science Data System (SDS). The Mission Operations Center maintains responsibility for the control and monitoring of the SORCE *spacecraft*. The Instrument Operations system has responsibility for all *instrument* operational activities. All science data production and management responsibilities are provided by the SORCE Science Data System (SDS), which resides at the SORCE Science Operations Center (SOC) at LASP. Data processing is performed automatically with the production of data through Level 3 commencing 2–5 days after the time of data reception from the spacecraft, allowing for telemetry retransmissions and receipt of definitive spacecraft orbital ephemerides from NORAD. The SORCE SDS processes all levels of scientific data products and manages them using a relational database system. The SORCE SDS delivers Level 0 and Level 3 data, including algorithms and associated software packages, metadata, production histories, ancillary data and Quality Assessment (QA) data to the NASA GSFC Earth Sciences (GES) Distributed Active Archive Center (DAAC) for archival and distribution (*http://daac.gsfc.nasa.gov/upperatm/sorce/*). Scientific data products are also available from the SORCE web site at *http://lasp.colorado.edu/sorce*.

The science data processing software is implemented using an object-oriented design, which isolates algorithms that are independent of one another. In any science processing system, requirements are expected to change over time; this design permits modification – and even replacement of algorithms – with minimal impact on other processing system elements. Furthermore, a base library of data configuration management functionality is built into the system to facilitate version management and run-time discovery of system changes. This will be discussed in more detail later.

2. Data Products and Availability

The SORCE Science Data System (SDS) has full responsibility for all science data production activities, and consists of both the hardware and software components necessary to generate, manage, analyze, validate, and distribute all standard science data products. The SDS produces two principal science data products: total solar irradiance data and spectral solar irradiance data. Measurements made by the TIM instrument (Kopp and Lawrence, 2005; Kopp, Heuerman, and Lawrence, 2005; Kopp, Lawrence, and Rottman, 2005) are combined to produce representative daily and 6-hourly values of the TSI. Measurements made by the SOLSTICE (McClintock, Rottman, and Woods, 2005; McClintock, Snow, and Woods, 2005; Snow *et al.*, 2005), SIM (Harder *et al.*, 2005a,b,c), and XPS (Woods, Rottman, and Vest, 2005; Woods and Rottman, 2005) instruments are combined into merged daily and 6-hourly solar spectra, each containing representative irradiances reported from 1 to 2700 nm (excluding 34–115 nm, which is not covered by the SORCE measurements, but is available from the TIMED SEE project (Woods *et al.*, 2005)) on a fixed wavelength scale, which varies in spectral resolution from 1 to 34 nm over the entire spectral range. At the time of this writing, the SDS is not routinely producing data products containing spectral measurements from 1600 to 2700 nm. Measurements in this spectral interval are being made by the instrument on a routine basis, and corresponding data products will be available in the future.

2.1. Data Level Definitions

The following data level definitions describe the SORCE data products. The different data levels are generally processed in ascending order; e.g., Level 2 products are produced using Level 1 data products, as shown in Figure 1. Each processing level is associated with the same level as its data output; e.g., processing Level 2

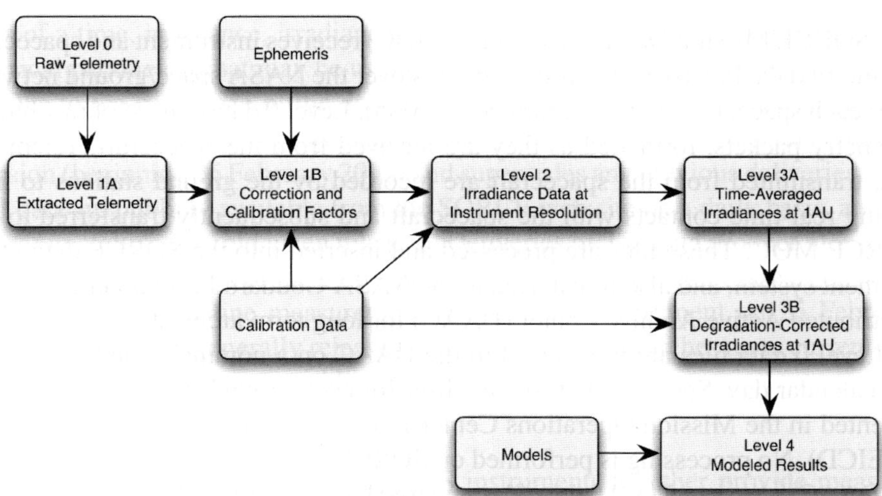

Figure 1. Data Processing Flow Diagram. This figure depicts the high-level flow of data through the SORCE data processing system, whereby less refined data products are used to produce more refined data products. Processing of data levels generally proceeds in ascending order, from Level 0 through Level 4.

resolution varying from 1 to 34 nm (Harder *et al.*, 2005a,b), and the XPS instrument measures six broadband samples from 0.1 to 34 nm and also at Lyman α (121.6 nm) (Woods, Rottman, and Vest, 2005; Woods and Rottman, 2005). Measurements from these instruments are combined into merged daily and 6-hourly spectra (6-hourly data is currently available for the XPS instrument only and may be available from other SORCE instruments in the future), each containing representative irradiances reported on a standard wavelength scale, with spectral intervals that vary from 1 to 34 nm in size. Irradiances are reported at a mean solar distance of 1 astronomical unit (AU) and zero relative line-of-sight velocity with respect to the Sun.

The SORCE spectral data products consist of daily and 6-hourly representations for each calendar day (universal time), resulting in two data files per day, each representing 24 h of data beginning at midnight universal time (UT). Delivery nominally occurs daily and the size of each delivered data file remains approximately constant throughout the mission. Each of these two Level 3 products contains a single solar spectrum constructed using measurements from the SOLSTICE, SIM, and XPS instruments. The spectral irradiance data files are delivered in HDF version 5 (HDF5) format, a platform-independent self-documenting file format developed by the National Center for Supercomputing Applications (NCSA). Each includes the following elements, stored and annotated as HDF data structures:

- The time-annotated solar irradiance spectrum, presented on a standard wavelength scale. One spectrum in the daily average file, four spectra in the 6-hourly file.
- Time-annotated spectral irradiance measurements for six XPS channels.
- Ancillary solar, geophysical, or other physical parameters, such as the mean Sun–Earth distance and solar Carrington latitude and longitude.

Gaussian fits and integrations for selected spectral emission and absorption line profiles will be included in a future version. Also note that SORCE spectral data products may be obtained interactively via the SORCE interactive data access web pages at *http://lasp.colorado.edu/sorce/sorce_data_access*.

The SORCE HDF-formatted files are binary compatible with many computer platforms, including Intel-based PCs running Windows or Linux, MacOS 9, MacOS X, Sun Solaris, etc. Numerous tools are available to permit access to the data contained in these files, and the SORCE SSI web site (*http://lasp.colorado.edu/sorce/ssi_data.html*) presents some of the more common and convenient methods, including a customized file reader for users of the IDL data analysis environment.

2.4. Data Availability

Routine science data processing is performed automatically within 2–5 days of data reception from the spacecraft, and data products are nominally made available to the public shortly thereafter, following preliminary data inspection. This preliminary data quality assessment typically occurs within 24 h of data processing, and products are made available for public access immediately thereafter.

Scientific data products, along with associated metadata and documentation, are available from two locations: the GES DAAC and the SORCE web site. The GES DAAC maintains a SORCE data portal, which may be accessed via the URL *http://daac.gsfc.nasa.gov/upperatm/sorce*. While the GES DAAC archives and distributes Level 3B science products, the SORCE web site (*http://lasp.colorado.edu/sorce*) provides access to a slightly larger selection of data products, including selected Level 2 and analysis products. Visitors may either download the standard data files, or perform custom selections of data directly from the SORCE database. Additionally, data are available from the SORCE web site slightly sooner than they are available at the DAAC, typically 6 days after the instruments make the measurements.

Some time-dependent corrections (such as instrument degradation) require periodic in-flight calibration data several months into the future. As a result, subsequent (and periodic) *reprocessing* is required in order to maintain the quality of the science products. Updates to algorithms also occur occasionally, themselves warranting a reprocessing of data. Based on past experience, such reprocessing activities typically take place 1–2 times per year for each instrument's processing system. When reprocessing of data is needed, it almost always requires reprocessing of the complete mission. Changes to algorithms alone usually warrant a full-mission reprocessing, but for more practical reasons it is convenient to ensure a uniform version of data for the entire mission. The SORCE SDS has been designed to facilitate reprocessing of the complete mission within a period no greater than 30 days, after which the new versions of standard data products are delivered to the public.

2.5. Algorithm Releases

Occasionally, the SORCE program will release a reference archive of the SORCE science processing software to the DAAC. These Delivered Algorithm Packages (DAPs) are provided as a matter of record only, and are not intended to easily install or directly execute within computing environments outside of LASP. Nevertheless, the SORCE team takes a sensible approach to software development and incorporates reasonable coding standards and portability considerations. At the time of this writing, the SORCE science processing software is known to execute successfully on the Solaris, Windows, and MacOS X platforms. The SORCE Delivered Algorithm Packages will take the form of a single archive that contains all relevant SORCE science software, basic design documentation, a description of the intended computing environment, the SORCE Algorithm Theoretical Basis Document (ATBD) and documentation of the science products and related usage information. It is expected that there will be only limited interest in accessing this software archive, and the content and format are chosen accordingly. Following delivery, these DAPs will be available from the GES DAAC.

3. System Architecture and Design

The SORCE Science Data System consists of several subsystems, each of which utilizes a common centralized database, as depicted in Figure 2. The core of the system consists of a commercial relational database system, in which all telemetry, calibration data, scientific data products, and ancillary information are stored. Public users access the SORCE scientific data products stored in the database management system via a variety of web-based user interfaces, each of which directly queries the database on-demand. In order to support routine operations associated with the SORCE mission, local SORCE project personnel interact with the same database server as part of their routine activities. Routine planning of on-orbit activities, including instrument science experiments and spacecraft maneuvers, is facilitated using data stored in the database, and all planned activities are themselves managed

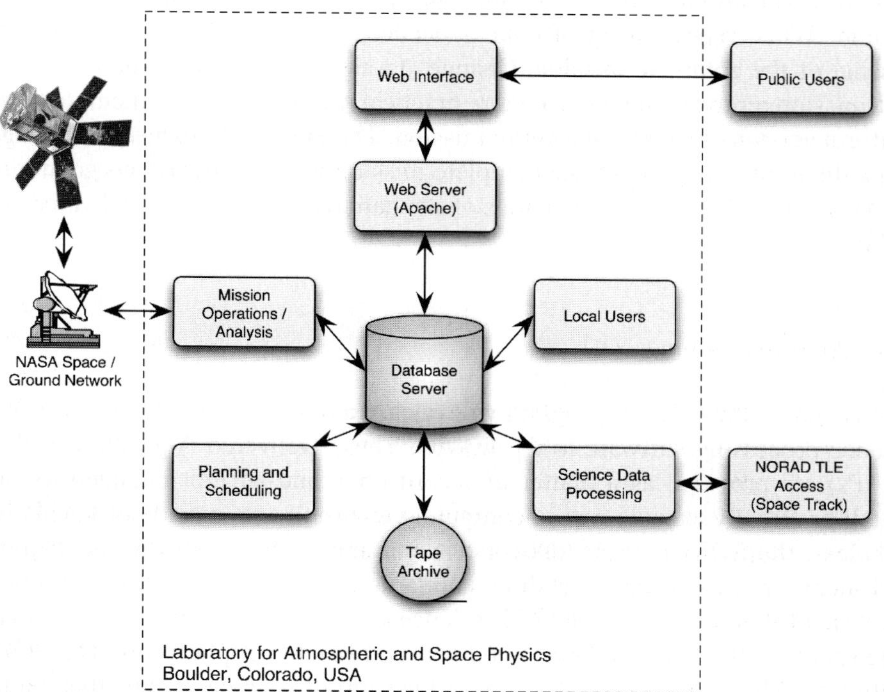

Figure 2. SORCE Science Data System (SDS) Architecture. A centralized commercial relational database management system is used to manage all SORCE data, distributing these data to elements within the SORCE project and to the general public.

by the database. The mission operations segment of the SORCE mission, which is also located at LASP, uses the database to manage a variety of data, such as operational procedures, spacecraft contact schedules, and raw telemetry data. The science processing system stores scientific data products in the database, which are subsequently accessed by local project personnel and the public for analysis purposes.

At the time of this writing – 2 years after launch of the SORCE – the database system manages slightly less than 1 terabyte of data. All telemetry and science data are stored in the database as individual time-referenced points to provide direct and rapid access to each datum received from the spacecraft or instruments or those data that are subsequently processed. Certain file cataloging and archiving activities are also required to manage these data, for instance, design documentation and raw telemetry data as received from the ground stations following spacecraft contacts.

The data processing and calibration data management software are tightly coupled with the SORCE project database. The data processing component of the system interacts directly with the time-referenced data stored in the database in order to provide efficient data utilization. Input data are obtained from the database system and generated data products are stored to the database. With this technique, processing steps do not produce science product data *files*, but rather store each science datum in one or more normalized database tables, along with all metadata necessary for complete traceability.

Algorithms are implemented using a data-centric object-oriented design philosophy, in which the software design parallels the logical relationships between primary and intermediate data products. Software objects each have a scope limited to the production of a specific datum, and relationships between objects are minimized to reduce overall system complexity. This design has proven very successful, and is discussed in more detail in subsequent sections.

In legacy data processing approaches, distinct data products are identified (Level 1, Level 2, Level 3, etc.) and are each associated with a specific program executable that utilizes lower-level or ancillary data products to produce a new, higher-level, data product. The initiation of each executable occurs by way of a process that can be initiated manually or automatically. In such an approach, if even a single calibration datum changes the complete executable for that level of processing must again be executed to produce a new version of the data product, thereby performing all intermediate computations as well. If an algorithm change occurs (even a minor one), the complete executable must be rebuilt and re-run in order to produce new data, and every calculation must again be performed. This data processing approach has inefficiencies considering the fact that added and unnecessary CPU cycles are consumed to reproduce redundant data, and added data storage space is also required. One merit of this approach, however, is a simpler conceptual design, in that all data and process flows are linear and predictable. This linear approach for data processing is only used for the SORCE XPS products as its software is reused from the TIMED XPS data processing system (Woods *et al.*, 2005).

The approach adopted for the SORCE SDS (except for SORCE XPS) eliminates these difficulties by associating each primary and intermediate data product with a single software package that is executed independently of the packages associated with other, unrelated, data types. Each software package contains the implementation of a single algorithm, is responsible for generating a single type of data, and is itself responsible for initiating the generation of other intermediate results on which it depends. Any change to the data or algorithms used by that subsystem does not affect other independent subsystems.

For example, consider the SOLSTICE instrument processing algorithms (Pankratz *et al.*, 2000; McClintock, Rottman, and Woods, 2005; McClintock, Snow, and Woods, 2005), which generate several intermediate quantities in order to produce an irradiance datum. These include wavelength, filter transmission, detector temperature gain, photometric calibration, and several others, all of which are combined using the instrument measurement equation to produce an irradiance value. If a new photometric calibration dataset is released, it will be used to recalculate the photometric calibration factor that is eventually applied in the irradiance measurement equation (Pankratz *et al.*, 2000). A new photometric calibration will have no affect on other independent calibration factors, such as the wavelength calibration, filter transmission, etc. As a result, a new version of these independent quantities is not required and new data need not be generated. This approach is facilitated by the data-centric object-oriented design, and integration of data configuration management capabilities within the software and data management system. Additionally, in this manner, the task of generating a science data product (like Level 2 SSI) can take place completely independently of the processes that generate the calibration parameters, gain factors, etc.

4. Algorithm and System Implementation

The SORCE Science Data System produces selected EOS standard products (King *et al.*, 2003) using data acquired from the SORCE spacecraft, instruments, pre-flight and in-flight calibration data, and spacecraft orbital ephemerides obtained from NORAD. Algorithms to produce these products are implemented in accordance with the SORCE Algorithm Theoretical Basis Document (ATBD) (Pankratz *et al.*, 2000), which presents the theoretical and mathematical basis for the algorithms utilized in the production of scientific data products. Primary motivating factors in the design of the SORCE SDS included increased *reliability*, such that software bugs are more easily identified and repaired; *extensibility*, to accommodate changes in requirements; *reusability*, so software components are usable for other instruments on SORCE and on future projects; and *portability*, to permit execution on other computer platforms.

The SDS was designed with the expectation that algorithm requirements were likely to change over time, sometimes significantly, even after launch and commissioning of the instruments. This scenario can lead to an unpredictable elevation in long-term software maintenance costs, and mitigation of this risk on the SORCE program was paramount. This design objective strongly suggested the use of an object-oriented software design, with software objects limited in scope to the implementation of a single algorithm, which produces a unique type of data. Such a data-centric object-oriented design minimizes the coupling between software elements and thus minimizes the scope and impact of future modifications.

The SDS design needed to accommodate convenient accessibility of data products to members of the SORCE science team, who utilize a variety of computer platforms. The second design objective suggested the use of a centralized and network-accessible repository for the data products. Other requirements, including data traceability and flexibility in creating customized datasets, led to the selection of a system design which relies on a centralized relational database management system (RDBMS).

The SORCE science data processing software is tightly coupled with the SORCE project database at LASP and is implemented in a variety of programming languages, including Java, IDL, FORTRAN, C, and UNIX shell scripts. Software objects are developed using the Java programming language for portability, with some Java methods implemented in other languages, such as C or FORTRAN. This permits reuse of legacy code from other projects and improves overall system performance by utilizing the appropriate programming language for each particular task. Each programming language is selected for specific needs based on its suitability, performance potential, and the ability to make use of existing software libraries. Software developed for the SORCE mission conforms to generally accepted software coding standards and best practices. Useful object-oriented design philosophies were used from Wirfs-Brock, Wilkerson, and Wiener (1990), Meyer (1997), and Coad and Mayfield (1999) and several very useful specific object-oriented design patterns were taken from Gamma *et al.* (1995) and Grand (1998), which apply to languages other than Java, as well.

All software is configuration managed at the SORCE SOC using a commercial configuration management system and software versions and configurations are fully traceable to each produced datum.

4.1. Data management, versioning, and traceability

All data used in the generation of the SORCE scientific data products are managed within a relational database, which maintains, under configuration control, raw instrument and spacecraft telemetry data, calibration data, science data products, operational plans, and other ancillary data. The SORCE SDS also establishes and maintains audit trails that provide full traceability between data produced by the data processing system and the original source code, operation plans, calibration data, and other relevant information. In addition the versions of each relevant software module, values of processing job control parameters, and the versions of other related data products are also associated with each produced datum.

A key objective of the SORCE data model is to facilitate full traceability and reproducibility of each and every datum that is produced and managed by the system. As the algorithms described in the SORCE ATBD typically require the generation of several intermediate types of data, to maintain full traceability the SDS must formally manage these data in addition to the standard products. The software components that implement the algorithms that produce data to be managed are referred to as *data servers* and the data they produce are referred to as *managed datasets*. Each *managed dataset* consists of one or more tuples of data, which are stored in the SORCE database. Each has a pedigree of dependencies, including the versions of software modules used in the calculation, job control parameters that specify, for instance, a tolerance criterion, and other managed datasets that may themselves be other managed datasets. This pedigree represents the state of the data processing system used to generate a particular data product. In this way, each variable that appears in the measurement equation is calculated and stored individually in the database as a managed dataset, and each managed dataset has a unique version assignment that distinguishes it from those generated with a different state of the data processing system.

There are two classifications of managed datasets: *data products* and *primitive datasets*. Data products represent those data that are generated by applying algorithms to produce data, possibly making use of other managed datasets. Data servers, in most cases, depend on data returned by other data servers, including both data products and primitive datasets. Primitive datasets, on the other hand, represent those data that are not generated by the data processing system. These data usually originate outside of the science data system, possibly in the calibration laboratory or from another outside source. Examples of primitive data include telemetry data, certain physical constants, and calibration datasets.

Each distinct algorithm produces data and stores those data into one or more specific relational database tables, where each record includes a time stamp, version, instrument channel identifier, and the generated data. The internal database tables are normalized and designed for ease of maintenance, maximum data integrity, and optimal performance. The specific data structures and the interfaces associated with these tables are designed for flexible and appropriate scientific "views" of the data. A key benefit of using a relational database system to store all of the science data is the ability to store data in a logically optimized fashion, while maintaining more conceptual views of the data to various categories of users. Strict file-based systems can only achieve this by redundantly storing data in files for each desired "view", or by reading and reformatting the data at the time the user requests the data.

To facilitate data versioning and traceability, metadata are generated at run-time by the software itself and are also stored in normalized relational database tables. Metadata associated with each datum produced by the system include the versions of all software modules, control parameters, and other data products that were used.

Some algorithms are common to multiple SORCE instruments, for instance the TIM (Kopp, Heuerman, and Lawrence, 2005) and SIM (Harder *et al.*, 2005b) ESR detectors. This commonality leads to the need for an identifier to distinguish managed datasets according to the instrument, channel, and operational mode for which the data are associated. In the SORCE SDS, this identifier is referred to as an *instrument mode*. Each managed dataset produced by the system is therefore created for and identified with a particular instrument mode. For a given project, instrument modes are predefined and uniquely identify the project (e.g., SORCE), the instrument, and any designated instrument configurations or operational modes that impact the data production activities.

Data version numbers are commonly used in the production of scientific datasets and require little introduction. For the SORCE SDS, the use of a data version indicator implies the existence of associated traceability (dependency) information and therefore the existence of a traceability tree establishing relationships between all data products for the mission. A data version is determined based on the unique union of three types of dependencies: (1) the software version of all modules used in the calculation; (2) job control parameters that specify, for instance, tolerance criteria; (3) the data versions of other managed datasets that may themselves be other data products.

4.1.1. *Inherent Data Traceability Determination*

One unique aspect of the SORCE SDS is that the logic to determine data version numbers is embedded within the data processing software itself. At run-time, the system instantiates software objects, determines software versions, obtains run-time control parameters, and analyzes them in relation to previous executions of the processing software. As differences are identified, the system selectively assigns new data version numbers to the managed datasets that are directly or indirectly affected by the changes. Data versions exist as integers that are incremented by one for each data version promotion that occurs. For a given managed dataset, if any dependency changes, a new version results and any data generated by the system during that process will incorporate this new version number.

The embedded data traceability algorithms are implemented in a common code base that is inherited by each data server, such that software developers may implement algorithms without the burden of any explicit awareness of the data traceability system. In this way, the data traceability software itself is decoupled from the algorithms, and its presence does not adversely impact the implementation of new algorithms, or the modification of existing algorithms. This is another major benefit of choosing an object-oriented software design.

Each data server, when instantiated at run-time, performs a one-time check with the database to establish its data version (and other metadata) as appropriate for the current run-time state of the data processing system. The ensuing process is performed "behind the scenes", such that the data servers need not explicitly identify their dependencies in source code. That is, programmers do not need to explicitly identify the dependencies when they implement an algorithm. The software design inherently facilitates this capability, minimizing the otherwise present risk of a programmer forgetting to explicitly identify a particular dependency. If an existing data version is not found in the data management system to match the current software, parameter, and dependency sets, then the managed dataset and its entire client tree are notified that any existing data within the database are out-of-date and may not be delivered to client data servers. Each data server then prompts a new data version record to be established in the database and ensures that all subsequent client requests for data result in new data generation. If the data server object does have a data version record that matches the current software, parameters, and dependencies, any data for that managed dataset are simply retrieved from the database, and no new calculations are performed. By eliminating the need to perform redundant computations, overall performance of the data processing system is improved for reprocessing efforts, in which only a few changes are typically made; targeted changes to software or calibration data need not affect the system as a whole.

Safety checks are employed in the data traceability subsystem to prevent dependencies from inadvertently being missed. For instance, data servers are prevented from accessing control parameters after the data version check takes place; otherwise the data version check would be incomplete.

A conceptually representative database table schema demonstrating the SORCE data model is shown in Figure 3. The data values produced by the processing algorithms would be stored in database tables designated ProductData1 and ProductData2, a particular calibration dataset would be stored in CalibrationData, and the other tables facilitate storage of the version and traceability information. The

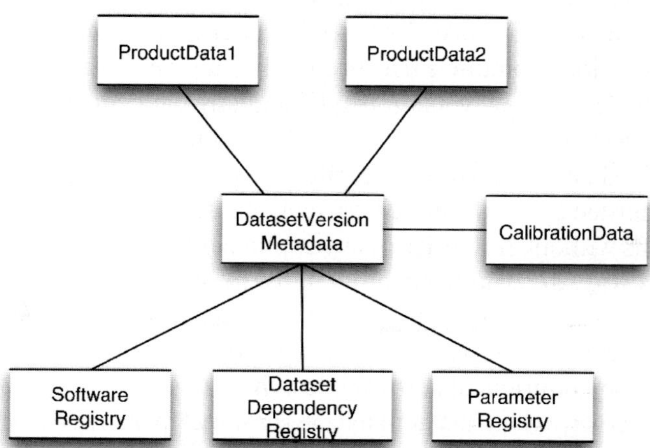

Figure 3. SORCE SDS Conceptual Data Model. This conceptually representative data model resembles the database schema actually used to store SORCE scientific data products and their associated traceability metadata.

DatasetVersionMetadata table stores the version for each unique managed dataset, as well as the date the version was first created, and references to the information that describes the three classifications of dependencies associated with each data product: software, parameters, and dataset dependencies. For a given data product, these three fundamental dependency types are cataloged in the SoftwareRegistry, ParameterRegistry, and DatasetDependencyRegistry. The SoftwareRegistry consists principally as a listing of software modules and their associated version. The ParameterRegistry contains a listing that identifies any parameters used to control the behavior of the software and the values specified for those parameters. The DatasetDependencyRegistry contains a listing of references to tuples in the DatasetVersionMetadata table, thereby associating a data product with its use of other product servers and primitive servers, whose own dependencies are also managed. In this way, a hierarchical tree of dependencies can be constructed that identifies the relationships between every data product managed by the SORCE SDS.

Figure 4 shows a representative object-oriented design model illustrating the generation of SORCE data products for the SOLSTICE instrument. Each object identified in the dependency diagram represents a particular type of data that must

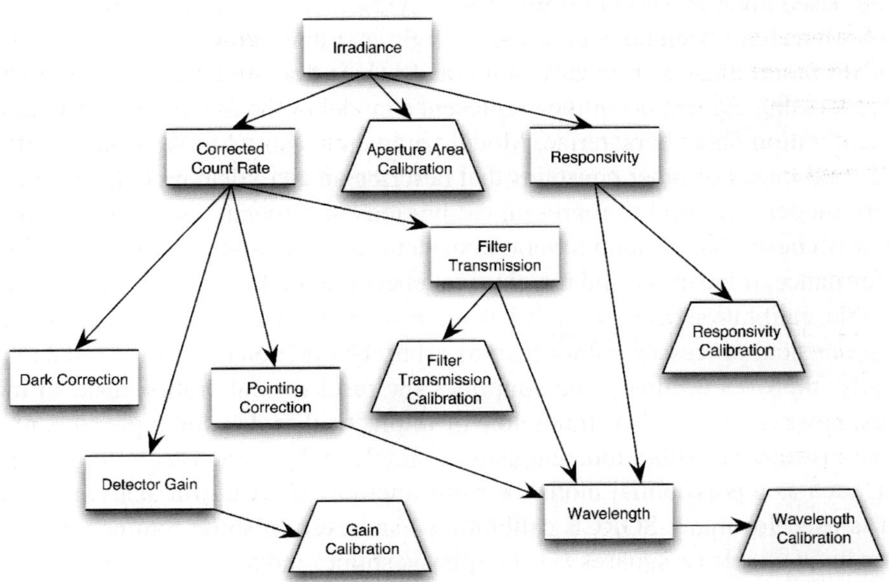

Figure 4. Dependency Diagram Excerpt for SOLSTICE Irradiance Processing Software. This excerpt of a much larger dependency diagram shows the relationships between variables in the SOLSTICE measurement equation, and also illustrates the relationships between data servers and data products. Product servers are represented by rectangles, and primitive (calibration) servers are represented by trapezoids. The direction of the arrows represents dependencies between measurement equation variables, rather than data flow. Relationships can be read in the direction of the arrows as "depends upon" or "requires"; for example "Irradiance depends upon Responsivity".

be produced as part of the generation of a standard scientific data product. Each of the objects in the diagram corresponds to a data server, and also to a type of data that is managed in the database as a managed dataset. In this diagram, data servers are categorized according to their data type and/or how these data are produced. Data for each of these data servers are stored individually in the science database, and are assigned a data version and linked to other objects on which they depend. During the execution of the data processing system, when a data server accesses another data server to obtain data, the resulting dependency is recorded, along with associated software and data versions. Data servers are also tasked with propagating uncertainties at the time calculations are performed.

4.1.2. *Calibration Data Management*

Calibration data provide a transfer between a standard and an instrument feature (e.g., ground aperture calibration, thermistor calibration). As mentioned previously, calibration data are classified and managed as primitive datasets, which are data that have no explicit dependencies within the scope of the data processing system, but can serve as dependencies for product datasets. Some calibration data are based on ground-based measurements, and other calibration data originate with in-flight calibration measurements. The SORCE mission launched with all calibration datasets being based upon pre-flight ground-based measurements; however, many of these calibrations have been updated using in-flight measurements.

Calibration data are typically numerical values that are either used directly in the processing algorithms or they represent a model of the functional behavior that the calibration data characterize. Model parameters can take the form of polynomial coefficients or other constants that describe an analytical model. In the cases where models are used to represent calibrations, the specific analytical representation is chosen based upon several considerations, including numerical stability, performance, robustness, and the physical phenomenon being modeled. Whenever possible, calibration data are analyzed and managed as parameterizations of underlying physical processes, rather than as arbitrary functional fits to raw data. This usually improves accuracy and simplifies the resulting calibration data. In many cases, however, it is either impossible or impractical to develop a physical model for an instrument calibration, and using a simple and generic function representation, such as a polynomial model, is most appropriate. A useful analytical model that is used for many SORCE calibrations, and even in some routine processing algorithms, is a least-squares *b-spline* (basis spline) model (Lawson and Hanson, 1974; de Boor, 1978). A *b-spline* model is comprised of a series of cubic polynomials, such that its zeroth, first, and second derivatives are piecewise continuous. A *b-spline* function is analytic; it can be evaluated, integrated, and differentiated analytically, with uncertainties analytically propagated. As an example, the calibration data associated with the "Filter Transmission Calibration" object shown in Figure 4 exist as coefficients to a generic analytical model of the underlying data. The measured filter transmission calibration data vary with wavelength, and a single least-squares *b-spline* model provides a convenient and flexible way to represent these data.

When a calibration needs to be applied in a data processing algorithm, the application of this calibration is facilitated by one or more data servers that are dedicated to this task. If the managed calibration data are to be used directly in a processing algorithm, only a single primitive server is required. This is the case with the "Aperture Area Calibration" object shown in Figure 4. However, when calibration data represent a model, two separate data servers are required, as with the aforementioned filter transmission example. One server provides access to the managed primitive calibration data, and the other server contains an implementation of the algorithm that permits evaluation of the representative model. Two servers are required in order to provide separation between primitive data, which have their own inherent version, and algorithms, which are implemented by product servers.

All calibration data are stored in database tables along with associated metadata and version information. Metadata are provided with each new version, providing documentation of the calibration data, including the date of release, the person responsible for producing the data, the rationale for the new version, a description of the data and how they were obtained, as well as the date on which they should take effect. Because calibration data are regarded as managed datasets, data products that are derived from calibration data are fully traceable to these primitive datasets.

4.1.3. *Job Control Parameters*

Job control parameters are keywords that are used to supply external information to a data processing system for the purpose of controlling the behavior and application of the science algorithms. The intent of these control parameters is to allow some degree of flexibility in regulating the processing without requiring recompilation of the software. Some dataset classes require parameters, while others do not. Parameters fall into two categories, those that impact the quality of data and those that do not. This designation is known in the SDS as *major* or *minor* parameters, respectively. For example, two parameters used by the SORCE SDS are the FilterWidthInCycles *major* parameter and the WriteToDatabase *minor* parameter. The FilterWidthInCycles parameter is used to specify the number of shutter cycles used in the TIM and SIM phase sensitive detection processing algorithms (Kopp and Lawrence, 2005; Harder *et al.*, 2005a). A change to its value affects the associated algorithm, which results in a change to the generated data products, and thus qualifies it as a *major* parameter. The WriteToDatabase parameter is used to prevent the data processing system from storing the data it generates into the database, which is occasionally useful during system testing. A change to its value does not result in a change to data quality, which qualifies it as a *minor* parameter.

Parameters are name-value pairs that are directed to specific data servers in the system. The parameter handling mechanism in the software is designed to allow the scope (or target server(s)) associated with a particular parameter to be regulated, such that the parameter information can be used by a designated algorithm, or even by all algorithms in the system. Each parameter may be directed as job-wide, limited to a particular product, or limited to a particular instance of a data server. To accomplish this, each parameter optionally includes keys attaching it to a specific data server, a specific instrument mode, or no targeted designation at all. For example, a specific product server uses the aforementioned FilterWidthInCycles parameter, so it is designated for use with only this specific server. By contrast, the aforementioned WriteToDatabase parameter is commonly used to render the database read-only and is typically applied with global scope to facilitate system testing.

4.2. Development process

Software development of any system component follows a methodology that encompasses planning, requirements analysis to capture both static and dynamic requirements, design and development, and testing and integration. The evolution of a software project is commonly viewed at a high-level by a lifecycle model. Lifecycle models provide both management and development levels of guidance through the software development process. During development, SORCE personnel referenced accepted standards for software development, including IEEE 12207 (1996), which accommodates the selection of various lifecycle models. The SORCE SDS development effort utilized a simplified software development methodology, based on both the Evolutionary and the Incremental lifecycle models, both of which are referenced in MIL-STD-498 (1994) and NASA software standard documents, including the NASA Software Management Guidebook (NASA, 1996). This methodology was adopted because it is easy to follow, is well suited for an object-oriented design, accommodates the notion of changing (or evolving) requirements, and caters to a relatively small development team, as was the case for SORCE.

Using the SORCE methodology, requirements were categorized according to related data types (e.g., orbit, pointing, gain, irradiance) resulting in *requirement sets* that define independent packages of functionality. The term *package* is used when referring to logical breakdown of the system-level problem into partitions of narrower and related scope. The package associated with a given requirement set corresponded to a software development schedule milestone, which typically entailed the development of several software modules. Each requirement set was assigned to an individual developer and that developer maintained full responsibility for completing the implementation of the requirements associated with that requirement set. The scope and complexity associated with each requirement set were intentionally limited in order to keep software development cycles relatively short (2–4 weeks, on average). Six milestones were informally tracked in the development of each requirement set: Requirements definition, package specification, design review, code walkthrough, qualification review, and release.

The initial specification for a given package provided the starting point for the developer, including the specification of all associated package-level requirements. The design review followed the initial software development, and was intended to ensure consistency with the design of other packages and to identify and resolve any ambiguous requirements. Code walkthroughs involved one or more people in addition to the developer, and provided a forum for suggesting alternative implementation possibilities and generalization options. Code walkthroughs were typically performed for only those algorithms that possessed a significant level of complexity. Qualification reviews entailed the presentation of testing criteria and results, and resulted in the ultimate consent for release.

Configuration management procedures govern the development of all software. At the heart of the SORCE configuration management system is the capability for tracking software change requests (SCRs), anomaly or trouble reports and to associate them with the software components that they ultimately affect. These software change requests are key for software configuration management, facilitating documentation of code and any associated changes. The configuration management system is also used to create a manifest of the software versions that comprise each software build. These manifests are stored in the database and are used in the establishment of science product traceability "pedigrees", which were discussed earlier.

4.3. System metrics

In total, the SORCE SDS consists of roughly 200 000 lines of computer source code, with elements implemented in the Java, C, C++, FORTRAN, and Interactive Data Language (IDL) programming languages. This includes approximately 1300 Java class files, 20 C functions, 100 FORTRAN subroutines, and 100 IDL files. Numerous scripts are also implemented in the C-shell, *bash*, perl, and Ruby scripting languages to control various elements of data processing operations. The SORCE database consists of approximately 220 database tables, and manages approximately 1 terabyte (TB) of data.

5. Software and Data Quality Assurance

Configuration management, testing, and validation provide the basis for both software and data quality assurance in the SORCE SDS. Both software and data are carefully configuration-managed, such that software modules and data have assigned versions that are fully traceable to one another. Before software modules are released, they are extensively tested, such that they behave as expected and produce the anticipated results. Furthermore, the principal irradiance data products are validated against similar measurements made by other instruments, as well as theoretical and proxy solar irradiance models as a check for internal consistency.

Software configuration management is facilitated by adherence to defined configuration management procedures, and is supported by several software development and release tools. The SORCE SDS uses the Razor commercial configuration management system, which tracks software change requests (SCRs), versions, and build manifests. This system is used to track SCRs, anomaly or trouble reports and to associate them with particular staff members, as well as the specific software modules they affect. These SCRs, or issues, are essential for software configuration management, facilitating documentation of code and associated changes. The same configuration management system also tracks the versions of each software module checked-in to the system, associating each version with one or more issues. The third capability, build management, collectively assigns versions to groups of software modules, facilitating combined versioning of software packages (multiple software modules that together map to a single managed dataset) in order to accommodate data traceability requirements, as discussed previously. The SORCE SDS personnel also make use of various software development tools to minimize risk from programming errors, bugs, or other anomalies. In particular, object modeling standards and tools, as well as code analyzers, are used to inspect code for problems that could otherwise remain undiscovered.

As has been discussed previously, data configuration management capabilities are actually built-in to the SDS, such that all data (calibration data, production parameters, intermediate data, and final science data) are maintained under configuration control in the SORCE SDS relational database management system. Each datum managed in the SORCE database is linked (e.g., by foreign key references) to other data on which it depends or from which it is derived. With this configuration management design the data management system contains all of the relevant *metadata* concerning any tuple of data produced by the data processing software, a capability that provides full accounting and ensures the integrity of all data products.

5.1. Data Quality Assessment

Assessment of the quality of SORCE data products begins with each product server performing automated quality checks at the time data are generated. At this initial stage in the quality assessment process, each product server knows best how to judge the quality of the data it produces, and thus most product servers return quality information with their data, which is determined based upon predetermined quality criteria. This quality information is used subsequently in processing as the basis for automated flagging or rejection of data. For example, in the TIM data processing system, an indicator is set within the Level 2 irradiance data for times when the Sun was not within the instrument's field of view. This flag helps to distinguish between solar measurements and dark characterization measurements.

Post-processing preliminary data quality assessment is accomplished by building high-level trend and limit checking into the data production software and analysis tools. Long- and short-term trends undergo comparison with expected values, and data are flagged if expected limits are exceeded. These trends and processing logs are reported automatically to data production staff for inspection. In the event that trends or other criteria exceed predefined limits, automated public releases of new data are temporarily halted in order to prevent release of suspect data. When the system detects the presence of suspect data, automated data releases are prevented and cognizant staff members are automatically notified.

During more rigorous subsequent validation of SORCE data products, calibrations are scrutinized, data are compared with other instruments (when possible) and theoretical predictions, and changes in instrument sensitivity are analyzed. This data validation process is coordinated by the SORCE science team and is beyond the scope of this paper.

6. Processing Operations

SORCE data processing operations nominally take place in two distinct modes: routine daily processing and occasional reprocessing. Since the data processing requirements are sufficiently unique for each of the four SORCE instruments (TIM, SIM, SOLSTICE, and XPS), the SDS is managed as four separate processing systems, each operating independently of the others. The routine execution of each processing system is performed by unique scripts, each of which facilitates the generation of each instrument's respective data products.

On a daily basis, the routine processing of all Level 1 through Level 3 science products takes place. Scripts written for the UNIX *bash* shell execute automatically at prescribed times, initiating the generation of the Level 2 and Level 3 data products. Each script determines the day to process, establishes the appropriate settings for the process, and begins the processing task by running the main function (known as the main *method*) in the appropriate Java product server class, formally initiating the generation of data. As the methods in this top-level product server object execute, other required product servers are instantiated and begin generating their own data. In this manner, data products are generated in an event-driven fashion in which higher-level servers prompt the generation of lower-level data products on an as-needed basis. Data products produced during a given processing job are inserted into the SORCE database and are available for immediate access by subsequent processing jobs, as well as end users for analysis purposes.

Certain product servers require spacecraft position and velocity information that is provided in the form of NORAD Two Line Element (TLE) sets, which are acquired from the Space Track web site (*http://www.space-track.org*). These TLEs are nominally available on a daily basis, but occasionally contain erroneous data, the presence of which requires quality checking prior to use. In order to maximize the quality of data products, it is best to avoid using predictive ephemerides, which are generated by propagating TLE measurements forward in time. Instead, SORCE data processing for a particular day is postponed (usually for 5 days) to allow for the receipt TLEs that were measured after the day being processed. This permits the use of definitive measurements of spacecraft position and velocity by interpolating the results of the two nearest TLEs.

In addition to routine daily processing of incoming data, the SDS provides full and partial mission *reprocessing* capability, as discussed earlier. This reprocessing capability is designed to support the SORCE mission requirement to complete full-mission reprocessing activities in less than one calendar month. The SORCE mission was designed for a life expectancy of at least 5 years, and this requirement was intended to ensure that, at the end of the SORCE mission, the entire dataset could be reprocessed in a reasonable amount of time. In actuality, the SORCE SDS is presently able to reprocess data at a 100$\times$ rate (100 days processed per 24-h period), easily achieving the design requirement. Once a full-mission reprocessing has been completed, the data are inspected and then released to the public, superseding previous data versions.

7. Science Analysis Support

In addition to generating and publicly releasing standard Level 3 data products, the SDS must also provide convenient access to all managed information for members of the SORCE science team and other SORCE program personnel. Users utilize a variety of computing platforms and require access locally from LASP in Boulder, Colorado, as well as from remote locations. The performance of modern computer networks makes it possible to support these requirements using a centralized, network-accessible data repository, and a centralized database server lends itself well to this approach.

Members of the SORCE science team typically use the commercially available Interactive Data Language (IDL) for analysis, visualization, and offline processing of SORCE science and in-flight calibration data. As a result, a key function of the SORCE SDS is to provide IDL-based tools to facilitate convenient access to the SORCE database, which stores all SORCE data, including instrument and spacecraft telemetry and science products. A fundamental objective in the design of this software component of the SORCE SDS was the desire to conceal from end-users the use of a relational database system to store the data they are interested in obtaining. This considerably simplifies the user experience and eliminates the need for broad training of personnel in the use of relational database systems.

As illustrated in Figure 5, the analysis software subsystem is implemented as a three-tier architecture, which isolates end-users from the data storage details and access protocols. The User Interface Layer provides both graphical and non-graphical interfaces, which are conceptually convenient for users. The Data Access

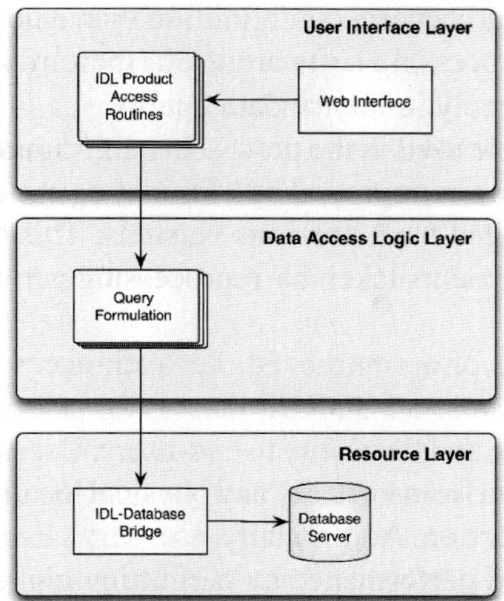

Figure 5. Analysis Software Architecture. A three-layer approach is used to simplify software maintenance and avoid requiring users to have a detailed knowledge of low-level data storage aspects and data access protocols. Public users access data products via the web interface, while SORCE project personnel can access data products directly from IDL on their respective computers.

Logic Layer facilitates the mapping between user-based views of the data and the correct logical representation of the data in the data management system. The Resource Layer includes the services that provide the low-level protocol support for interfacing with the database server, as well as the database server itself. Access routines are available to provide users a simplified way of accessing a particular data product. These routines identify the desired data products to the routines present in the Data Access Logic Layer, which then formulate a specific database query that can be used to provide the desired data. This database query is executed within the Resource Layer, transferring the desired data back to the user's process.

In addition to providing access for SORCE project personnel, a selection of the aforementioned IDL library routines are themselves used to facilitate the data access capabilities available to the public on the SORCE web site. In this manner, only one set of analysis data access functionality needs to be maintained.

8. Summary

The SORCE Science Data System is designed to bring advanced capabilities and modern software engineering principles to the production, management, and access of scientific data products. The use of a rigorous data-centric object-oriented software design facilitates many of the unique capabilities of the SORCE SDS, including automatic data version determination, data configuration management, and tolerance for changing requirements and the inevitable evolution of algorithms.

Version determination and general configuration management of data products are embedded in the data processing software itself, thereby minimizing the potential for human error to adversely influence data quality.

With each data variable used in the processing algorithms managed individually, reprocessing performance is improved by eliminating the need to recalculate variables that have not changed from previous versions. This minimizes the execution time and storage requirements taken by reprocessing activities, improving overall processing efficiency.

Additionally, the use of a centralized, network-accessible relational database management system to store all SORCE data – not just metadata – has brought a high level of convenience and flexibility to end-users. Users can conveniently access every type of SORCE data from virtually any physical location, as long as they have a computer network connection. Additionally, user services to access the data provide significant flexibility and performance by permitting highly customizable queries. Users can conveniently access a wide variety of cross-sections of the SORCE data while, at the same time, the system minimizes the need to transfer superfluous data by returning only the data that are requested.

Acknowledgements

Several professional software engineers and many students have contributed to the design, development, testing, and operation of the SORCE SDS. The dedication, expertise, and insight provided by these individuals over the years have produced a functionally unique and robust data system, and their efforts are gratefully acknowledged. Certain commercial equipment, materials, or software are identified in this paper to foster understanding. Such identification does not imply recommendation or endorsement by the Laboratory for Atmospheric and Space Physics (LASP), nor does it imply that the equipment, materials, or software identified are necessarily the best available for the purpose. This research is supported by NASA contract NAS5-97045 to the University of Colorado.

References

Coad, P. and Mayfield, M.: 1999, *JAVA Design*, Yourdon Press, New Jersey.

De Boor, C.: 1978, *A Practical Guide to Splines*, Springer-Verlag, New York.

Gamma, E., Helm, R., Johnson, R., and Vlissides, J.: 1995, *Design Patterns: Elements of Reusable Object-Oriented Software*, Addison-Wesley, Boston, Massachusetts.

Grand, M.: 1998, *Patterns in Java*, Vol. 1, Wiley, New York.

Harder, J., Lawrence, G., Fontenla, J., Rottman, G., and Woods, T.: 2005a, *Solar Phys.*, this volume.

Harder, J., Fontenla, J., Lawrence, G., Woods, T., and Rottman, G.: 2005b, *Solar Phys.*, this volume.

Harder, J., Fontenla, J., Rottman, G., Woods, T., White, O., and Lawrence, G.: 2005c, *Solar Phys.*, this volume.

Institute of Electrical and Electronics Engineers (IEEE)/Electronic Industries Association (EIA) 12207.0-1996 Standard Industry Implementation of International Standard ISO/IEC12207: 1995 and (ISO/IEC 12207), *Standard for Information Technology – Software life cycle processes*, IEEE Product No. SS94581.

King, M., Closs, J., Spangler, S., and Greenstone, R. (eds.): 2003, *EOS Data Products Handbook*, Vol. 1, NASA Goddard Space Flight Center, Greenbelt, Maryland.

Kopp, G. and Lawrence, G.: 2005, *Solar Phys.*, this volume.

Kopp, G., Heuerman, K., and Lawrence, G.: 2005, *Solar Phys.*, this volume.

Kopp, G., Lawrence, G., and Rottman, G.: 2005, *Solar Phys.*, this volume.

Lawson, C. L. and Hanson, R. J.: 1974, *Solving Least Squares Problems*, Prentice Hall, New Jersey.

McClintock, W., Rottman, G., and Woods, T.: 2005, *Solar Phys.*, this volume.

McClintock, W., Snow, M., and Woods, T.: 2005, *Solar Phys.*, this volume.

Meyer, B.: 1997, *Object-Oriented Software Construction*, Prentice Hall, New Jersey.

Military Standard (MIL-STD)-498: 1994, *Software Development and Documentation*, December 5.

NASA: 1996, *Software Management Guidebook*, National Aeronautics and Space Administration Software IV&V Facility, Fairmont, West Virginia.

Pankratz, C. K. *et al.*: 2000, *SORCE Algorithm Theoretical Basis Document*, Laboratory for Atmospheric and Space Physics, Boulder, Colorado. http://eospso.gsfc.nasa.gov/eos_homepage/for_scientists/atbd/viewInstrument.php?instrument=16

Snow, M., McClintock, W., Rottman, G., and Woods, T.: 2005, *Solar Phys.*, this volume.

Sparn, T. *et al.*: 2005, *Solar Phys.*, this volume.

Wirfs-Brock, R., Wilkerson, B., and Wiener, L.: 1990, *Designing Object-Oriented Software*, Prentice Hall, New Jersey.

Woods, T. N. and Rottman, G.: 2005, *Solar Phys.*, this volume.

Woods, T. N., Rottman, G., and Vest, R.: 2005, *Solar Phys.*, this volume.

Woods, T. N., Eparvier, F. G., Bailey, S. M., Chamberlin, P. C., Lean, J., Rottman, G. J., Solomon, S. C., Tobiska, W. K., and Woodraska, D. L.: 2005, *J. Geophys. Res.* **110**, A01312, doi: 10.1029/2004JA010765.